The Astrophysics of Emission-Line Stars

Astrophysics and Space Science Library

Recently Published in the ASSL series

Volume 344: *Space Weather: Research Toward Applications in Europe*, edited by Jean Lilensten. Hardbound 1-4020-5445-9

Volume 343: *Organizations and Strategies in Astronomy: Volume 7*, edited by A. Heck. Hardbound 1-4020-5300-2

Volume 342: *The Astrophysics of Emission Line Stars*, by Tomokazu Kogure, Kam-Ching Leung. Hardbound ISBN: 0-387-34500-0

Volume 341: *Plasma Astrophysics, Part II: Reconnection and Flares*, by Boris V. Somov. Hardbound ISBN: 0-387-34948-0

Volume 340: *Plasma Astrophysics, Part I: Fundamentals and Practice*, by Boris V. Somov. Hardbound ISBN 0-387-34916-9, September 2006

Volume 339: *Cosmic Ray Interactions, Propagation, and Acceleration in Space Plasmas*, by Lev Dorman. Hardbound ISBN 1-4020-5100-X, August 2006

Volume 338: *Solar Journey: The Significance of Our Galactic Environment for the Heliosphere and the Earth*, edited by Priscilla C. Frisch. Hardbound ISBN 1-4020-4397-0, September 2006

Volume 337: *Astrophysical Disks*, edited by A. M. Fridman, M. Y. Marov, I. G. Kovalenko. Hardbound ISBN 1-4020-4347-3, June 2006

Volume 336: *Scientific Detectors for Astronomy 2005*, edited by J. E. Beletic, J. W. Beletic, P. Amico. Hardbound ISBN 1-4020-4329-5, December 2005

Volume 335: *Organizations and Strategies in Astronomy 6*, edited by A. Heck. Hardbound ISBN 1-4020-4055-5, November 2005

Volume 334: *The New Astronomy: Opening the Electromagnetic Window and Expanding our View of Planet Earth*; edited by W. Orchiston. Hardbound ISBN 1-4020-3723-6, October 2005

For other titles see www.springer.com/astronomy

The Astrophysics of Emission-Line Stars

Tomokazu Kogure
Emeritus Professor of Astronomy
Kyoto University, Japan

Kam-Ching Leung
Professor of Physics and Astronomy
University of Nebraska-Lincoln, USA

and

Institute of Astronomy and Astrophysics
Academia Sinica, Taiwan, China

Springer

Tomokazu Kogure
Kyoto University
Personal: 1–10, Toganoo Hashimoto,
Yawata, Kyoto 614-8322
Japan
tkogure@paz.so-net.ne.jp

Kam-Ching Leung
Department of Physics & Astronomy
University of Nebraska-Lincoln, Lincoln, NE
USA
Lincoln 68588-0111
1186 Brace Laboratory
kleung@unlserve.unl.edu

Library of Congress Control Number: 2007922569

ISBN-10: 0-387-34500-0
ISBN-13: 978-0-387-34500-0
e-ISBN-10: 0-387-68995-8
e-ISBN-13: 978-387-68995-1

Printed on acid-free paper.

9 8 7 6 5 4 3 2 1

springer.com

Contents

II Emission-Line Stars

Preface

Many types of stars show conspicuous emission lines in their optical spectra. These stars are broadly referred to as emission-line stars, but, in the past, they were considered a type of peculiar stars, because emission lines were thought to be an indication of behaviors "peculiar" from the normal stellar atmospheres. Prior to 1950s, early-type emission-line stars such as Wolf-Rayet stars, Be stars, and P Cygni stars were called the early-type peculiar stars. With the advance in theories of emission-line formation, the name "emission-line stars" has been widely adopted to include both early- and late-type stars, and the name of "peculiar-stars" has been only used for chemically peculiar stars. Some stars that have no particular names as emission-line stars, such as cataclysmic variables and Mira variables, are also included in the category of emission-line stars. In closer examination most of stars on the Hertzsprung–Russell (HR) diagram show somehow evidence of emission lines formed in a less-developed form as in case of the Sun. In this book, however, we confine our examination of emission-line stars to the stars having strong lines in the optical region.

In the later half of the twentieth century, the physics of emission-line stars has been surprisingly developed under collaborations between ground-based and space observations. Wide wavelength observations have opened a new era of understanding the active stellar envelopes in various forms, such as stellar winds, accretion flows, flare activities, and binary interaction.

In this book an attempt is made to outline the physics of emission-line stars that are widely located on the HR diagram. Particular attention is paid to the spectral analysis of emission lines mainly in the optical region.

Although intended mainly for the use of graduate student and teachers of stellar astronomy, the present work should also provide a useful reference for practicing astronomers, particularly, for small-telescope users in institutions and public or private observatories. For these observers, emission-line stars may be an attractive choice to observe/monitor by their mysterious and often violent variable behaviors.

This book consists of a brief historical review in Chapter 1 followed by two major parts.

In Part I, first two chapters review the basic concepts on the spectroscopic processes (Chapter 2) and gas dynamical processes (Chapter 3) in stellar atmospheres. Chapter 4 is devoted to the mechanisms of emission-line formation in static and moving envelopes and in nonthermal atmospheres. Readers who are already familiar with basic astrophysics can skip Chapters 2 and 3 and move to Chapter 4.

In Part II, broad overviews of emission-line stars are given in three chapters. Chapter 5 deals with the early-type emission-line stars (WR, Of, Oe, Be, LBV), mostly forming developed expanding envelopes. Chapter 6 yields the late-type stars (dMe, flare stars, Mira variables) and close binary systems (Algol, RS CVn, cataclysmic variables, symbiotic stars), where the nonthermal processes are prevailing. In Chapter 7 the pre-main sequence stars (HES and TTS) are considered as activities in the contracting phase of stellar evolution. Thus Part II as a whole will show an amazing variety of emission-line stars.

Though we have confined the topics to the stars in this book, there are numerous objects showing strong and active emission-line phenomena in and out of the Galaxy. This book is expected to be useful for these related fields. Considering remarkable progress in the field of the physics of emission line stars since submission of the manuscript, we have added a self-contained Supplement at the end of the book to bring it up to 2006.

For further reading, books, and review articles are prepared for each chapter. The cited references are also given. It should be noticed that this book owes a great debt to the numerous works cited for their thoughts, data, and figures. The authors express sincere gratitude to those who have gone before. "We stand on the shoulders of giants." One of the authors (T.K.) published a book entitled *Emission Line Stars* (2002) written in Japanese. The present book is realized on the basis of this original book. On this occasion T.K. expresses his thanks to the late Mr. Mitsuo Goto, who encouraged him for a long time and passed away just before the publication of the original book. K-C.L. would like to thank Professor Sun Kwok for the invitation to spend a semester at the Institute of Astronomy and Astrophysics, Academia Sinica, R. O. C., Fall, 2006. He would also like to acknowledge the support of the Department of Physics and Astronomy, University of Nebraska for releasing him from teaching to complete the manuscript of this book. The authors are grateful to Yvonne Norton Leung for her editing assistance.

Tomokazu Kogure
Kam Ching Leung

Chapter 1

Introduction

1.1 Emission-line stars

Stars having remarkable emission lines in the optical spectra are called emission-line stars and widely distributed on the Hertzsprung-Russell (HR) diagram including various stellar types:

(1) Early type stars—Wolf-Rayet (WR) stars, Of, Oe/Be/Ae stars, Luminous Blue Variables (LBV).
(2) Late type stars—dMe stars, flare stars, red giants, Mira variables.
(3) Close binaries—Algol stars, cataclysmic variables (CV), symbiotic stars.
(4) Pre-main sequence stars—Herbig Be/Ae stars, T Tauri stars.

The locations of main types of emission-line stars on the HR diagram are schematically shown in Figure 1.1. It is seen that the emission-line stars are, roughly speaking, concentrated in the region of early- and late-type stars. Although main-sequence and giant stars in the intermediate F–G region also exhibit weak emission components in CaII H, K, as in case of the Sun, we do not refer to them as emission-line stars in this book.

Emission-line stars reveal characteristic lines according to their spectral types. For O-type stars, ionized helium and highly ionized ions of CNO elements show remarkable emission lines. For stars later than B type, the Hα and Hβ lines are the most conspicuous, and CaII H and K lines become ubiquitous in late-type stars. These emission lines often show variability of long or short time scales in intensity and/or line profiles. Sometimes emission lines become weaker and even disappear so that the term "emission-line phenomena" is occasionally used instead of "emission-line stars."

It is now known that emission lines originate mainly in the three following ways:

(1) Stellar envelopes or outer stellar atmospheres. This type includes expanding envelopes or strong stellar winds (e.g., WR, Of, LBV), rotating disks or rings (Be, Ae stars), accretion disks (cv, pre-main sequence stars), chromosphere-corona structure (red giants), and pulsating atmospheres (Mira variables).

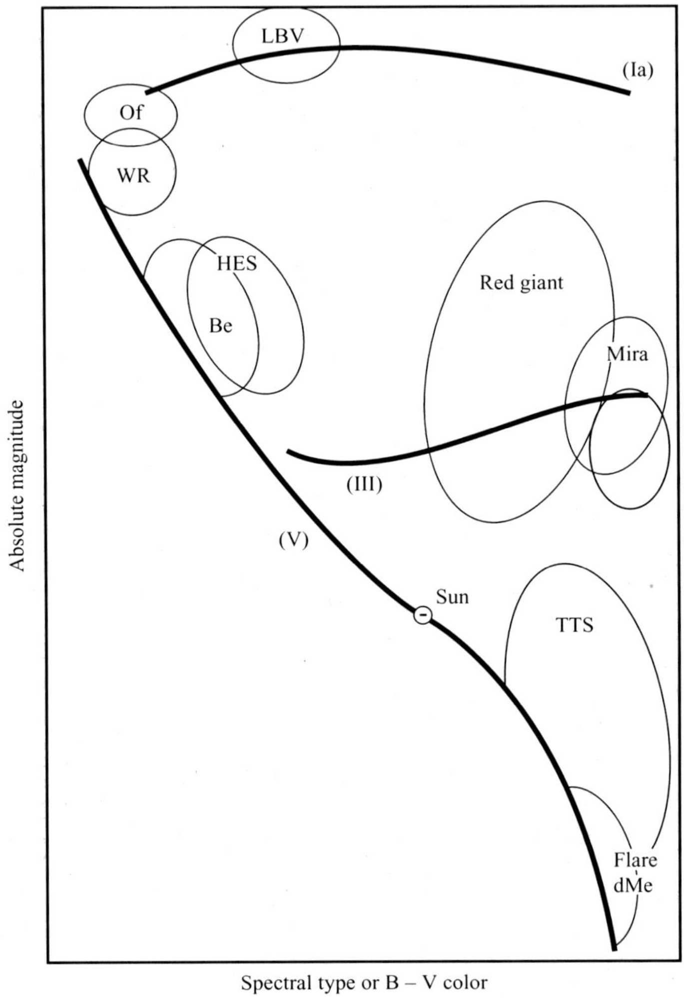

Figure 1.1: HR diagram and distribution of main emission-line stars.

(2) Stellar activities. Flare outbursts are most remarkable in this type. Others are prominence (RS CnV) and dark or bright star spots (RS CnV, CV).
(3) Binary interaction. This includes Algol binaries, CV, and symbiotic variables, where the mass exchange and accretion flows are the prominent processes.

Studies of emission-line stars offer the tools for understanding the physical state and dynamic structure of the envelopes and the active regions of stars.

By combining ground-based and space observations, it has become possible to explore the fine structure of emission phenomena and to build up reliable models of stellar activities. Our knowledge of emission-line stars has thus been dramatically advanced in recent decades.

1.2 Early history of stellar spectroscopy and discovery of emission-line stars

1.2.1 Early days of stellar spectroscopy

In around 1814 J. von Fraunhoffer, found numerous dark lines on the bright continuous spectrum of the Sun, now known as Fraunhoffer lines of A, a, B, C, D, E, b, F, G, H, etc. He also observed the spectra of Sirius, Betelgeuse, and some other bright stars. However, the stellar spectroscopy was not actually initiated until the 1860s by two pioneers, W. Huggins and A. Secchi.

At this time, G. Kirchhoff in Heiderberg developed the spectroscopic method of chemical analysis of matter and applied it to interpret the Fraunhoffer lines. Excited by this discovery, Huggins started spectroscopic observations of bright stars in 1863, using the spectrograph attached to the 20-cm refractor at his Tulse Hill Observatory, London. His main interest was the chemical analysis of bright stars by using the Fraunhoffer lines in stars. He found the existence of elements H, Na, Mg, Ca, and Fe in their spectra.

In 1864 he first pointed his telescope to a planetary nebula and unexpectedly found only emission lines at λ 5007, 4959 Å, and Hβ, in opposition to what was found in ordinary stellar absorption spectra. Contrary to the current view of the time, he found that the planetary nebulae and some diffuse nebulae are not the aggregates of faint stars but are gaseous clouds. Huggins attributed two green lines at λ 5007, 4959 Å to an unknown element "nebulium," though its identification remained unsolved until the twentieth century.

Spectroscopic observations of stars were also promoted by A. Secchi, at the Roman College Observatory by using a direct vision spectrograph attached to the 24-cm refractor. In 1862, he started his spectroscopic observations of stars brighter than around eighth magnitude for the purpose of classification of stellar spectra. He observed visually that the stellar colors appeared conspicuous in conjunction with spectral line behaviors. He thus classified the stars first into two groups: (a) yellow or red stars including α Ori, α Tau, and β UMi and (b) white stars with a few spectral lines such as Sirius, Rigel, and α Lyr. Thereafter, he extended his classification into three and finally into four groups. By 1872, Secchi had carried out spectroscopic observations for about 4000 stars and classified them according to his four-group system. Table 1.1 shows Secchi's spectral classification and prototype stars, proposed in 1866.

Table 1.1: Secchi's spectral classification

Type	Stellar color	Spectral lines	Prototype stars
I	White or blue	Hydrogen absorption metallic lines weak or absent	Sirius Vega
II	Yellow	Metallic lines rich	Sun, α Aur α Boo
III	Orange or red	Many absorption bands (faded toward red side)	α Ori, α Sco o Cet (often variable)
IV	Deep red	Many absorption bands (faded toward blue side)	19 Pis (carbon stars)

1.2.2 Early discoveries of emission-line stars

In B-type stars, emission lines were observed in γ Cas and β Lyr by Secchi in 1866. γ Cas exhibited strong emission components in Hα $\sim$ Hδ superimposed on the wide absorption lines, and this is now recorded as the first discovery of Be stars. In 1895, W. W. Campbell also observed γ Cas at Lick Observatory and found what we know as the characteristic feature of the Be stars in bright hydrogen lines. At that time Campbell listed around 30 Be stars, mostly discovered by using the Harvard objective prism for observations. The spectroscopic binary β Lyr was observed by Huggins in 1897–1998. He noticed the existence of emission in the Balmer and helium lines. Thus β Lyr becomes the first example of emission-line stars in close binary systems.

A "new star" (nova) appeared in May 1866 in the constellation Coronae Borealis. Huggins carried out spectroscopic observations of this nova (Nova T CrB) when it was as bright as second magnitude, and detected several emission lines including Hα, Hβ, and Hγ, along with some absorption lines such as Na D line. He was the first observer of a nova spectrum. The nova was fading rapidly reaching the ninth magnitude by the beginning of June 1866. Another nova (Nova Aurigae) erupted in December 1891. This nova was observed by Huggins, Campbell, and others. It showed emission-line spectrum with absorption lines displaced to the short-wavelength side. Ultraviolet spectrum showing the higher members of hydrogen Balmer lines was photographed by Mr. and Mrs. Huggins. This nova entered the nebular stage in the following year and exhibited the characteristic nebular spectrum with the two green lines at λ 4959 and λ 5007 Å.

In 1867, C. Wolf and G. Rayet at Paris Observatory found three faint stars (8–9 magnitude) with very broad and strong emission lines during a spectral survey with direct spectrograph in the Cygnus region. Emission lines were due to ionized helium and carbon/nitrogen. The line widths reached several 1000 km s^{-1} in Doppler velocity. These stars are now known as HD 191765,

Table 1.2: Comparison of spectral types

Secchi type	Draper memorial types (Pickering 1890)
I	A, B, (C), D
II	(E), F, G, (H), (I), K, (L) (J is omitted)
III	M
IV	N
—	O (WR star), P (Planetary nebula), Q (others)

Note: The types given in brackets are later omitted by Fleming herself.

HD 192103, and HD 192641 and called Wolf-Rayet stars (or WR stars) honoring the names of their discoverers.

In 1888, E. W. Maunder observed the star P Cygni and found a number of strong emission lines with absorption components at their violet edges. These line profiles are called P Cyg-type profile, and P Cygni is the prototype of these peculiar stars. Actually, P Cygni itself is a very slow nova that erupted in 1600 and is still slowly varying its brightness and spectrum.

1.2.3 Spectral classification and emission-line stars

In 1885, E. C. Pickering at Harvard College Observatory started the project of spectral classification of stars for the whole sky by using the objective prism (apex 13°) attached to the 20- or 28-cm telescope at Harvard College Observatory for the northern sky and at Boyden Station in Arequipa, Peru, for the southern sky. Stellar spectra were recorded on photographic plates of 20 × 25 cm size for stars brighter than about 7 magnitudes.

The classification of stellar spectra at Harvard College Observatory was an improvement of the Secchi's system, and three different classification schemes were successively proposed. The first one was made in 1885 by W. Fleming who revised Secchi's classification into 13 new types. This classification adopted the Roman letters A, B, G, K, M, etc., and is now called Pickering-Fleming classification. A total of 10,351 stars are classified and contained in the Draper Memorial Catalogue published in 1890. Table 1.2 gives a comparison of the Pickeing-Fleming classification with Secchi's classification.

The second classification at Harvard was made in 1897 by A. C. Maury, whose classification consisted of 22 types represented by the Roman numerals I to XXII, instead of the letters B, A, F, G, K, and M, in the Draper Memorial Catalogue. The last two types are denoted as XXI for carbon stars (Secchi's type IV) and XXII for WR stars.

The third classification was that of Annie J. Cannon. She started her work as an extension of Maury's classification for southern stars. In her classification, however, she did not use Maury's scheme, but adopted the existing Draper Memorial Catalogue system with some revisions. She used the letters

Table 1.3: Type and number of emission-line stars observed up to 1916 (from Cannon 1916)

Type	Number of stars
P (nebula)	150
O (W-R and other ionized He stars)	107
P Cyg type	10
Novae (including supernovae)	20
Emission-line B stars (including β Lyrae)	99
Md (Mira variables)	364
Total	750

O, B, A, F, G, K, M (in this order), together with P for planetary nebulae and Q for three peculiar stars with bright lines. Cannon's classification system was formally adopted at the general assembly of the IAU in 1922 and was called the Harvard classification. Cannon classified 225,300 stars and published the Henry Draper Catalogue in 1918–1924.

In Cannon's time, atomic theory as the physical basis for spectral sequence was still in its infancy. Nevertheless, she recognized in her classification a temperature sequence of stars which are thought to be somehow related to the evolution of stars. The direction of evolution was not yet clear as to whether stars evolve from hot stars to cool stars or vice versa.

The appearance of emission lines in stars such as WR and Be stars were thought as a sign of star's peculiarity in the Harvard classification. Because of their remarkable features many attempts have been made to observe emission-line stars. In 1912, W. Fleming carried out spectroscopic observations of a number of variable stars, including Algols (134 stars), β Lyr stars (10), short-period variables (168), and long-period variables (629). She noted that many of variable stars, particularly Mira variables, exhibit emission lines in their spectra. Revising Fleming's work, Cannon (1916) published a list of emission-line stars divided into six types as shown in Table 1.3.

1.2.4 Additional discoveries of emission-line stars

After Cannon's classification, new kinds of emission-line stars were discovered in the first half of the twentieth century.

Symbiotic stars

In 1933, P. W. Merrill at Mt. Wilson Observatory observed three peculiar stars of CI Cyg, RW Hya, and AX Per, which exhibited bright lines of ionized helium (characteristic of early type stars), in combination with dark bands of titanium oxide (TiO) ordinarily appearing in late-type stars. Merrill called them the "stars of combination spectra." Since the word "combination" involves other types of stars such as typical long-period variables which exhibit

M-type spectra, plus bright hydrogen and metal lines, Merrill (1958) proposed the name of "symbiotic stars" for the M-type stars that show bright lines including highly excited ions such as HeII, [OIII], [NeIII], etc., and this name has become widely accepted.

T Tauri stars

In 1945, A. H. Joy at Mt. Wilson Observatory defined a new class of variable stars as T Tauri type. His original definition was (1) irregular variations of about 3 magnitudes, (2) spectral types F5–G5 with emission lines resembling the solar chromosphere, (3) low luminosity, and (4) association with bright or dark nebulosity. Joy listed 11 stars of this type including T Tau, RW Aur, RY Tau, etc. These stars exhibit strong emission lines in CaII H and K, Hα, and many bright lines of low excitation in FeI, FeII, CaI, and others. The presence of irregular variables associated with nebulosity had been known already in the nineteenth century, but it was Joy who found the remarkable features of emission lines. Since these stars are generally fainter than 9 magnitude, spectroscopic observations with sufficient spectral resolution were carried out only after 1940s when more efficient equipment became available.

Two alternative possibilities on the nature of T Tauri stars had been proposed: one was that the stars were newly born inside nebulosities and the other was that normal stars encountered and interacted with nebulosities (Herbig 1952). Supporting the youth of these stars, Herbig (1958) redefined the T Tauri stars from spectroscopic criteria. Basic features are the presence of emission lines in CaII H and K and hydrogen lines, in addition to low-excitation forbidden lines such as [SII] and [OI]. The existence of strong absorption line of Li at λ 6708 Å is also an indication of the youth. Since this time the pre-main sequence nature of T Tauri stars has become widely accepted.

Late-type stars and chromospheric emission lines

The presence of weak emission components in CaII H and K of the solar spectrum was first noticed by Deslandres in 1892 and Eberhard and Schwarzschild in 1913, and it was attributed to the chromospheric origin. Since then, emission components of CaII have been widely observed in late-type stars. In particular, Wilson and Bappu (1957) observed conspicuous emission lines in G, K, M type stars with a wide range of luminosities and found a relationship between the width of emission components and stellar luminosity. This is known as the Wilson-Bappu relation. Since then, it has been recognized that the red giants and supergiants possess developed chromospheres accompanied with strong CaII H and K emission.

Late-type dwarf stars (dM) exhibit more or less emission components in CaII H and K lines. When emission is strong in these lines, the star often shows emission in Hα. Those stars with Hα emission are called red-dwarf emission-line stars (dMe). Since intensities of emission lines are generally variable, no clear separation is seen between dM and dMe stars.

Flare stars

Flare stars are the group of variable stars that exhibit irregular sudden brightening over a short time scale. In 1924, E. Hertzsprung observed that DH Car suddenly brightened about 2 magnitudes. In 1940, A. van Maanen observed the faint dM6 star Lalande 21258B, which showed temporally increase of brightness by 1.5–1.8 magnitudes over a short time period. These observations, however, received little attention at the time of their discovery. In 1949, A. H. Joy and M. L. Humason observed the eruptive brightening of the dMe star L726-8, which increased its intensity of emission lines and continuum about 1 magnitude in a few minutes and thereafter gradually declined. Since then, these stars have been called flare stars. By 1955, 12 flare stars were observed (Roques 1955), and all of them were dMe stars with strong Balmer line emission in the quiescent stage.

Herbig Ae/Be stars

After the pre-main sequence nature of T Tauri stars was accepted, the next step was to inquire as to whether some newly formed stars of an early type could be identified. Herbig (1960) started his spectroscopic observations with this inquiry and found 26 plus 7 candidate stars that each satisfied the following conditions: (a) the spectral type is A or earlier, with emission lines, (b) the star lies in an obscured region, and (c) the star illuminates fairly bright nebulosity in its immediate vicinity. These stars were confirmed as pre-main sequence stars and called Herbig Ae/Be stars.

Emission-line stars in 1948

The number of emission-line stars and their types discovered by 1948 are summarized by Joy (1948) as given in Table 1.4. The stellar types are arranged roughly in the order of decreasing temperature. The names of these stellar types are somewhat different from those of Cannon as well as those adopted in this book. For example, the type "stars in dark lanes" has the stars embedded in dark lanes, which were thought to be a different type from T Tauri variables due to their spectral features. They have spectra earlier than F type or they show the bright lines of HeI, resembling the Herbig Ae/Be stars. Remarks are added in brackets in the type column of Table 1.4, though not exactly corresponding to the respective type.

Table 1.4 shows a global state of emission-line star observations made by 1948 and can be compared to the data in Table 1.3 in the epoch of 1916. Identification of the varieties of emission-line stars and their numbers increased remarkably in the 30 years between the publications of Cannon and Joy.

Observations of emission-line stars have been greatly expanded since 1950s, and the results of new observations will be considered in their respective chapters.

Table 1.4: Type and number of emission-line stars observed up to 1948 (from Joy 1948)

Type	Number of Stars
WR stars	80
Stars of combination spectra (symbiotic stars)	20
P Cygni stars	20
Be and Ae stars	700
Eclipsing stars (Algols and close binaries)	20
W Virginis variables (Cepheids with CaII emission)	5
SS Cygni stars (dwarf novae)	40
T Tauri variables	12
RV Tauri variables (Pulsating supergiants)	30
Stars with bright H and K (stars with chromospheric activity)	150
Stars in dark lanes (T Tau-like stars)	40
Faint M stars (dMe)	40
Long-period variables (Mira variables)	3000
Total	4157

Note: Additional explanation is given in the brackets.

The first international meeting on emission-line stars was held in Liège in 1957 (Etoiles à Raies d'Emission, Inst. D'Astrophysique, Belgique, 1958). A wide field of emission-line stars from early to late spectral type was discussed, mainly based on the optical observations. After 1960s, the topics of meetings have been more specific reflecting the development of observations and theories in each field of emission-line stars. In Part II, we shall introduce and give an overview for each of the main type of emission-line stars.

1.3 Development of theoretical approach

Theoretical approach in astrophysics is closely connected with the development of physical theories particularly on quantum mechanics and hydrodynamics from the late nineteenth century to the middle of twentieth century. In this section, we briefly review the historical development of physical theories in this period from the view point of stellar astrophysics and formation of spectral lines.

1.3.1 Formation of emission lines

In 1859, G. Kirchhoff and R. Bunsen introduced the spectroscopic method of chemical analysis of matter, showing that each element discloses characteristic emission lines at distinct wavelengths in a flame and so can be identified. They supposed that the same analysis could be applied to the Sun. By comparing solar and laboratory spectra, Kirchhoff found that, though the solar spectrum appears in absorption (Fraunhoffer lines), as compared to the bright spectral lines from the laboratory, many lines showed essentially identical wavelengths between them. Thus it became clear that terrestrial elements do exist in the solar spectrum. Using the same reasoning for the Fraunhoffer lines appearing in absorption, he inferred that a cool gas exists outside the glowing solar photosphere and absorbs the light of that particular wavelength. He also noticed that stronger emission in laboratory spectrum tends to appear as stronger absorption in the Fraunhoffer spectrum. Based on this evidence he arrived at the general law of emission and absorption that states

> In thermodynamic equilibrium the radiant energy at any wavelength emitted by a body equals the radiant energy absorbed by that body.

This is now known as Kirchhoff's law and has a major influence as a basic concept for the interpretation of solar and stellar spectra. During Kirchhoff's time, however, there was no theory on the origin of emission lines in stars and nebulae.

A new era opened with the development of modern atomic theories and quantum mechanics in early twentieth century. Planck introduced the concept of thermodynamic equilibrium and derived the spectral energy distribution emitted from matter in a thermodynamic equilibrium (black body). The Sun and stars are shining roughly like black bodies, which enable us to determine the effective temperature and colors of the stars. On closer inspection, however, stars are not the ideal black bodies because of the existence of atmospheres where the selective absorption or scattering operates to form absorption line spectra. The formation of absorption lines in stellar spectra has become one of the principal research problems in stellar astrophysics. The first model, Schuster-Schwarzschild model (1905–1914), stated that a stellar atmosphere can be simplified into two layers: a deep black-body emitting layer and an outer rarefied line-forming layer. This model was later replaced by the Milne–Eddington model (1916–1930), in which the local thermodynamic equilibrium (LTE) is assumed everywhere inside the atmosphere instead of dividing the atmosphere in two (see Menzel 1966). In circumstellar envelopes, however, deviations from the LTE become remarkable in an un-isotropic radiation field due to dilution effects, and some large-scale motion of gas in envelopes. Hence the introduction of non-LTE treatment is essential in any theories on the formation of emission-line spectra.

Early theoretical studies of emission lines in planetary and emission nebulae by Zanstra (1927) marked a successful application of quantum mechanics to astrophysics. These nebulae show the conspicuous emission in the Balmer series of hydrogen, allowing him to simplify the nebula to consist wholly of hydrogen and to consider the transition cycle from ground state to ionized state followed by recombination, cascade transitions toward the original ground state. He called this process the fluorescence mechanisms. In order to calculate the intensities of emission lines, one needs to calculate the transition probability of each transition based on quantum mechanics. This cycle of ionization–recombination, caused by ultraviolet radiation from hot stars, has been widely known as the basic process of emission-line formation in emission nebulae and was fully formulated in 1930s by Menzel and coworkers (see Menzel 1962).

Emission lines are generally classified into two types: permitted and forbidden lines. Lines of hydrogen and helium belong to the permitted lines, which are produced by transition from upper to lower energy levels through the electric dipole radiation. For some energy levels, downward transitions are prohibited by the selection rules of transition and only possible through the electric quadrupole or magnetic dipole radiation with very small transition probabilities. Such levels are called the metastable levels. The lines from metastable levels are called the forbidden lines and are rarely observable in laboratory conditions. It was Bowen (1928) at California Institute of Technology who first identified the forbidden lines of NII, OII, and OIII in nebular spectra that were previously attributed to unknown nebulium by Huggins. Forbidden lines are now designated as [NII], [OII], and [OIII] and provide an efficient diagnostic method for rarefied gas. Forbidden lines are widely observed in nebular objects and sometimes observable in stellar envelopes.

The conditions of emission-line formation in stellar spectra were considered by Rosseland (1926) at Mt. Wilson Observatory. By solving the equations of radiative transfer under the condition of radiative equilibrium in stellar atmosphere, he showed that the main stellar disk will show a continuous spectrum with absorption lines, while the atmosphere, as seen at the limb where the total optical depth along the line of sight is less than unity, will show a spectrum principally of emission lines. This suggests that the strength of emission lines depends on the relative extension of its outer atmosphere, where the cyclic transitions of ionization–recombination play an important role in the formation of emission lines. This cyclic transitions, initially considered in three-level atom, is now called Rosseland cycle.

At the same time, Rosseland pointed out that this cyclic transition will not operate in late-type stars due to the lack of sufficient exciting or ionizing radiation. In order to form high-excitation emission lines such as Balmer series, some additional energy input is required. He supposed the possibility of either local formation of hot regions in the atmosphere or corpuscular streams of high-energy particles as in case of terrestrial aurora. In any case, he recognized the necessity of nonthermal heating in the atmospheres of late-type stars.

By around 1930, it has become clear that early-type, emission-line stars are surrounded by extended envelopes accompanied by expanding or rotational motions. Beals (1929) first suggested that the large width of emission lines in the spectra of WR stars reveals the existence of an envelope expanding with a prodigious velocity. He also showed that the expanding envelope itself produces absorption lines on the violet edge of the emission band as seen in P Cygni type stars. Similarly, Struve (1931) proposed the rotation hypothesis for Be stars to explain single- or double-peaked emission-line profiles superimposed on broad photospheric absorption lines. These emission-line profiles could be explained by a different inclination angle of a rotating disklike envelope.

The non-LTE treatment of the radiation field of stellar envelopes needs to solve the equations of radiative transfer and that of statistical equilibrium, simultaneously. Since a direct solution to these sets of equations is difficult to obtain, numerous works have been devoted to approaching the problem with some simplification depending on the special circumstance of the envelopes in question. There are basically two approaches: the static envelope theory and the moving envelope theory.

Static envelope theory treats the non-LTE problem more precisely without considering the effects of large-scale gas motions. The general formulation of an atom with a finite energy level was given by L. G. Henyey (1938) at Yerkes Observatory, who applied it to a three-level problem in nebular case (small dilution factor of the order of 10^{-10}). S. Miyamoto at Kyoto University (1949–1952) solved the 3–5 level problems of hydrogen atoms in Be star envelope and derived the Balmer decrements.

Moving envelope theory takes into consideration the effects of gas motion by avoiding the solution of radiative transfer. In early days, emission-line profiles, formed in expanding or rotating envelopes, have been calculated by assuming that the envelopes are transparent for the lines in question (see Rosseland 1936). Sobolev (1947) at Leningrad State University introduced the concept of escape probability from moving envelopes and solved the equations of statistical equilibrium without using the equations of radiative transfer.

These theories are mostly concerned with the radiation fields of early-type stars. In late-type emission-line stars, nonthermal process is a basic requirement to excite hydrogen and helium atoms as already suggested by Rosseland (1926). This is closely related to the development of hydrodynamic approach given in the next section.

1.3.2 Hydrodynamic approach

Hydrodynamics has a long history from the early 1800s. Basic concepts that are now popular and widely used in astrophysics have their origin in such early research.

Supersonic flows and shock waves are prevalent phenomena in many types of stars and interstellar media. The Mach number which expresses the flow velocity in unit of sound velocity was introduced by Ernst Mach in 1877 in his works on supersonic phenomena. Establishing the basic relationship between the physical parameters before and after shock passage of the stationary shock waves was done by Rankine in London (1879) and Hugoniot in Paris (1889). In turbulent motion, the Leynolds number, which gives the nondimensional value at critical change from laminar to turbulent flows, was introduced by

O. Leynolds in 1883. Numerous concepts and fundamental equations that now describe astrophysical phenomena originated in the nineteenth century.

It was, however, only in the middle of the twentieth century that hydrodynamics became essential to astrophysics. The symposium held in Paris in 1949 was a declaration of the opening of a new era of the close relationship between astrophysics and hydrodynamics (Burgers and van de Hulst 1951). The aims of the symposium were to exchange information between astrophysicists and physicists and discuss hydrodynamical phenomena in a cosmic scale. Particular attention was focused on the problems of turbulence, expansion phenomena, and shock waves. Subsequently, a series of these symposia has continued every few years up to 1969, when hydrodynamics fully settled into astrophysics with a resulting major impact on the history of astrophysics.

Hydrodynamical phenomena appear in gas flows, which are often supersonic, and in the form of wave propagations, which are often of finite amplitude causing shock waves in stellar envelopes.

Structure and stability of stellar envelopes

In early-type stars, the formation of emission lines in envelopes is usually attributed to the ionization–recombination mechanism caused by the ultraviolet radiation from hot stars. Hydrodynamical interest is mostly concentrated on the formation and stability of circumstellar envelopes, and much research has been carried out beginning in the 1940s. For example, Struve (1942) explained the formation of expanding shells in terms of radiation pressure from the Lyman α flux. He suggested that the difference between stationary and expanding shells could be explained by the difference of optical depth of the shell in the Lyman continuum radiation. On the other hand, Miyamoto (1953) examined the stability of envelopes of P Cygni stars in terms of radiation pressure by the Lyman continuum and electron scattering. He derived a criterion parameter and showed that P Cygni stars are distributed near the instability line on the effective temperature—surface gravity diagram. It is also noted that Miyamoto (1943) solved the equations of motion and found the existence of two types of flows: accelerating and decelerating. Both are intersected at a singular point, now known as the X-type singular point. In the time of Miyamoto, however, it was not clear that the singular point expresses the sonic point in flows changing from subsonic to supersonic or vise versa. The basic physical theory of stellar winds was initiated by E. N. Parker (1958).

Shock waves

Attention was first drawn to shock waves in stellar atmospheres by the heating source of chromospheres and coronae. In late-type stars, developed convection layers beneath the photospheric surface produce the sound waves that propagate outward growing progressively up to finite-amplitude waves, and finally transform into shock waves. The mechanical energy of shock waves is dissipated into heat and used as the heating source of outer atmospheres.

Schatzman (1949) considered the efficiency of dissipated energy of a saw-toothed train of shock-waves propagating in the solar chromosphere. His results showed that the temperature increases rapidly in the outer chromosphere and reaches 10^6 degrees at 1.05 solar radius, which might be sufficient to maintain the isothermal corona. Later development showed that a simple shock-wave model is not adequate to form the chromosphere-corona structure, but shock-wave heating was a novel idea in the 1940s.

The structure of shock waves under cosmic conditions is characterized by strong interaction with a radiation field. Remarkable theoretical developments in this field were made in the 1950s. For the interstellar medium, Pikel'ner (1954) considered the structure of shock waves behind the shock front and calculated the relative intensities of emission lines in the case of optical filaments in supernova remnants. In stellar atmospheres, Deutsch and Merrill (1959) analyzed the spectrum of the long-period variable R Cygni and showed the combined formation of absorption and emission lines in the shock-heated atmosphere. Similarly, Wallerstein (1959) applied the shock-wave model to the interpretation of emission-line spectrum in the population II Cepheid W Virginis. Since 1960, numerous research publications have been devoted to the understanding of shock-wave structure in stellar atmospheres.

Further reading

Hearnshaw, J. B. (1986). *The Analysis of Starlight, One Hundred and Fifty Years of Astronomical Spectroscopy.* Cambridge University Press, Cambridge.

Leverington, D. (1995). *A History of Astronomy from 1890 to the Present.* Springer-Verlag, Berlin.

Tassoul, J.-L. and Tassoul, M. (2004). *A Concise History of Solar and Stellar Physics.* Princeton University Press, Princeton, NJ.

References

Beals, C. S. (1929). On the nature of Wolf-Rayet emission. *MNRAS*, **90**, 202–212.

Bowen, I. S. (1928). The origin of the nebular lines and the structure of the planetary nebulae. *Ap. J.* **67**, 1–15.

Burgers, J. M. and van de Hulst, H. C. (1951). Problems of cosmical aerodynamics. *Proceedings of the Symposium on the Motion of Gaseous Masses of Cosmical Dimensions*, Paris, August 16–19, 1949.

Cannon, A. J. (1916). Spectra having bright lines. *Ann. Harvard Coll. Obs.* **76**, 19–42.

Deutsch, A. J. and Merrill, P. W. (1959). Gross differences between the R Cygni spectra at two successive maxima. *Ap. J.* **130**, 570–577.

Henyey, L. G. (1938). The theory of cyclic transitions. *Ap. J.* **88**, 133–163.

Herbig, G. H. (1952). Emission-line stars in galactic nebulosities. *JRAS. Canada*, **46**, 222–233.

Herbig, G. H. (1958). Stars of F, G, and K with emission lines: Introductory report. Etoiles a raies d'emission. *Liège Conference*, July 1957, 251–270.

Herbig, G. H. (1960). The spectra of Be and Ae type stars associated with nebulosity. *Ap. J. Suppl.* **4**, 337–368.

Hertzsprung, E. (1924). Note on a peculiar star or nova of short duration. *Bull. Astr. Inst.* **2**, 87–88.

Joy, A. H. (1945). T Tauri variable stars. *Ap. J.* **102**, 168–195.

Joy, A. H. (1948). Emission lines in stellar spectra. *Ap. J.* **53**, 107.

Joy, A. H. and Humason, M. L. (1949). Observations of the faint dwarf star L726-8. *PAS. Pacific*, **61**, 133–134.

Menzel, D. H. (ed.) (1962). *Selected Papers on Physical Processes in Ionized Plasmas*, Dover, New York.

Menzel, D. H. (ed.) (1966) *Selected Papers on the Transfer of Radiation.* Dover, NY.

Merrill, P. W. (1933). Four stars whose spectra have bright helium lines. *Ap. J.* **77**, 44–50.

Merrill, P. W. (1958). Symbiosis in astronomy: Introductory report. Etoiles a raies d'emission. *Liège Conference*, July 1957, 436–448.

Miyamoto, S. (1943). On the envelopes of peculiar stars (in Japanese). *Tenmon-gaku Iho* **2**, 157–186.

Miyamoto, S. (1953). The atmospheres of the P Cygni stars. *PAS. Japan*, **5**, 55–73.

Parker, E. N. (1958). Dynamics of interplanetary gas and magnetic fields. *Ap. J.* **128**, 664–676.

Pickering, E. C. (1890). The Draper Catalogue of stellar spectra photographed with the 8-inch Bache telescope as a part of Henry Draper Memorial. *Harvard College Obs. Ann.*, **27**, 1–388.

Pikel'ner, S. B. (1954). Spectrophotometric investigation on the mechanisms of excited filamentary nebulae. *Izv. Krimk. Astrofiz. Obs.* **12**, 93.

Roques, P. (1955). A search for flare stars. *PASP*, **67**, 34–38.

Rosseland, S. (1926). On the Origin of bright lines in stellar spectra. *Ap. J.*, **63**, 218–235.

Rosseland, S. (1936). Stars with extensive envelopes. *Theoretical Astrophysics.* Chapter 20. Oxford University Press, Oxford.

Schatzman, E. (1949). The heating of the solar corona and chromosphere. *Ann. d'Ap.*, **12**, 203–218.

Sobolev, V. V. (1847). Moving Envelopes of Stars (in Russian, English translation is published from Harvard Univ. Press in 1960).

Struve, O. (1931). On the origin of bright lines in spectra of stars of class B. *Ap. J.*, **73**, 94–103.

Struve, O. (1942). Extended stellar atmospheres: A review of the problems of gaseous shells. *Ap. J.*, **95**, 134–151.

van Maanen, A. (1940). The photographic determination of stellar parallaxes with the 60- and 100-inch telescopes *Ap. J.*, **91**, 503–506.

Wallerstein, G. (1959). The shock-wave model for the population II Cepheids. *Ap. J.*, **130**, 560–569.

Wilson, O. C. and Bappu, M. K. V. (1957). H and K emission in late type stars: Dependence of line width on luminosity and related topics. *Ap. J.*, **125**, 661–683.

Zanstra, H. (1927). An application of the quantum theory to the luminosity of diffuse nebulae. *Ap. J.*, **65**, 50–70.

Part I

Stellar Atmospheres and Formation of Emission Lines

Chapter 2

Stellar Spectra and Radiation Fields

2.1 Basic properties of stars

2.1.1 Photometric system

Apparent magnitude m of a star is defined by the logarithms of energy I received above the Earth's atmosphere as follows:

$$m = -2.5 \log I + c, \tag{2.1.1}$$

where c is the constant to be determined by comparison with a standard star.

Stellar energy I is usually measured as an integrated energy in a specific wavelength range called the photometric band. Historically the most basic photometric bands were the UBV system introduced by Johnson (1963), which is composed of three bands of U (ultraviolet, ~3600 Å), B (blue, ~4300 Å), and V (visual ~5500 Å). Let the magnitudes $m_{\rm U}, m_{\rm B}$, and $m_{\rm V}$ of a star in these bands be denoted as U, B, and V, respectively, then each of the following magnitude differences,

$$U - B = m_{\rm U} - m_{\rm B}, \tag{2.1.2}$$

$$B - V = m_{\rm B} - m_{\rm V}, \tag{2.1.3}$$

defines the color index in the UBV system. The sensitivity of each color is adjusted so as to give $U - B = 0, B - V = 0$ for Vega, the standard star in this system. Hence a star with $B - V > 0, U - B > 0$ is redder than Vega and ultraviolet light is weaker than Vega. This color system has been extended later toward the infrared region in many different photometric systems as summarized in Table 2.1.

The photometric observations in the UBV system have long been made using a photoelectric multiplier or photographic plate, both of which are sensitive in blue light. In the 1980s, the charge coupled device (CCD), a CCD detector, sensitive in redder wavelengths was introduced so that the V, G, R, I bands in the spectral range 5000–10000 Å prevailed in photometric observations.

Table 2.1: Main photometric systems

Photometric system	Name of bands	Central wavelength $\lambda_0(\mu m)$	Effective bandwidth $\Delta\lambda(\mu m)$	Note
Johnson–Morgan's three-color system				
	U	0.36	0.04	
	B	0.44	0.10	
	V	0.55	0.08	
Johnson's infrared system				
	R	0.70	0.21	
	I	0.88	0.22	
	J	1.25	0.3	
	H	1.62	0.2	
	K	2.2	0.6	
	L	3.5	0.9	
	M	5.0	1.1	
	N	10.4	6.0	
Stebbins's six-color system	U B V G R I			(1)
Strömgren–Crawford system	u	0.35	0.034	(2)
	b	0.41	0.020	
	v	0.47	0.016	
	y	0.55	0.024	
Crowford's Hβ system	β	0.486	0.00315	(3)

(1) The names of each band and its central wavelength (μm) are U (0.353), B (0.422), V (0.488), G (0.570), R (0.710), I (1.030) (Stebbins and Kron 1964).
(2) Photometric system with intermediate bandwidths (Strömgren 1963).
(3) Narrow band system for measuring the strength of Hβ absorption (Crawford 1958).

2.1.2 Stellar parallax and stellar luminosity

Stellar parallax and absolute magnitude

Stellar parallax is defined by $\pi = 1/d$ in arcsecond, where d denotes the distance to star in parsec (1 pc = 3.08 $\times 10^{18}$ cm). The ground-based trigonometric parallax is given in the New Yale Catalogue by van Altena et al. (1992), which contains 7881 stars with accuracy typically 15 milliarc second (mas). The Hipparcos satellite launched in 1989 revised the accuracy up to 2 mas for bright stars (V < 9) and 4–5 mas for fainter stars (V < 12.5). The Hipparcos data contain more than 100,000 stars, mostly closer than about 500 pc (Gömez 1993).

To express the intrinsic brightness of stars, the absolute magnitude is defined as the brightness observed at a distance of 10 pc from the star. If we take into account the interstellar extinction, A, the relationship between absolute magnitude M and apparent magnitude m of a star is given by

$$M = m + A - 5 \log d + 5. \tag{2.1.4}$$

This relation holds for every color band such as for M_V and m_V.

Stellar luminosity

The total radiative energy $I(\text{erg cm}^{-2}\, s^{-1})$ of a star received above the Earth's atmosphere is written as $I = \pi f$, where f is called the total flux of radiation and given as the integration of the flux of radiation, f_λ, at the wavelength λ per unit wavelength, over the whole wavelength range. That is, we have

$$f = \int_0^\infty f_\lambda \, d\lambda. \tag{2.1.5}$$

Flux f depends on the distance from the star.

The luminosity of a star L (erg s^{-1}) is defined as the total radiative energy emitted from the whole stellar surface, and connected with the total flux of radiation πf at the stellar surface by the relationship

$$L = 4\pi R^2 \pi F, \tag{2.1.6}$$

where R is the star's radius. If there is no interstellar absorption, luminosity L can be expressed by using the total flux πf at the Earth as

$$L = 4\pi\, d^2 \pi f. \tag{2.1.7}$$

From these relations we have $f = F \times (R^2/d^2)$, where R/d is the angular radius of the star seen from the Earth.

The total radiative energy of the Sun received at the Earth's orbit is called the solar constant and designated as $S_\odot$. The observed value is

$$S_\odot = 1.38 \times 10^6 \,\text{erg cm}^{-2}\,\text{s}^{-1} = \pi f_\odot.$$

Since the distance to the Sun is $1\,\text{AU} = 1.49 \times 10^{13}$ cm, then we have from Equation (2.1.7)

$$L_\odot = 3.96 \times 10^{33}\,\text{erg}\,\text{s}^{-1}.$$

On the other hand, the mass of the Sun is known as $M_\odot = 1.98 \times 10^{33}$ g. It is interesting to see that the energy productivity of the Sun per unit mass is $L_\odot/M_\odot \sim 2$ erg s^{-1} g^{-1}. This productivity is called luminosity-mass ratio and takes a large variety among stars. For example, the luminosity-mass ratio is around $L/M \sim 10^4$ for a supergiant with $L \sim 10^6\, L_\odot$ and $M \sim 100\, M_\odot$, whereas it is around $L/M \sim 10^{-5}$ for a dwarf star with $L \sim 10^{-6}\, L_\odot$ and $M \sim 0.1\, M_\odot$, Thus this ratio has a large variety of values depending on the mass of stars and their evolutionary states.

The stellar magnitude defined by the total radiation over the entire wavelength range is called the bolometric magnitude and is designated as m_{bol} and M_{bol} for the apparent and absolute magnitudes, respectively. The correction term yielding the bolometric magnitude from V magnitude is called the bolometric correction (BC) and defined by the relationship

$$m_{\text{bol}} = m_{\text{V}} - BC \quad \text{or} \quad M_{\text{bol}} = M_{\text{V}} - BC. \tag{2.1.8}$$

Since the surface temperatures of normal stars lie in a range of 2000–30,000 K, and the maximum brightness falls in the spectral region from ultraviolet to near infrared (see next section), then the values of BC are relatively small around 0–2. Bolometric magnitude is a basic parameter related to the energy production rate of a star at each epoch of its evolution.

2.1.3 Spectral classification

Harvard classification

The Harvard classification arranges the spectral types based on the relative strengths of absorption lines as follows:

```
                      R - N
                     /
O - B - A - F - G - K - M
                     \
                      S
```

Each spectral type has 10 subtypes signified by numbers 0 to 9, except O type which starts from O3 subtype. Besides the main types O to M, stars with different chemical compositions are designated as R, N, S. Spectral characteristics of the main types and its sample stars are shown in Table 2.2.

The R to N sequence is characterized by CN, CO bands (R type) and by C_2, NaI D (N type). Since R-N stars contain abundant carbon or carbon compounds, this sequence is now called the carbon series ranging from C0 to C7. On the other hand, S type stars are conspicuous by their strong bands of TiO, ZrO, and other metal oxides.

The Harvard classification yields mainly a temperature sequence from high (O type) to low (M type). The special features of stars are shown by a combination of spectral type and prefix or suffix, as shown below:

Prefix	c: supergiant (Note 1)
	g: giant
	d: dwarf
	sd: subdwarf
Suffix	n: broad absorption line (“nebulous” line)
	s: sharp absorption line
	e: stars with emission lines
	f: a kind of emission-line stars in O type (Note 2)
	q: stars showing P Cygni type profiles (Note 3)
	v: stars with time variations
	p: stars with peculiar spectra
	m: stars showing conspicuous metallic lines

Notes: (1) In Maury and Pickering’s (1897) classification, stars are divided into three steps according to line widths as a (broad), b (mediate), and c (sharp), among which the stars with step c have been used to express supergiants. (2) Of star is a kind of O type emission-line star, distinguished from Oe stars. (3) q: A kind of peculiar star, p, first designated as Q type by Maury.

Table 2.2: The Harvard classification

Spectral type	Main feature	Subtype	Classification criteria	Sample stars
O	Predominant HeII,	O3	HeIλ 4471/HeII 4541 = 0.1	HD 93205
	Intensive blue	O4	= 0.2	HD 96715
	continuum	O7	= 0.9	HD 48099
		O8	= 1.3	λ Ori
		O9	= 1.8	ι Ori
B	Conspicuous HeI	B0	CIIIλ 4650 max. HeI > HeII	τ Sco, ε Ori
		B3	HeI max. HI(Balmer):A0 ∼ 1/2	π^4Ori, α Pav
		B5	Siλ 4128 > Heλ 4121	19Tau, φ Vel
		B8	HeIλ 4471 = MgIIλ 4481. Metal lines appear	β Per, $\bar{\text{O}}$ Cru
A	Dominant Balmers	A0	Balmer lines max. MgII conspicuous	α CMa
	max. at A0	A3	CaII K = 0.8 Hδ	α PsA, τ ^{3}Eri
	grad. decline	A5	CaII K = 0.9 (CaIIH + Hε) > Hδ	β Tri, α Pic
F	Metal lines	F0	HI(Balmer):A0 ∼ 1/2, CaII K = H + Hε	δ Gem, α Car
			Many metal lines	
		F5	G band = 0.6 Hγ, CaI4227 = 0.5 Hγ	α CMi, ρ Pup
G	Solar type	G0	Caλ 4227 = Hβ, G band = 2 Hγ	α Aur, β Hya
		G5	FeIλ 4325 > Hγ	κ Gem, α Ret
K	Neutral metals	K0	H, K line max. Balmer lines weak	α Boo, α Phe
		K5	G band dissolved in lines,	
			CaI and CaII dominant, TiO bands.	α Tau
M	TiO bands dominate	M0–2	TiO dominate, CaI 4226 max.	α Ori, β And
		M3–5	Molecular bands dominate	ρ Per, π Aur
		M6–10	Abundant strong molecular lines	W Cyg, ρ Per
			M0e–M10e H_γ, H_δ in emission	$\dot{\chi}$ Cyg, o Cet

Note: Classification criteria (Keenan 1963, Lang 1991), equality or inequality denotes the relative intensity of the respective absorption lines.

Additional designations are used such as cA (A-type supergiant), dM (M-type dwarf star), Be (B-type emission-line star), Bn (B-type star with broad spectral lines), Kv (K-type stars with time variations), Ap (A-type peculiar star), and so on.

The original HD Catalogue contains 225,300 stars, as stated in Chapter 1. This catalogue was later extended and called HDE Catalogue (Extended HD Catalogue) with 86,932 stars listed with HDE numbers (*Ann. Harvard Coll. Observ., Vol.* 100, 1926–1936; Vol. 105, 1937; and Vol. 112, 1949).

Recently, new spectral types L and T for dwarf stars later than M type have been introduced on the basis of infrared spectroscopy by Kirkpatrick et al. (1999) and Reid (1999). The TiO, and VO bands that are dominant spectral features in M type stars, rapidly weaken toward the end of M type. Replacing TiO, L type is characterized by the presence of metallic hydrides such as CaH, FeH, CrH, and neutral alkali metals NaI, KI, RbI, etc., making a sequence L0 to L8 toward lower temperature in the range 2000–1300 K. On the other hand, T type dwarfs are characterized by strong CH_4 bands in the infrared spectral region and make a tail of the L type sequence in the temperature range 1200–1000 K. The stars later than M type are an aggregation of substellar objects and contain a number of brown dwarfs that are considered intermediate objects between stars and planets. In these latest types of stars there exist some emission lines offering interesting spectroscopic problems.

MK classification

Morgan (1937) paid attention to the importance of surface gravity and introduced a sequence of luminosity classes from dwarfs to supergiants, as follows:

Luminosity class	Stellar type
I	Supergiants (Ia, Iab, Ib)
II	Bright giants
III	Giants
IV	Subgiants
V	Dwarfs

The class I (supergiants) is divided into three subclasses Ia, Iab, and Ib in the order of decreasing stellar brightness. Stars having luminosity lower than Class V are called subdwarfs and designated as Class VI.

The luminosity class is determined based on the degree of Stark broadening and/or the relative intensities of absorption lines sensitive to the surface gravity. Thus the luminosity class is essentially a parameter of surface gravity, g, of stars.

Morgan and Keenan have introduced a two-dimensional spectral classification combining the HD scheme in temperature sequence and the luminosity class in gravity sequence. This is the so-called Morgan–Keenan (MK) classification scheme. This classification was originally based on the spectrograms (dispersion 125 Å mm^{-1}) taken with the slit spectrograph attached to the 100-cm reflector at Yerkes Observatory. MK classification was first published in 1943 by the name of MKK Atlas (Morgan et al. 1943) and contains the spectrograms of around 200 spectroscopic standard stars.

BCD classification

Both HD and MK classifications are essentially based on the visual inspection of spectrograms or spectral intensity tracings of stars; thus systematic personal bias on the classification is inevitable. To avoid this, several quantitative or automatic classifications have so far been proposed. Among them we show the BCD classification by Barbier, Chalonge, and Divan (Chalonge, and Divan 1952, 1973), mostly applicable to B and A stars. This classification will be referred to sometimes in this book.

These stars exhibit strong absorption lines in the Balmer series, and also remarkable discontinuity at the series limit, called Balmer discontinuity or Balmer jump. The BCD classification is based on the quantitative measurements of this Balmer jump. A low-dispersion spectrum near the Balmer limit is shown in Figure 2.1, where the curves AB and DE denote the continuous

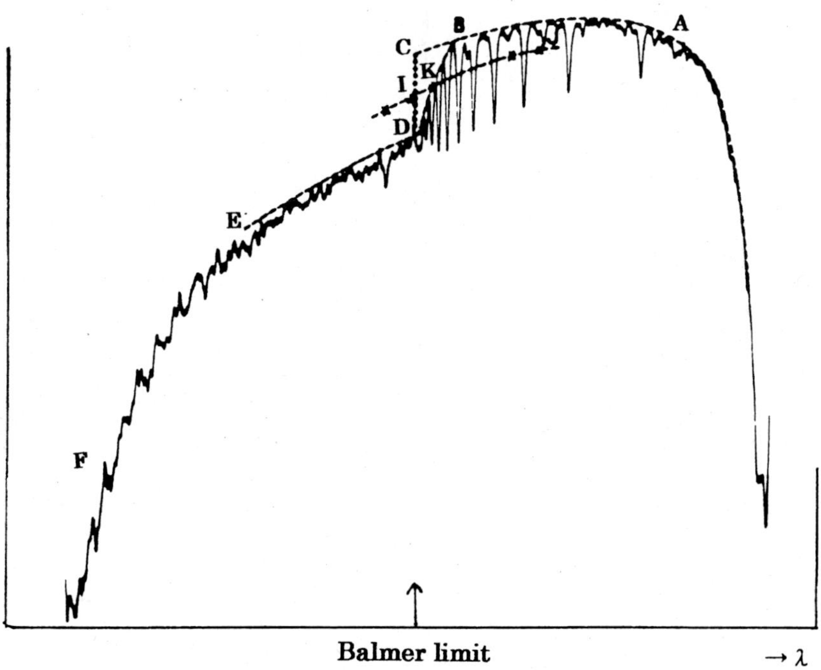

Figure 2.1: The definition of the Balmer jump. AB, continuous spectrum in blue region; BC: extrapolation of AB toward the Balmer limit; CD, the Balmer discontinuity (Balmer jump); DE, the Balmer continuum; IJ, the curve parallel to AB passing the central point of DE; K, the cross point of the curves BD and IJ. The Balmer jump D is defined as the intensity difference of points C and D. (See text)

spectrum in the red and violet sides of the series limit. The Balmer jump is defined by the following two parameters (D, λ_1).

D: A measure of the Balmer jump. Let I_c and I_d be the spectral intensities on the longward and shortward sides of the wavelength of the series limit, respectively, then the Balmer jump D is defined by the logarithmic difference as

$$D = \log I_c - \log I_d. \tag{2.1.9}$$

λ_1: effective wavelength of the jump, is defined at cross point K of two curves, one is parallel to AB passing through the midpoint of CD and the other is the part of stellar spectrum BD.

Parameters (D, λ_1) both depend upon the spectral type and the luminosity class, and we can draw a grid of the theoretical curves with constant $T_{\rm eff}$ and L on the ($D - \lambda_1$) plane. By plotting the observed values of these parameters on this grid, we can determine the two-dimensional spectral type. Figure 2.2 illustrates an example of the grid on the ($D - \lambda_1$) plane and the positions of observed stars (Chalonge and Divan 1952).

As seen in Figure 2.2, the Balmer jump D takes the maximum value at about A2 and declines toward both sides of early and late-types. If we can previously judge whether the stellar type is earlier or later by some eye estimation or other method, the BCD classification will provide a useful two-dimensional classification scheme.

2.1.4 HR diagram and stellar parameters

The HR diagram was originally introduced to express the relationship between spectral type (abscissa) and absolute magnitude (ordinate) of stars. The abscissa can be converted to color index ($B - V$), or logarithms of stellar effective temperature ($T_{\rm eff}$), and the ordinate to the bolometric magnitude ($M_{\rm bol}$) or stellar luminosity (L). For distant stars, color index is more easily accessible up to fainter objects. The HR diagram converted to ($\log T_{\rm eff} - L$) can be compared with the theories of stellar evolution.

In Figure 2.3, we illustrate the HR diagram for bright stars (V $<$ 2.5 mag) and nearby stars (distance $<$ 5 pc), along with approximate curves of luminosity class (V, III, Ib, Ia).

Physical parameters of stars are closely related to the spectral types. In Table 2.3 we summarize the colors and some physical parameters for stars of main-sequence stars, giants, and supergiants, separately.

In this table, the color index ($B - V$)is defined in (2.1.3), and is often adopted as stellar color instead of stellar spectral type. Absolute magnitude in V band, M_V, is derived from (2.1.4) with suitable correction for the interstellar absorption A. The bolometric correction BC is defined by (2.1.8) to derive the bolometric magnitude. The effective temperature $T_{\rm eff}$ denotes the temperature expression of the total luminosity of the star (see section 2.5.3) and declines with advancing spectral type. The surface gravity g, stellar mass M, and stellar radius R are given as the relative values to the Sun.

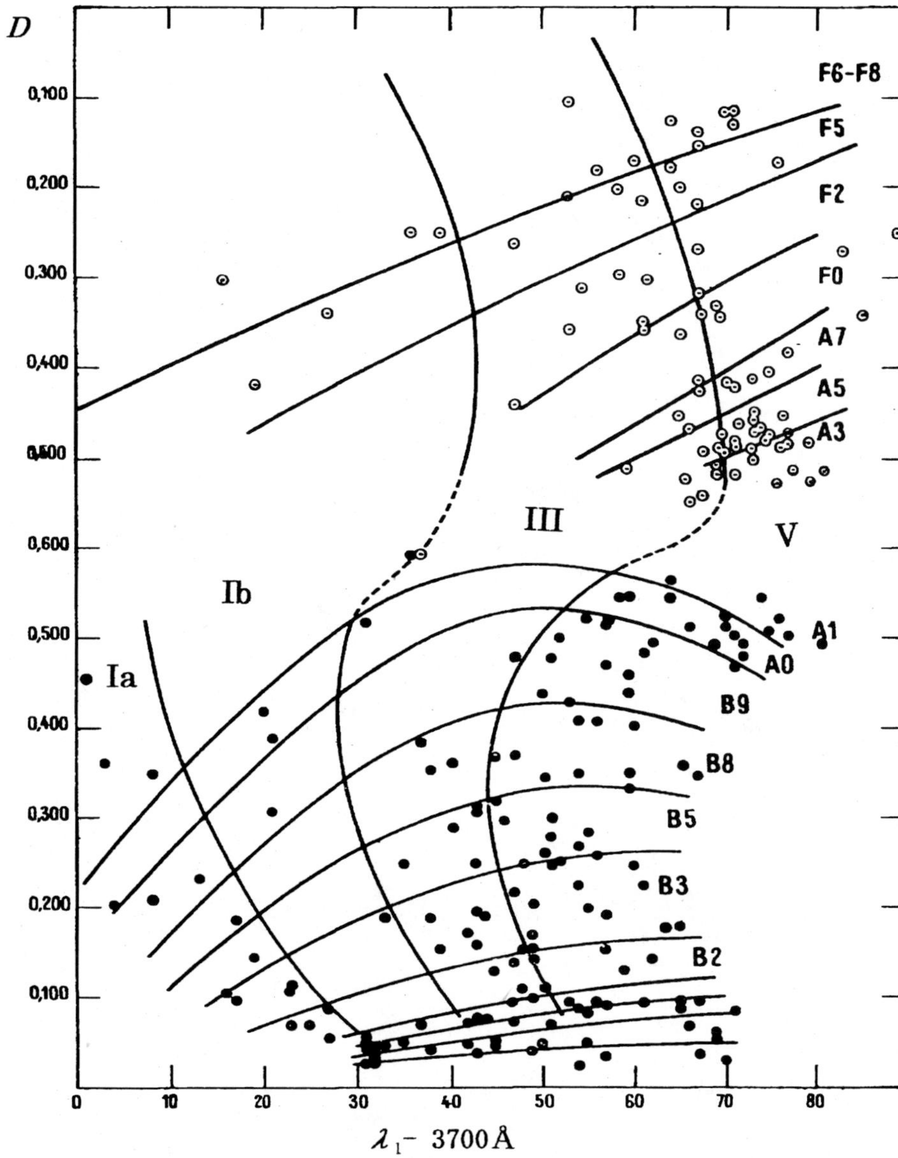

Figure 2.2: BCD classification. A grid of theoretical curves with constant values of T_{eff} (spectral type) and luminosity. Observed points are also plotted. The abscissa is $(\lambda_1 - 3700)$Å, and the ordinate is D. Note that this figure is a combination of two diagrams for stars later than A2 stars (upper half) and for stars earlier than A1 stars (lower half). The value of D takes its maximum at the center of the ordinate and declines to both upper and lower sides. (From Chalonge and Divan 1952)

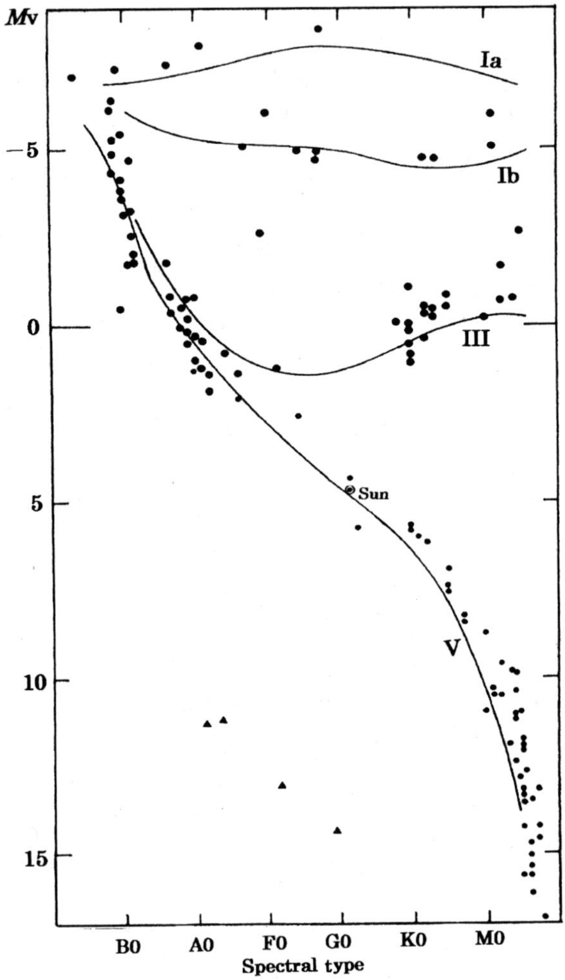

Figure 2.3: HR diagram for bright and nearby stars. Bright stars (V < 2.5 mag) (large filled circles), nearby stars (less than 5 pc) (small filled circles) and nearby white dwarfs (triangles) are plotted on the (spectral type, M_V) diagram. Approximate position of the luminosity classes Ia, Ib, III, and V are shown.

2.2 Atomic spectrum

Basic properties of atomic spectra are briefly introduced in this section, particularly for spectral lines familiar in emission-line stars such as hydrogen, helium, and calcium.

2.2.1 Hydrogen and ionized helium

Hydrogen atoms and spectral series

A hydrogen-like atom with nuclear charge Z consists of a nucleus and an electron connected with a total energy E. If E is positive, the atom is in the

Table 2.3: Physical parameters along spectral type

Spectral type	B-V	M_V	BC	T_{eff}	$\log g/g_\odot$	$M/M_\odot$	$R/R_\odot$
Main sequence stars (V)							
O5	−0.33	−5.7	4.40	42000	−0.4	60	12
B0	−0.30	−4.0	3.16	30000	−0.5	17.5	7.4
B5	−0.17	−1.2	1.46	15200	−0.4	5.9	3.9
A0	−0.02	+0.65	0.30	9790	−0.3	2.9	2.4
A5	+0.15	+1.95	0.15	8180	−0.15	2.0	1.7
F0	+0.30	+2.7	0.09	7300	−0.1	1.6	1.5
F5	+0.44	+3.5	0.14	6650	−0.1	1.4	1.3
G0	+0.58	+4.4	0.18	5940	−0.05	1.05	1.1
G5	+0.68	+5.1	0.21	5560	+0.05	0.92	0.92
K0	+0.81	+5.9	0.31	5150	+0.05	0.79	0.85
K5	+1.15	+7.35	0.72	4410	+0.1	0.67	0.72
M0	+1.40	+8.8	1.38	3840	+0.15	0.51	0.60
M5	+1.64	+12.3	2.73	3170	+0.5	0.21	0.27
Giant stars (III)							
G0	—	—	—	—	−1.5	1.0	6
G5	+0.86	+0.9	0.34	5050	−1.9	1.1	10
K0	+1.00	+0.7	0.50	4660	−2.3	1.1	15
K5	+1.50	−0.2	1.02	4050	−2.7	1.2	25
M0	+1.56	−0.4	1.25	3690	−3.1	1.2	40
M5	+1.63	−0.3	2.48	3380	—	—	—
Supergiant stars (I)							
O5	—	—	—	—	−1.1	70	30
O9	−0.27	−6.5	3.18	32000	—	—	—
B0	—	—	—	—	−1.6	25	30
B2	−0.17	−6.4	1.58	17600	—	—	—
B5	−0.10	−6.2	0.95	13600	−2.0	20	50
A0	−0.01	−6.3	0.41	9980	−2.3	16	60
A5	+0.09	−6.6	0.13	8610	−2.4	13	60
F0	+0.17	−6.6	0.01	7460	−2.7	12	80
F5	+0.32	−6.6	0.03	6370	−3.0	10	100
G0	+0.76	−6.4	0.15	5370	−3.1	10	120
G5	+1.02	−6.2	0.33	4930	−3.3	12	150
K0	+1.25	−6.0	0.50	4550	−3.5	13	200
K5	+1.60	−5.8	1.01	3990	−4.1	13	400
M0	+1.67	−5.6	1.29	3620	−4.3	13	500
M5	+1.80	−5.6	3.47	2880	—	—	—

Note: $g_\odot = 2.739 \times 10^4$ cm s^{-2}, $M_\odot = 1.989 \times 10^{33}$ g, $R_\odot = 6.959 \times 10^{10}$ cm (Cox 1999).

state of ionization. If E is negative, it takes discrete values E_n given by

$$E_n = -\frac{2\pi^2 m e^4}{h^2}\frac{Z^2}{n^2} \quad (n = 1, 2, \ldots), \tag{2.2.1}$$

where m is the mass of electron, e the electron charge, h the Planck constant.

The number n is called the principal quantum number. When $n = 1$, the atom is called in the ground state, and when $n > 1$, the atom is in excited state. Line emission or absorption by atom is expressed as a transition of electron between energy level n and m and follows Bohr's frequency condition

$$h\nu_{nm} = |E_n - E_m| \,, \tag{2.2.2}$$

where ν_{nm} is the corresponding frequency and can be converted to the wavelength by the relation $\lambda_{nm} = c/\nu_{nm}$.

From Equations (2.2.1) and (2.2.2) and inserting the values of physical constants, we have the following spectral series ($m > n$)

$$\lambda_{nm} = \frac{911.5069}{Z^2} \frac{n^2}{1 - (n/m)^2} \text{ Å}, \tag{2.2.3}$$

where $Z = 1$ for hydrogen and $Z = 2$ for ionized helium.

In hydrogen, the spectral series are named by the combinations of n and m as Lyman series ($n = 1, m = 2, 3, \ldots$), Balmer series ($n = 2$, $m = 3, 4, \ldots$), Paschen series ($n = 3, m = 4, 5, \ldots$), Bracket series ($n = 4$, $m = 5, 6, \ldots$), Pfund series ($n = 5, m = 6, 7, \ldots$), etc. The line wavelengths of each series and the name of lines are given in Table 2.4 for hydrogen and in Table 2.5 for ionized helium.

Table 2.4: Spectral series of hydrogen atom

Series m	Lyman $n=1$		Balmer $n=2$		Paschen $n=3$		Bracket $n=4$		Pfund $n=5$	
2	Lα	1215.24		—		—		—		—
3	Lβ	1023.44	Hα	6562.85		—		—		—
4	Lγ	972.27	Hβ	4861.37	Paα	18750.99		—		—
5	Lδ	949.48	Hγ	4340.51	Paβ	12818.06	Brα	40511.1		—
6	Lε	937.54	Hδ	4101.78	Paγ	10938.08	Brβ	26251.3	Pfα	74577.8
7		930.49	Hε	3970.11	Paδ	10049.36	Brγ	21655.1	Pfβ	46524.8
8		925.97		3889.09	Paε	9545.96	Brδ	19445.4	Pfγ	37395.1
9		922.90		3835.43		9229.0	Brε	18174.0	Pfδ	32960.7
10		920.71		3797.94		9014.90		17362.0		30383.5
11		919.10		3770.67		8862.77		16806.4		28721.9
12		917.88		3750.20		8750.46		16407.1		27574.9
13		916.93		3734.41		8665.02		16109.2		26743.8
14		916.18		3721.98		8598.38		15880.4		26119.2
15		915.57		3712.01		8545.37		15700.5		25636.1
16		915.08		3703.90		8502.47		15556.3		25253.8
17		914.67		3697.20		8467.24		15438.8		24945.5
18		914.32		3691.60		8437.94		15341.7		24692.9
19		914.03		3686.87		8413.31		15260.4		24483.1
20		913.79		3682.85		8392.39		15191.7		24306.8
Limit		911.50		3646.85		8203.56		14585.1		22787.6

Table 2.5: Spectral series of ionized helium

Series m	Lyman-1 $n = 1$	Lyman-2 $n = 2$	Fowler $n = 3$	Pickering $n = 4$	$n = 5$
2	303.7	—	—	—	—
3	256.3	1640.1	—	—	—
4	243.0	1214.9*	4686.1	—	—
5	237.3	1087.7	3203.4	10124.4	—
6	234.3	1025.0*	2733.6	6562.8*	18638
7	232.5	992.2	2511.4	5413.8	11637
8	231.4	971.9*	2385.6	4861.3*	9345.6
9	230.6	958.5	2306.4	4543.5	8237.4
10	230.1	949.1*	2252.9	4340.5*	7593.3
Limit	227.8	911.2*	2050.2	3646.0*	5695.0

*The wavelength coincident with that of the hydrogen Lyman or Balmer series.

The values of E_n given in (2.2.1) represent the eigenvalues of the Schrödinger equation, and the solutions having the eigenvalue are called eigenfunctions. In hydrogen, eigenfunction for the principal quantum number n is not unique but has some different forms distinguished by different set of quantum numbers (l, m), where l is the azimuthal quantum number, m the magnetic quantum number. For a given value of n, l takes the values $0, 1, \ldots (n-1)$, and the electron corresponding to the values of $l = 0, 1, 2, 3, 4, \ldots$ is called s, p, d, f, g, ... electron, respectively. Then the state of electron in hydrogen atom is expressed as shown in Table 2.6. The magnetic quantum number expresses the situation that each energy levels of hydrogen atoms in (n, l) state splits into $(2l+1)$ sublevels, from $m = -l, -l+1, -l+2, \ldots, +l-1, +l$, under the existence of magnetic field.

When the energy levels are distinguished by the state of electron, the levels are called terms such as 2p, 3d term. In the hydrogen-like atoms energy levels are determined by the principal quantum number regardless the difference in the state of electron. The graphical representation of the terms (energy levels) and possible spectral lines is called an energy level diagram or a Grotrian

Table 2.6: State of electron in hydrogen atom

l electron state	0 s	1 p	2 d	3 f	4 g
$n = 1$	1s				
2	2s	2p			
3	3s	3p	3d		
4	4s	4p	4d	4f	
5	5s	5p	5d	5f	5g
—	—	—	—	—	—

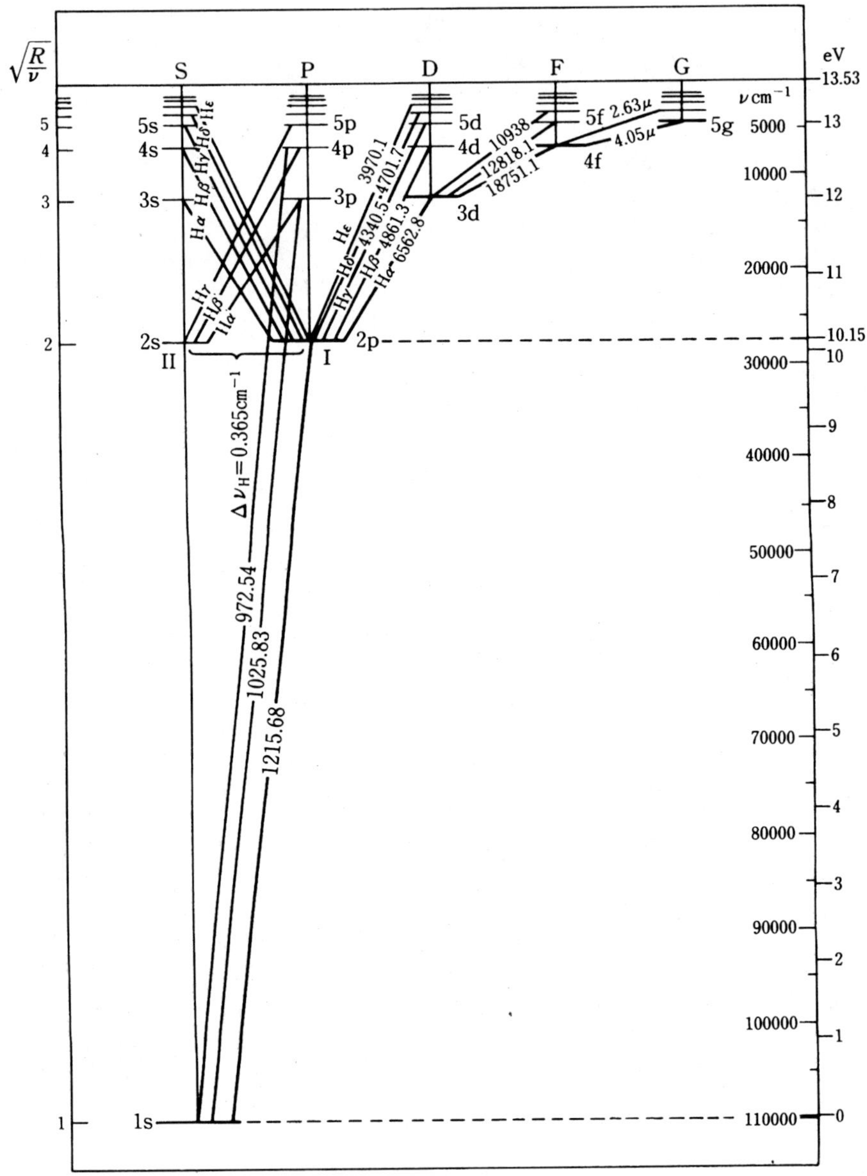

Figure 2.4: Grotrian diagram of HI.

diagram. Figure 2.4 illustrates the Grotrian diagram of HI atom. In Grotrian diagram, energy levels are typically expressed by electron volts (eV) and/or wave number ($1/\lambda$ in cm^{-1}). The Grotrian diagrams for a number of atoms and ions are given by Merrill (1958), along with wide descriptions on spectroscopic features of emission lines in stars.

2.2.2 Spectra of general atoms

General atoms

An atom or ion generally consists of a nucleus with charge Z and bounded electrons of number Z (atom) or less than Z (ion). These electrons are classified by the existence of the so-called shells of electrons. All electrons belonging to the same shell are characterized by the same principal quantum number n. The first shell $n = 1$, the second shell $n = 2$, the third shell $n = 3$, etc., are called closed, when they contain 2, 8, 18, 32, etc., electrons, respectively. These numbers are given by the formula $N = 2n^2$. The electrons in the outermost shell, which is not closed, act to absorb or emit radiation and are called the valence electrons. Atoms are called one-electron atoms, two-electron atoms, or multielectron atoms, according to the number of valence electrons, one, two, or more than two, respectively. We show some examples of each atom and its spectroscopic designation as

One-electron atom HI ($Z = 1$), HeII ($Z = 2$), LiI ($Z = 3$), NaI ($Z = 11$), CaII ($Z = 20$)
Two-electron atom HeI ($Z = 2$), CI ($Z = 6$), NII ($Z = 7$), MgI ($Z = 12$), CaI ($Z = 20$)
Multielectron atom NI ($Z = 7$), OI ($Z = 8$), NeII ($Z = 10$), SIII ($Z = 16$), FeI ($Z = 26$)

where the roman letter I, II, III denotes the neutral, once-ionized, twice-ionized state of the element, respectively.

One-electron atoms

Alkali atom (LiI, NaI, KI, etc.) and CaII belongs to one-electron atom and the configuration of electrons in the ground state is described as

LiI : $\underline{1s^2}2s$
NaI : $\underline{1s^2 2s^2 2p^6}3s$
CaII : $\underline{1s^2 2s^2 2p^6 3s^2 3p^6}4s$

where the underline denotes the closed electron shell.

Consider NaI atom. Though valence electron is one, the nuclear force is shielded by inner electron shells in some way so that the spectral series appears more complex than hydrogen-like atoms, and several series are named as shown in Table 2.7. The Grotrian diagram is given in Figure 2.5. The

Table 2.7: Spectral series of NaI

Electronic transition	Upper level	Name of the series
3p – ns	$n = 4, 5, 6, \ldots$	S (Sharp series)
3s – np	$n = 3, 4, 5, \ldots$	P (Prinsipal series)
3p – nd	$n = 3, 4, 5, \ldots$	D (Deffuse series)
3d – nf	$n = 4, 5, 6, \ldots$	F (Fundamental series)

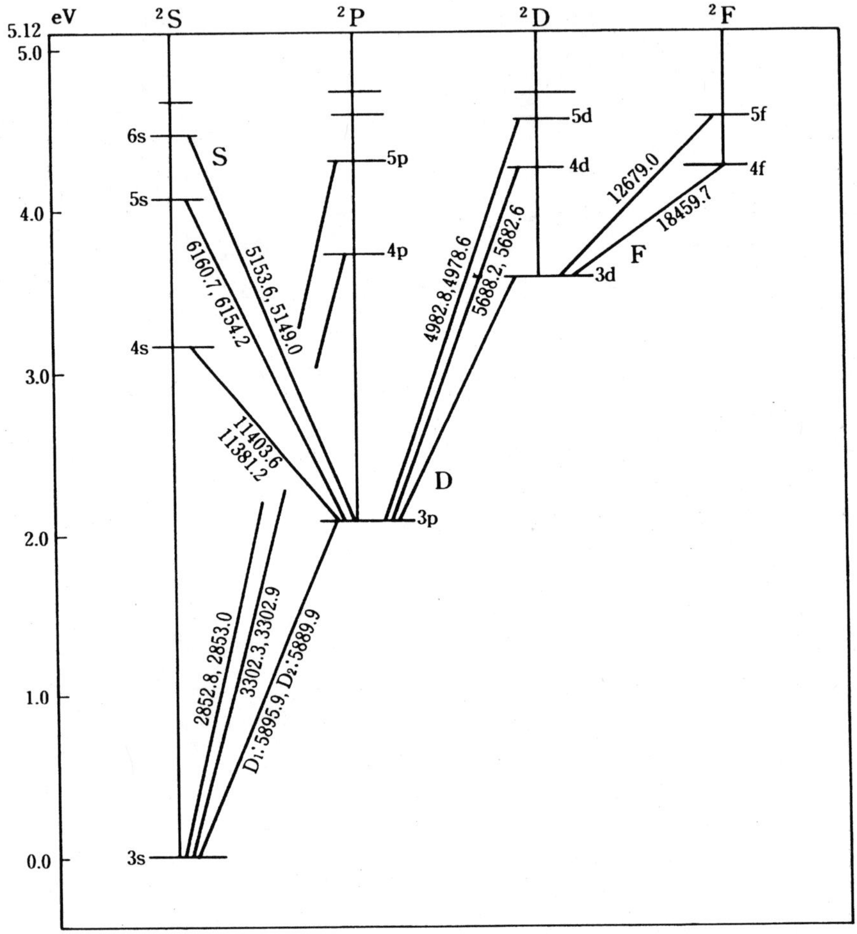

Figure 2.5: Grotrian diagram of NaI.

most conspicuous lines of NaI in optical region are the Fraunhofer D lines at λ 5895.923, λ 5889.953 Å corresponding to 3s–3p transition in Principal series. These lines make a doublet by the difference in electron spin quantum number $s = 1/2$ and $s = -1/2$. The lines by transitions between the ground state and excited levels are called the resonance lines, and the lines by transitions between excited levels are called the subordinate lines. NaI D lines are the resonance lines. In general, resonance lines yield the strongest lines in stellar spectra among the lines of the same atom (or ion).

Next, consider CaII ion. This is also one-electron ion but has much more complex than NaI by the large number of Z. The Grotrian diagram is shown in Figure 2.6. CaII H and K lines at λ 3968 and 3933 Å are the resonance lines and most conspicuous in stellar lines among CaII lines. The infrared CaII lines λ 8498, 8542, 8662 Å are next to H and K in importance in intermediate and late-type stars. The lower level 3d ^{2}D is a metastable level, then the lines λ 7291, 7323 Å are forbidden lines by the selection rules (see below).

Two- and multielectron atoms

For atoms with two or more valence electrons, we need a summation procedure of the energy states of individual electrons by quantum mechanics.

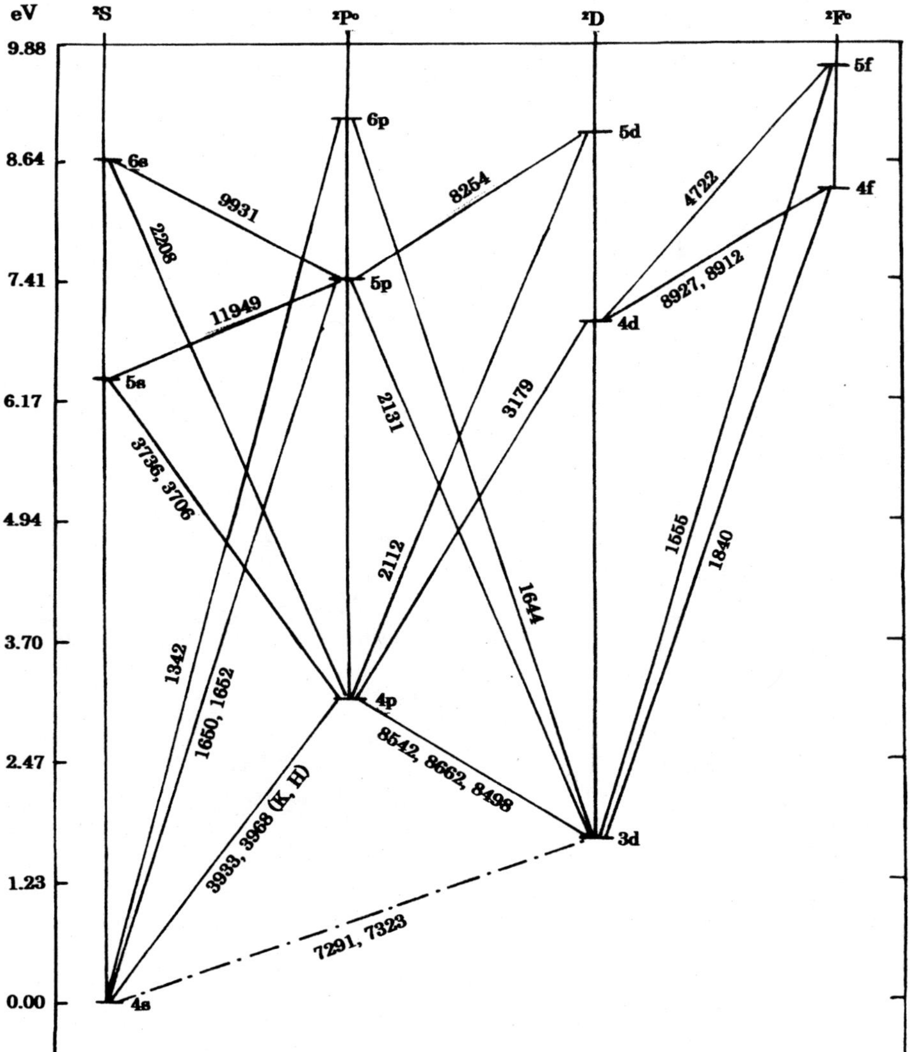

Figure 2.6: Grotrian diagram of CaII.

Consider HeI as a two-electron atom. The energy states of both electrons are designated by a set of quantum numbers (n_1, l_1, s_1) and (n_2, l_2, s_2). Though two electrons are mutually interacted electromagnetically, interaction (l_1, l_2) and (s_1, s_2) can be supposed stronger than the interaction (l_1, s_1) and (l_2, s_2). Then (l_1, l_2) and (s_1, s_2) are appropriately combined making the resultant orbital angular momentum L and the resultant spin quantum state S as

$$L = l_1 + l_2, l_1 + l_2 - 1, \ldots, \mid l_1 - l_2 \mid \tag{2.2.4}$$

$$S = s_1 + s_2, s_1 + s_2 - 1, \ldots, \mid s_1 - s_2 \mid . \tag{2.2.5}$$

In addition, L and S are summed to make the total angular momentum J as

$$J = L + S, L + S - 1, \ldots, \mid L - S \mid . \tag{2.2.6}$$

Since spin quantum number of each electron is 1/2, then $S = 1/2 \pm 1/2$, that is, $S = 0$ or 1. In this way we have

$$\begin{array}{ll} J = L & \text{when } S = 0 \\ J = L + 1, \;\; L, \;\; L - 1 & \text{when } S = 1. \end{array}$$

These relations indicate that the energy level is split into three components when $S = 1$ making a triplet, while the level is single when $S = 0$ making a singlet. As the results, HeI spectrum is separated into two groups of singlet and triplet. Transitions between singlet and triplet is prohibited by the selection rules of transition (see below) so that HeI forms separate spectral series: parahelium (singlet series) and orthohelium (triplet series). The Grotrian diagram of HeI is shown in Figure 2.7.

Multielectron atoms or ions, such as FeI, FeII ($Z = 26$), TiI, TiII ($Z = 22$), need more complex rules of electron summation though not shown here (see Herzberg 1944 or Aller 1963). In solar and stellar spectra they exhibit numerous absorption lines, often forming emission lines in some types of emission-line stars.

For the electronic transitions in atoms and ions, there are several rules to be obeyed, called the selection rules. Important rules contain the followings:

(1) There is no restriction in the transitions between principal quantum numbers. That is, transitions from level n to $n + \Delta n$ ($\Delta n = 0$, $\pm 1, \pm 2, \ldots$) can be made without restriction.
(2) The orbital quantum momentum l can change only for $\Delta l = \pm 1$.
(3) The change of total angular momentum J may occur for $\Delta J = \pm 1$ and 0, except the transition from $J = 0$ to $J = 0$.
(4) Transitions between different multiplicity such as between singlet and triplet are generally prohibited.

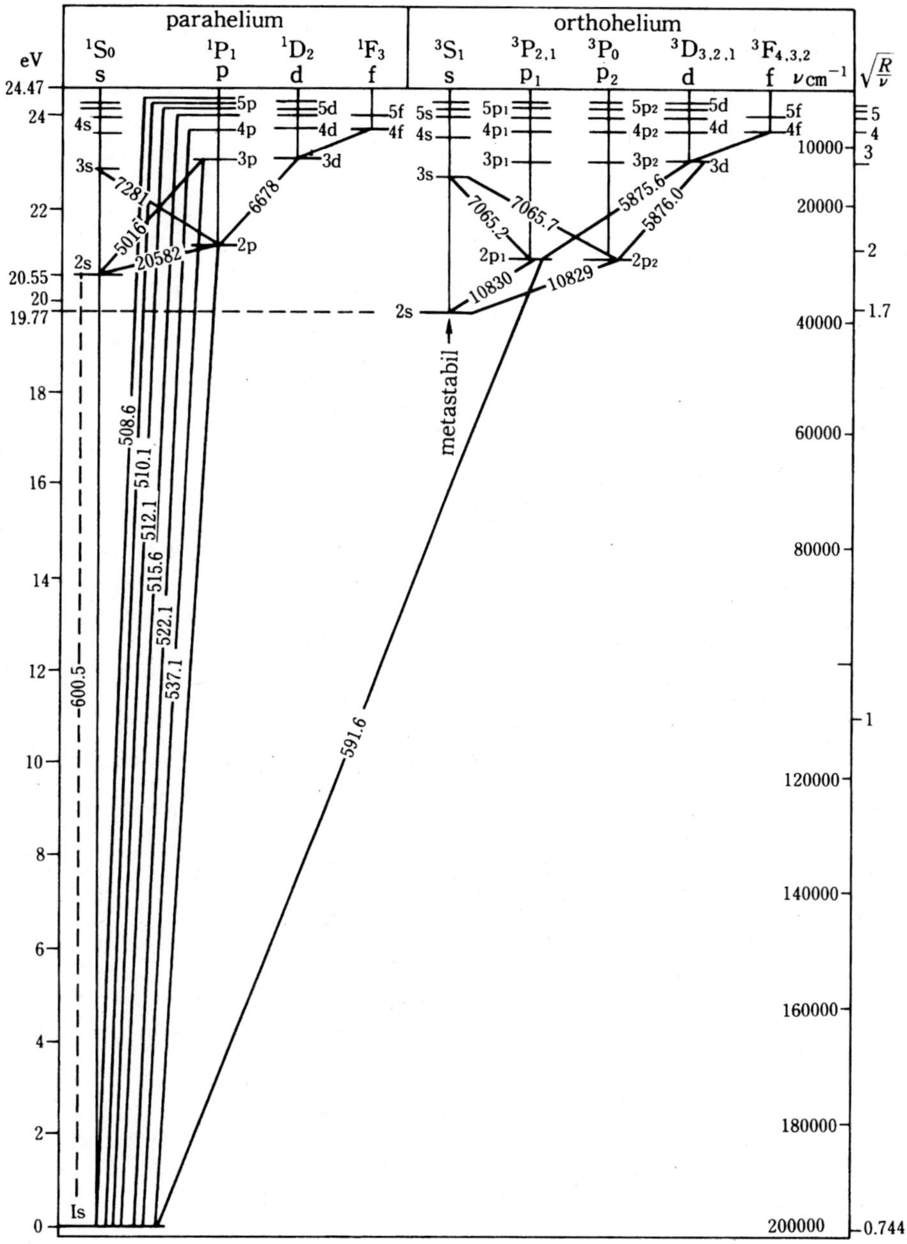

Figure 2.7: Grotrian diagram of HeI.

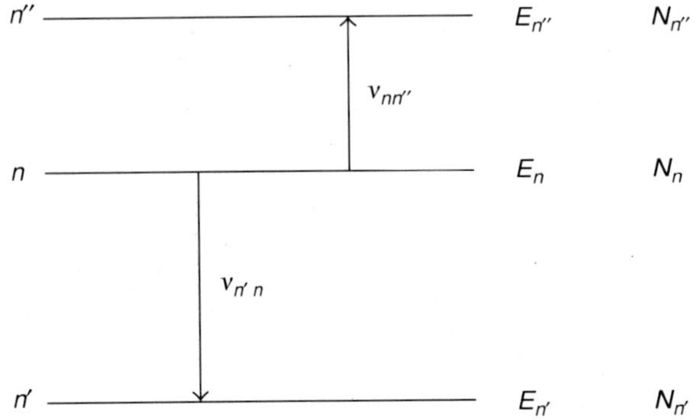

Figure 2.8: Energy levels and transitions. Level energy E and level population N are shown for levels n', n, and n''.

2.2.3 Emission and absorption processes in atoms or ions

As stated in Section 2.2.1, atoms or ions emit or absorb radiation by the transition between two energy levels as given in Equation (2.2.2). Let us consider electronic transitions from level n to a lower level n'(emission) and to an upper level n'' (absorption), as illustrated in Figure 2.8. These levels have the respective energies E_n, $E_{n'}$, $E_{n''}$, and level populations N_n, $N_{n'}$, $N_{n''}$. Any transitions can be performed as probabilistic process; hence the transition can be expressed by the transition probabilities. There are three types of transitions: spontaneous emission, induced emission, and absorption. They can be expressed by the Einstein coefficients as follows:

(i) Spontaneous emission. Atoms (or ions) excited to level n can emit the radiation $\nu_{nn'}$ by the spontaneous transition n to n' (spontaneous means transition without interaction with surrounding radiation field). The number of transitions of this type per unit volume per unit time are proportional to the level population and they can be expressed as

$$A_{nn'}N_n,$$

where $A_{nn'}$ is the coefficient for spontaneous emission.

(ii) Induced emission. If the atom is exposed to radiation, the atom in level n can emit the radiation with a frequency proportional to the radiation intensity $J_{nn'}$ and level population N_n; hence we can write the number of transitions as

$$B_{nn'}N_nJ_{nn'}.$$

$B_{nn'}$ is the coefficient for the induced emission. Hence the total number of atoms leaving level n for n' in unit time, in unit volume is given by

$$N_n(A_{nn'} + N_{nn'}J_{nn'}).$$

(iii) Absorption. When the atom is in the radiation field, the atom can absorb the radiation $\nu_{nn''}$ by transition to upper level n'', and the number of transitions is proportional to $J_{nn''}$ and N_n and given by

$$B_{nn''} N_n J_{nn''},$$

where $B_{nn''}$ is the coefficient for line absorption.

In later sections, we consider the radiation field in line radiation of frequency ν (in Section 2.4 and thereafter) and define the absorption coefficient κ_ν (cm –1) per frequency, and volume emissivity ε_ν (erg s^{-1}) per frequency. It is noticed that Einstein's coefficients are concerned with the total spectral line in concern (not per frequency), and then both the definitions are connected by the following relationship.

Absorption coefficient:

$$\int \kappa_\nu d\nu = \frac{1}{4\pi} h\nu_{nn''} \left(N_n B_{nn''} - N_{n''} B_{n''n}\right), \tag{2.2.7}$$

where the integration covers the whole range of line, and the coefficient $1/4\pi$ denotes the flow of radiation which is defined in unit of solid angle. The second term of the right-hand site indicates the contribution of induced emission which corresponds to the negative absorption.

Emission coefficient:

$$\int \varepsilon_\nu d\nu = \frac{1}{4\pi} h\nu_{nn'} N_n A_{nn'}, \tag{2.2.8}$$

The range of integration is the same as Equation (2.2.7).

Einstein coefficients $A_{nn'}$, $B_{n'n}$, and $B_{nn'}$ are not independent; mutual relation will be derived in Section 2.3.2.

2.2.4 Line intensities in spectral sequence

Spectral sequence of stars in the Harvard classification, given in Table 2.2, can be roughly understood in terms of the ionization potential (IP) of atoms or ions concerned. In Table 2.8, the IP of representative atoms and ions are shown.

Ionized helium lines, HeII λ 4541, λ 5411, λ 4686 Å, etc. are most conspicuous in the earliest spectral types O3–O4 ($T_{\rm eff} > 40,000$ K). In order to form these absorption lines, ionized helium should be excited to the third or fourth energy levels higher than 48 eV. By these high ionization and excitation potentials, HeII lines only appear in the earliest type and rapidly decline toward B type.

Neutral helium lines, such as HeI λ 3888, λ 5875 Å (triplet), λ 3964, λ 5015, λ 6678 Å (singlet), take the maximum absorption intensities in around B3 type ($T_{\rm eff} \sim$ 20,000 K.) Triplet lines are discomposed into three lines in high-dispersion spectra. By the requirement of sufficient exciting energies, these lines almost disappear in B9 or A0 type.

Table 2.8: IP of selected elements

Atom	Z	Stage of ionization I	II	III
		eV	eV	eV
H	1	13.6	—	—
He	2	24.6	54.4	—
Na	11	5.14	47.3	71.6
Ca	20	6.11	11.9	51.2
Ti	22	6.82	13.6	27.5
Fe	26	7.87	16.2	31.6

Hydrogen Balmer lines are widely observable along the spectral sequence because of high relative abundance of hydrogen and show the maximum intensities in around A0 type ($T_{\rm eff} \sim$ 10,000 K). The weakening of Balmer lines from B to O stars is caused by the ionization of hydrogen leading to the decrease of neutral hydrogen and then the decrease of the population in the second energy level. In contrast, the weakening of Balmer lines from A to late-type stars is caused by the shortage of energy to excite atoms to the second level (10.2 eV above the ground state).

Heavy elements are generally characterized by the low IP as seen in Table 2.8. Among these, NaI (and also KI) is a one-electron atom with low IP and the NaI D lines are conspicuous in late-type stars. Meanwhile, NaII (and KII) has closed electron shells so that the IP is high. This infers that NaII and KII lines are very rare in stellar spectra. In contrast, Ca, Ti, Fe, and other metallic elements have rather low IP for both neutral and once-ionized state so that both of CaI, TiI, FeI, and CaII, TiII, FeII, etc. are observable in a wide range of intermediate and late-type stars. In this way, metallic elements form a number of absorption lines, and though their abundance is low as compared to hydrogen and helium; they play an important role in the spectra of intermediate and late-type stars. In K and M type star, band structure by diatomic molecules such as TiO, CN, MgH become noticeable. In Table 2.2, G band around λ4300 Å is a mixture of atomic and molecular bands including CN band.

The atlas of representative stellar spectra by Yamashita et al. (1977) shows the stellar spectrograms obtained by the 91-cm reflector at the Okayama Astrophysical Observatory. It contains 127 standard stars and 42 peculiar stars with spectral dispersion 73 Å mm^{-1} Hγ in the wavelength range λ 3800–4900 Å. Major lines are identified. This atlas also presents the partial Grotrian diagrams for around 20 atoms or ions, including CII, CIII ($Z = 6$), NII, NIII, NIV ($Z = 7$), OII, OIII ($Z = 8$), SiII, SiIII, SiIV ($Z = 14$), TiII ($Z = 22$), and FeI, FeII ($Z = 26$).

2.3 Thermodynamic equilibrium and black-body radiation

2.3.1 Planck function

Color of a heated matter depends solely on its temperature and not on the specific substance of the matter. This radiation is called the thermal radiation. The source of an ideal thermal radiation is called the black body.

Physically, a black body is a light source in the state of thermodynamic equilibrium (TE) in which there is no internal flow of heat and radiation, anywhere within an enclosure. An ideal light source is best realized by a volume of gas enclosed by a perfectly insulated box with a tiny hole in its wall. The gas inside this box finally reaches to the state of TE, and radiation going out through the tiny hole is the black-body radiation. Speaking at the atomic level, TE is a state of detailed balancing, where any emission transition from one energy level to another occurs at the exact frequency with its opposite absorption transition. When such balancing is realized in all of the possible transitions in atoms, ions, or molecules, this gas is "in detailed balancing." Due to its detailed balancing, any atomic properties such as spectral lines are lost from the radiation, and then the black-body radiation is purely continuous. The wavelength distribution of black-body radiation is called the Planck function, $B_\lambda(T)$, and expressed as a function of temperature in the form

$$B_\lambda\,(T)\;d\lambda = \frac{2\pi hc^2}{\lambda^5}\,\frac{1}{e^{hc/\lambda kT}-1}\,d\lambda, \tag{2.3.1}$$

where the physical constants are light velocity $c = 2.998 \times 10^{10}$ cm s^{-1}, Boltzmann constant $k = 1.380 \times 10^{-16}$ erg K^{-1}, and Planck constant $h = 6.626 \times 10^{-27}$ erg s^{-1}. The spectral energy distributions for some sample temperatures are shown in Figure 2.9.

Based on Equation (2.3.1), we can derive some well-known radiation laws from the Planck function. As seen in Figure 2.9, the Planck function has a maximum intensity at a wavelength $\lambda_{\rm m}$ which is obtained by the condition of $\partial B_\lambda/\partial\lambda = 0$. As the result we have

$$\lambda_{\rm m}\,T = 2890\ (\mu{\rm m\,K}). \tag{2.3.2}$$

This is Wien's displacement law indicating that the wavelength at maximum intensity is inversely proportional to the Kelvin temperature.

Next, we derive the total radiation energy E by integrating $B_\lambda\,(T)$ in Equation (2.3.1) over the whole wavelength range as follows.

$$E = \int B_\lambda\,d\lambda = \frac{2k^4T^4}{h^3c^2}\int_0^\infty \frac{x^3}{e^x-1}\,dx = \sigma T^4 \quad ({\rm erg\ cm^{-2}s^{-1}}), \tag{2.3.3}$$

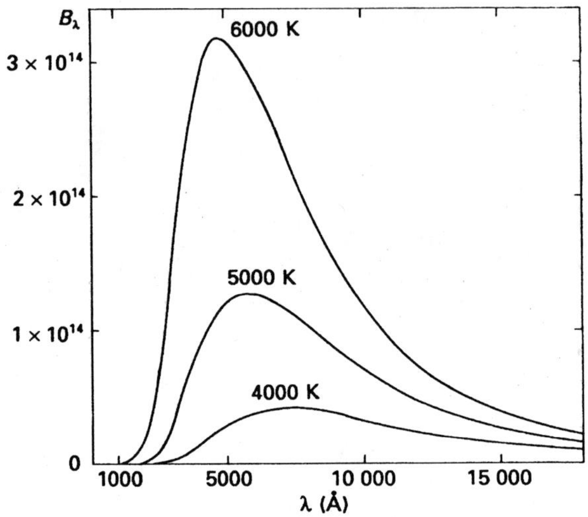

Figure 2.9: Spectral energy distribution of the Planck function. The ordinate gives the radiation energy per unit wavelength (cm), B_λ (erg cm^{-2} s^{-1} cm^{-1}), and the abscissa is the wavelength (Å).

where $\sigma = 2\pi^5 k^4/(15\ c^2\, h^3) = 5.67 \times 10^{-5}$ erg cm^{-2} s^{-1} K^{-4} is the Stefan-Boltzmann constant. This is the Stefan-Boltzmann's law stating that the total energy emitted from the black body is proportional to the fourth power of the Kelvin temperature.

Finally we give two approximate formulae for shorter and longer wavelength regions with respect to the wavelength λ_m at maximum intensity.

(i) Short wavelength side ($\lambda \ll \lambda_m$, or $hc/(\lambda kT) \gg 1$). We have Wien's approximation as follows

$$B_\lambda \approx \frac{2\pi hc^2}{\lambda^4}\, e^{-hc/(\lambda kT)}. \tag{2.3.4}$$

This is the approximation useful in the ultraviolet and X-ray regions.

(ii) Long wavelength side ($\lambda \gg \lambda_m$ or $hc/(\lambda kT) \ll 1$). We have Rayleigh-Jeans' approximation as

$$B_\lambda \approx \frac{2\pi c}{\lambda^4}\, kT. \tag{2.3.5}$$

This is appropriate for the infrared and radio spectral regions.

2.3.2 Boltzmann's law and Einstein coefficients

In the state of TE, the fundamental relationship is Boltzmann's law (or Boltzmann's distribution) which states that the relative number of atoms in the lower levels $E_{n'}$ and higher level E_n is given by

$$\frac{N_n}{N_{n'}} = \frac{g_n}{g_{n'}} e^{-\chi_{nn'}/kT}, \tag{2.3.6}$$

where $\chi_{nn'} = E_n - E_{n'}$ and g_n and $g_{n'}$ are the statistical weight for level n and n', respectively. The statistical weight of a level expresses the total number of sublevel. In general atom at level n, g_n is given by the total angular quantum number $2J_n + 1$. In case of hydrogen atom, level n is split by a set of (l, m), where l takes the values 0, 1, ... $(n-1)$, and the m takes $(2l+1)$ levels so that the number of sublevels, that is, the statistical weight g_n is given by

$$g_n = 2 \times \sum_{l=0}^{n-1} (2l+1) = 2n^2, \tag{2.3.7}$$

where factor 2 is caused by the split of electron spins to $-1/2$ and $1/2$. Thus in hydrogen atoms, level population relative to the ground state is

$$\frac{N_n}{N_1} = n^2 e^{-\frac{\chi_{n1}}{kT}}. \tag{2.3.8}$$

This is the case of TE. In non-TE state, deviation from this distribution becomes remarkable and plays important roles, particularly, in the formation of emission lines as we shall see in Chapter 4.

Now, we consider the detailed balancing in the TE state. The number of downward transition from n to n' must equal the number of upward transition from n' to n. This balancing can be written by making use of Einstein coefficient as

$$N_n \left(A_{nn'} + B_{nn'} I_{n'n}\right) = N_{n'} B_{n'n} I_{n'n}. \tag{2.3.9}$$

Combining Equations (2.3.6) and (2.3.9), and recalling that under conditions of TE $I_{nn'}$ is given by the Planck function (2.3.1) expressed by wave frequency, we have, after some manipulation,

$$\frac{g_n}{g_{n'}} A_{nn'} = \frac{2h\nu^3}{c^2} B_{n'n} \frac{\left[e^{h\nu/kT} - \frac{g_n}{g_{n'}} \frac{B_{nn'}}{B_{n'n}}\right]}{e^{h\nu/kT} - 1}. \tag{2.3.10}$$

In this Equation, $A_{nn'}$ is a constant of the atom only, independent of the temperature. Therefore, in order for this equation to hold at any temperature we must have

$$g_n B_{nn'} = g_{n'} B_{n'n} \tag{2.3.11}$$

and

$$\frac{g_n}{g_{n'}} A_{nn'} = B_{n'n} \frac{2h\nu^3}{c^2}. \tag{2.3.12}$$

Notice that the Einstein coefficients are atomic constants. Although the relations (2.3.11) and (2.3.12) are derived under the conditions of TE, these relations hold in any case of non-TE state.

2.4 Concepts of spectral-line formation

2.4.1 Equations of radiative transfer

Suppose a volume element with length ds and cross section $d\sigma$ in a gas layer. The radiation passing through this volume perpendicularly to the surface $d\sigma$ suffers some absorption and reduces its intensity. The absorbed energy dE_λ is supposed to be proportional to the intensity of radiation and the distance ds, and can be written as

$$dE_\lambda = -\kappa_\lambda \, E_\lambda \, ds = -\kappa_\lambda \, ds \, I_\lambda \, d\omega \, d\lambda \, d\sigma, \tag{2.4.1}$$

where the constant κ_λ (cm^{-1}) is called the absorption coefficient. At the same time, the radiation intensity is increased by emission from this volume element. The amount of increase is written as

$$dE_\lambda = \varepsilon_\lambda \, d\omega \, d\lambda \, ds \, d\sigma, \tag{2.4.2}$$

where ε_λ (erg cm^{-3} s^{-1}) is the energy of radiation emitted from unit volume per unit time and called the volume emissivity. The net increase (or decrease) of radiation, dE_λ, is given by the sum of absorption and emission as follows

$$dE_\lambda \;=\; dI_\lambda \, d\sigma \, d\lambda \, d\omega \;=\; -\,\kappa_\lambda \, I_\lambda \, d\omega \, d\sigma \, ds \;+\; \varepsilon_\lambda \, d\omega \, d\lambda \, ds \, d\sigma.$$

Then we have

$$\frac{dI_\lambda}{ds} = -\kappa_\lambda \, I_\lambda + \varepsilon_\lambda. \tag{2.4.3}$$

If we define the optical thickness $d\tau_\lambda$ as

$$d\tau_\lambda = \kappa_\lambda \, ds, \tag{2.4.4}$$

then Equation (2.4.3) takes the form

$$\frac{dI_\lambda}{d\tau_\lambda} = -I_\lambda + S_\lambda, \tag{2.4.5}$$

where S_λ is called the source function and given by

$$S_\lambda = \frac{\varepsilon_\lambda}{\kappa_\pi}. \tag{2.4.6}$$

Equation (2.4.5) is the equation of radiative transfer for a gas layer.

2.4.2 Absorption versus emission

Let us consider one-dimensional flow of radiation through a gas tube in a laboratory. Distance s and the optical thickness τ_λ in a tube are taken as shown in Figure 2.10. Then the equation of the radiative transfer in this tube is given by Equation (2.4.5).

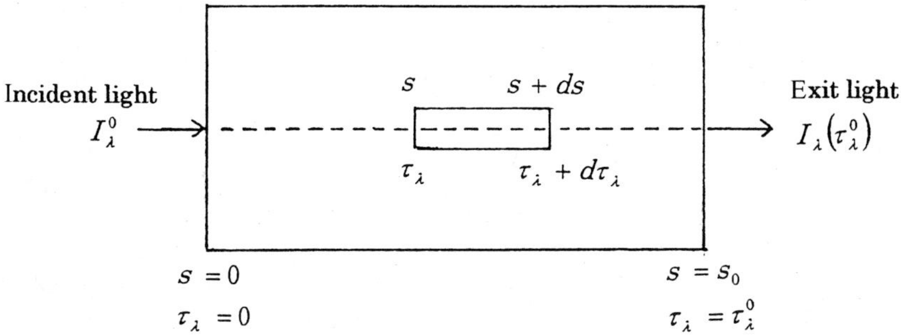

Figure 2.10: Flow of radiation in a gas tube.

If the physical state of gas is uniform, the source function S_λ is constant and then the transfer equation can easily be solved as

$$I_\lambda(\tau_\lambda) = I_\lambda^0 \, e^{-\tau_\lambda} + S_\lambda \left(1 - e^{-\tau_\lambda}\right), \tag{2.4.7}$$

where I_λ^0 is the incident radiation coming from a continuum source. Let τ_λ^0 be the optical thickness of the gas tube, the emergent intensity at the tube exit can be obtained from Equation (2.4.7). We shall consider the solutions in some special cases.

In case of $I_\lambda^0 = 0$ (i.e., no incident continuum radiation) emergent intensity is given by

$$I_\lambda\left(\tau_\lambda^0\right) = S_\lambda \left(1 - e^{-\tau_\lambda^0}\right). \tag{2.4.8}$$

We have two limiting cases.

(a) $\tau_\lambda^0 \ll 1$, i.e., gas tube is optically thin, we have in this case

$$I_\lambda\left(\tau_\lambda^0\right) \approx \tau_\lambda^0 \, S_\lambda. \tag{2.4.9}$$

If S_λ is constant inside the tube, we have

$$I_\lambda\left(\tau_\lambda^0\right) \approx \tau_\lambda^0 \, S_\lambda \approx \kappa_\lambda \, s_0 \, S_\lambda. \tag{2.4.10}$$

We may see that, since s_0 and S_λ were taken as constant, the emergent intensity is proportional to the absorption coefficient. For a spectral line, κ_λ has a sharp maximum at some wavelength, then, the emergent intensity also exhibits a maximum at the same wavelength. In this case, the light leaving from this gas tube forms an emission line spectrum.

(b) $\tau_\lambda^0 \gg 1$, i.e., gas tube is optically thick. In this case we have

$$I_\lambda\left(\tau_\lambda^0\right) \approx S_\lambda \approx \text{constant}, \tag{2.4.11}$$

and the emergent intensity exhibits no wavelength dependence. Gas in the tube approaches toward TE state, i.e., $S_\lambda \approx B_\lambda$, the spectrum takes the form of approximate Planck function.

In case of $I_\lambda^0 \neq 0$, we have also two limiting cases.

(a) $\tau_\lambda^0 \ll 1$, i.e., gas tube is optically thin; the emergent intensity is written as

$$\begin{aligned} I_\lambda\left(\tau_\lambda^0\right) &\approx I_\lambda^0\left(1-\tau_\lambda^0\right)+\tau_\lambda^0 S_\lambda \\ &= I_\lambda^0+\tau_\lambda^0\left(S_\lambda-I_\lambda^0\right). \end{aligned} \tag{2.4.12}$$

It is seen that if $I_\lambda^0 > S_\lambda$, the last term in the right-hand side is negative, and the radiation intensity reduces by the absorption proportional to the optical thickness. This is the formation of absorption line spectrum and is realized when the temperature of gas in the tube is lower than the temperature of the continuum light source. On the other hand, if $I_\lambda^0 < S_\lambda$, the emergent intensity is stronger than the incident radiation, and we see the formation of emission lines superimposed on the continuous spectrum.

(b) $\tau_\lambda^0 \gg 1$, i.e., gas tube is optically thick, we have in this case

$$I_\lambda\left(\tau_\lambda^0\right) \approx S_\lambda.$$

Emergent intensity exhibits no wavelength dependence. When gas in the tube approaches the TE state, i.e., $S_\lambda \approx B_\lambda$, the spectrum takes the form of approximate Planck function.

There is another process to form emission lines superimposed on the continuous spectrum. That is when a continuum light source and an optically thin gas tube are placed in parallel along a line of sight. If this system is located at a sufficiently large distance, the observer will not resolve the two sources and see the composite spectrum showing emission lines on a continuous radiation. This is analogous to a star surrounded by an extended envelope, where the star acts as a continuous light source and the envelope as an optically thin gas tube.

Based on the above formulation, we can summarize the situation for the formation of emission-line spectrum.

(1) An optically thin volume of gas with no background light emits emission-line spectrum as in emission nebulae.
(2) An optically thin volume of hot gas in front of background cooler TE source emits emission lines superimposed on a continuous spectrum. Stellar chromosphere above the cooler photosphere may be of this type.
(3) Two parallel sources composed of an optically thin volume of gas and an optically thick TE source, seen from a distant observer. This is analogous to a star surrounded by an optically thin extended envelope.

2.4.3 Source function and black-body radiation

When a gas medium is in the state of TE, there exists no flow of radiation in the medium as stated in Section 2.3. In this case we can put $dI_\lambda/d\tau_\lambda = 0$ in

Equation (2.4.5) and hence we get

$$I_\lambda = S_\lambda. \tag{2.4.13}$$

On the other hand, since radiation from TE gas is given by the Planck function B_λ in Equation (2.3.1), we have

$$I_\lambda = B_\lambda. \tag{2.4.14}$$

Combining these equations we get

$$S_\lambda = I_\lambda = B_\lambda. \tag{2.4.15}$$

This indicates that the source function in TE gas is the Planck function and only depends on temperature. From Equations (2.4.6) and (2.4.15) we have

$$\varepsilon_\lambda = \kappa_\lambda B_\lambda. \tag{2.4.16}$$

This relationship is the so-called Kirchhoff's law stating that the volume emissivity of a medium at a temperature is proportional to the absorption coefficient of this medium.

In stellar atmospheres, the source function is an unknown function of the depth from the surface. So we often assume the TE state at the depth τ_λ and put $S_\lambda = B_\lambda(\tau_\lambda)$ in the equation of radiative transfer. In this case, stellar atmospheres are called in the state of local thermodynamic equilibrium (LTE).

2.5 Stellar atmospheres and formation of absorption lines

2.5.1 Radiation fields of stellar atmospheres

The stellar atmosphere is the field of outflow radiation from inner to outer layers, and we observe the radiation emitted in the line-of-sight direction (Figure 2.11). We now consider the radiation field of a stationary atmosphere viewed by an outside observer.

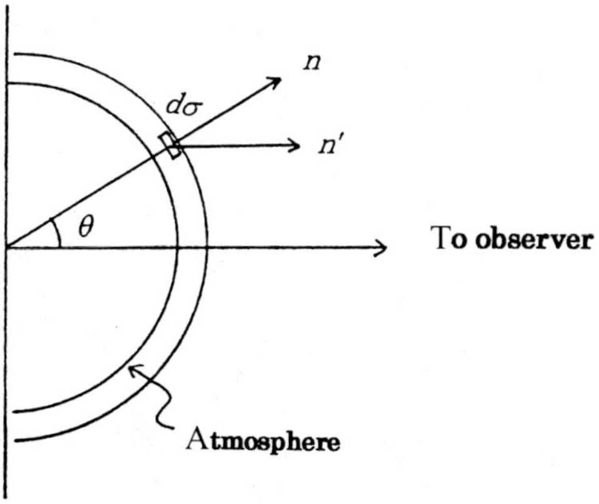

Figure 2.11: Stellar atmosphere observed by an observer.

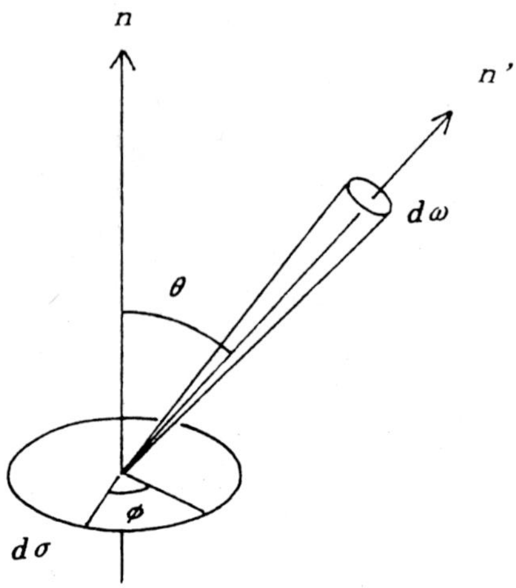

Figure 2.12: Definition of the specific intensity of radiation.

The radiation field is defined as the spatial function of the specific intensity of radiation $I(\theta,\ \varphi)$. Consider a surface element $d\sigma$ and the light passing through a narrow cone of solid angle $d\omega\,(\theta,\ \varphi)$ toward a direction n' which makes an angle θ to the normal n of the surface (Figure 2.12). Let dE_λ be the radiative energy passing through this narrow cone in the wavelength range λ and $\lambda + d\lambda$ in unit time. Then dE_λ can be written by using the specific intensity $I_\lambda\,(\theta,\ \varphi)$ as follows:

$$dE_\lambda = I_\lambda\,(\theta, \phi)\ d\lambda\ \cos\theta\, d\sigma\, d\omega. \tag{2.5.1}$$

The specific intensity $I_\lambda\,(\theta,\ \varphi)$ expresses the amount of radiation energy passing through an unit area of the surface $d\sigma$ per unit solid angle, per unit wavelength, and per unit time, to the direction $(\theta,\ \varphi)$ from the normal of the surface $d\sigma$.

The total amount of energy leaving $d\sigma$ in all direction per wavelength per second is written as $\pi F_\lambda\, d\sigma$, where F_λ defines the flux of radiation, and it is given as an integral of the specific intensity over the whole solid angle as follows.

$$\begin{aligned}\pi F_\lambda &= \int I_\lambda\ \cos\theta\, d\omega \\ &= \int_0^{2\pi}\int_0^{\pi} I_\lambda\ \cos\theta\ \sin\theta\, d\theta\, d\phi.\end{aligned} \tag{2.5.2}$$

The flux of radiation obviously vanishes when the radiation field is isotropic. In the stellar atmospheres, however, there is a general flow of radiation toward the outer space and so the flux does not vanish in general. The amount of net flux is defined as the difference of outward flux F_λ^+ and inward flux F_λ^-, each of which is defined by

$$\pi F_\lambda^+ = \int_0^{2\pi} \int_0^{\pi/2} I_\lambda \cos\theta \sin\theta \, d\theta \, d\phi \tag{2.5.3}$$

$$\pi F_\lambda^- = \int_0^{2\pi} \int_{\pi/2}^{\pi} I_\lambda \cos\theta \sin\theta \, d\theta \, d\phi.$$

Thus the net flux is given by

$$F_\lambda = F_\lambda^+ - F_\lambda^-. \tag{2.5.4}$$

In contrast to the Sun, stars are generally too distant to resolve their surfaces. We only observe the total radiation integrated over the hemisphere facing us (Figure 2.11). This is just the flux of radiation defined in Equations (2.5.2) and (2.5.3). Therefore, when we treat the stellar spectrum, we should make use of the flux of radiation F_λ, instead of the specific radiation I_λ.

We also define the mean intensity of radiation J_λ as the specific intensity averaged over all directions as

$$J_\lambda = \frac{1}{4\pi} \int I_\lambda \cos\theta \, d\omega. \tag{2.5.5}$$

The total intensity, total flux, and total mean intensity of radiation, integrated over the whole wavelength range, are respectively defined by the following integrals:

$$I = \int I_\lambda \, d\lambda, \qquad F = \int F_\lambda \, d\lambda, \qquad J = \int J_\lambda \, d\lambda. \tag{2.5.6}$$

2.5.2 Radiative transfer and limb darkening

Let us consider the radiation field in a plane–parallel stellar atmosphere. Suppose a radiation beam passing through a volume elements $d\sigma\, ds$ along the path s which makes an angle θ from the normal. The geometry is shown in Figure 2.13. Geometrical depth t and optical depth τ_λ are measured downward from the upper surface of the atmosphere, whereas the path s is directed toward the observer. Then we have

$$\begin{aligned} d\tau_\lambda &= \kappa_\lambda \, dt \\ ds &= -\, dt \sec\theta. \end{aligned} \tag{2.5.7}$$

The equation of radiative transfer along the path s for the radiation at a wavelength λ can be written similarly with Equation (2.4.3):

$$\frac{dI_\lambda(\tau_\lambda, \theta)}{ds} = -\,\kappa_\lambda I_\lambda(\tau_\lambda, \theta) + \varepsilon_\lambda. \tag{2.5.8}$$

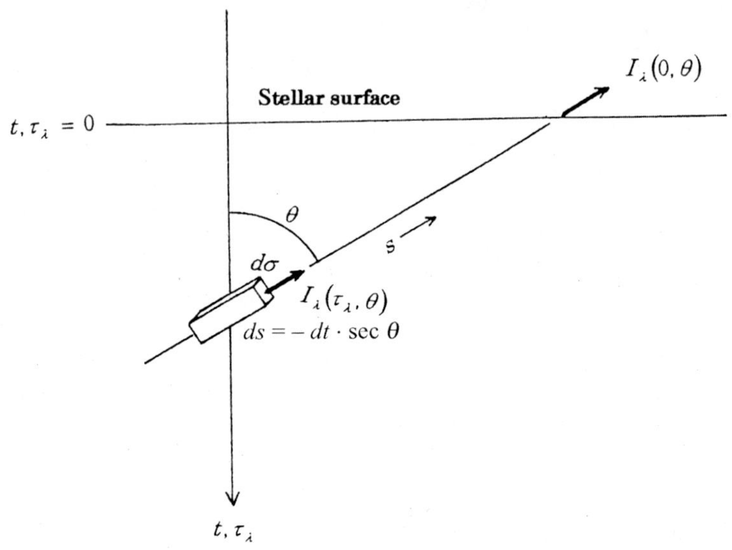

Figure 2.13: Geometry of a light beam passing through a volume element $d\sigma\ ds$ with an angle θ with respect to the normal to the surface.

By making use of the Relation (2.5.7), we have

$$\cos\theta \frac{dI_\lambda(\tau_\lambda, \theta)}{d\tau_\lambda} = I_\lambda(\tau_\lambda, \theta) - S_\lambda. \tag{2.5.9}$$

This equation differs from Equation (2.4.5) in the sign of the right-hand side. This is due to the definition of the optical depth, which is usually taken downward from the surface in the case of stellar atmospheres.

In stellar atmosphere, the source function S_λ is an unknown function of the optical depth since the temperature distribution along the depth is unknown. For simplicity, we first assume that the source function is a given function, then Equation (2.5.9) is formally solved to give the surface intensity of the radiation $I_\lambda\ (0,\ \theta)$ as follows:

$$I_\lambda(0, \theta) = \int_0^\infty S_\lambda(\tau_\lambda)\ \exp\left(-\tau_\lambda \sec\theta\right)\ d\left(\tau_\lambda \sec\theta\right). \tag{2.5.10}$$

This equation represents the solar limb darkening which is the declining of surface brightness of the solar disk from the disk center to the limb. In Equation (2.5.10), the surface intensity is $I_\lambda\ (0,\ 0)$ at the disk center $(\theta = 0^0)$, and $I_\lambda(0, \theta)$ at angle θ from the center. Let us consider a simple case where the source function is given as a linear equation of the optical depth as given

$$S_\lambda(\tau_\lambda) = a_\lambda + b_\lambda\, \tau_\lambda, \tag{2.5.11}$$

where a_λ and b_λ are constants to be determined. Inserting this relation to Equation (2.5.10), we have

$$\begin{aligned} I_\lambda(0,\theta) &= a_\lambda \int_0^\infty \exp(-\tau_\lambda \sec\theta)\, d(\tau_\lambda \sec\theta) \\ &\quad + b_\lambda \int_0^\infty \tau_\lambda \exp(-\tau_\lambda \sec\theta)\, d(\tau_\lambda \sec\theta) \\ &= a_\lambda + b_\lambda \cos\theta. \end{aligned} \tag{2.5.12}$$

This is the equation where τ_λ of (2.5.11) is replaced by $\cos\theta$. Combining these two equations we have

$$I_\lambda(0,\theta) = S_\lambda(\tau_\lambda = \cos\theta) = a_\lambda + b_\lambda \cos\theta. \tag{2.5.13}$$

This expresses the law of limb darkening of the solar photospheric disk, that is, we have

$$\begin{aligned} &I_\lambda(0,\ 0) = a_\lambda + b_\lambda &&\text{at the disc center } (\theta = 0) \\ &I_\lambda(0, \pi/2) = a_\lambda &&\text{at the limb } (\theta = \pi/2), \end{aligned}$$

and thus the surface intensity decreases from $a_\lambda + b_\lambda$ to a_λ according to the law of darkening in Equation (2.5.13). This law can be compared with observed darkening to determine the constants a_λ, b_λ and then to deduce the depth dependence of the source function for the atmospheric layer that the radiation at wavelength λ is emitted.

2.5.3 Radiative flux and effective temperature

As stated in Section 2.5.1, the observed radiative energy from the stellar surface is the flux of radiation given by (2.5.2). Suppose a plane parallel atmosphere as in Figure 2.12, and assume that there is no incident radiation from outside, then the flux of radiation at the stellar surface is given by the upper Equation of (2.5.3), i.e.,

$$\begin{aligned} \pi F_\lambda^+(0) &= \int_0^{2\pi}\int_0^{\pi/2} I_\lambda(0,\theta)\cos\theta \sin\theta\, d\theta\, d\phi \\ &= 2\pi \int_0^1 I_\lambda(0,\theta)\cos\theta\, d(\cos\theta). \end{aligned} \tag{2.5.14}$$

We again consider a simple case where the source function is given by (2.5.11). Then, inserting Equation (2.5.12) for I_λ $(0,\ \theta)$ in Equation (2.5.14), we have

$$\begin{aligned} \pi F_\lambda(0) &= 2\pi \int_0^1 (a_\lambda + b_\lambda \cos\theta)\cos\theta\, d(\cos\theta) \\ &= \pi\left(a_\lambda + \frac{2}{3} b_\lambda\right). \end{aligned} \tag{2.5.15}$$

By comparing this with the source function (2.5.11), we get

$$F_\lambda(0) = S_\lambda\left(\tau_\lambda = \frac{2}{3}\right). \quad (2.5.16)$$

This is the so-called Eddington–Barbier relation and shows that the flux coming out of the stellar surface equals the source function at the optical depth $\tau_\lambda = 2/3$. Though this is accurate only to the degree that the source function can be approximated by a linear equation as in (2.5.11), this relation is useful for the understanding of the formation of stellar spectra.

We now consider the stellar atmosphere under the state of LTE, i.e., $S_\lambda(\tau_\lambda) = B_\lambda(\tau_\lambda)$. We then have

$$F_\lambda(0) = B_\lambda(\tau_\lambda = 2/3). \quad (2.5.17)$$

For simplicity, let us consider a gray atmosphere where the absorption coefficient κ_λ is independent of λ. By putting $\kappa_\lambda = \kappa$, $\tau_\lambda = \tau$, we have

$$F_\lambda(0) = B_\lambda(T(\tau = 2/3)). \quad (2.5.18)$$

Thus the emergent flux is given by a black-body radiation at the optical depth $\tau = 2/3$. Let T_1 be the temperature at this depth, i.e., $T_1 = T(\tau = 2/3)$, then the total flux of radiation $F(0)$ is given as

$$\pi F(0) = \pi\int_0^\infty F_\lambda(0)\, d\lambda = \pi\int_0^\infty B_\lambda(T_1)\, d\lambda, \quad (2.5.19)$$

where the last term can be written as σT_1^4 by Stefan–Boltzmann's law.

We now define the effective temperature of a star T_{eff} as a measure of the total emergent energy as follows:

$$\pi F(0) = \sigma T_{\text{eff}}^4. \quad (2.5.20)$$

Then Equation (2.5.19) gives the relation

$$T_{\text{eff}}^4 = T_1^4 = [T(\tau = 2/3)]^4. \quad (2.5.21)$$

We can see that the effective temperature of a star equals the temperature of the atmosphere at the optical depth $\tau = 2/3$.

2.5.4 Radiative equilibrium and temperature gradient

Suppose the total energy flux πF passing through the unit area on the sphere with radius r from the star's center. The flux F is generally the sum of the fluxes in radiative (F_{r}), convective (F_{conv}), and conductive (F_{cond}) energies. For stars in a stationary state, the total energy flux through the whole sphere should be kept constant at any radius. In this case the system is called in the state of thermal equilibrium. By this definition, the condition of thermal equilibrium can be given by

$$\frac{d}{dr}\left(r^2 \pi F\right) = 0. \quad (2.5.22)$$

If the energy transport is confined only to radiation, the above equation expresses the conservation of total radiative flux πF_r, and can be written as

$$\frac{d}{dr}\left(r^2 \pi F_r\right) = \frac{d}{dr}\int_0^{\infty} \pi r^2 F_r\left(\lambda\right) d\lambda = 0. \tag{2.5.23}$$

We call this state the radiative equilibrium (RE). The difference between RE and TE should be noted. As explained in Section 2.3, the system in TE has no flow of radiation so that the temperature should be constant everywhere in the system. In contrast, RE only requires the stationary flow of radiative energy, and then the temperature gradient can exist inside the system.

In the case of a plane-parallel atmosphere the condition of RE is the conservation of flux of radiation passing through the unit area in the horizontal plane. In this case we have

$$\frac{d}{dt}\left(\pi F_r\right) = 0, \tag{2.5.24}$$

where t denotes the geometrical depth from the stellar surface.

Now we shall derive the temperature gradient in a plane-parallel atmosphere in RE. Multiplying $\cos\theta$ on both sides of the Equation (2.5.9) and integrating over the whole solid angle, we get

$$\frac{d}{d\tau_\lambda}\int_{4\pi} \cos^2\theta\, I_\lambda\left(\tau_\lambda,\theta\right) d\omega = \int_{4\pi} \cos\theta\, I_\lambda\left(\tau_\lambda,\theta\right) d\omega - \int_{4\pi} \cos\theta\, S\left(\tau_\lambda\right) d\omega,$$

where the last term on the right-hand side vanishes when S_λ is assumed to be isotropic. The first term denotes πF_λ by definition. On the left-hand side we define the function $K_\lambda\left(\tau_\lambda\right)$ for the integral part as

$$4\pi\, K\left(\tau_\lambda\right) = \int_{4\pi} \cos^2\theta\, I_\lambda\left(\tau_\lambda,\theta\right) d\omega. \tag{2.5.25}$$

Then the above equation can be written as

$$\frac{dK_\lambda\left(\tau_\lambda\right)}{d\tau_\lambda} = \frac{1}{4} F_\lambda\left(\tau_\lambda\right). \tag{2.5.26}$$

In order to derive the temperature gradient, the flux must be integrated over the whole wavelength range. Since wavelength dependence of τ_λ is not known, we again assume a gray atmosphere for simplicity. Then Equation (2.5.26) for the integrated radiation can be written as

$$\frac{dK}{d\tau} = \frac{1}{4}F, \qquad K = \int_0^{\infty} K_\lambda\, d\lambda, \tag{2.5.27}$$

where $d\tau_\lambda$ is replaced by $d\tau$ for the optical depth of total radiation. In case of RE, the total flux F is kept constant by Equation (2.5.24), then we get

$$K = \frac{1}{4} F\,\tau + c, \tag{2.5.28}$$

where c is an integration constant.

By definition, functions K, K_λ are the second moment of specific intensity $I_\lambda(\tau_\lambda, \theta)$ with respect to cos θ. Physically, K, K_λ are the quantities proportional to the radiation pressure at the point in question. Then Equation (2.5.28) yields the relation that the radiation pressure increases linearly with the increase of optical depth in the atmospheres in RE.

In the stellar atmospheres, the specific intensity $I_\lambda(\tau_\lambda, \theta)$ is getting isotropy with the increase of optical depth. Hence it is useful to replace the intensity $I_\lambda(\tau_\lambda, \theta)$ by mean intensity $J_\lambda(\tau_\lambda)$ for the calculation of K_λ as follows:

$$\begin{aligned} 4\pi K_\lambda(\tau_\lambda) &= \int_{\omega=4\pi} I_\lambda(\tau_\lambda, \theta)\cos^2\theta\, d\omega \\ &= J_\lambda(\tau_\lambda)\int_{\omega=4\pi}\cos^2\theta\, d\varpi = \frac{4\pi}{3}J_\lambda(\tau_\lambda). \end{aligned} \quad (2.5.29)$$

Or, simply we have both for K_λ and K,

$$K_\lambda = \frac{1}{3}J_\lambda \qquad \text{and} \qquad K = \frac{1}{3}J. \quad (2.5.30)$$

This relation is called the Eddington approximation and provides sufficient accuracy inside the stellar atmosphere except for its outermost part. Inserting this relationship into the Equation (2.5.28) we get the total radiation of

$$J(\tau) = \frac{3}{4}F\tau + \text{constant} = S(\tau), \quad (2.5.31)$$

where the last equality denotes the case of RE and gray atmosphere, i.e., by integrating both sides of the Equation (2.5.9) over a solid angle and inserting the condition of RE in (2.5.24), we immediately get the relation $J(\tau) = S(\tau)$.

In a plane-parallel atmosphere under LTE, the source function can be replaced by the Planck function as

$$S(\tau) = B(\tau) = \frac{\sigma}{\pi}T^4(\tau), \quad (2.5.32)$$

then we have

$$\sigma T^4(\tau) = \frac{3}{4}\pi F\,(\tau + \text{constant}) = \pi S(\tau). \quad (2.5.33)$$

Since total flux can be expressed in terms of effective temperature as

$$\pi F = \sigma T_{\text{eff}}^4,$$

we get

$$\sigma T^4(\tau) = \frac{3}{4}T_{\text{eff}}^4\,(\tau + \text{constant}), \quad (2.5.34)$$

where the integration constant is determined by a boundary condition at the surface $\tau = 0$ as follows. We first define the surface temperature T_0 using the mean intensity at the atmospheric surface as

$$J(0) = B(0) = \frac{\sigma}{\pi}T_0^4. \quad (2.5.35)$$

Near the surface, the outward flow of radiation is predominant against the inward flow. Then we assume a one-dimensional flow and designate the outward and inward intensities by I^+ and I^-, respectively. In this case we have

$$J = \frac{I^+ + I^-}{2}, \qquad F = I^+ - I^- . \tag{2.5.36}$$

As a boundary condition we assume that there is no incident radiation from outside, i.e., $I^-(0) = 0$. In this case we have

$$J(0) = \frac{1}{2} I^+, \qquad F(0) = I^+ \tag{2.5.37}$$

and we get, using Equation (2.5.20)

$$J(0) = \frac{1}{2} F(0) = \frac{1}{2} \left(\frac{\sigma}{\pi} T_{\text{eff}}^4 \right) . \tag{2.5.38}$$

Combining with Equation (2.5.35) and putting $\tau = 0$ in Equation (2.5.34), we obtain constant $= 2/3$ in Equation (2.5.34). Thus the temperature distribution is given as

$$T^4(\tau) = \frac{3}{4} T_{\text{eff}}^4 \left(\tau + \frac{2}{3} \right) . \tag{2.5.39}$$

This relationship was derived by using the Eddington approximation. It represents an approximation of the linear flow of radiation in the outermost part of the atmosphere. According to more accurate derivation without employing the Eddington assumptions, the temperature distribution is given as follows

$$T^4(\tau) = \frac{3}{4} T_{\text{eff}}^4 \left(\tau + q(\tau) \right) , \tag{2.5.40}$$

where $q(\tau)$ is a slowly varying function. It varies with τ in a small range of values such as

$$q(0) = 0.577, \qquad q(\infty) = 0.710.$$

Comparing the approximate value of $q(\tau) = 2/3$ in Equation (2.5.39) to the above values, one will see that the approximation is reasonable.

2.5.5 Formation of absorption lines

The temperature of the stellar atmosphere increases with increasing depth. According to Kirchhoff–Bunsen's experiments, the continuous radiation emitted from the inner hotter layer suffers selective absorption by atoms or ions in the outer cooler layer. In this way absorption lines are formed in stellar atmospheres. Let us consider this process in some detail.

Let τ_c, τ_λ be the optical depths for the continuum and an absorption line, respectively. Then we have from Equation (2.5.7),

$$d\tau_c = \kappa_c \, dt, \qquad d\tau_\lambda = \kappa_\lambda \, dt, \tag{2.5.41}$$

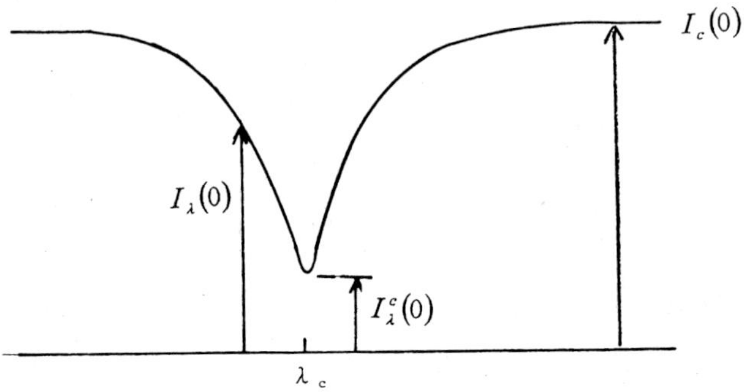

Figure 2.14: Schematic profile of an absorption line, where the intensity at typical point is designated as I_c (0), surface intensity of the continuum; I_λ (0), surface intensity inside the absorption line at wavelength λ; I^c_λ (0), surface intensity at the line centerλ_c. In case of stellar atmospheres, observed quantities are the emergent fluxes, $F_c(0), F_\lambda(0), F^c_\lambda(0)$; respectively.

where κ_c, κ_λ (both cm^{-1}) denote the absorption coefficients for the continuum and absorption line, respectively. For simplicity, suppose the radiation propagating vertically outward in a horizontal layer, and let $I_c\,(0)\,,\, I_\lambda\,(0)$ be the surface intensities of the continuum and absorption lines, respectively. The profile of absorption line is expressed as in Figure 2.14.

For an absorption line in Figure 2.14, the line depth R_λ and residual intensity A_λ are defined as

$$R_\lambda = \frac{I_c\,(0)\; - \; I_\lambda\,(0)}{I_c\,(0)} = 1 - \frac{I_\lambda\,(0)}{I_c\,(0)}$$
$$A_\lambda = 1 - R_\lambda. \tag{2.5.42}$$

The surface intensities $I_c(0)$ and $I_\lambda(0)$ are given by putting $\theta = 0$ in Equation (2.5.10) as

$$I_c\,(0) = \int_0^\infty S_c\,(\tau_c)\,\exp\,(-\,\tau_c)\;d\tau_c, \tag{2.5.43}$$

$$I_\lambda\,(0) = \int_0^\infty S_\lambda\,(\tau_\lambda)\,\exp\,(-\,\tau_\lambda)\;d\tau_\lambda. \tag{2.5.44}$$

In these equations the integrands give the contribution to the surface intensity from gas element at the depth τ_c or τ_λ.

Now, let us consider a strong absorption line ($\kappa_c \ll \kappa_\lambda$) and assume LTE in the atmosphere. Then the source functions in the above equations can be replaced by the Planck function $B_\lambda\,(\tau_c)$ and $B_\lambda\,(\tau_\lambda)$, respectively. The values of these functions increase with optical depth. On the other hand the exponential factor in the integrand decreases rapidly so that the contribution to the surface intensity takes a maximum value at some optical depth. According to Eddington–Barbier's relationship given in (2.5.16), the maximum contribution to the surface intensity comes from the optical depth $\tau = \frac{2}{3}$ for both continuum and line. The geometrical depth to that point differs considerably for the continuum and line center. In the continuum, the effective depth ($\propto \kappa_c^{-1}$) is large and corresponds to a hotter deep layer, whereas, in the line center, the effective depth ($\propto \kappa_\lambda^{-1}$) is small and located in cooler outer layer. That is, the continuum is formed effectively in the deeper hot layer, while the absorption line is formed in the cooler layer near the surface. This explains the formation of absorption lines.

In order to see the line depth of an absorption line, we shall consider a simple one-dimensional flow of radiation near the surface by denoting outward intensity I_λ^+ and inward intensity I_λ^- as in the previous section. Then the flux and mean intensity at the surface can be given by

$$F_\lambda(0) = I_\lambda^+(0), \qquad J_\lambda(0) = \frac{1}{2} I_\lambda^+(0). \tag{2.5.45}$$

For the continuum, similarly with Equation (2.5.17), we have

$$I_c(0) = F_c(0) = B_c(\tau_c = 2/3) \tag{2.5.46}$$

and, since atmospheric temperature at $\tau_c = 2/3$ equals to the effective temperature as seen in Equation (2.5.21), we have approximately

$$I_c(0) \approx B(T_{\text{eff}}). \tag{2.5.47}$$

For the line center, Equation (2.5.44) can be written as

$$I_\lambda(0) = \int_0^\infty B_\lambda(T(\tau_\lambda)) \exp(-\tau_\lambda)\, d\tau_\lambda. \tag{2.5.48}$$

Again, by making use of Eddington–Barbier's relation we get

$$I_\lambda(0) = B_\lambda\left(T\left(\tau_\lambda = \frac{2}{3}\right)\right) = B_\lambda(T_0), \tag{2.5.49}$$

where $T_0 = T(\tau_\lambda = 2/3, \tau_c \approx 0)$ is the temperature at $\tau_\lambda = 2/3$.

Inserting these relationships to Equation (2.5.42), we get

$$R_\lambda = 1 - \frac{I_\lambda(0)}{I_c(0)} \approx 1 - \frac{B_\lambda(T_0)}{B_c(T_{\text{eff}})}. \tag{2.5.50}$$

This relationship implies that the depth of an absorption line is determined through the difference between T_0 and T_{eff}. In a weak absorption line, κ_λ may not so much differ from κ_c, and the difference between $B_\lambda(T_0)$ and $B_c(T_{\text{eff}})$ may also be small, then the depth R_λ is small. In contrast, strong absorption lines have large absorption coefficients ($\kappa_c \ll \kappa_\lambda$) and $B_\lambda(T_0) \ll B_c(T_{\text{eff}})$. We have deep central depths of $R_\lambda \sim 1$.

In this way we can understand the process of formation of absorption lines in terms of the difference in the values of absorption coefficients. We have so far considered the plane-parallel atmosphere, In stellar atmospheres, we can also put forward the similar process by making use of the flux of radiation instead of radiation intensity.

2.6 Spectral-line profiles

2.6.1 Profiles of absorption lines

The profiles of spectral lines are basically determined by the form of absorption coefficients. In this section we first summarize the three basic profiles of absorption coefficients: the damping profile caused by a finite width of energy levels of atoms; the Doppler profile caused by the thermal motion of gas; and the Voigt profile as a composition of the damping and Doppler profiles. Thereafter we consider the broadening of spectral lines by the Stark effect and by stellar rotation, both of which are important in the spectral analysis of emission-line stars.

One may note that the line absorption coefficient per unit volume κ_L (cm^{-1}) and per atom $\kappa_{L,s}$ (cm^{-2}) are connected by the relation

$$\kappa_L = N_L \, \kappa_{L,a}, \tag{2.6.1}$$

where N_L (cm^{-3}) denotes the number density of atoms.

Damping profile or Lorentz profiles

A spectral line of an atom is formed by a transition of electron between two energy levels, whose difference yields the frequency of the line by the relation $E_2 - E_1 = h\nu_{12}$, where E_2 and E_1 are the upper and lower energy levels, respectively. Energy levels of atoms are intrinsically broadened due to the uncertainty principle so that the spectral lines are also broadened. This type of broadening is called natural broadening and the profile of the absorption coefficient takes the form

$$\kappa_{L,a} = \frac{\pi \, e^2}{m \, c} \frac{\gamma}{(\Delta\omega)^2 + \left(\frac{\gamma}{2}\right)^2} f, \tag{2.6.2}$$

where $\Delta\omega = \omega - \omega_0$ denotes the offset circular frequency from the central frequency ω_0 $(= 2\pi\nu_0)$, and γ is the constant called the damping constant and is given, according to the classical electrodynamics, as

$$\gamma = \frac{8\pi^2 \, e^2}{3mc \, \lambda_0^2}, \tag{2.6.3}$$

where λ_0 is the center wavelength of the line. In addition, f is a parameter proportional to the absorption coefficient and called the oscillator strength. Equation (2.6.2) gives the damping profile or the Lorentz profile of absorption coefficient.

Doppler profile

The profile broadened by thermal motion is expressed by the following exponential function

$$\kappa_{L,a} = \frac{\sqrt{\pi}\, e^2}{mc^2} \frac{\lambda^2}{\Delta\lambda_D} \exp\left[-\left(\frac{\Delta\lambda}{\Delta\lambda_D}\right)^2\right], \tag{2.6.4}$$

where $\Delta\,\lambda_D$ denotes the Doppler width given by the thermal velocity ξ_0 as

$$\Delta\lambda_D = \lambda\frac{\xi_0}{c}, \quad \text{where} \quad \xi_0 = \sqrt{\frac{2\,RT}{\mu}}, \tag{2.6.5}$$

R and μ denote the gas constant and mean molecular weight of gas, respectively.

Voigt profile

The Voigt profile is composed of the Damping and Doppler profiles. In a strong absorption line, the Doppler profile is predominant in the line center, while the damping profile dominates in the wing region. This is because the Doppler profile decreases exponentially from the line center, and the damping profile declines only as $1/(\Delta\lambda)^2$. The schematic Voigt profile, composed of Doppler

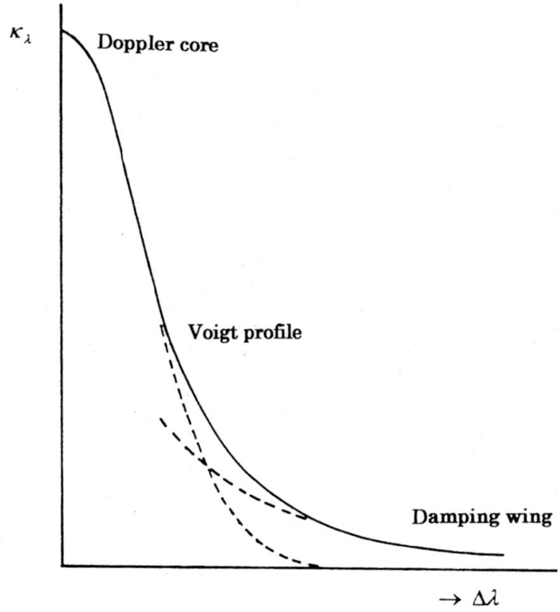

Figure 2.15: Schematic Voigt profile of a line absorption coefficient. The abscissa is the wavelength difference from the line center. The Voigt profile is a composite of Doppler and damping profiles. The central part is called a Doppler core and the outer skirt is called a damping wing.

core and the Damping wing, is shown in Figure 2.15. The actual form of the profile is determined by a parametric ratio α of damping width $\gamma/2$ versus Doppler width $\Delta\lambda_D$ in frequency unit as given

$$\alpha = \frac{\gamma}{2\,\Delta\omega_D}. \tag{2.6.6}$$

This ratio is generally smaller than unity and takes a value in a range of 0.1–0.001.

2.6.2 Line broadening by the Stark effect

When a line–emitting light source is placed in a magnetic field, the spectral lines split into several components. This is the Stark effect and is classified into two cases as:

(a) Linear Stark effect. The size of the splitting of spectral lines $\Delta\nu$ is proportional to the applied electric field F. This effect appears in hydrogen and hydrogen-like atoms.

(b) Quadratic Stark effect. The splitting $\Delta\nu$ is proportional to the square of the field, F^2. This effect is seen in heavy elements and the amount of the splitting is small as compared to the linear Stark effect.

In view of its importance in emission-line stars, we now consider the linear Stark effect of hydrogen atoms. When an electric field exists, excited levels of hydrogen split into several components. The energy difference $\Delta\,E$ from their original level is given for the principal quantum number n as

$$\Delta E = h\Delta\nu = \frac{3}{8\pi^2}\frac{h^2}{em_e}Fnn_F, \tag{2.6.7}$$

where n_F $(= 0,\ \pm1, \ldots,\ \pm(n-1))$ denotes the magnetic quantum number. One may see from this equation that the number of components increases with the principal quantum number. In Table 2.9, we show the number of components and magnetic quantum numbers for some lower principal quantum numbers.

Now we consider the hydrogen atmosphere in RE. Hydrogen atoms respond to the effects of an electric field of surrounding charged particles. Let N (cm^{-3})

Table 2.9: Components of energy levels split by Stark effect in hydrogen atom

n	nF	Number of components
1	0	1
2	0, ±1	3
3	0, ±1, ±2	5
4	0, ±1, ±2, ±3	7
5	0, ±1, ±2, ±3, ±4	9

be the number density of charged particles, their average distance r_0 is given by

$$\frac{4\pi}{3} r_0^3 N = 1, \tag{2.6.8}$$

then the average electric field F_0 acting for hydrogen atoms is

$$F_0 = \frac{e}{r_0^2} \propto N^{2/3}, \tag{2.6.9}$$

where e denotes the electric charge of charged particles. By this Coulomb force, energy levels of hydrogen atoms split into Stark components given by Equation (2.6.7).

Through the thermal motion of atoms and ions, both the electric field and the Stark components perpetually change. As a result, spectral lines are broadened by the Stark effects depending on the charge density N. In hydrogen atoms, the number of Stark components and then the line widths of the Balmer lines increase with the quantum number n. On the other hand, the separation between adjacent energy levels in the Balmer series decreases with the increase of n. Hence the Balmer lines mutually overlap and become indistinguishable above a critical quantum number n_{m}. This merging of adjacent lines depends upon the charge density of the gas, and the critical quantum number is lower in denser gas.

Based on the calculation of the Stark effect, Inglis and Teller (1939) derived the relation between the critical quantum number n_{m} and the charge density N as

$$\log 2N = 23.26 - 7.5 \log n_{\mathrm{m}}. \tag{2.6.10}$$

This is called the Inglis–Teller formula, which enables us to estimate the charge density of the Balmer-line forming region, by using a simple procedure to find the critical quantum number of the Balmer series. In Equation (2.6.10), the factor $2N$ in the left-hand side is the revised one from the original N (Aller 1963). Although the charged particles are the sum of ions and electrons, the effect of electrons on the line broadening is different in higher and lower electron temperatures as shown

$$N = Ni + Ne \quad \text{for} \quad T < 10^5/n_{\mathrm{m}}$$
$$N = N_{\mathrm{i}} \quad \text{for} \quad T > 10^5/n_{\mathrm{m}},$$

where N_{i} and N_{e} denote the ion and electron number densities, respectively. This difference is caused by the thermal velocity of electrons that is too high to the broadening of the spectral lines in the hotter gases.

In main sequence stars, the critical quantum number of the Balmer lines is around $n_{\mathrm{m}} \sim 14 - 16$ so that, $N = N_i \sim (2 - 0.8) \times 10^{14}$ cm^{-3} for stars hotter than $T \sim 7000$ K. In supergiant stars, we usually observe $n_{\mathrm{m}} \sim 20$–23, then we have $N = N_i \sim (1 - 0.5) \times 10^{13}$ cm^{-3} for early-type stars. Among Be stars, Be-shell stars exhibit high values of $n_{\mathrm{m}} = 25$–35 which correspond to $N = N_i \sim N_e \sim 10^{11}$ according to Inglis–Teller formula. However, it should be mentioned that the disappearance of Balmer absorption lines (shell lines) in these stars can be different from the Stark effect (See Section 5.3.4).

2.6.3 Line broadening by turbulence

Thermal motions yield the Doppler profiles in spectral lines as seen in Section 2.6.1. If turbulent motions exist in the medium, spectral lines will show additional broadening. In stellar atmospheres, there are two types of turbulent motion: macroturbulence and microturbulence. They cause different effects on the broadening of absorption lines.

Macroscopic turbulence

The example of this type is seen in the solar spectrum. When we observe the Sun with a long-slit spectrograph, we can see some jagged structures along the lines. These are supposed to be the Doppler shifts due to the solar granulation motion. If we observe the Sun from very far away, the solar radiation would be integrated over the whole disk, and then, the jagged structure will disappear and some broadened spectral line will appear instead. This is due to different parts of the stellar disk having different velocities in macroscopic scale. If the distribution of radial velocities of macroscopic turbulent motion can be expressed by the Gaussian distribution as in case of thermal motion, then the distribution of the resulting velocities will also yield the Gaussian distribution. Then the reference velocities of the resulting motion ξ_0 can be written as

$$\xi_0^2 = \xi_{\rm th}^2 + \xi_{\rm turb}^2, \qquad (2.6.11)$$

where $\xi_{\rm th}$ and $\xi_{\rm turb}$ denote the reference velocities of thermal and turbulent motions, respectively. The Doppler width of an observed spectral line, $\Delta\lambda_0$, is given by Equation (2.6.5) with replacement of ξ_0 by the value given above.

Microscopic turbulence

As stated in Section 2.3, the main contributor to the formation of an absorption line comes from the outer layer down to the optical depth of around $\tau_\lambda = 2/3$, according to the Eddington–Barbier relation. If the atmosphere is in a turbulent state and there exist several turbulent elements in this outer layer, the radiation passing through this layer will suffer the Doppler shifts caused by the different line-of-sight velocities through different turbulent elements. Therefore, an absorption line with central wavelength λ_0 will have a width $\Delta\lambda$ caused by thermal velocity (temperature T) and the turbulent velocity ($v_{\rm turb}$) given as

$$\Delta\lambda = \frac{\lambda_0}{c}\left(\frac{2kT}{\mu m} + v_{\rm turb}^2\right)^{1/2}, \qquad (2.6.12)$$

where m denotes the proton mass and μ the mean molecular weight. This means that the line is broadened as if the apparent gas temperature is increased by the effect of turbulent motion. Since the sizes of turbulent elements are supposed to be smaller than the mean free path of a photon, the turbulence in this case is called microscopic. The microscopic turbulence thus broadens the line as in case of macroscopic turbulence. However, in contrast to the latter, the microscopic turbulence broadens the absorption coefficients, and then gives some changes in the ratio of line to continuous absorption coefficient. As the results, the observed absorption lines suffer some changes in their equivalent widths and profiles. These changes can be quantitatively measured by the detailed study of the curves of growth (Section 2.7.1). In macroscopic turbulence, the effects of turbulent motion from different parts of the stellar disk are mutually independent so that the observed line profiles are only broadened without changes of equivalent widths.

2.6.4 Line broadening by stellar rotation

Since stars are generally rotating (unless viewed pole-on), one-half of the stellar disk moves away from us, while the other half moves toward us. Thus the spectral lines are broadened by the Doppler shifts by different radial velocities from different parts of the stellar disk due to stellar rotation. Now we consider the line profiles in rotating stars (Unsöld 1955).

Suppose a rigidly rotating star, with its angular velocity vector being $\boldsymbol{w}$ (rotation velocity at the equator is $w = |\boldsymbol{w}|$), and the axis of rotation makes an angle i toward the observer. Let us take the rectangular coordinates (ξ, η, ζ) centered at the star's center and ζ-axis being directed toward the observer as shown in Figure 2.16. Since the velocity component of vector $\boldsymbol{w}$ in η direction is $w \sin i$, the star appears as if rotating with $w \sin i$ for the observer equator-on.

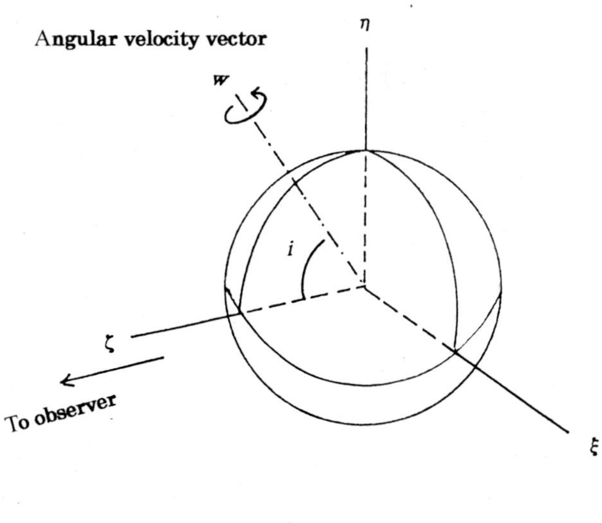

Figure 2.16: Coordinate system of a rotating star.

Let $v\,(\xi,\,\eta)$ be the line-of-sight velocity of stellar rotation at a point $(\xi,\,\eta)$ on the disk. After some geometrical manipulation, we obtain

$$v\,(\xi,\eta) = R\,w\,\sin\,i\,\left(\frac{\xi}{R}\right) = V\,\sin\,i\,\left(\frac{\xi}{R}\right)\cdot\ (\xi \le R)\,, \tag{2.6.13}$$

where R denotes the stellar radius and V the rotational velocity at the stellar equator. This relationship indicates that the line-of-sight velocity of rotation takes the same value for a stripe across the stellar disk parallel to η axis.

Now, let $\Delta\lambda$ be the Doppler shift for the spectral line λ at the distance ξ, we have

$$\Delta\lambda = \frac{\lambda}{c}\xi\,w\,\sin\,i. \tag{2.6.14}$$

This shift takes a maximum value, say b, at the stellar radius, $\xi = R$. By putting $b = \Delta\lambda\,(\xi = R)$, we get

$$b = \frac{\lambda}{c}R\,w\,\sin\,i = \frac{\lambda}{c}V\,\sin\,i. \tag{2.6.15}$$

We shall adopt the new variables x, y for the wavelength shift and coordinate as

$$\frac{\Delta\lambda}{b} = \frac{\xi}{R} = x, \qquad \frac{\eta}{R} = y, \tag{2.6.16}$$

then the spectral line intensity in the wavelength range $\Delta\lambda$ is determined by the contribution from the stripes $(x, x + dx)$ on the stellar disk.

Let β be the limb-darkening coefficient, that is, we write the intensity distribution on the stellar disk as follows

$$I = \text{constant}\ (1 + \beta\,\cos\theta)\,, \tag{2.6.17}$$

where θ denotes the angular distance from the disk center as given by

$$\sin\,\theta = \frac{\sqrt{\xi^2 + \eta^2}}{R} = \sqrt{x^2 + y^2} \tag{2.6.18}$$

then we have

$$I = \text{constant}\ \left\{1 + \beta\sqrt{1 - (x^2 + y^2)}\right\}. \tag{2.6.19}$$

We now normalize the surface intensity I by the condition

$$\int_{-1}^{+1} I\,dx = 1, \tag{2.6.20}$$

then, the relative intensity $A\,(x)$ along the stripe at x can be written

$$\begin{aligned} A\,(x) &= \frac{\displaystyle\int_0^{\sqrt{1-x^2}} I\,(x,y)\,dy}{\displaystyle\int_{x=-0}^{+1}\int_{y=0}^{\sqrt{1-x^2}} I\,(x,y)\,dy\,dx} \\ &= \frac{3}{3+2\beta}\left[\frac{2}{\pi}\sqrt{1-x^2} + \frac{\beta}{2}\,(1-x^2)\right]. \end{aligned} \tag{2.6.21}$$

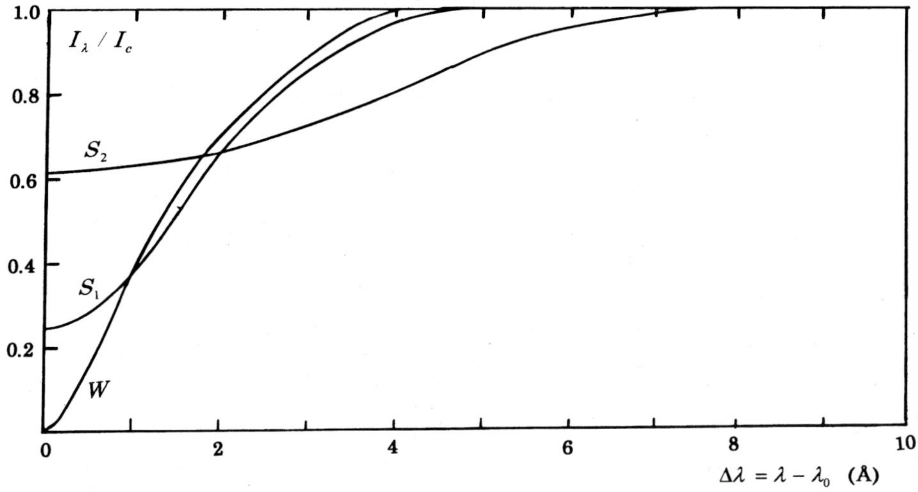

Figure 2.17: Rotationally broadened line profiles. W denotes the profile with no broadening, S_1 and S_2 are the broadened profiles for $V \sin i = 100$, 300 km s^{-1}, respectively. (From Unsöld 1955)

This yields the profile of the rotationally broadened line and serves as a broadening function. If there is no limb darkening, $A(x)$ gives an elliptic profile.

Let W(x) be the fictitious line profile not broadened by rotation, then the absorption line profile $S(x)$ formed from the whole stellar disk can be derived by the convolution of W(x) with the broadening function A(x) as given

$$S(x) = \int_{-\infty}^{+\infty} W(x, y)\, A(y)\, dy. \qquad (2.6.22)$$

In Equation (2.6.22) it is evident that $S(x)$ will become $A(x)$ when the ratio of the width of $W(x)$ to that of $A(x)$ is sufficiently small, whereas $S(x)$ will approach the form of $W(x)$ when the same ratio is sufficiently large. The fictitious profile $W(x)$ can be obtained from model-atmosphere calculation or from the observed profile for slowly rotating stars. Figure 2.17 illustrates some line profiles $S(x)$ broadened by rotation, along with the unbroadened profile $W(x)$, where the broadening is calculated in the case of no limb darkening ($\beta = 0$).

The procedure of determination of $V \sin i$ was first given by Huang and Struve (1953). As spectral lines of photospheric origin, He Iλ 4471 Å and Mg Iλ 4481 Å are usually adopted in the optical spectral region. Chauville et al. (2001) calculated the broadened theoretical profiles of He Iλ 4471 Å line, based on their non-LTE model stellar atmospheres by taking into account the full limb darkening of spectral lines. They thus derived the values of $V \sin i$ for 116 Be stars, by fitting with observed profiles.

Rotational velocity is one of the basic quantities of stars, closely related to the structure of stellar envelopes and chromospheric activities, as seen in

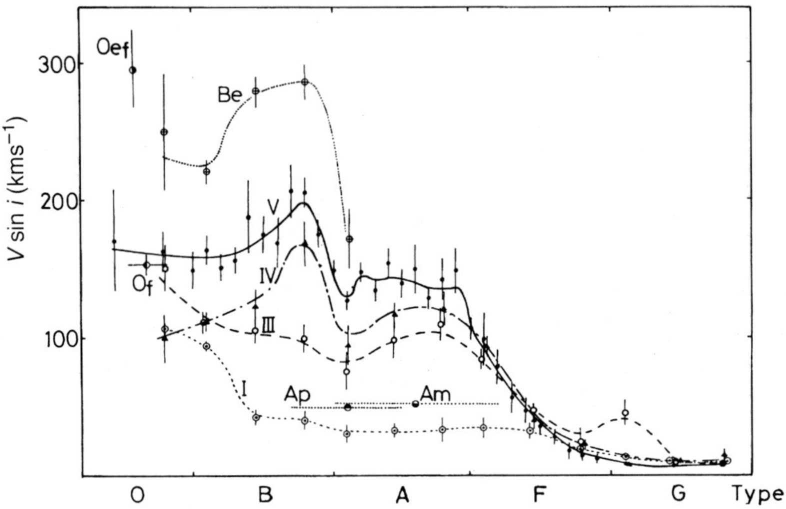

Figure 2.18: Average rotational velocities of stars $\langle V \sin i \rangle$. Stars are designated by ● Main sequence stars (V), ▲ Subdwarfs (VI), ○ Giants (III), ⊚ Supergiants (I), ⊕ Be stars, ◐ Of stars, ◑ Oef stars, ◒ Am stars, and ◓ Ap stars. (From Fukuda 1982)

Part II of this book. The measurements of $V \sin i$ have so far been carried out for a number of stars. Generally speaking, rotational velocity is higher in early-type stars and lower in late-type stars. Figure 2.18 exhibits the average velocities along the spectral type for several kinds of stars (Fukuda 1982).

When rotational velocity is lower than several $\mathrm{km\,s^{-1}}$, the separation between rotation and turbulence becomes important. Several methods have so far been proposed; all of them are based on the detailed quantitative analysis of absorption profiles (see Gray and Turner 1987, Takeda 1995).

2.7 Absorption lines and model atmospheres

2.7.1 Curve of growth

The intensity of an absorption line increases with increasing amount of absorbing gas, but not proportionally to the amount of gas.

We define the equivalent width W_λ of an absorption line by

$$W_\lambda = \int_{\text{line}} R_\lambda \, d\lambda, \qquad (2.7.1)$$

where R_λ denotes the depth of absorption line given in Equation (2.5.42) at wavelength λ. The integration covers the whole absorption line.

In order to clarify the growing process of an absorption line, we imagine a laboratory system, where a gas tube is placed in front of a continuum light source. In the tube, a gas having an absorption coefficient κ_λ at central wavelength λ_0 is enclosed. If the gas density N (cm^{-3}) in the tube gradually increases, the equivalent width W_λ of the line will also increase.

Let the radiation emitted from the continuum light source enters the gas tube with incident intensity I^0_λ and exits the tube with the intensity I_λ. Then we have

$$I_\lambda = I^0_\lambda \, e^{-HN\kappa_\lambda}, \tag{2.7.2}$$

where H denotes the geometrical length of the tube.

When gas density N is low and the tube is optically thin, Equation (2.7.2) can be expanded as

$$I_\lambda \approx I^0_\lambda \, (1 - HN\kappa_\lambda)\,. \tag{2.7.3}$$

Then we have

$$W_\lambda = \int \frac{I^0_\lambda - I_\lambda}{I^0_\lambda}\, d\lambda = HN \int \kappa_\lambda \, d\lambda, \tag{2.7.4}$$

where the integration of absorption coefficient over the whole line can be expressed by making use of the oscillator strength f as

$$\int \kappa_\lambda \, d\lambda = \frac{\pi\, e^2}{m\, c^2}\, \lambda^2 f. \tag{2.7.5}$$

Hence Equation (2.7.4) can be written as

$$W_\lambda = \frac{\pi\, e^2}{m\, c^2}\, \lambda_0^2 \, HNf. \tag{2.7.6}$$

This relationship implies that the equivalent width increases proportionally with the column density *NH* during the optically thin stage of the gas tube.

With the increase of gas density, the line profile changes as shown in Figure 2.19. In the optically thin stage, the equivalent width is proportional to *NH*, as stated above, and the line shows the Doppler profile as given in Section 2.6.1. In optically thick stage, the central part of the line first saturates and then the wing part gradually develops with increasing column density. The relation between logW_λ and logNH is schematically shown in Figure 2.20 and called the curve of growth.

Atomic or ionized gas generally has many absorption lines with different values of f_λ, and the curve of growth can be constructed as the relation between (logW_λ and logf_λ) or (log W_λ and log NHf_λ). In every case, the curve

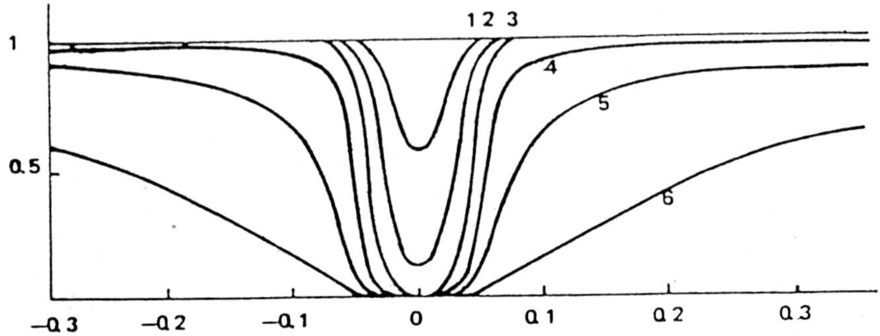

Figure 2.19: Growth of an absorption line. The line profile changes in the order of number by the increase of absorbing gas.

of growth is composed of the following three parts in the order of increasing f_λ or NHf_λ as schematically shown in Figure 2.20.

(1) The linear part. This appears when the gas is optically thin, as given in Equation (2.7.6).
(2) The flat part. When the central part of the line approaches to its saturation, the line equivalent width increases very slowly.
(3) The damping part or square root part. The curve begins to increase due to the effect of damping wings.

The curve of growth can be derived theoretically based on some suitable model atmospheres. Theoretical curves of growth are usually expressed by the relationship between log (W_λ/λ) and log (NHf), while observational curves of growth are expressed by the relationship between log (W_λ/λ)

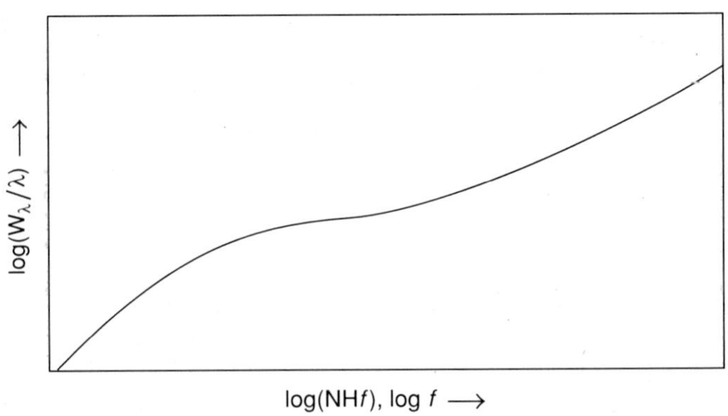

Figure 2.20: Schematic curve of growth on the log (W_λ/λ) versus log (NHf) or log f diagram.

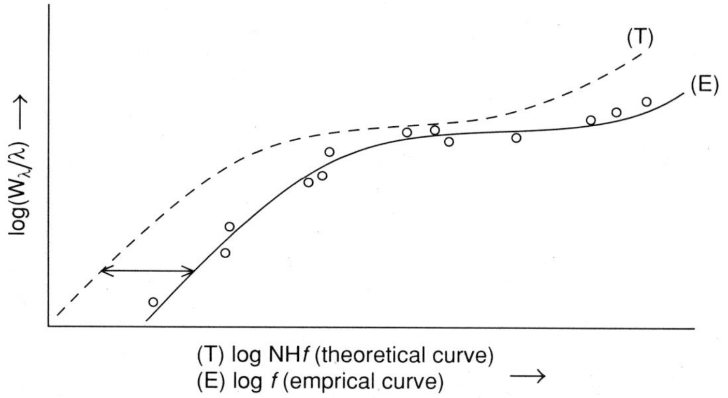

Figure 2.21: Procedure of measurement of the column density NH on the curves of growth. The abscissa is log (NHf) or log f for theoretical (T) and empirical (E) curves of growth, respectively, in the same scale. The ordinate is log (W_λ/λ) and solid circles denote fictitious observed points. The horizontal distance $(\longleftrightarrow)$ gives the value of log NH.

and log f, since we have no knowledge on the values of NH in advance in observations.

The curves of growth provide a useful method in the spectral analysis of stars. By comparing the theoretical and observed curves of growth in lines of different atoms or ions, or in lines of different excitation levels, we can deduce the physical parameters such as (a) chemical abundance of the atmosphere, (b) excitation degree and excitation temperature, (c) ionization degree and ionization temperature, and (d) turbulent velocity.

As an example, we consider the derivation of the relative abundance. Suppose an atomic or ion gas having absorption lines in wavelengths $\lambda_1, \lambda_2, \lambda_3, \ldots$ with known oscillator strength f_1, f_2, $f_3, \ldots$. First, measure the equivalent widths of these lines, W_1, W_2, $W_3, \ldots$ and draw the empirical curve of growth in $\log(W_\lambda/\lambda)$ versus log f diagram. Next, draw the theoretical curve of growth in $\log(W_\lambda/\lambda)$ versus log (NHf) diagram and compare both curves in the same scale as shown in Figure 2.21. Since the difference in both curves lies in the value of log (NH), we can obtain this value by horizontal shift of the theoretical curve. Repeating the same procedure for the atomic, excited, and ionized states of an element, we obtain the total column density of this element. If the geometrical height H can be assumed as the same for all absorption lines, the total column densities of different elements yield the relative abundance of elements considered. This is the principle of the measurement of chemical abundance of stellar atmospheres.

In addition, it is mentioned that the theoretical curves of growth also depend upon the turbulent velocity (microturbulence) and γ value in Equation (2.6.3). This allows us to estimate these parameters by the detailed fitting of curves of growth (see Aller 1963).

2.7.2 Model atmosphere

Assumptions and basic equations

Theories of model atmospheres deal with the calculation of the structure of the outer layers of a star knowing the total radiation flux, radius, and surface gravity. The purpose of model atmospheres is to calculate the reliable theoretical profiles of absorption lines and the continuous stellar spectrum, and to provide the basic data for interpreting the observed stellar spectra.

The model calculations are generally based on the following assumptions:

(a) plane-parallel geometry
(b) hydrostatic equilibrium
(c) radiative equilibrium (RE)
(d) chemical homogeneity

The geometrical thickness of stellar atmosphere is sufficiently small as compared to the stellar radius. In case of the Sun, geometrical thickness of the photosphere is around 550 km against the solar radius of 700,000 km, i.e., less than 0.1% in ratio. Plane-parallel assumption may then be appropriate for most main-sequence and giant stars, but may fail for the supergiants. Hydrostatic equilibrium may also be assured for main-sequence and giant stars. The atmospheres of these stars are mechanically stable without showing large-scale gas motions such as expansion or contraction. RE states that the bolometric flux in a plane-parallel atmosphere is constant. In some stars, particularly in late-type stars, convective energy transportation becomes important and we need to take into account for (see Chapter 3 for convection). Chemical homogeneity assumes that the elements are completely mixed throughout the atmosphere. Each of the above four assumptions can be dropped according to the conditions of stellar atmospheres.

The basic equations of the model atmospheres can be written under the above assumptions as follows:

(i) Equation of energy transfer. The equation of radiative transfer is given in Section 2.5.2. If we distinguish the continuous and absorption line radiation, it takes the form

$$\cos\theta \frac{dI_v(\theta)}{\rho dx} = -(\kappa_c + \kappa_L) I(\theta) + \kappa_L J_v + \kappa_c B(v, T), \qquad (2.7.7)$$

where the second and last term in the right-hand side indicate the emission term by scattering process and by the LTE radiation, respectively, in the atmosphere.

(ii) Equation of hydrostatic equilibrium. From the balance of pressure gradient and gravity, we have

$$dP = \rho g \, dz, \qquad (2.7.8)$$

where z denotes the vertical geometrical depth, which is connected with the optical depth as

$$d\tau_\lambda = (\kappa_c + \kappa_L)\,\rho\,dz. \tag{2.7.9}$$

Combining these equations we get

$$\frac{dP}{d\tau_\lambda} = \frac{g}{\kappa_c + \kappa_\lambda}. \tag{2.7.10}$$

(iii) Equation of state. Stellar atmosphere can be supposed as the combination of perfect gas and photon gas, of which the latter gives rise to the radiation pressure P_r and becomes effective in early-type stars. The equation of state is then written as

$$P = \frac{\rho k T}{\mu m_H} + P_r. \tag{2.7.11}$$

In model atmosphere calculation, we solve these basic equations simultaneously. Since absorption coefficients κ_c, κ_L, and mean molecular weight μ are complicated functions of gas state (ρ, T, A), where A stands for chemical composition, the calculation requires a large amount of computing time.

The basic parameters in the solutions of basic equations are

(a) Stellar effective temperature $T_{\rm eff}$ as defined by (2.5.20)

$$\sigma T_{\rm eff}^4 = \pi \int F_\lambda(0)\,d\lambda. \tag{2.7.12}$$

(b) Stellar surface gravity

$$g = \frac{GM}{R^2}.$$

Surface gravity takes the values of $g \sim 10^4$ for main-sequence stars, and $g \sim 10^2$ for supergiants (see Table 2.3).

(c) Chemical abundance A. Taking the solar abundance as the standard, the abundance of heavy elements is usually assumed as 1/10, 1/100 times of the solar abundance.

LTE and non-LTE models

We show two cases of model calculations for LTE and non-LTE models.

(a) LTE models. Kurucz (1979) has calculated the LTE model atmospheres for stars earlier than G type, under the basic assumption stated above, and taking into account the effects of convective energy transfer and of radiation pressure. For atomic absorption coefficients, 10^6 spectral lines are included. Adopted parameters are $\tau_{\rm eff} = 5500 - 50{,}000$ K, $\log\, g = 4.5 - 1.0$, and the chemical abundance of heavy elements $A = 1,\ 1/10,\ 1/100$ solar abundance.

The results of the calculation are given for the temperature structure $T(\tau)$, emergent flux F_λ, color index of UBVuvby, bolometric correction BC, and

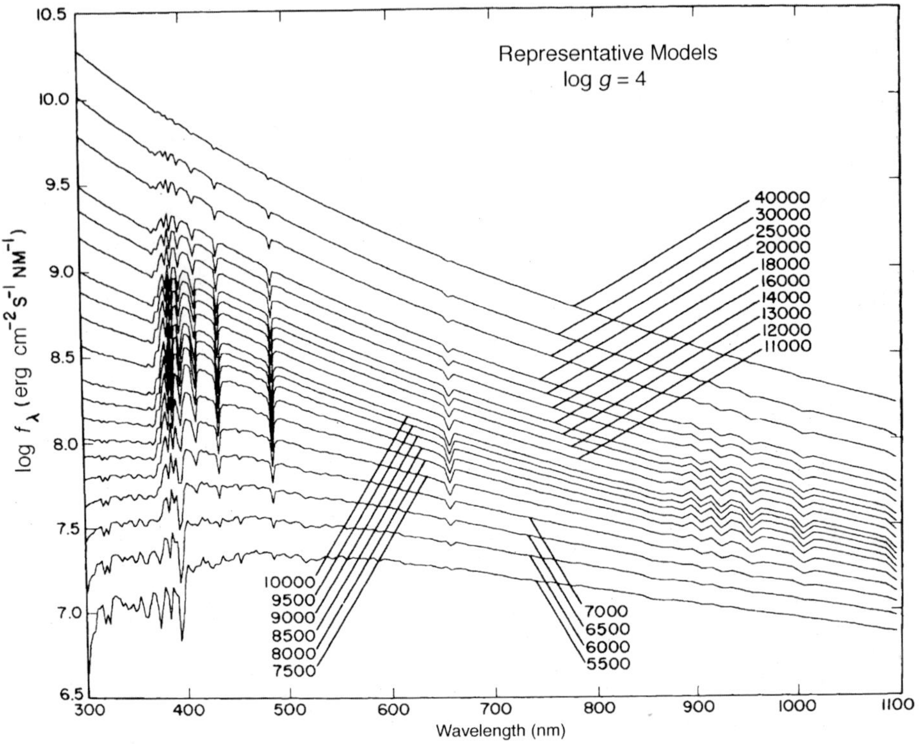

Figure 2.22: The emergent flux as a function of $T_{\rm eff}$ in the Kurucz's LTE model. Surface gravity log $g = 4$, and Solar abundance for heavy elements are adopted. (From Kurucz 1979)

the Balmer line profiles. Among these results, Figure 2.22 exhibits flux spectra in the wavelength range $\lambda 3000 - 11{,}000$ Å for the effective temperature from 5500 to 40,000 K with a fixed surface gravity log $g = 4$. Figure 2.23 shows the profiles of $H\gamma$ line for the same parameter range as in Figure 2.22. The profiles are symmetric with respect to the line center.

Kurucz's (1979) model has been revised and extended in a series of published ATLAS with improvements in the adopted number of absorption lines, improvement of absorption coefficients etc. ATLAS12, the most recent atlas was published in 1993 (Kurucz 1993).

(b) Non-LTE models. In the state of TE the detailed balancing holds in all energy levels of atoms and ions (Section 2.3), and the level population is given by the Boltzmann distribution. LTE models assume the TE state with the local temperature in the atmosphere so that the level populations are given by the Boltzmann distribution. In actual stellar atmospheres, there is always the flow of radiation causing the deviation of level populations from the Boltzmann distribution. Non-LTE models are the models that

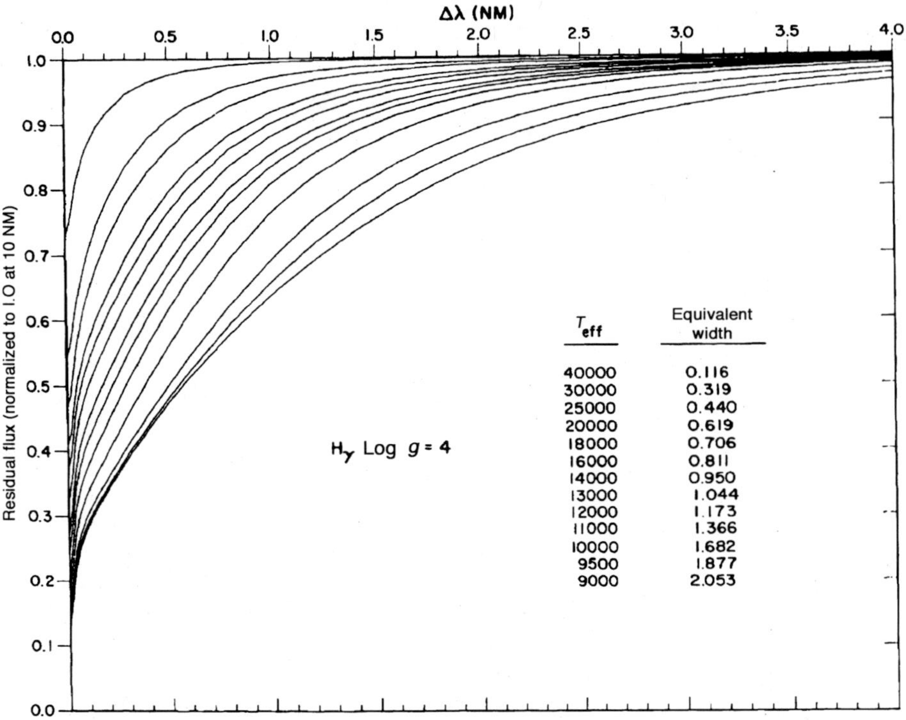

Figure 2.23: Hγ profiles in early-type stars in the LTE models. The profiles are arranged in the order of decreasing effective temperature from top to bottom. The decrease in equivalent width with increasing $T_{\rm eff}$ is seen. (From Kurucz 1979)

treat the radiation fields without assuming the Boltzmann distribution. This inevitably requires a new set of equations, i.e., the equations of statistical equilibrium, expressing the stationary state of electronic transitions between energy levels (see Chapter 4). Therefore, simultaneous solutions of the equation of radiative transfer and of statistical equilibrium are needed in the non-LTE models, making the problem more complicated. Generally speaking, non-LTE models treat the electronic transitions between energy levels more precisely than LTE models. This difference mostly appears in the calculation of line intensities.

For example, Kudritzki (1973, 1976) and Mihalas (1974) pointed out a remarkable increase of absorption line-intensities in HI, HeI, HeII in early-type stars when compared with the LTE models. Figure 2.24 illustrates a comparison between model calculations and observations of the equivalent widths of $H\beta$ line in stars earlier than B0. One may see that the calculations in non-LTE model generally yield better coincidence with observed values. The difference from the LTE calculation is remarkable as one move toward the higher temperature range, particularly for stars earlier than around O6.

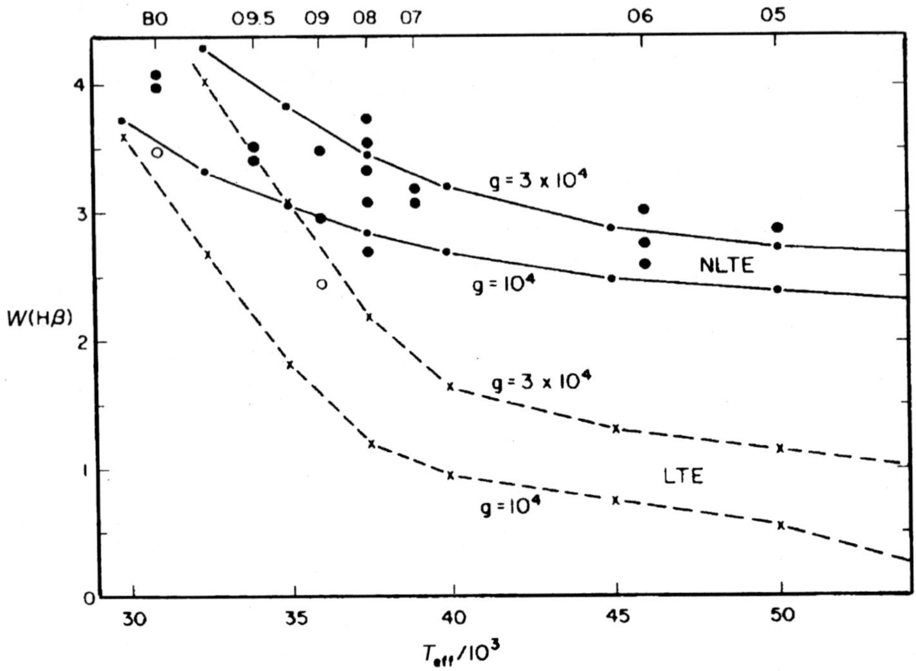

Figure 2.24: Equivalent widths of the Hβ line in O stars. Solid curves, non-LTE calculation; dashed curves; LTE calculation; dots, observations by Zinn (1970). (From Mihalas 1974)

Extended atmospheres

When the geometrical thickness of stellar atmospheres or its scale height in gas pressure is comparable to the stellar radius, the assumption of plane-parallel atmosphere fails to be applicable. Such atmospheres are called extended atmospheres. Supergiants in the upper end of the HR diagram are the examples. A new approach based on spherical symmetry becomes inevitable. In these atmospheres we usually observe large-scale outflow of gas, or strong stellar winds (see Chapter 3), linking the radiation fields closely to the velocity fields. Therefore, simultaneous treatment of hydrodynamic equations and equations of radiative transfer is important for the precise calculation of the stellar atmospheres. Currently three approaches have been proposed as follows.

First approach is to treat the radiation process as precise as possible for static atmospheres. For static spherical atmospheres Chandrasekhar (1934) and Kosirev (1934) have already obtained the approximate solution for gray atmospheres and showed that the spherical symmetric atmospheres yield more flattened continuous spectra in a sense to redden the stellar color as compared to the plane-parallel atmospheres.

More recently Cassinelli (1971) and Castor (1974) recognized the same effects based on the LTE-model calculations. Further more, Mihalas and Hummer (1974) developed the theory of extended atmospheres to the calculation of non-LTE models for a star with $M = 60\ M_{\odot}$, $T_{\rm eff} = 39{,}500$ K, $R = 24\ R_{\odot}$, and log $g = 3.45$, corresponding to a spectral type near O6. Their results, as compared with LTE models, showed the followings:

(a) The models predict an intrinsic reddening of the colors of extended envelopes relative to plane-parallel atmospheres.
(b) The Balmer discontinuity appears in absorption, as observed. The weakening of the UV side becomes more evident when the atmosphere becomes more extended. This is in contrast with both the spherical and planar LTE models that predict the discontinuity to be in emission.
(c) The profiles and equivalent widths for the Lα, Hα, Hβ, and Hγ are computed and compared with other models. The Lα line is strongly in emission for all of the spherical models, while for all planar and the LTE spherical models this line is in absorption, or only weakly in emission. The first three Balmer lines remain in absorption but weaken as the atmosphere becomes more extended, while the LTE spherical models predict increasing strength in emission.

These results show that spherical treatments become important in some types of stars, particularly in the calculation of continuous spectra. For the line spectra, since intensities and profiles are strongly dependent upon the velocity field in the atmospheres, it is necessary to take the kinematic state of the atmosphere into account.

The second approach is to consider the hydrodynamical processes more directly and the radiative processes supplementarily. This approach mainly aims at an analysis of the stellar winds. Cassinelli and Castor (1973) and Castor et al. (1975) have considered the stellar winds in early-type stars by combining the conservation of radiative momentum and energy with the equations of gas flows. Among them, Cassinelli and Castor assumed the optically thin winds, whereas Castor et al. have adopted the escape probability method for line radiation (see Section 4.2.2).

The third is an approach to unified models. The extended atmospheres of early-type stars are generally accompanied by stellar winds, where the strong emission lines are often formed. The atmosphere and wind have usually been connected by some artificial boundary condition. Gabler et al. (1989) proposed a new approach for non-LTE atmospheres of hot stars to avoid such artificial condition. They combined the radiation-driven stellar winds with a non-LTE model atmosphere for spherical geometry. This model has applied to the O4f star ζ Puppis and following results were obtained:

(a) This model can reproduce wind-contaminated photospheric lines as well as wind lines.

(b) The observed infrared excess of ζ Puppis is reproduced.

(c) The emergent flux shortward of the HeII-edge at 228 Å is increased by a factor of a thousand relative to the plane-parallel models. This is caused by the presence of the wind outflow in the region where the continuum is formed.

Further reading

Aller, L. H. (1963). *Astrophysics: The Atmospheres of the Sun and Stars*, 2nd edition. Ronald.

Böhm-Vitense, E. A. (1989). *Introduction to Stellar Astrophysics,* Vols.1 and 2. Cambridge University Press, Cambridge.

Kaler, J. B. (1989). *Stars and Their Spectra.* Cambridge University Press, Cambridge.

Kudritzki, R. P. and Hummer, D. G. (1990). Quantitative spectroscopy of hot stars. *Ann. Rev. A.A*, **28**, 303–345.

References

Aller, L. H. (1963). *Astrophysics: The Atmospheres of the Sun and Stars*, 2nd edition. Ronald.

Cassinelli, J. P. (1971). Extended model atmospheres for the central stars of planetary nebulae. *Ap. J.*, **165**, 265–284.

Cassinelli, J. P. and Castor, J. I. (1973). Optically thin stellar winds in early-type stars. *Ap. J.*, **179**, 189–207.

Castor, J. I. (1974). The effect of sphericity on stellar continous energy distributions. *Ap. J.*, **189**, 273–283.

Castor, J. I., Abbott, D. C., and Klein, R. I. (1975). Radiation driven winds of Of stars. *Ap. J.*, **195**, 157–174.

Chalonge, D. and Divan, L. (1952). Recherche sur les spectres continues stellaires. V. *Ann. d'Ap.*, **15**, 201–236.

Chalonge, D. and Divan, L. (1973). La classification stellaire BCD: Paramètres caractéristiques du type spectral calibration en magnitudes absolues. *A.A.*, **23**, 69–79.

Chandrasekhar, S. (1934). The radiative equilibrium of extended stellar atmospheres. *M. N. R. A. S*, **94**, 444–466.

Chauville, J., Zorec, J., Ballereau, D., Morrell, N., Didal, L., and Garcia, A. (2001), High and intermediate-resolution spectroscopy of Be stars 4481 lines. *A. A.* **378**, 861–882.

Cox, A. N. (1999). *Allen's Astrophysical Quantities*, 4th edition. Springer Verlag, Chapter 15, Normal stars.

Crawford, D. L. (1958). Two-dimensional spectral classification by narrow-band photometry for stars in clusters and associations. *Ap. J.*, **128**, 185–206.

Fukuda, I. (1982). A statistical study of rotational velocities of the stars. *Pub. A.S.P.*, **94**, 271–284.

Gabler, A., Gabler, R., Kudritzki, R. P., Puls, J., and Pauldrach, A. W. A. (1989). Unified NLTE model atmospheres including spherical extention and stellar winds: Method and first results. *A. A*, **226**, 162–182.

Gömez, A. E. (1993). Stellar distances and Hipparcos, in *Inside the Stars*, IAU Coll. 137, A.S.P. Conf. Ser. Vol. **40**, 324–332.

Gray, D. F. and Turner, C. G. (1987). An analysis of the photospheric line profiles in F, G, and K supergiants. *Ap. J.*, **322**, 360–367.

Herzberg, G. (1944). *Atomic Spectra and Atomic Structure*, Dover Book Publ., N.Y.

Huang, S. S. and Struve, O. (1953). A study of line profiles: The spectrum of Rho Leonis. *Ap. J.*, **118**, 463–476.

Inglis, D. R. and Teller, E. (1939). Ionic depresssion of series limits in one-electron spectra. *Ap. J.*, **90**, 439–448.

Johnson, H. L. (1963). Vol.3, Chapter 8. *Basic Astronomical Data, Stars and Stellar Systems*, K. A Strand (ed.), University of Chicago Press, IL.

Keenan, P. C. (1963). Classification of stellar spectra. Vol. 3. *Basic Astronomical Data, Stars and Stellar Systems*, Strand K. A. (ed.), University of Chicago Press, IL.

Kirkpatrick, J.D., Reid, I. N., and 8 co-authors (1999). Dwarfs cooler than M. The definition of spectral type L using discovering from the 1-micron all-sky survey (2MASS). *Ap. J.*, **519**, 802–833.

Kosirev, N. A. (1934). Radiation equilibrium of the extended photosphere. *M. N. R. A. S.*, **94**, 430–443.

Kudritzki, R. P. (1973). Non-LTE effects and influence of helium abundance in A0 Ia supergiant-atmospheres. *A.A.*, **28**, 103–107.

Kudritzki, R. P. (1976). Non-LTE model atmospheres of subluminous O-stars. *A.A.*, **52**, 11–21.

Kurucz, R. L. (1979). Model atmospheres for G, F, A, B, and O stars. *Ap. J. Suppl.*, **40**, 1–340.

Kurucz, R. L. (1993). A new opacity-sampling model atmosphere program for arbitrary abundances, in *Peculiar Versus Normal Phenomena in A-type and Related Stars*, A.S.P. Conf. Vol. **44**, 87–97.

Lang, K. R. (1991). Section 9.5, Spectral classification of the stars, Section 9.7, Stellar temperature and luminosity. *Astrophysical Data : Planets and Stars.* Springer-Verlag, Berlin.

Maury, A. C. and Pickering, E. C. (1897). Spectra of bright stars photographed with the 11-inch Draper telescope as a part of the Henry Draper Memorial and discussed by Antonia C. Mauary under the direction of E. d. Pickering *Annals of the Astronomical Observatory of Harvard College*, **28** (Part 1), 1–128.

Merrill, P. W. (1958). *Lines of the Chemical Elements in Astronomical Spectra.* Carnegie Institute of Washington Publication, Washington, DC.

Mihalas, D. (1974). Progress towards an interpretation of stellar spectra. *A. J.*, **79**, 1111–1121.

Mihalas, D. and Hummer, D. G. (1974). Theory of extended stellar atmospheres. I. Computational method and first results for static spherical models. *Ap. J. Suppl.*, **28**, 343–372.

Morgan, W. W. (1937). On the spectral classification of the stars of types A to K., *Ap. J.*, **85**, 380–397.

Morgan, W. W., Keenan, P. C., and Kellman, E. (1943). *An Atlas of Stellar Spectra.* Astrophysical Monograph, Chicago, IL.

Reid, I. N. (1999). M dwarfs, L dwarfs, T dwarfs and subdwarfs : Ψ(M) at and below the hydrogen-burning limit, *Proc. of Star Formation 1999*, Nakamoto, T. (ed.), Nobeyama Radio Observatory, Nagano, Japan, 327–332.

Stebbins, J. and Kron, G. E. (1964). Six-color photometry of stars. XI. Black-body color temperature of 25 stars. *Ap. J.*, **139**, 424–434.

Strömgren, B. (1963). Vol. **3**, Chaper 9, Quantitative classification methods. *Basic Astronomical Data, Stars and stellar Systems*, K. Strand (ed.), Universtiy of Chicago Press, IL.

Takeda, Y. (1995). Self-consistent multi-parameter fitting of stellar flux spectra. *Pub. A.S. Japan*, **47**, 287–298.

Unsöld, A. (1955). *Physik der Aternatmosphären*, Springer-Verlag, Berlin. p. 509, Figure 168.

van Altena, W. F., Lee, J. T., and Hoffleit, E. D. (1992). *The General Catalogue of Trigonometric Stellar Parallaxes.* Yale University Observatory, IL.

Yamashita, Y., Nariai, K., and Norimoto, Y. (1977). *An Atlas of Representative Stellar Spectra.* University of Tokyo Press, Tokyo.

Zinn, R. J. (1970). The temperature dependence of Hβ strength in O stars. *Ap. J.*, **162**, 909 912.

Chapter 3

Dynamic Processes in Stellar Atmospheres

3.1 Convection layers and atmospheric structure

3.1.1 Convection layers and the Schwarzschild criterion

Convection is a form of energy transport. In stellar interiors, energy can be transported by convective current in addition to radiative flow. Convection layers appear in the intermediate- to late-type stars and play important roles as the source of mechanical energy that heats up outer atmospheres of these stars.

We first consider the principle of convection in a plane parallel atmosphere. Suppose that the temperature $T_{\rm a}$ and the gas density $\rho_{\rm a}$ (subscript 'a' stands for atmosphere) increase with the geometrical depth t from the surface. If a gas bubble at a depth is heated (by any method or means) in the surrounding atmosphere, the temperature and gas density are changed to the values of $T_{\rm b}$ and $\rho_{\rm b}$, respectively, (subscript 'b' stands for bubble), under the pressure balance between bubble and atmosphere. When $T_{\rm b} > T_{\rm a}$ and $\rho_{\rm b} < \rho_{\rm a}$, this bubble becomes buoyant and begins to rise up until the bubble temperature is no longer higher than the surrounding temperature. Falling motion of the bubble occurs just in the opposite case. Therefore, whether the gas bubble can continuously rise or not depends on the temperature gradients of the bubble and atmosphere. The condition for rising and falling of a gas bubble, i.e., the onset of the convective instability, can be expressed as follows:

$$\frac{dT_{\rm b}}{dt} < \frac{dT_{\rm a}}{dt}. \qquad (3.1.1)$$

If we assume that the expansion of the bubble during its rising motion is adiabatic, the temperature variation of the bubble follows the adiabatic temperature gradient, that is, $dT_{\rm b}/dt$ is replaced by $(dT/dt)_{\rm ad}$, where 'ad' stands for adiabatic.

Then, by making use of the relation

$$\frac{dT}{dt} = \frac{dT}{dP_g}\frac{dP_g}{dt},$$

we can rewrite Equation (3.1.1) as

$$\left(\frac{dT}{dP_g}\right)_{ad} < \left(\frac{dT}{dP_g}\right)_{a}, \tag{3.1.2}$$

since dP_g/dt is common for the bubble and atmosphere. In addition, if the atmosphere is in radiative equilibrium, the above equation is written as

$$\left(\frac{dT}{dP_g}\right)_{ad} < \left(\frac{dT}{dP_g}\right)_{rad}, \tag{3.1.3}$$

where rad stands for radiative. If we multiply Equation (3.1.2) by P_g/T, we get

$$\nabla_{ad} = \left(\frac{d\ln T}{d\ln P_g}\right)_{ad} < \left(\frac{d\ln T}{d\ln P_g}\right)_{rad} = \nabla_{rad}, \tag{3.1.4}$$

where the nabla ∇ denotes the logarithmic gradient. Relation (3.1.4) is called the Schwarzschild criterion for convective instability. From this criterion we can see that the logarithmic gradient ∇_{ad} should be small, or ∇_{rad} should be large, or both, for the onset of convective instability. In the next section we consider these conditions.

3.1.2 Convective instability

Radiative temperature gradient

In what case does the radiative temperature gradient ∇_{rad} become larger? Suppose that the atmosphere is in LTE and consider that the total mean intensity J and total flux F_r are connected by Equations (2.5.27) and (2.5.30) as

$$\frac{dJ}{d\tau} = \frac{3}{4}F_r, \tag{3.1.5}$$

where $d\tau = \kappa\, dt$ denotes the mean optical depth. In the LTE atmosphere, we have $J(\tau) = S(\tau) = B(\tau) = (\sigma/\pi)\, T^4(\tau)$ from Equations (2.5.31) and (2.5.32). Then, inserting these relations into (3.1.5), we get

$$\frac{d\ln T}{dt} = \frac{3\pi\kappa F_r}{16\,\sigma\, T^4}. \tag{3.1.6}$$

Since the pressure gradient in the static equilibrium is given by

$$\frac{d\ln P}{dt} = g\,\frac{\rho}{P_g}, \tag{3.1.7}$$

we have the pressure gradient as

$$\nabla_{\rm rad} = \frac{(d\ln T)/dt}{(d\ln P)/dt} = \frac{3\pi F_{\rm r}\kappa P_{\rm g}}{16\sigma T^4 \rho g}. \tag{3.1.8}$$

The radiative gradient can be very large if the flux $F_{\rm r}$ is very large or if κ becomes very large at some level of the atmosphere.

A large flux can happen in the interior of massive stars. In these stars the energy production is sharply concentrated near the star's center, due to the very high sensitivity of energy production rate to temperature in CNO cycle. Since the flux is given by $\pi F = L/(4\pi r^2)$ and r is small, we have a very large flux, causing the convective instability. This explains why massive stars have convection cores.

A large absorption coefficient can cause the formation of convection layers in cool stars having surface temperature less than about 7000 K. When the temperature exceeds 6000–7000 K, the mean absorption coefficient κ increases rapidly with temperature. This is due to the effect of continuous absorption by negative ions of hydrogen as the main cause of opacity in the atmospheres of cool stars.

Negative ions can be formed by the capture of an electron by a hydrogen atom with the binding energy of 0.7 eV. The bound-free transitions of these negative ions produce the continuous absorption for $\lambda < 8200$ Å. Near the stellar surface cooler than $\sim$6000 K, free electrons are supplied by the ionization of metallic atoms like Na, Ca. When the temperature exceeds 6000–7000 K, hydrogen starts to be ionized. Since hydrogen abundance is so high as compared to metals, only 0.1% ionization of hydrogen gives 10 times the electron density and produces a remarkable increase in the continuous absorption.

In addition, the bound-free transitions from hydrogen atoms become conspicuous with the increase of temperature, and contribute to the continuous absorption. For these two reasons, the total absorption coefficient κ increases very steeply. In the upper part of hydrogen ionization layers of cool stars, the gradient $\nabla_{\rm rad}$ can exceed even 1000 times due to the very steep increase of the absorption coefficient.

Adiabatic temperature gradient

The second condition for the convective instability is the smallness of gradient $\nabla_{\rm ad}$. This condition is derived as follows.

Let $C_{\rm p}$ and $C_{\rm v}$ be the specific heats constant pressure and constant volume, respectively, and their ratio $\gamma = C_{\rm p}/C_{\rm v}$ is also defined. In adiabatic change, the gas pressure P and the density ρ follow the relation

$$P = P_0 \left(\frac{\rho}{\rho_0}\right)^{\gamma}, \tag{3.1.9}$$

and the equation of state for an ideal gas is given by

$$P = \frac{R\rho T}{\mu}, \tag{3.1.10}$$

where R and μ denote the gas constant and mean molecular weight, respectively.

By eliminating ρ from (3.1.9) and (3.1.10), and calculating the logarithmic differential, we can easily determine $\nabla_{\rm ad}$ for an ideal gas as

$$\nabla_{\rm ad} = \left(\frac{d\ln T}{d\ln P_{\rm g}}\right)_{\rm ad} = 1 - \frac{1}{\gamma}. \tag{3.1.11}$$

For a monoatomic gas, since $\gamma = 5/3$, we have $\nabla_{\rm ad} = 0.4$. The value of γ generally exceeds unity, with the limiting case of $\gamma = 1$ for isothermal changes at its lowest value. According to (3.1.11), $\nabla_{\rm ad}$ decreases when γ decreases. Now we have by definition

$$\gamma = \frac{C_{\rm p}}{C_{\rm v}} = \frac{C_{\rm v} + R}{C_{\rm v}} = 1 + \frac{R}{C_{\rm v}}, \tag{3.1.12}$$

then, γ approaches 1 if $C_{\rm v}$ becomes very large. This happens if a large amount of energy is needed to heat the gas, such as latent heat in ionization of gas. In that case $\nabla_{\rm ad}$ becomes small and convection can set in.

In cool stars, as atmospheric temperature increases inward, hydrogen atoms start to ionize at a level of the temperature 6000–7000 K, and enter the convection layer by small value of $\nabla_{\rm ad} = 0.2$–0.1. In much deeper layers where temperatures exceed around 20,000 K, $\nabla_{\rm ad}$ recovers to around 0.4 so that the convection layers only occur in this temperature range, i.e., from commencement of ionization up to fully ionized state in stellar atmospheres. This is called the hydrogen convection layer because of the important role of hydrogen. Beneath this layer there is the helium convection layer, in the temperature range 20,000–50,000 K, where ionization of helium plays the important role. However, the contribution of the helium convection layer to the supply of mechanical energy to the outer atmosphere is small, since helium abundance is small and the layer lies so deep under the stellar atmosphere. In hot stars with surface temperature higher than 10,000 K, the hydrogen ionization layer already reaches the surface so that there appears to be no hydrogen convection layer. The helium convection layer might have some roles in early-type stars.

3.1.3 Convection layers and mechanical energy

The convection layer is generally in a turbulent state, and mechanical energy is produced in the form of sound waves or magnetic sound waves by turbulent activity. Proudman (1952) considered the generation of sound waves by homogeneous and isotropic turbulence and derived the acoustic power $\varepsilon_{\rm A}$

(erg cm^{-3} s^{-1}) in the form

$$\varepsilon_{\rm A} = 38\rho \frac{u^8}{c_{\rm s}^5 L} \tag{3.1.13}$$

where ρ is the gas density, u the mean turbulent velocity, $c_{\rm s}$ the sound speed, and L the length scale of turbulence. From this acoustic power, the total emerging acoustic flux F_A (erg cm^{-2} s^{-1}) from a stellar convection zone is given by

$$F_{\rm A} = \frac{1}{2} \int_{\Delta z} \varepsilon_{\rm A}\, dz, \tag{3.1.14}$$

where Δz is the thickness of convection layer and the factor 1/2 takes into account that only half of the acoustic flux is emitted in the outward direction.

This theory was extended later by Stein (1968) to include the turbulent medium with temperature and density gradients. He then derived the acoustic flux spectrum. Based on this theory, Ulmschneider et al. (1996) considered the acoustic wave generation by turbulent convection in late-type stellar atmospheres and computed both the acoustic spectrum and total acoustic flux for stars in the range of effective temperature $T_{\rm eff}$ = 2000–10,000 K and gravity log g = 0–8. According to them, the acoustic spectrum $F(\omega)$, where ω (= $2\pi\nu$) is circular frequency of sound, is expressed as

$$F(\omega) = \frac{dF_{\rm A}}{d\omega} = 2\pi \int_{\Delta z} dz \int_0^{\pi/2} z^2 F(z,\omega) \cos\theta \sin\theta \, d\theta, \tag{3.1.15}$$

where z is the vertical distance, Δz the thickness of convection layer, and the energy flux per unit frequency $F(z,\omega)$ is calculated after Stein's expression. The total acoustic energy flux F_A can be obtained by

$$F_{\rm A} = \int_{\omega_1}^{\infty} F(\omega)\, d\omega, \tag{3.1.16}$$

where ω_1 denotes the critical frequency at the low-frequency end, below which sound waves are trapped in the convection layer. The critical frequency is estimated by using the scale height $H = kT/(\mu m_{\rm H} g)$ of the atmosphere as

$$\varpi_1 = \frac{c_{\rm s}}{2H}, \tag{3.1.17}$$

where $c_{\rm s}$ is the sound velocity. The acoustic flux spectrum thus obtained is shown in Figure 3.1, where the parameters are taken for two stellar types of giants (log g = 3) and dwarfs (log g = 5) in the range of $T_{\rm eff}$ = 2000–8000 K or 9500 K. Parameter α is the so-called mixing length of convection layer and takes a value between 1 and 2. In Figure 3.1, α is fixed to 2.

The acoustic spectrum generally has a broad maximum depending on log g and $T_{\rm eff}$ as seen in Figure 3.1. Acoustic energy flux and the frequency at maximum are much higher for dwarf stars than for giant stars at the same effective temperature. The temperature dependence of the total acoustic flux

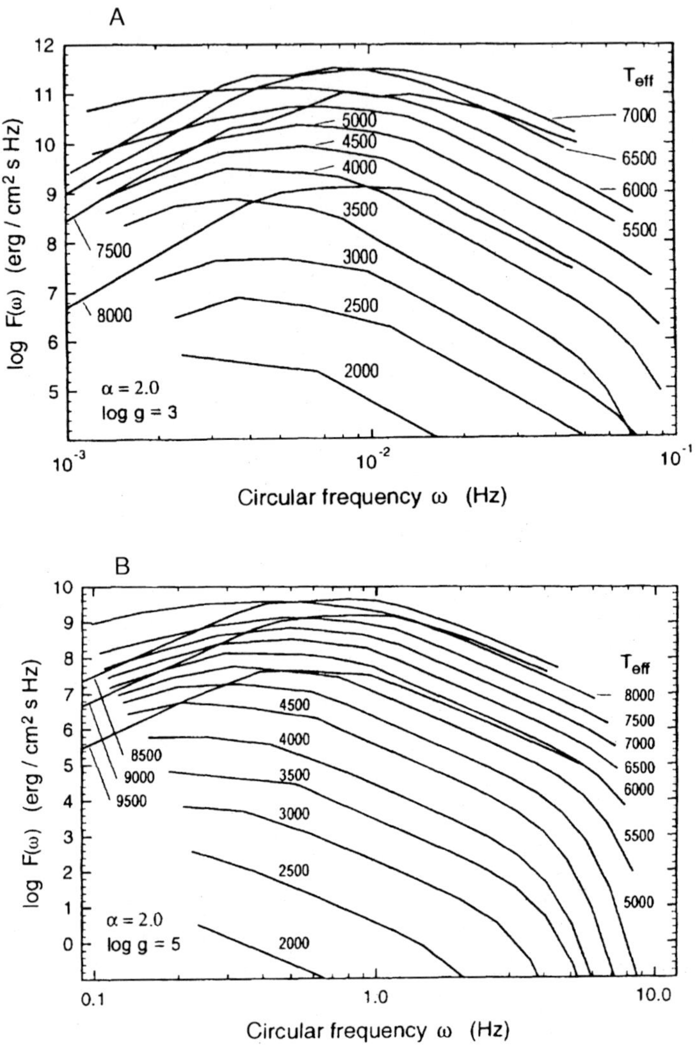

Figure 3.1: Acoustic spectrum $F(\omega)$ versus circular frequency $\omega = 2\pi\,\nu$ for giants (upper panel) and dwarfs (lower panel). $T_{\rm eff}$ is taken as parameter. (From Ulmschneider et al. 1996)

is exhibited in Figure 3.2, where the surface gravity is taken in a range log $g = 0$–8. The calculated total fluxes by Ulmschneider et al. (1996) (solid lines) are compared with those generated using classical theory of Proudman (1952) (dotted lines). It is seen that the maximum points of the total flux appear in around $T_{\rm eff} = 6000\,{\rm K}$ for giants (log $g = 3$) and $T_{\rm eff} = 8000$ K for dwarfs (log $g = 5$). The classical theory generally yields the total flux several factors higher than Ulmschneider et al.'s theory.

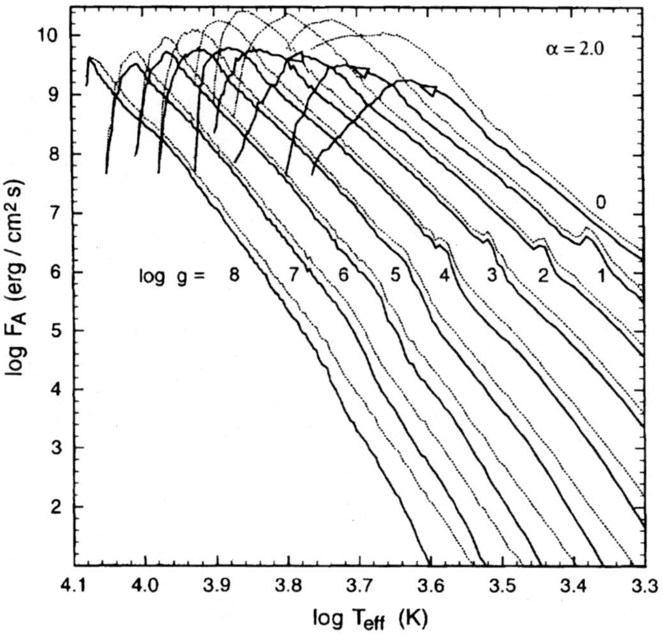

Figure 3.2: Total acoustic fluxes F_A versus T_{eff} for different gravity (solid lines). For comparison, calculations by the classical Lighthill–Proudman formula are also shown (dotted lines). (From Ulmschneider et al. 1996)

3.1.4 Stellar evolution and chromospheric activities

Kinetic energy flux generated in the convection layer can be obtained from the stellar model by taking the stellar effective temperature and surface gravity as the parameters. Kippenhahn (1973) derived the fluxes and drew the lines of equal mechanical energy flux F_{mech} (erg cm^{-2} s^{-1}) on the ($\log g - \log T_{eff}$) plane as seen in Figure 3.3. He thus considered the chromospheric activities and evolution of stars from pre-main sequence to post-main sequence stages. Figure 3.3 A illustrates the loci of the equal mechanical flux $\log F_{mech}$ (erg cm^{-2}), along with the approximate positions of dwarfs, giants, and supergiants. Panel B shows the evolutionary paths of the stars with mass of 1 and 5 $M_\odot$ on the same diagram. As seen in panel A a peak of maximum flux runs approximately in vertical direction at around abscissa $\log T_{eff} \sim 3.8$. This peak corresponds to the instability stripe for pulsational instability on the HR diagram. On both sides of the stripe acoustic flux declines, particularly rapidly in the hotter side.

According to this figure, a star having mass nearly $1 M_\odot$ passes regions of high mechanical energy flux during most of its evolutionary life, whereas a massive star of around $5 M_\odot$ passes such regions only in the pre-main sequence star stage and in evolved post-main sequence stage. The difference between

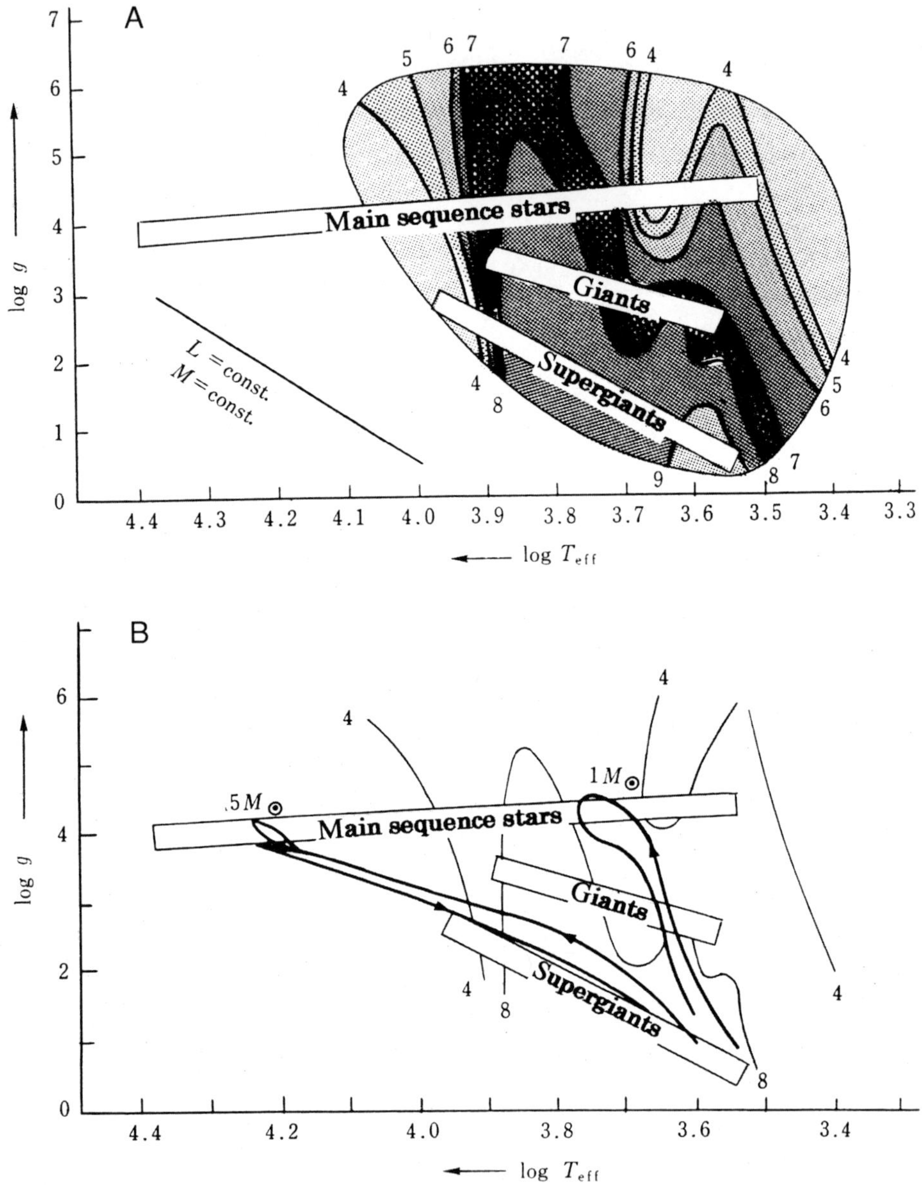

Figure 3.3: The curves of equal mechanical energy flux and evolutionary paths of stars in log $T_{\rm eff}$ versus log g plane. Panel A denotes the loci of the equal mechanical flux log $F_{\rm mech}$ and the approximate locations of dwarfs, giants, and supergiants. Panel B shows the evolutionary paths of the stars of 1 and $5M_\odot$ on the same plane. (From Kippenhahn 1973)

early- and late-type stars in the structure of outer atmospheres and wind structure can be understood by the difference of the generation of mechanical energy in the convection zones.

Returning to Equation (3.1.13), one may see that the rate of sound wave generation strongly depends on the turbulent velocity. At the same time, it also depends on the stellar temperature. For the same condition, the generation rate of sound waves is higher for cooler stars (by smaller c_s). This may explain the general tendency that chromospheric and flare activities are increasingly stronger in stars of lower temperatures (see Chapter 6).

3.2 Stellar winds

3.2.1 Basic concepts of stellar winds

In most stars, outflows of gas from the surfaces can occur steadily or unsteadily in various forms. Relatively steady outflows are called the stellar winds. We first consider the basic form of the solar winds in steady, spherically symmetric, and isothermal gas flow. The driving force is the pressure gradient in hot corona.

Let M and R be the mass and radius of the star and ρ, p, v the density, pressure, and flow velocity of gas at distance r from the star's center, respectively. The basic hydrodynamic equation is written as follows:

(a) Equation of mass conservation

$$4\pi\rho v r^2 = \text{constant}. \tag{3.2.1}$$

(b) Equation of momentum conservation

$$v\frac{dv}{dr} + \frac{1}{\rho}\frac{dp}{dr} + \frac{GM}{r^2} = 0. \tag{3.2.2}$$

Equation of energy conservation is not needed under isothermal condition. From the above equations we get

$$\frac{r}{v}\frac{dv}{dr} = -\frac{GM}{r}\frac{1-(r/r_c)}{v^2 - c_s^2}, \tag{3.2.3}$$

where

$$c_s^2 = RT, \quad \text{(sound velocity)} \tag{3.2.4}$$

$$r_c = \frac{GM}{2c_s^2}. \tag{3.2.5}$$

We now rewrite Equation (3.2.3) as

$$\frac{1}{2}\frac{dv^2}{dr} = -\frac{GM}{r^2}\frac{N}{D}, \tag{3.2.6}$$

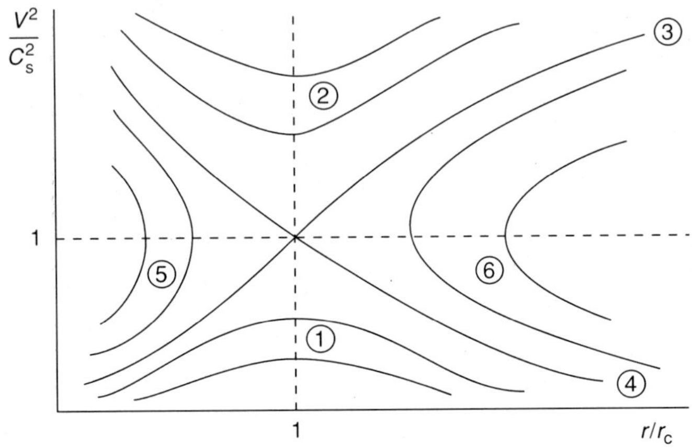

Figure 3.4: Topological solutions in corona-type stellar winds. The abscissa is the distance from the star's center, in units of the distance to the singular points r_c. The ordinate is the velocity of flow expressed by the square of the Mach number v/c_S. For topological types ①-⑥ (see text).

where

$$N = 1 - \frac{r}{r_c}, \qquad D = v^2 - c_s^2. \tag{3.2.7}$$

The signs of numerator (N) and denominator (D) for isothermal flow can be read as

$$N > 0 \quad \text{for } r < r_c \qquad \text{and} \qquad N < 0 \quad \text{for } r > r_c$$

for N, and

$$D > 0 \text{ (supersonic)} \quad \text{for } v > c_s \qquad \text{and} \qquad D < 0 \text{ (subsonic)} \quad \text{for } v < c_s$$

for D. Thus Equation (3.2.6) has a singular point at $r = r_c, v = c_s$, and its solution topologically takes the following six types in Figure 3.4.

① Subsonic flow: Always subsonic with a maximum velocity at $r = r_c$.
② Supersonic flow: Always supersonic with a minimum velocity at $r = r_c$.
③ Transonic accelerating flow: Transition occurs from subsonic to supersonic at $r = r_c$.
④ Transonic decelerating flow: Transition occurs from supersonic to subsonic at $r = r_c$.
⑤, ⑥ Meaningless solution with double roots for velocity.

As seen in Equation (3.2.5), the position of any singular point depends on the stellar mass and the temperature of gas flow, and the type of flow is determined by the boundary conditions at the stellar surface and at infinity. The singular point is located on the crossing point of ascending and descending flows, and this type is called the X-type singularity.

In case of the sun, we have

$$r_{\rm c} = \frac{GMm}{4kT} = \frac{5.76 \times 10^6}{T} R_\odot, \quad c_{\rm s}^2 = \frac{2kT}{m} = 1.66 \times 10^8\, T, \qquad (3.2.8)$$

then, putting $T \sim 2 \times 10^6$ K for the coronal temperature, we get

$$r_{\rm c} \sim 3R_\odot, \qquad c_{\rm s} \sim 180 \text{ km s}^{-1}.$$

In solar winds, there is no supersonic flow in the corona while the wind velocity reaches fully supersonic ($v \sim 500$ km s^{-1}) at around 1 AU. Thus the wind type is the transonic accelerating flow (type ③). In reality, the solar wind is not isothermal, since wind temperature at the earth's orbit is measured as $T \sim 2 \times 10^5$ K, which is much lower than the coronal temperature of 2×10^6 K. Therefore the equation of energy conservation should be simultaneously solved in order to get the realistic solution.

The mass loss rate due to the solar wind, which can be estimated from ρ and v at suitable distance, is around $10^{-14} M_\odot$ per year. The total mass lost from the sun in its life ($\sim 10^{10}$ year) is then $10^{-4} M_\odot$. This is sufficiently smaller than the present mass of the sun, implying that the effect of mass loss to the evolution of the sun should be negligibly small.

For stars having coronae, the wind type is the same as the sun (type ③). Since the radius of any singular point $r_{\rm c}$ increases with the decrease of the coronal temperature as seen in (3.2.5), the stellar atmosphere will approach the state of static equilibrium if the coronal temperature declines to the level of the stellar effective temperature.

The solar wind and similar stellar winds originating from coronae are called the coronal-type winds. In early-type stars, the winds are usually powered by radiation pressure and called the radiation-driven winds. In late-type stars, some different types occur such as dust-driven winds, pulsationally driven winds, and wave-driven winds. We now proceed to the main types of stellar winds (see Lamers and Cassinelli 1999).

3.2.2 Radiation-driven winds in early-type stars

In stars earlier than around B0 type, the driving force of the wind is mainly coming from the radiation pressure of the stellar radiation. The equation of motion in this case (by adding the radiation pressure term $g_{\rm R}$ in Equation (3.2.2)) is given as

$$v\frac{dv}{dr} + \frac{1}{\rho}\frac{dp}{dr} + \frac{GM}{r^2} - g_{\rm R} = 0, \qquad (3.2.9)$$

where $GM/r^2 = g$ is the gravitational acceleration and proportional to r^{-2}. Since radiation pressure term $g_{\rm R}$ is proportional to the radiation flux πF, which is also proportional to r^{-2}, its ratio

$$\Gamma_{\mathrm{R}} = \frac{g_{\mathrm{R}}}{g} \tag{3.2.10}$$

can be regarded as a distance-independent constant. The radiation pressure term is generally given as

$$g_{\mathrm{R}} = \frac{4\pi}{c} \int_0^\infty \kappa_\nu \, F_\nu \, d\nu = \frac{k_{\mathrm{c}} \, L_{\mathrm{c}}}{4\pi c \, r^2} + \sum_L g_{\mathrm{L}}, \tag{3.2.11}$$

where k_{c} is the mean absorption coefficient ($\mathrm{cm}^2 \, \mathrm{g}^{-1}$), L_{c} the total luminosity of the star, g_{L} the radiation pressure term by a spectral line, and the summation gives the sum of contributions from all effective absorption lines. The equation of motion is rewritten by using (3.2.10) as

$$v \, \frac{dv}{dr} + \frac{1}{\rho} \frac{dp}{dr} + \frac{GM}{r^2} \, (1 - \Gamma_{\mathrm{R}}) = 0. \tag{3.2.12}$$

This corresponds to Equation (3.2.2), and the flow is specified by the parameter Γ_{R}.

Now consider the stellar winds of OB stars powered by the radiation pressure on a number of spectral lines, particularly, of resonance lines. In the outer atmospheres, if an ion is accelerated outward by absorbing a resonance line radiation which usually has a large absorption coefficient, its line center shifts to the blue side becoming able to absorb another radiation of the same ion. By repeating the same process, this ion can be accelerated further. When two lines are very close in wavelength, the blanketing effect will strengthen the acceleration of the ion. Thus, if there are a large number of such ions and atoms available, and the absorption–acceleration process is effectively repeated, the gas will be finally accelerated up to supersonic velocity. This is the principle of radiation-driven stellar winds in hot stars.

The radiation pressure term g_{L} for a line in Equation (3.2.11) is expressed as

$$g_{\mathrm{L}} = \frac{\Delta\nu_D \, F}{c} \, \frac{\kappa_{\mathrm{L}}}{\tau_{\mathrm{L}}} \, \left(1 - e^{-\tau_{\mathrm{L}}}\right), \tag{3.2.13}$$

where κ_{L} is the absorption coefficient for the line radiation L, τ_{L} the optical thickness of the envelope outside radius r. In the case of accelerating envelopes, τ_{L} is approximately given by the escape probability method (see Section 4.2) as

$$\tau_{\mathrm{L}} = \int_0^\infty \rho \, \kappa_{\mathrm{L}} \, dr \approx \frac{\rho \, v_{\mathrm{th}} \, \kappa_{\mathrm{L}}}{dv/dr}, \tag{3.2.14}$$

where v_{th} is the thermal velocity and dv/dr denotes the acceleration of gas flow at radius r.

The theory of radiation-driven stellar winds was developed by Lucy and Solomon (1970) and subsequently by Castor et al. (1975). It is known as the CAK theory. Later, Abbott (1980, 1982) modified and extended this theory

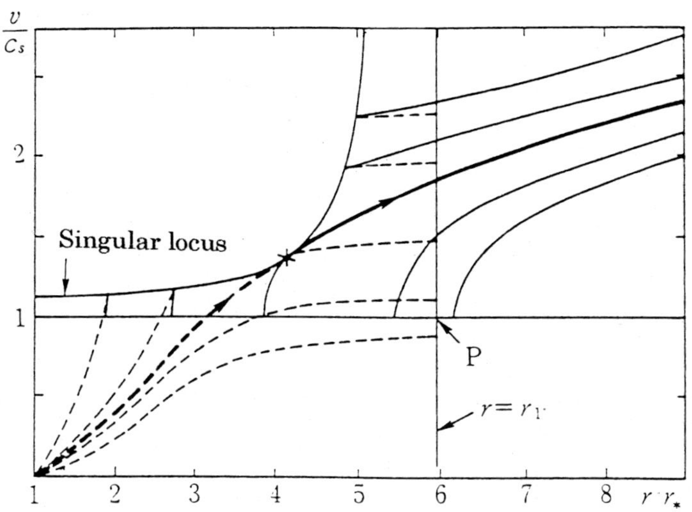

Figure 3.5: A topological solution of a radiatively driven wind. Heavy solid and broken lines show the solution curve of the accelerating flow, which contacts with the singular locus at point x in the upper left corner. By contrast, the sonic point of a solar-type wind P is located at $r = r_1$. (From Abbott 1980)

by increasing the number of atoms and ions up to 30 elements along with the physical interpretation of the critical point (MCAK theory).

In radiation-driven winds, we can also obtain the topological solutions connected from stellar surface to interstellar medium, passing through the critical point as seen in coronal-type winds. The difference is that, in radiative-driven winds, the critical point is not a point but a singular locus and the transonic point is separately located from this singular locus. An example of the solution topology by Abbott (1980) is shown in Figure 3.5, where the abscissa is the relative distance r/r^* in unit of stellar radius and the ordinate is the Mach number v/c_s. The solution curve for accelerating flow is shown by the curve with arrows whereas the dashed line gives the subcritical branch and the solid line the supercritical branch. The solution curve contacts the singular locus at a point given by a cross. Other topological solutions that are not realistic are also shown by thin solid or dashed lines. The singular point P corresponding to the solar wind is located at the cross of radius r_1 and sonic velocity, where radius r_1 is corresponding to Equation (3.2.5) and given here as

$$r_1 = \frac{GM}{2c_s^2}\,(1 - \Gamma_R)\,. \tag{3.2.15}$$

If the radiation pressure term disappears, the solution reduces to the normal corona-type winds shown in Figure 3.4.

In CAK or MCAK theory, the stellar wind regions are assumed to be in radiative equilibrium, i.e., electron temperature T_e is nearly equal to the star's effective temperature T_{eff} (~10,000–50,000 K). By this reason these theories are called the cool radiation pressure model. Since higher temperature phenomena prevail in the stellar winds of early-type stars, Hearn (1975) proposed a different model. He assumed that the principal cause of the wind is the same as that of the solar wind, namely, the presence of a hot corona, which has a large escape velocity for the stellar gravity. This model is called the hot coronal model with the coronal temperature as high as 3.5×10^6 K as needed to explain the high mass-loss rate of hot stars. In expanding envelopes of early-type stars, often highly ionized ions such as OVI, NV, and SiIV are observed. Lamers and Snow (1978) assumed the ionization equilibrium in the envelope and proposed the stellar winds, called the warm radiation pressure model to explain the existence of these ions. In this model, the envelopes are heated mechanically up to the temperature of the order of $0.7 \sim 4 \times 10^5$ K needed to attain the ionization equilibrium.

It should be noted that the stellar winds in early-type stars are not stationary flows as treated above but are subject to various kinds of instabilities that can cause wind gas to heat to coronal temperature. Mechanical heating may explain the various types of high-temperature phenomena in stellar winds.

3.2.3 Magnetic rotator model of the solar wind

It is well known that the active phenomena in solar chromosphere and corona are closely connected with the magnetic field, which, in turn, can affect the motion of highly conductive fluid such as the solar wind. In order to understand the basic properties of the magnetic solar wind, we now consider the theory developed by Weber and Davis (1967), called the magnetic rotator model (Lamers and Cassinelli 1999).

The motion of the solar wind can be described by means of the magnetohydrodynamic equations for a fluid with an infinite conductivity, no viscosity, and a scalar pressure. Suppose a steady-state solar wind confined in the equatorial plane. The two-dimensional velocity $\vec{v}\,(u, v_\phi)$ and magnetic field $\vec{B}\ (B_r, B_\phi)$ are taken as shown in Figure 3.6 and given as

$$\vec{v} = u\,\vec{e}_r + v_\phi\,\vec{e}_\phi, \tag{3.2.16}$$

$$\vec{B} = B_r\,\vec{e}_r + B_\phi\,\vec{e}_\phi, \tag{3.2.17}$$

where $\vec{e}_r$, $\vec{e}_\phi$ denote the unit vectors in the radial and longitudinal directions, respectively. The roots of magnetic lines of force are vertically fixed at the surface and rigidly rotating with the sun.

By mass conservation law we have

$$\rho\,u\,r^2 = \text{constant} \tag{3.2.18}$$

Figure 3.6: Velocity and magnetic field vectors defined in a rotating star, and unit vectors in radial and horizontal directions on the equatorial plane.

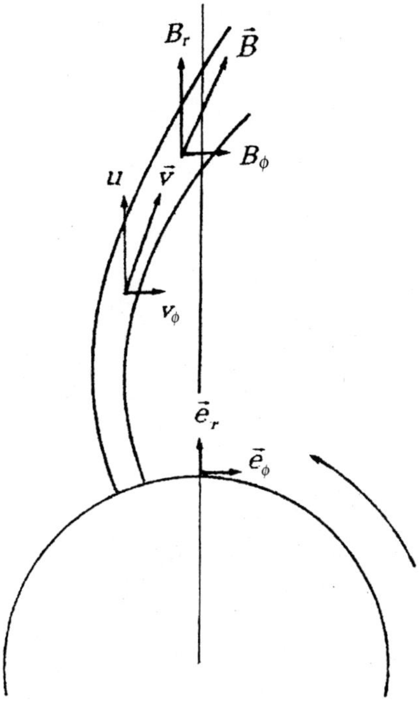

where ρ is the gas density (g cm^{-3}). Since the solar wind is assumed to be a perfect conductor, the electric-field vector is given as $\vec{E} = -\vec{v} \times (\vec{B}/c)$ by Ohm's law, and the gas motion is parallel to the magnetic vector (frozen-in state) in a frame rotating with the sun. In addition, if the wind is in a steady state, we have, from Maxwell's equation,

$$\frac{\partial B_\phi}{\partial t} = c\left(\text{rot}\,\vec{E}\right)_\phi = \frac{1}{r}\,\frac{d}{dr}\,[r\,(u\,B_\phi - v_\phi\,B_r)] = 0. \tag{3.2.19}$$

Prior to integration we consider the wind near the solar surface. First, the magnetic line of force is nearly vertical so that we can put $B_\phi \ll B_r$, while the velocity of gas is prevailing in azimuthal motion by rotation as compared to the vertical wind motion. Hence we have $v_\phi = r\Omega \gg u$, where Ω is the angular velocity of the roots of the lines of force. Under these conditions, Equation (3.2.19) can be integrated as

$$r\,(u\,B_\phi - v_\phi\,B_r) = \text{constant} = -\,\Omega r^2 B_r. \tag{3.2.20}$$

Also, since div$\vec{B} = 0$, the conservation of magnetic flux in a nearly vertical flux tube can be written as

$$r^2 B_r = r_0^2 B_{r0}, \tag{3.2.21}$$

where 0 refers to an arbitrary reference level, $r = r_0$, near the surface.

The azimuthal equation of motion in an axi-symmetric flow is given by

$$\rho \frac{u}{r} \frac{d}{dr} (r v_\phi) = \frac{1}{c} \left(\vec{J} \times \vec{B}\right)_\phi = \frac{1}{4\pi} \left[\left(\mathrm{rot}\vec{B}\right) \times \vec{B}\right]_\phi = \frac{B_r}{4\pi r} \frac{d}{dr} (r B_\phi), \tag{3.2.22}$$

where J is the electric current density. By combining Equations (3.2.18) and (3.2.21) we have

$$\frac{B_r}{4\pi\rho u} = \frac{B_r r^2}{4\pi\rho u r^2} = \text{constant}.$$

This allows us to integrate Equation (3.2.22) as

$$r\, v_\phi - \frac{B_r}{4\,\pi\,\rho\,u}\, r\, B_\phi = \text{constant} = L, \tag{3.2.23}$$

where L is a constant to be determined later. The first term of this equation is the ordinary angular momentum per unit mass and the second term represents the torque associated with the magnetic stresses. The constancy of their sum indicates that the total angular momentum is carried away from the sun per unit mass loss through magnetic solar wind.

Now we introduce the Alfvénic velocity c_{A} and Alfvénic Mach number M_{A} in radial direction as

$$c_{\mathrm{A}}^2 = \frac{B_r^2}{4\pi\rho}, \qquad M_{\mathrm{A}}^2 = \frac{4\,\pi\,\rho\,u^2}{B_r^2} = \frac{u^2}{c_{\mathrm{A}}^2}. \tag{3.2.24}$$

Solving Equations (3.2.20) and (3.2.23) for azimuthal velocity v_ϕ by making use of new variable M_{A}, we get

$$v_\phi = \Omega\, r\, \frac{\left(M_{\mathrm{A}}^2\, L r^{-2}\, \Omega^{-1} - 1\right)}{\left(M_{\mathrm{A}}^2 - 1\right)}. \tag{3.2.25}$$

The radial Mach number M_{A} is much smaller than 1 near the surface of the sun, but M_{A} is approximately 10 at 1 AU. Thus there should exist a point taking the value of $M_{\mathrm{A}} = 1$ between the sun and the earth. Let the radius and radial velocity at this point be r_{a} and u_{a}, respectively. This point is called the Alfvénic critical point. As seen in Equation (3.2.25), the denominator goes to zero at this point, then we require that the numerator should also vanish at the same point in order to keep the expression for v_ϕ finite. By this condition the value of L must be arranged to have the value

$$L = \Omega\, r_{\mathrm{a}}^2. \tag{3.2.26}$$

By substituting Equations (3.2.18) and (3.2.21) into the second equation of (3.2.24), we can see that $M_{\mathrm{A}}^2/u\,r^2$ is a constant along the radius. Using the values r_{a} and u_{a} at the critical point where $M_{\mathrm{A}} = 1$, we have

$$M_{\mathrm{A}}^2 = \frac{u\,r^2}{u_{\mathrm{a}}\,r_{\mathrm{a}}^2} = \frac{\rho_{\mathrm{a}}}{\rho}. \tag{3.2.27}$$

With these relations, v_ϕ and B_ϕ can be written as

$$v_\phi = \Omega\, r\, \frac{u_{\mathrm{a}} - u}{u_{\mathrm{a}}\,(1 - M_{\mathrm{A}}^2)}, \tag{3.2.28}$$

$$B_\phi = -\,B_r\, \frac{\Omega\, r}{u_{\mathrm{a}}}\, \frac{r_{\mathrm{a}}^2 - r^2}{r_{\mathrm{a}}^2\,(1 - M_{\mathrm{A}}^2)}. \tag{3.2.29}$$

The approximate behavior of these functions can be described: for $r \ll r_{\mathrm{a}}$ and $u \ll u_{\mathrm{a}}$, we have $v_\phi \propto r$, $B_\phi \propto 1/r$, since $B_r \propto r^{-2}$ from Equation (3.2.21), and for $r \gg r_{\mathrm{a}}$, the radial velocity u becomes almost constant and thus $M_{\mathrm{A}} \propto r$ and both of v_ϕ and B_ϕ vary as $1/r$.

The radial velocity u can be derived from the solution of the equation of motion that includes the effect of magnetic field and rotation of gas,

$$\rho\, u\, \frac{du}{dr} = -\frac{dp}{dr} - \rho\, \frac{GM_0}{r^2} + \frac{1}{c}\left(\vec{J} \times \vec{B}\right)_r + \rho\, \frac{v_\phi^2}{r}, \tag{3.2.30}$$

where p is the gas pressure and M_0 the mass of the sun.

For simplicity, we suppose that the solar wind is composed of fully ionized hydrogen and assume the perfect gas law and polytrope law, instead of solving the equation of energy conservation. Thus we have

$$p = \frac{2kT}{m}\,\rho, \qquad p = p_0\left(\frac{\rho}{\rho_0}\right)^\gamma = p_{\mathrm{a}}\left(\frac{\rho}{\rho_{\mathrm{a}}}\right)^\gamma, \tag{3.2.31}$$

where γ is the polytrope index. If we take the specific heat ratio for γ, the polytropic law yields the adiabatic change as seen in Equation (3.1.9).

The magnetic force term in Equation (3.2.30) can be written as

$$\frac{1}{c}\left(\vec{J} \times \vec{B}\right)_r = -\,\frac{1}{4\,\pi\, r}\, B_\phi\, \frac{d}{dr}\,(r\, B_\phi)\,. \tag{3.2.32}$$

Inserting these relations, (3.2.30) can be transformed to the following wind equation after some algebraic manipulation,

$$\frac{du}{dr} = \frac{u}{r}\left\{\left(\frac{2\,\gamma\, p_{\mathrm{a}}}{\rho_{\mathrm{a}}\, M_{\mathrm{A}}^{2(\gamma-1)}} - \frac{GM_0}{r}\right)\left(M_{\mathrm{A}}^2 - 1\right)^3 + \Omega^2\, r^2\left(\frac{u}{u_{\mathrm{a}}} - 1\right)\right.$$
$$\left.\times\left[\left(M_{\mathrm{A}}^2 + 1\right)\frac{u}{u_{\mathrm{a}}} - 3\, M_{\mathrm{A}}^2 + 1\right]\right\}. \tag{3.2.33}$$

The topological family of the solutions is shown in Figure 3.7. It is seen that there exist two critical points: a standard X-type singularity at $r = r_{\mathrm{c}}$, and the so-called Alfvén-type singularity at $r = r_{\mathrm{a}}$. There are four solutions passing through the critical points designated $u_{\alpha 1}$ and $u_{\alpha 2}$ with zero pressure at infinity, and $u_{\beta 1}$ and $u_{\beta 2}$ with nonzero pressure at infinity.

We have considered the basic properties of the magnetic solar wind based on the theory of Weber and Davis. With this theory it has been found that there exist two types of singularity: Alfvén-type and the standard X-type.

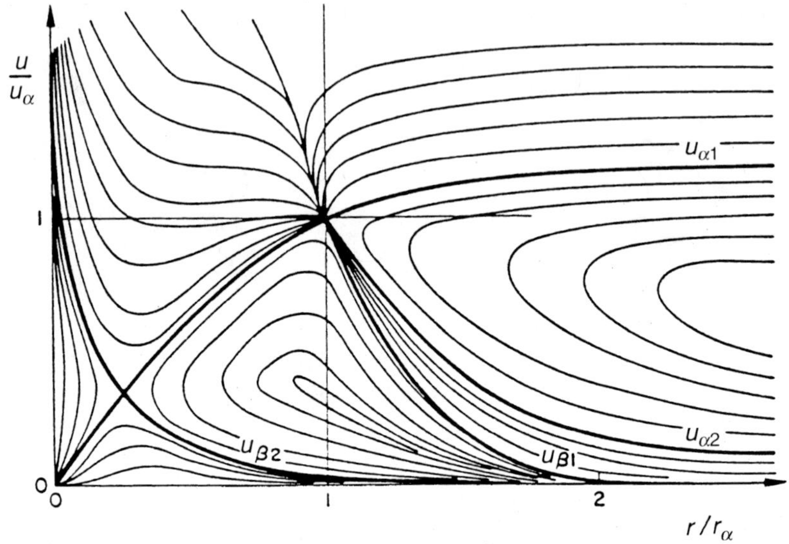

Figure 3.7: Family of solutions of Equation (3.2.33) for a given γ and r_a. The solutions passing through the critical points are designated as $u_{\alpha 1}$, $u_{\alpha 2}$ (with zero pressure at infinity) and $u_{\beta 1}$, $u_{\beta 2}$ (with non-zero pressure at infinity). (From Weber and Davis 1967)

Remarkable angular-momentum loss can also be derived along with the mass loss. In case of the sun, the Alfvén-type singularity is located at around 15 to 50 solar radius, and the time scale of angular momentum loss is 7×10^9 years, much shorter than the time scale of mass loss around 10^{14} years.

Further development has been made by Brandt et al. (1969) by taking into account the energy equation instead of assuming the polytropic gas. Belcher and MacGregor (1976) applied it to solar-type stars.

3.2.4 Stellar winds in late-type stars

Late-type stars, from dwarfs to supergiants, mostly exhibit some forms of mass-loss phenomena, some eruptively and some relatively quietly. Among these, G–K dwarfs show the coronal-type stellar winds as in the sun, while red giants and supergiants give rise to different forms of stellar winds. We have seen the coronal-type winds in Section 3.2.1, and we now consider the winds prevailing in red giants and supergiants. The mechanisms that cause the mass loss and drive the winds are not yet definitively known. Essentially three mechanisms have been proposed: dust-driven winds, pulsationally driven winds, and wave-driven winds. We now consider main types of these winds.

Dust-driven winds

In the outer atmospheres of red giants, dust grains are formed by condensation of outstreaming gas when the temperature drops below the equilibrium temperature for grain formation. The grains can absorb stellar radiation over a broad range of wavelength and gain momentum by radiation pressure. Through collisions between grains and gas, the stellar winds can be effectively accelerated. This is the principle of the dust-driven winds. The necessary conditions to power the winds of this type are the efficient condensation of gas to form grains and the existence of sufficiently high stellar luminosity as the source of radiation pressure.

Although the possibility of radiation pressure acting on molecules and grains was already suggested in the early 1960s (Weymann 1963), it is only since Kwok (1975) that hydrodynamic models have been constructed for the stellar winds in cool giants. It became clear that the formation and growth of grains is the key factor for the dust-driven winds (Gail and Sedlmayer 1987, Gail 1990).

Let us consider a simple case of spherically symmetric and isothermal winds in steady state. Mass conservation law gives

$$4\pi r^2 \rho v = \dot{M} = \text{constant}. \tag{3.2.34}$$

The isothermal sound velocity c_s is

$$p = c_\mathrm{s}^2\, \rho. \tag{3.2.35}$$

The equation of motion can be written, similarly with Equation (3.2.3), as

$$\left(v^2 - c_\mathrm{s}^2 \right) \frac{1}{v}\frac{dv}{dr} = \frac{2\,c_\mathrm{s}^2}{r} - \frac{GM_*}{r^2}\,(1-\alpha)\,, \tag{3.2.36}$$

where M_* is the stellar mass and α is the ratio of the radiative acceleration to gravitational deceleration. Let $\langle \kappa_d \rangle$ be the mean absorption coefficient for dust, then α is given by

$$\alpha = \frac{L_*}{4\,\pi\, GM_*\, c}\;\frac{\langle \kappa_d \rangle}{\rho}. \tag{3.2.37}$$

The family of solutions of Equation (3.2.36) is topologically similar to the solar wind with an X-type singular point as shown in Figure 3.4. We can obtain the transonic wind flow passing through the singular point as before. The position of singular point r_c is given from (3.2.36) by putting

$$v^2 - c_\mathrm{s}^2 = 0, \qquad \frac{2c_\mathrm{s}^2}{r} - \frac{GM_*}{r^2}\,(1 - \alpha) = 0,$$

thus

$$\frac{r_\mathrm{c}}{R_*} = \frac{v_\mathrm{e}^2}{c_\mathrm{s}^2}\,\frac{1 - \alpha\,(r_\mathrm{c})}{4}, \tag{3.2.38}$$

where R_* is the stellar radius and v_e the escape velocity at the stellar surface given by

$$v_e = \sqrt{\frac{2GM_*}{R_*}}. \tag{3.2.39}$$

Equation (3.2.38) is an implicit function of r_c, as α is a function of r_c. This wind differs from the coronal type at this point. Since $\langle\kappa_d\rangle$ is the product of dust density and absorption cross section, α becomes 0 in the vicinity of the photosphere where no dust exists, and the wind is reduced to the normal coronal type (3.2.3). As it goes out, dusts begin to form and α gradually increases and reaches to $\alpha = 1$ at some radius where the radiaton pressure force balances the gravity. By putting $\alpha = 1$ in Equation (3.2.37), we get

$$L_* = \frac{4\pi G M_* \rho}{\langle\kappa\rangle_H}. \tag{3.2.40}$$

When α exceeds 1, the right-hand side of (3.2.36) becomes positive and the gas is in the state of outward acceleration. Equation (3.2.36) has topological solutions similar to the solar wind, i.e., there is a transonic fast wind passing through an X-type singularity along with a slow breeze wind that is always subsonic and gradually decelerates after passing the maximum velocity point. The singular point is located near the point of $\alpha = 1$. The supersonic flow thus obtained is the dust-driven wind. A large amount of mass loss can be obtained with this wind, but the terminal velocity is several tens km s^{-1} much slower than the solar wind or radiation-driven winds. This is because the gas density decreases in the outer part of the wind where collisions between gas and dust decrease and the acceleration by dust becomes ineffective.

However, the conditions, under which the dust-driven winds can effectively be powered are rather severe. Equation (3.2.40) is the luminosity that corresponds to the Eddington luminosity limit (the stability boundary of atmospheres against the radiation pressure). This infers that the stars should have sufficient luminosity higher than the Eddington limit. In addition the stars should have sufficiently high mass-loss rate in order to form dust grains effectively. For stars with $T_{\rm eff} \sim 3000$ K and $\sim$2000 K, the Eddington limit is around 10^5 and $10^4 L_\odot$, respectively (Dominik 1990). Hence the stars that fulfill the requirements of dust-driven winds are limited to the bright red supergiants. In ordinary red giants, large mass-loss rate is also observed, but it is difficult to explain the winds of these stars as the dust-driven winds by the shortage of the luminosity of these stars (Gail and Sedlmayer 1987).

Pulsationally driven winds

It has been known that the long-period variables (Mira) show a high mass-loss rate along with the remarkable dynamical structure synchronized with the pulsational motion of atmospheres. Large mass outflow may originate from the stellar winds powered by pulsational motion. In the phase of expanding motion in Miras, large-scale shock waves propagate outward and gas bubbles suffer an impulsive force. These bubbles rise up along ballistic orbits in the atmosphere and again fall down. If the bubbles receive the impulsive force by next shock waves before returning to the original point, then the bubbles again rise up to much higher part of the atmosphere. Thus, the successive propagation of shock waves can transport the atmospheric matter effectively upward and produce the stellar winds. This is the principle of pulsationally driven winds. Hydrodynamical models were first proposed by Willson and Hill (1979) and Wood (1979), who were succeeded by Bertschinger and Chevalier (1985).

Consider a simple case of pulsationally driven winds following Willson and Hill (1979). Let the star pulsating with the period P, also the period of generation of shock waves at the base of the atmosphere. Suppose that a gas particle, ejected at radius r_0 with initial postshock velocity v_0 makes a ballistic motion. Then the particle velocity in a coordinate system at rest with respect to the star is given by

$$v = \pm \left[v_0^2 - v_e^2 \left(1 - \frac{r_0}{r} \right) \right]^{1/2}, \tag{3.2.41}$$

where $v_e = (2GM/r_0)^{1/2}$ is the escape velocity at radius r_0. The double sign in the right-hand side indicates that the gas particle first rise up ($v > 0$), then, after reaching its highest position ($v = 0$), the particle turns to fall ($v < 0$) toward the initial radius r_0. The ballistic time P_0 for the particle to return to the initial radius r_0 is given by

$$P_0 = \frac{2r_0}{v_e} \Pi_0 (v_0/v_e), \tag{3.2.42}$$

where Π_0 is a dimensionless gravitational period and a monotonically increasing function of v_0/v_e. For $v_0/v_e \propto v_0 r^{1/2}$, Π_0 tends to remain roughly constant as r increases. Therefore, P_0 is a monotonically increasing function of r, and there will be a point in the atmosphere above which the ballistic time exceeds the pulsational period ($P_0 > P$). Physically this happens because, as we move outward in r, the gravity decreases, increasing the ballistic time, while the pulsation period stays fixed. This point where $P = P_0$ is called the critical radius r_s.

In the outer atmosphere above this critical radius, we have $P_0 > P$ and gas particles accumulate kinetic energy from successive shock passage and finally exceed the escape velocity. As a whole, the gas flows appear as stellar winds powered by pulsational motion accompanied by successive shock waves. As an example of numerical calculation by Willson and Hill, particle paths above r_c for $v_0 = 20$ km s^{-1} are shown in Figure 3.8 for a star of 5 $M_\odot$ and a pulsation period of $P = 1.6 \times 10^7$ s. The abscissa denotes time t/P in unit of pulsation period and the ordinate is the distance from the star's center r/r_c in unit of critical radius. Thick lines give the position of shock waves and thin lines the loci of individual mass elements; the shock strength is arbitrarily assumed to

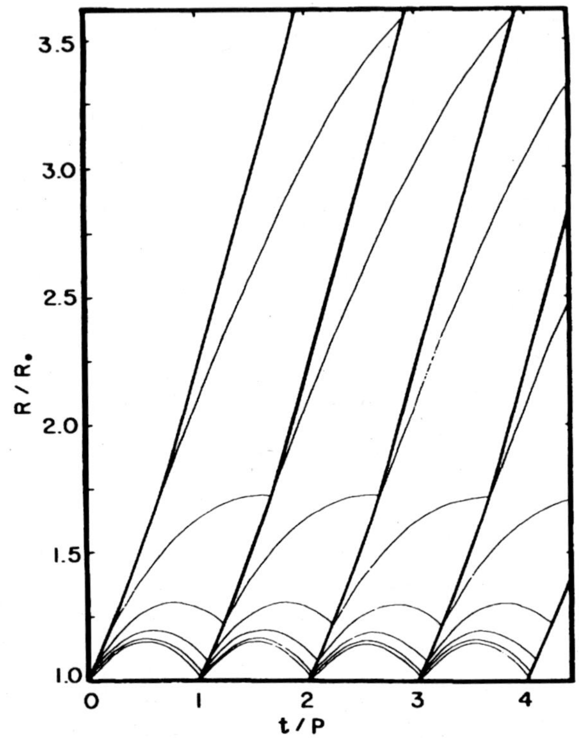

Figure 3.8: Sample calculations of pulsationally driven winds. Particle paths after shock passage is shown for a star of 5 $M_{\odot}$ and pulsation period of $P = 1.6 \times 10^7$ seconds. The heavy lines indicate the positions of shock waves. The paths of individual mass elements as a function of time between shocks are shown by thin lines. Different paths denote the different initial radius ejected by the first shock. (From Wilson and Hill 1979)

give the density jump of 10 at the shock front. It is seen in Figure 3.8 that gas particles can escape after several shock passages even slightly outside of the critical radius. For example, gas particles started at $r_{\rm initial}/r_{\rm c} = 1.001$ reach the escape velocity after five shock passages.

In this way, if the critical radius is located at a large distance from the stellar surface where the gas density is low, then the mass loss rate will be low. If the critical radius is sufficiently near the stellar surface where the density is high, the mass-loss rate may be high. Willson and Hill derived the mass-loss rate of Mira variables adopting suitable strength of shock waves.

Wave-driven winds

In convection layers of stars, various types of waves are generated and transport kinetic energy into the upper layers. These waves are either compressive

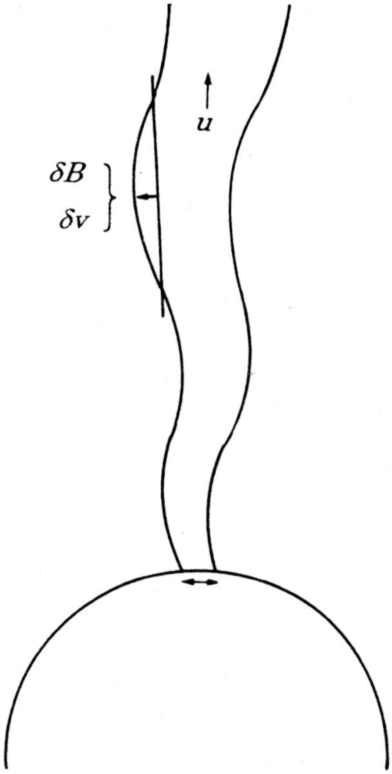

Figure 3.9: The magnetic lines of force and the propagation of Alfvén waves are schematically shown. Horizontal oscillation is given at the foot of the lines of force on the photospheric surface and propagates as the Alfvén waves. The propagation velocity of Alfvén waves, u, and amplitude (δB or δv) is assumed as functions of the radius r. Stellar rotation is ignored.

(e.g., acoustic waves or magnetoacoustic waves) or noncompressive (Alfvén waves). According to Hartmann and MacGregor (1980), the compressible waves grow up rapidly transforming into weak shock waves and dissipate in the lower part of the atmospheres. They are not able to lift a massive wind out the stellar gravitational field. In contrast, noncompressive Alfvén waves are not so rapidly damped and can reach the outer atmosphere and drive a quite massive wind. This type of winds is called the wave-driven winds.

Let us consider the steady and spherically symmetric wind. Suppose that the magnetic lines of force are extended in radial direction at its initial state. If we oscillate horizontally the roots of the magnetic lines of force, the oscillation propagates upward in the form of Alfvén waves. Figure 3.9 shows the magnetic field and the propagation of the Alfvén waves.

The basic magnetohydrodynamic equations are similar to those of the magnetic rotator model (Section 3.2.3), but here the stellar rotation is ignored and the momentum term D given by wave propagation is added as

$$u\,\frac{du}{dr} = -\frac{1}{\rho}\,\frac{dp}{dr} - \frac{GM_*}{r^2} + D. \qquad (3.2.43)$$

In order to consider the oscillation of the magnetic lines of force, we should solve the time-dependent equation. For simplicity, however, we consider the steady flows by taking a time average over several oscillation periods. Then the additional term D, which is called the gradient of wave pressure or magnetic pressure gradient, can be given as (Holzer 1987)

$$D = -\frac{1}{\rho}\frac{d}{dr}\left[\frac{\langle \delta B^2\rangle}{8\,\pi}\right] = -\frac{1}{\rho}\frac{d}{dr}\left[\frac{1}{2}\rho\,\langle \delta\, v^2\rangle\right], \tag{3.2.44}$$

where $\langle \delta B^2\rangle$ and $\langle \delta v^2\rangle$ are the mean-square magnetic field and velocity field of the wave, respectively. The Alfven wave energy flux density, f, is given by

$$f = \rho\,\langle \delta v^2\rangle\,\left(c_{\rm A} + \frac{3}{2}\,u\right), \tag{3.2.45}$$

where $c_{\rm A} = B/\sqrt{4\,\pi\rho}$ is the Alfvén speed.

For the undamped Alfvén waves, the equation of motion can be arranged to the wind equation given by

$$\frac{1}{u}\frac{du}{dr}\left[u^2 - v_{\rm T}^2 - \frac{1}{4}\left(\frac{1+3M_{\rm A}}{1+M_{\rm A}}\right)\langle\delta v^2\rangle\right] = \frac{2\beta}{r}\left[v_{\rm T}^2 + \frac{1}{4}\left(\frac{1+3M_{\rm A}}{1+M_{\rm A}}\right)\langle\delta v^2\rangle - \frac{1}{4}\frac{v_{\rm e}^2}{\beta}\right], \tag{3.2.46}$$

where $v_{\rm T}$ is the thermal velocity, $v_{\rm e}$ the escape velocity at radius r, and

$$\beta = \frac{r}{2A}\frac{dA}{dr}, \tag{3.2.47}$$

A is the cross sectional area for a radial magnetic tube and $A \propto B^{-1}$ for $u \times B = 0$. The mean amplitude of the Alfvén waves, $\langle \delta v^2\rangle$ or $\langle \delta B^2\rangle$, is a function of radius r and $v_{\rm e}$.

A critical point of the Equation (3.2.46) appears at $r = r_{\rm c}$ when the terms in brackets on the right and left sides of (3.2.46) both vanish. It is noticed that the flow speed at the critical point does not equal the thermal velocity but a sound velocity modified by Alfvén waves, $v_{\rm s}$, which is defined by

$$v_{\rm s}^2 = v_{\rm T}^2 + \frac{1}{4}\left(\frac{1+3M_{\rm A}}{1+M_{\rm A}}\right)\langle\delta v^2\rangle. \tag{3.2.48}$$

Though this is a case of undamped Alfvén waves, Equation (3.2.46) gives a transonic flow passing the critical point.

The case of stellar winds driven by the damped Alfvén waves has been numerically calculated by Holzer et al. (1983). The result is illustrated in Figure 3.10, which shows (A) radial profiles of density, ρ/ρ_0, (B) flow speed, $u/v_{\rm e0}$, (C) temperature T(K), and (D) wave energy flux $F_{\rm w}/F_{\rm w0}$, for four values of the Alfvén wave frequency in a hydrogen-helium atmosphere with 10% helium by number. The suffix 0 indicates the value at the stellar surface, $v_{\rm e0}$ the escape velocity at the surface. In the calculation, the Alfvén wave frequency ω is taken as the parameter related to the damping length L. L

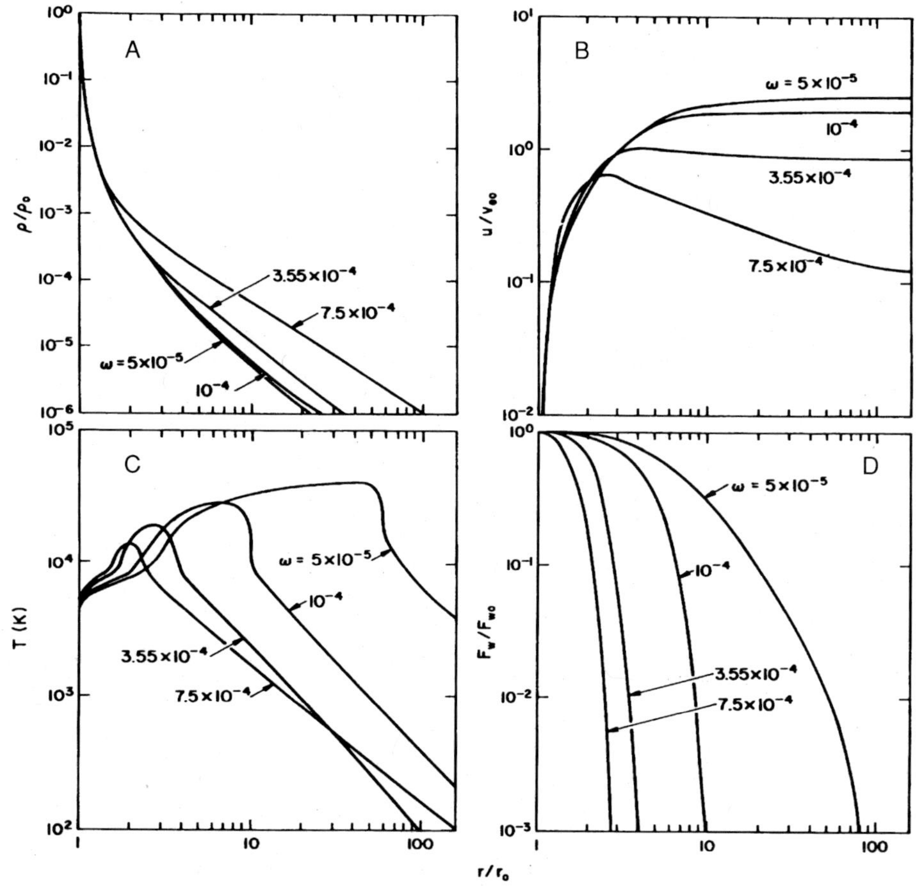

Figure 3.10: Model calculations of the wave-driven winds. The abscissa is the radial distance and the ordinates are the radial profiles of (A) density (ρ), (B) flow velocity (u/v_{e0}), (C) temperature (T), and (D) the Alfvén wave energy flux (F_w/F_{w0}) for four Alfvén wave frequencies. (From Holzer et al. 1983)

is smaller for higher frequency. In Figure 3.10, the value of $\omega = 3.55 \times 10^{-4}$ corresponds to the damping length equal to the stellar radius and the lower values of ω denote longer damping length by smaller damping effects.

As seen in Figure 3.10, the stellar winds powered by the Alfvén waves are quite different in some points from the solar type winds: (1) there are no high-temperature regions as high as solar corona—they are at most several ten thousand Kelvin; (2) the terminal velocities (given as the asymptotic curves of u/v_{e0}) are in the order of escape velocity at the stellar surface (around 100 km s^{-1}) or less, depending on the values of ω; (3) the effects of Alfvén wave damping is higher for higher Alfvén frequency (or smaller damping length) as shown in steep declining of Alfvén wave energy flux $F_\mathrm{w}/F_\mathrm{w0}$ for higher frequencies.

3.2.5 Stellar winds and mass-loss rates

As a consequence of stellar winds, most stars are losing mass, more or less steadily, into the interstellar space. The amount of mass loss is an important property of stars, and its effects on the evolution of stars become very significant in some types of stars.

The mass-loss rate $\dot{M}$ is defined as the amount of gas lost per year in unit of solar mass ($M_\odot$ per year), and expressed by the following general formula.

Consider a spherical surface at radius r from the star of radius R. Let $d\sigma$ be a surface element and $\rho(r)$ and $v(r)$ be the density and outward velocity of gas on this surface. If the gas outflow occurs from solid angle Ω, the mass-loss rate is given by

$$\dot{M} = \frac{dM}{dt} = \int_\Omega \rho\,(r)\; v\,(r)\; d\sigma, \tag{3.2.49}$$

where

$$d\sigma = r^2 \sin\theta\, d\theta\, d\phi. \tag{3.2.50}$$

In case of spherically symmetric outflow, mass-loss rate is given by

$$\dot{M} = 4\pi\, r^2 \rho\,(r)\; v\,(r)\,. \tag{3.2.51}$$

If the outflow is limited in equatorial region within the opening angle $\pm\theta_\mathrm{w}$, we have

$$\dot{M} = 4\pi\, r^2\, \rho\,(r)\; v\,(r)\; [1 \;-\; \cos\theta_\mathrm{w}]\,, \tag{3.2.52}$$

and if it is limited in high-latitude region higher than $\pm\theta_\mathrm{w}$, we have

$$\dot{M} = 4\pi\, r^2 \rho\,(r)\; v\,(r)\; \cos\theta_\mathrm{w}, \tag{3.2.53}$$

where $\rho(r)$, $v(r)$, and θ_w must be deduced from observations of stellar winds.

For most stars stellar winds are supposed to be spherically symmetric. In stars where the effect of rotation is important such as Be stars, equatorial and polar regions on the stellar surface may have different wind behaviors. Estimation of opening angle is often difficult and in many cases some suitable values are assumed.

We now consider the estimation of mass-loss rates for early- and late-type stars, separately.

Early-type stars

Stellar winds of early-type stars are generally the radiation-driven type having the transonic flows of ionized gas, which can usually be observed by emission lines, thermal radio emission, and infrared excess. The most prevailing emission line is the Hα, except WR stars, for which emission lines of ionized ions such as HeII, NIII, CIV, etc., play key roles. Generally, emission-line stars in early-type stars yield high mass-loss rates.

As an example, we consider the mass-loss rate from OB stars, derived by Leitherer (1988) from the intensity of Hα emission. In O stars and OB supergiants, Hα emission is supposed as having been formed in the stellar winds at a large distance from the stellar surface, where the envelope is generally assumed to be optically thin. In this case, the luminosity of Hα line, $L(\mathrm{H}\alpha)$ (erg s^{-1}), can be given by (see Chapter 4)

$$L(\mathrm{H}\alpha) = \int W N_3 A_{32} h\nu_{32} dV, \tag{3.2.54}$$

where the integration is carried out for all line forming region, and

$$dV = 4\pi r^2 \, dr, \tag{3.2.55}$$

W denotes the dilution factor (see Equation (4.1.4)), N_3 the population density of the third level of hydrogen, both the function of the radius r. In Equation (3.2.54), A_{32} is the transition probability for spontaneous emission for Hα, $h\nu_{32}$ is the energy of an Hα photon.

N_3 can be expressed in terms of the electron density N_e, proton density N_p, and electron temperature T_e in the form

$$N_3 = N_\mathrm{e} N_\mathrm{p} b_3 f(T_\mathrm{e}), \tag{3.2.56}$$

where $f(T_\mathrm{e})$ is derived from the ionization formula for hydrogen, i.e.,

$$f(T_\mathrm{e}) = \frac{9h^3}{(2\pi m_\mathrm{e} k)^{3/2}} T_\mathrm{e}^{-3/2} \exp\left(\frac{17530}{T_\mathrm{e}}\right) \tag{3.2.57}$$

and b_3 denotes the departure coefficient for the level $n = 3$ of hydrogen.

If we assume spherical symmetry and set $N_\mathrm{e} = N_\mathrm{p}$ (pure hydrogen envelope), equation of continuity gives the mass flux at radius r which is equal to the mass-loss rate by the relation

$$4\pi r^2 m_H \mu N_\mathrm{e} v(r) = \dot{M}, \tag{3.2.58}$$

where the velocity law $v(r)$ is parameterized as

$$v(r) = v_0 + (v_\infty - v_0)\left(1 - \frac{R}{r}\right)^\beta, \tag{3.2.59}$$

R is the stellar radius, v_0 the initial velocity on the stellar surface and v_∞ the terminal velocity of the wind, and β the parameter which governs the slope of $v(r)$. When $\beta < 1$, the flow velocity increases very rapidly, whereas, when $\beta > 1$, velocity gradually increases, both asymptotically approaching the terminal velocity.

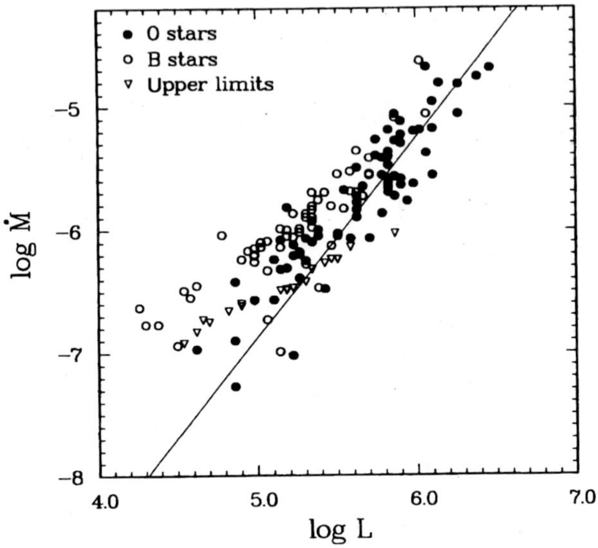

Figure 3.11: Mass loss rates for O and B stars derived from $L(\mathrm{H}\alpha)$ by using Equation (3.2.60). The solid line represents the predicted relation from radio and UV data by Garmany and Conti (1984). (From Leitherer 1988)

Combining the above equations, Leitherer derived the relation between $L(\mathrm{H}\alpha)$ and $\dot{M}$ as

$$\log L\,(\mathrm{H}\alpha) = 2\,\log\left|\dot{M}\right| - 2\log\,v_\infty - \log R + c\,(T_{\mathrm{eff}}) + I + 25.125, \tag{3.2.60}$$

where the units are taken as $L(\mathrm{H}\alpha)$ in $L_\odot$, $\dot{M}$ in $M_\odot$ per year, v_∞ in km s^{-1}, and R in $R_\odot$. The value of $c(T_{\mathrm{eff}})$ is given as a function of stellar effective temperature, and I is the quantity depending on the values of β and v_0/v_∞. The adopted value of β is 0.7 for O stars and 1.5 for B stars.

By using the Formula (3.2.60), Leitherer estimated the mass-loss rate for 150 OB stars. The correlation between derived mass-loss rate and stellar bolometric luminosity is shown in Figure 3.11, where O and B stars are distinguished by different symbols. The solid line indicates the predicted relation derived by radio and UV data by Garmany and Conti (1984). One may see that the mass-loss rate is an increasing function of luminosity as expected by the radiation-driven wind theory. On the average, $\dot{M}$ for O stars is in agreement with the predicted rate. In contrast, $\dot{M}$ for B stars shows some disagreement, indicating the deviation of the wind properties of B stars from those of hotter O stars.

Late-type stars

The stellar winds of late-type (G–M) stars are roughly classified into two types: solar type and red giant type.

The solar type (or coronal type) is characterized by hot, high-velocity winds flowing out from the hot coronae. Typical values in the solar wind are hot ($\sim$1–2 $\times 10^6$ K), high velocity ($\sim$500 km s^{-1}), with a low mass-loss rate (approximately a few of $10^{-14} M_\odot$ per year). Stars of this type are distributed along the main sequence and above it up to some boundary with giant stars in the HR diagram. Along the main sequence, chromospheric activities are strengthened toward K and M types. Mullan et al. (1992) carried out infrared spectroscopic observations and derived wind models that are powered by hot coronae with temperature of 3–4 $\times 10^6$ K, and the terminal velocity around 200–500 km s^{-1}. They thus estimated the mass-loss rate in the order of $10^{-10} M_\odot$ per year, several orders of magnitude higher than that of the Sun. This estimation was criticized by Lim and White (1996) based on aperture-synthesis observations at 3.5 mm of dMe stars. Since neither star did show detectable millimeter emission that is expected from Mullan et al.'s model, they put the upper limit of mass-loss rate to be around $10^{-12} M_\odot$ per year for these dMe flare stars.

Stars having the winds of the red giant type occupy the upper right part of the HR diagram, including red giants, Mira variables, asymptotic giant branch stars (AGB) and cool supergiants. The winds are generally cool and slow so that observations in molecular lines, infrared spectrum, and in some cases thermal radio emission, are the effective tools for the study of wind structure and mass-loss rates.

As a typical case, we consider the mass loss from cool stars following Knapp and Morris (1985). They have carried out observations of the molecular CO (J = 1–0) line at 2.6-mm wavelength and detected CO line emission for 50 stars including Mira variables, carbon stars, and supergiants among 105 program stars. The mass-loss rate was derived by matching the observed CO line profiles to those of model calculations in two cases of optically thick and thin envelopes. Optical thickness is determined by the line profile, i.e., the profile is round or parabolic for a thick envelope and sharp for a thin one. For mass-loss rate they derived the formula

$$\dot{M} = A \frac{T_{\mathrm{A}}^* v_0^2 D^2}{f} M_\odot \text{ per year,} \tag{3.2.61}$$

where A is a constant involving the envelope radius, f the abundance of CO relative to H_2 molecule, $T_A{}^*$ the peak temperature of CO line profile, v_0 the terminal velocity, and D the distance to the star in pc. Using this formula they estimated the mass-loss rates as

$\dot{M} = 4 \times 10^{-6} - 3 \times 10^{-4} M_\odot$ per year for optically thick envelopes and
$\dot{M} = 9 \times 10^{-8} - 8 \times 10^{-6} M_\odot$ per year for optically thin envelopes.

Thus the optically thick envelopes have well-developed winds giving higher mass-loss rates, as compared to the case of optically thin envelopes. The work of Knapp and Morris was further developed by Loup et al. (1993) and applied to 444 stars, mostly AGB stars. As the results, they have shown that the mass-loss rate is higher for stars with larger terminal velocities, in accordance with the Expression (3.2.61).

3.3 Accretion flows and accretion disks

Some types of stars show infalling flows of gas onto the stellar surface (accretion flow). Pre-main sequence stars (T Tauri stars, Herbig Ae/Be stars), interacting close binaries (Algol stars, Be binaries), and CV fit this category. Since accretion flows generally have some degree of angular momentum, the flows do not reach the stellar surface directly. Instead, they form the rotating disk or ring around the star, called the accretion disk, and then gradually fall onto the surface.

In this section, we first consider the spherical accretion flow (with no angular momentum) from the analogy of the stellar winds, and then proceed to the accretion flow with angular momentum and to the accretion disk.

3.3.1 Spherically symmetric accretion flows

For the spherically symmetric accretion, we can apply Equation (3.2.6) in the inverse direction of flows, i.e., $v < 0$. If we look for a solution which satisfies the boundary condition of $v = 0$ for $r \to \infty$ and have the negative velocity with inward acceleration, the solution is given by type ④in Figure 3.4. In a real solution, however, this accretion flow should approach the stellar surface with zero velocity; hence, the supersonic flow inside the critical point should have a shock front at some level through which the flow turns to subsonic. This is a basic process of star formation in spherical symmetry.

The numerical calculations by Winkler and Newman (1980) are an example of such spherical accretion process and are shown in Figure 3.12. They calculated the gravitational collapse of a homogeneous gas cloud of 1 $M_\odot$ with initial gas density $\rho = 10^{-20}$ g cm^{-3}. Figure 3.12A illustrates the density variation as a function of radius. Three different regions of the flow can be easily recognized: the free-falling envelope, shock front, and inner hydrostatic atmosphere. In Figure 3.12B, the velocity variation is shown as a function of radius. The velocity accelerates as $u \propto -r^{-1/2}$ in the envelope, sharply transits at the shock front, and then approaches zero in the hydrostatic core.

The formation of a protostar through the spherical collapse of isolated fragments was considered by Stahler et al. (1980). In the course of numerical calculations for the time-depending hydrodynamic equations, they particularly considered the main accretion phase in which a stable central core accretes matter from a distended envelope. The protostellar cloud in this phase

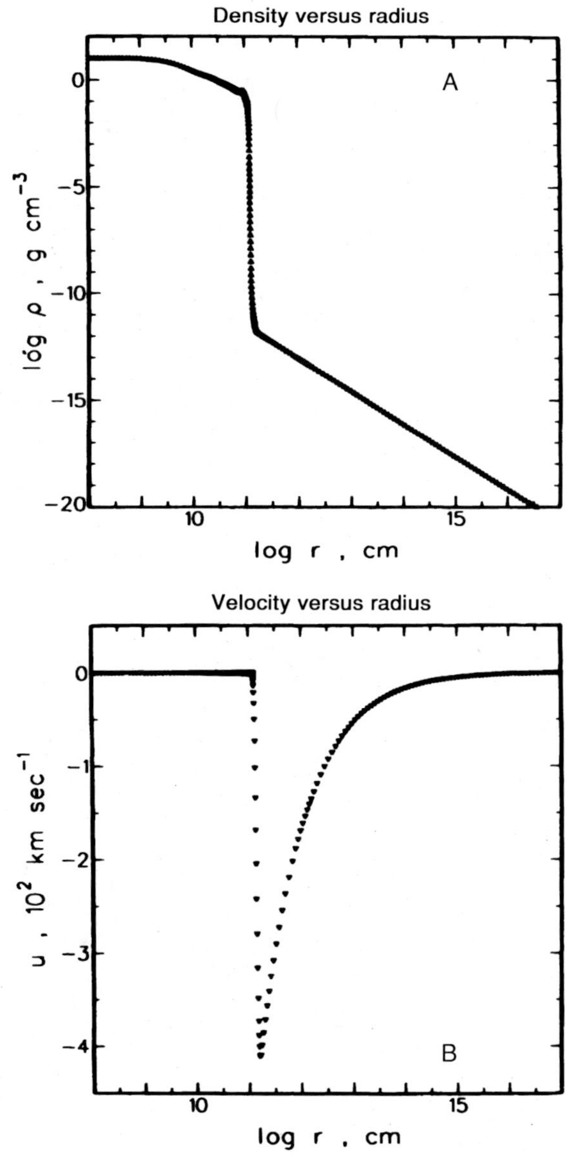

Figure 3.12: Density and velocity variations in spherically symmetric accretion flow; (A) density (B) velocity, both as functions of radius. (From Winkler and Newman 1980)

is divided into several layers with different behaviors that are illustrated in Figure 3.13. The outer envelope is an isothermal cool molecular gas layer under the free-falling state. The isothermal condition breaks down when densities become sufficiently high to trap the infrared radiation. The corresponding radius

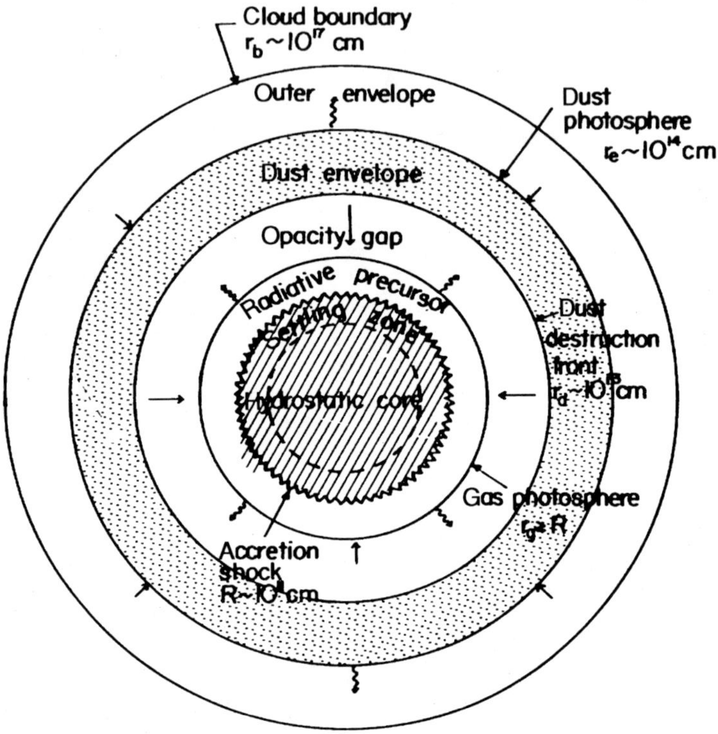

Figure 3.13: The structure of a protostar in the main accretion phase. Several layers with different features can be seen (see text). (From Stahler et al. 1980)

is denoted as the dust photosphere. Below this surface lies the dust envelope, where the temperature increases inward and eventually destroys the grains at the dust destruction front. The gas layer then becomes hotter and optically thicker and forms the gas photosphere shown in Figure 3.13. The gas will then cross the accretion shock region; precursor, shock front and relaxation zone, and finally reach the hydrostatic core. The interior core gradually develops, by accumulating the falling gas, and gravitationally contracting toward a main-sequence star that ignites nuclear burning at the very core.

3.3.2 Accretion disks of protostars

Formation of disks

The process of star formation starts from the gravitational collapse of a molecular cloud core when it becomes unstable. Since the collapsing velocity is higher in the denser part, the central part of the cloud core condenses in early stage into a hard nucleus, which gradually grows by the successive accretion flows from outer part of the cloud. Generally, the accretion flow is flattened by the angular momentum of the original molecular cloud, forming a large-scale molecular disk, often observable by radio molecular lines. Near the central nucleus, there also exists a fully condensed and flattened accretion disk. From the inner edge of the disk, gas falls onto the surface of the nucleus, steadily or irregularly, and sometimes accompanied by burst phenomena. Accretion processes in star formation is fully considered by Hartmann (1998).

Structure of disks

If the accretion flow preserves the angular momentum, the rotational velocity of the accretion disk will be considerably higher than that of the stellar surface. This forces the accretion flows to spin up the stellar rotation through its angular momentum. Usually, however, the angular momentum of the disk is supposed to be mostly transported outside by viscosity. Although viscosity is an important factor to determine the structure of the disk, its physical mechanisms are still poorly understood.

The accretion flow goes downward in a Keplerian disk, and its gravitational energy is converted into thermal energy by viscosity and must be radiated away from the disk surface mostly at infrared to millimeter wavelengths. At the same time, the disk is brightened by the irradiation from the central star. Hartmann (1998) showed that the irradiation exceeds the thermal radiation when stellar luminosity satisfies the condition

$$L_* > \frac{G M_*}{R_*} \dot{M}, \tag{3.3.1}$$

where M_* and R_* denote the mass and radius of the star and $\dot{M}$ the accretion rate for the disk. According to this criterion, disk irradiation becomes important for most of T Tauri stars.

The vertical thickness of the disk is sufficiently thin as compared to the equatorial dimension, since the disk is supposed to be in hydrostatic equilibrium. Its scale height is then determined by gas temperature provided the stellar mass and radius are known. The angular velocity of a steady accretion disk first increases from the outer edge inward, but its acceleration gradually declines through the increasing effects of viscosity. It reaches maximum velocity and then rapidly decreases through dissipation of angular momentum and kinetic energy. The distribution of angular velocity is shown in Figure 3.14.

The structure of the inner part of the accretion disk is distinguished by whether magnetic field exists or not. If the magnetic field is weak or absent, the boundary layer shown in Figure 3.14 becomes hot as high as 10,000 K by the viscous heating. The accretion disk in this case is shown in Figure 3.15.

When magnetic field exists making a magnetosphere around the star, the inner part of the disk is disrupted by magnetic field, and disk material flows onto the polar region of the stellar surface along the magnetic lines of force (Königl 1991, Ostriker and Shu 1995). The accretion of this type is called the magnetospheric accretion, and Figure 3.16 shows its schematic picture as a possible flow in low-mass pre-main sequence stars such as T Tauri stars. The disk part of the accretion flow is cold and emits the infrared and millimeter wave radiation. In contrast, the magnetospheric region is heated by the dissipation of kinetic energy, producing the Hα and other broad emission lines in the optical region. When the flow crashes into the stellar surface, a hot continuum is emitted from the region heated by a shock wave of accretion type.

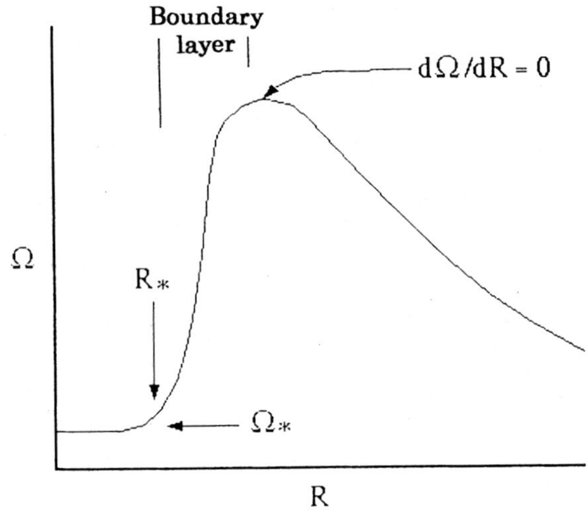

Figure 3.14: Schematic diagram of the angular velocity Ω as a function of radial distance R in the region where the disk reaches the stellar surface. The star with radius R_* is assumed to be rotating at a rate Ω_*, which is much less than the Keplerian velocity. The point of velocity maximum is located near the stellar surface, $\Delta R \ll R_*$. This region in the disk is called the boundary layer, where the disk material loses most of its rotational kinetic energy. (From Hartmann 1998)

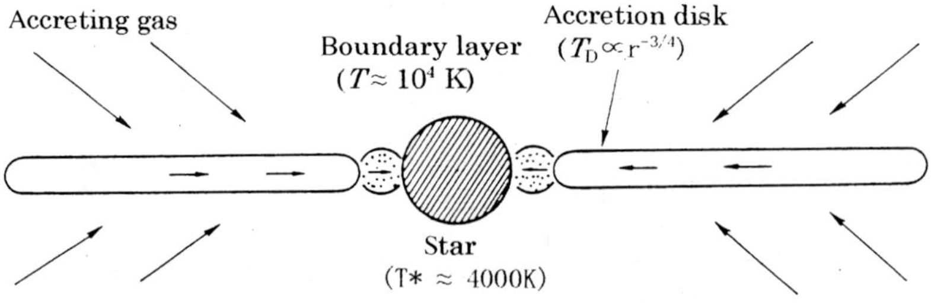

Figure 3.15: Schematic cross section of accretion disk when the effect of magnetic field is absent.

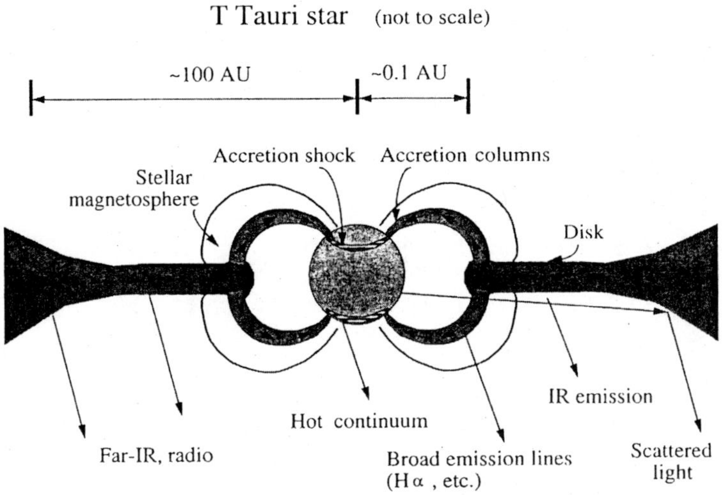

Figure 3.16: Schematic picture of accretion in T Tau type stars. The star is surrounded by an accreting circumstellar disk whose inner part is disrupted by the existence of magnetosphere. Accreting material falls onto the star along the magnetic lines of force. The picture shows that different parts of the accretion flows emit different radiation. (From Hartmann 1998)

3.3.3 Accretion disks of close binaries

Roche lobe

In close binary systems, an equipotential surface called the Roche lobe plays an important role in their dynamical structure. A small mass at a point in the binary system will experience three accelerations: gravitational attractions toward each of the stars and a centrifugal acceleration due to the orbital motion around the gravity center of the system. The loci of equal gravitational potential are known as the Roche equipotential surfaces. The critical surface within which material is bound to one or the other star defines the Roche lobe as shown in Figure 3.17, illustrating the equipotential curves on the orbital plane. The crossing point, L_1, in the Roche lobe is called the inner Lagrangian point. There is another critical equipotential surface which opens to outer space through a crossing point, L_2, called the outer Lagrangian point.

Roche lobe overflow and accretion disks

The accretion flows for one component of a binary system occur in two ways: the Roche lobe overflow and the stellar wind from a mass-losing star. We first consider the Roche lobe overflow, which was proposed by Paczynski (1971), and a number of hydrodynamic modeling has been carried out.

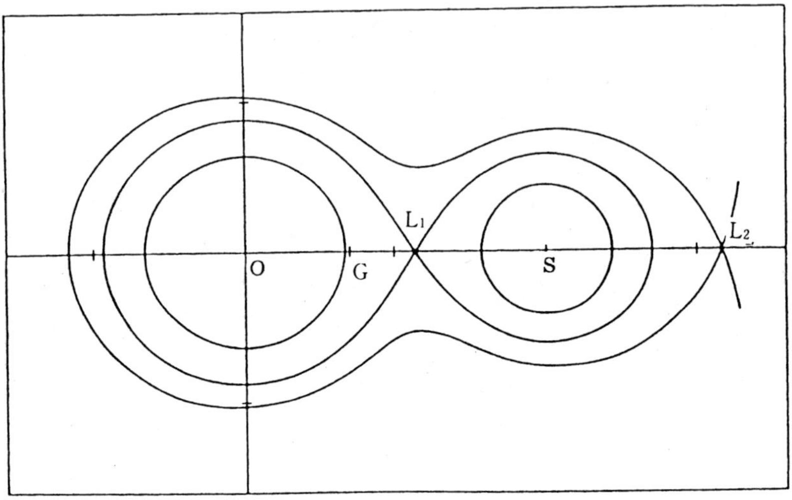

Figure 3.17: Roche equipotential curves on the orbital plane of a binary system. O, primary star with mass M_1; S, secondary star with mass M_2 (mass ratio M_1/M_2 is assumed to be 2 in this figure); G, gravity center of the system; L_1, the inner Lagrangian point; and L_2, the outer Lagrangian point.

In the evolution of close binary systems, if one star fulfills the Roche lobe, this star becomes a mass loser and the Roche lobe overflow sets in. That is, the mass flows through the inner Lagrangian point L_1 occur toward its companion, i.e., mass gainer, and the flow forms an accretion disk around the mass gainer. This type of accretion flow can be seen in Algol eclipsing binaries, CVs, and X-ray binaries (see Chapter 6).

To imagine the structure of accretion disks thus formed, we show a result of three-dimensional numerical calculations by Armitage and Livio (1996) in Figure 3.18. The Roche lobe overflow starts from the Lagrangian point located at the center of left-hand axis, and the disk is plotted on the orbital (x–y) and vertical (x–z) planes. The center of mass gainer is located at the origin of the coordinates. One may see a sufficiently flattened accretion disk, though actual scale relative to the Roche lobe is not shown in Figure 3.18.

The size of accretion disks depends on the types of binary system and mass ratio. Harrop-Allin and Warner (1996) have estimated the outer radius R_D of the accretion disks relative to the Roche lobe radius $R_{\rm RL}$ for 35 CVs showing the eclipses. The CVs are close binaries composed of a white dwarf (mass M_1, primary) and a companion (mass M_2) (see Figure 6.28), and the eclipse occurs when the accretion disk is masked by the companion. From the shape of the light curve of this eclipse, the size of the disk can be estimated. Harrop-Allin and Warner have thus found that nova-like variables tend to have $R_{\rm D}/R_{\rm RL}$ exceeding 0.61, and recurrent novae have very large disks with $R_{\rm D}/R_{\rm RL}$ exceeding 0.80, though there are some exceptions in both cases. Dwarf

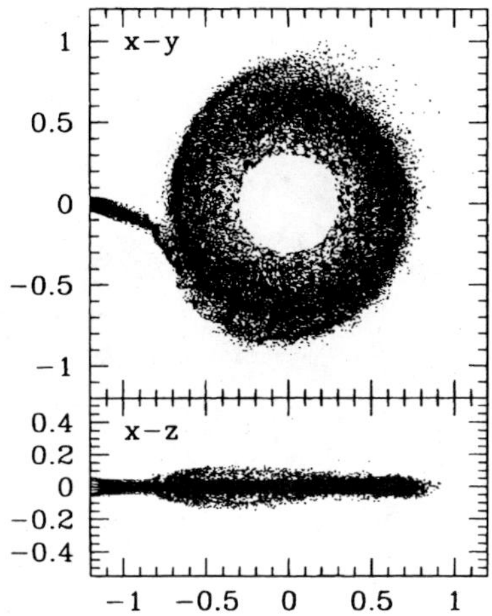

Figure 3.18: Model calculation of accretion disk in close binary in orbital plane (x–y), and vertical plane (x–z) in case of low-mass X-ray binary. The Lagrangian point L_1 of the binary system is given at the center of the ordinate in the left hand side. (From Armitage and Livio 1996)

novae have a disk ratio of 0.6 or higher during outburst making them resemble those of nova-like disks. In the quiescent phase the disk ratio becomes smaller, mostly $R_D/R_{RL} \sim 0.50$.

In Algol-eclipsing binaries, accretion disks have been detected in a wide range of orbital periods, from short (β Per, B8 + K2, $P = 2.86$ d, Richards et al. 1995) to long (RZ Oph, F5Ib + K5Ib, $P = 262$ d, Olson 1987). In short-period Algols ($P < 6$ d), the primary as a mass gainer is large, relative to the binary separation, so that the accretion flow directly hits the stellar surface through a shock front. A part of accretion flow can survive and form a transient accretion disk, though it may not be sufficient in scale to form emission lines. For long-period Algols ($P > 6$ d), the accretion disks are generally well developed and stable, favoring the formation of emission lines (see Section 6.4.2).

Stellar winds and accretion disks

When the mass loser in a binary system is a source of strong stellar wind such as OB stars, an accretion disk can be formed around the mass gainer fed by the stellar wind (Iben et al. 1995). Usually, the mass losers do not fulfill the Roche lobe. When the wind velocity is small, wind gas accretes onto the gainer through the Lagrangian point L_1. As the wind velocity increases, accretion occurs in a wider region around the Lagrangian point. In high-velocity wind only a part of the wind gas is captured by the gainer, and most of the gas flows out of the binary system. A numerical calculation, carried out by Theuns and

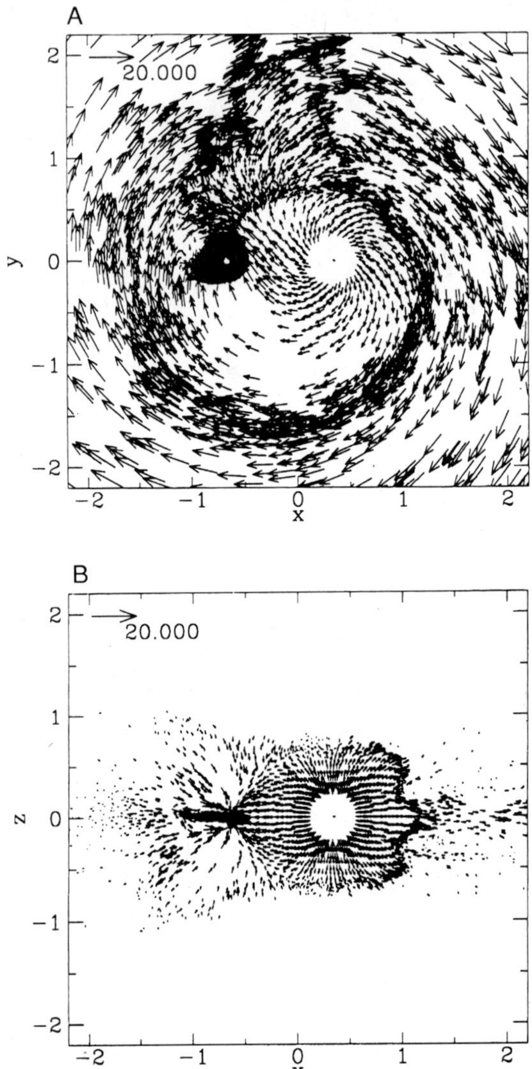

Figure 3.19: Structure of stellar wind and accretion disk in a binary system. The velocity structure is shown in orbital plane (A) and vertical plane (B). The mass losing star is at ($x = 0.33$, $y = z = 0$), and the accreting star at ($x = -0.66, y = z = 0$). The accretion disk is shown by black part around the accretion star. (From Theuns and Jorissen 1993)

Jorissen (1993) for the last case, is shown in Figure 3.19. Panel A illustrates the velocity structure in the orbital plane ($x - y$) for an isothermal model. The mass-losing star is at ($x = 0.33, y = 0, z = 0$), and the accreting star is at ($z = -0.66, y = 0, z = 0$). The black structure around the accreting star is the accretion disk. Panel (B) shows the same velocity structure in the vertical plane $(x - z)$. The spherically symmetric stellar wind flowed out from the mass-losing star turns to spiral flow due to the binary orbital motion, and a part of the gas is captured making an accretion disk around the gainer. Most of the wind gas flows away out of the binary system. One may see in the vertical plane that the accretion disk is sufficiently flattened even though the base of the wind is spherically symmetric. The actual structure of gas flow and accretion disk largely depends on the physical parameters such as initial wind velocity, gas density, orbital velocity, and thermal properties of gas.

3.4 Shock waves

3.4.1 Basic properties of shock waves

In the supersonic flow of gas, discontinuous jumps of physical parameters along the direction of gas motion can occur and are called the shock waves. In astrophysical processes in stars and interstellar media, shock waves are prevailing phenomena and play important roles, particularly in active phenomena. We now consider the basic properties of sound waves and shock waves.

Sound waves are the small amplitude waves of density variation $\Delta\rho/\rho$ ($\ll 1$), and their velocity c_s is determined by the gas temperature regardless the gas density, as given

$$c_S = \sqrt{\frac{\gamma RT}{\mu}}, \tag{3.4.1}$$

where R is the gas constant, and γ is the specific heat ratio with the value of $\gamma = 5/3$ for monoatomic gas and $\gamma = 7/5$ for diatomic molecular gas. The mean molecular weight μ is the constant depending on the element abundance, ionization, and/or dissociation degree.

Sound waves generated in a convection layer propagate in all directions. A part of the waves are directed to the upper layer where gas density is low and become the finite-amplitude waves as $\Delta\rho/\rho$ approaches finite. The propagating velocities of different points in the profile of a finite-amplitude wave are different: the points of compression move forward and those of rarefaction are left behind (Figurs 3.20). Finally, the profile may become such that the density $\rho(x)$ (for given t) is no longer one-valued: three different values of ρ correspond to the same x (the dashed line in Figure 3.20C). This is physically impossible. In reality, a shock wave is formed at this point and physical state of gas suffers a discontinuous jump. If the sound waves are generated as a wave train, a sequence of shock waves will be formed resembling teeth of a saw and is called the shock wave train.

In order to understand the jump of physical parameters, we consider the one-dimensional stationary shock wave propagating in an uniform medium. Imagine a long tube where a supersonic piston with constant velocity pushes the gas (Figure 3.21). At the head of the gas compressed by the piston, a

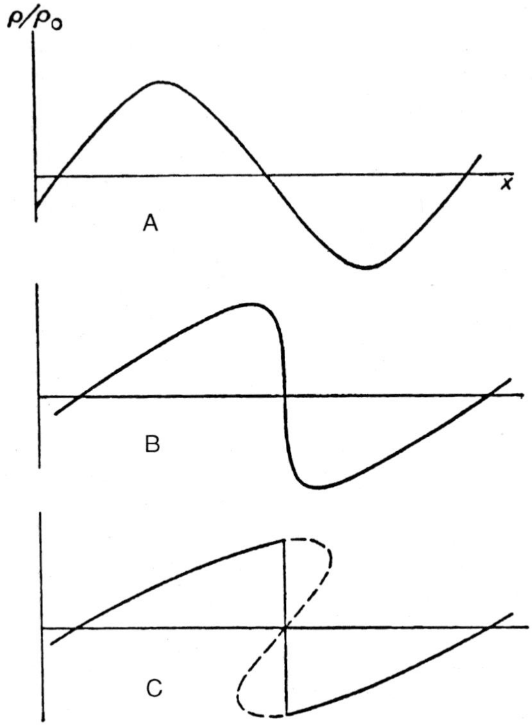

Figure 3.20: The variation of wave profile in the traveling waves. The ordinate is the gas density and the wave propagetes rightward. A, B, and C express time development of a treaveling wave. When the density becomes multivalued, a shock wave is set in at that place.

shock wave stands, and the gas passing through the shock front is heated and compressed. Shock front itself is actually a geometrically thin transition layer which is formed by the effects of viscosity and thermal conductivity of the gas, and its thickness is in the order of several times of the mean free path. This

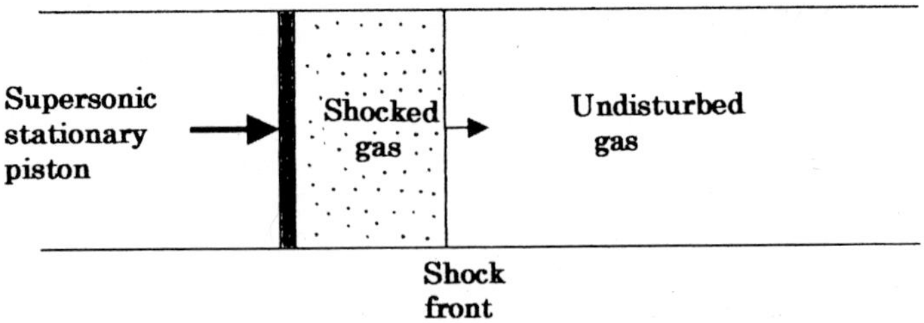

Figure 3.21: Generation of one-dimensional stationary shock wave in a long gas tube. The gas is compressed by a stationary supersonic piston, and a shock front is formed at the top of the compressed gas.

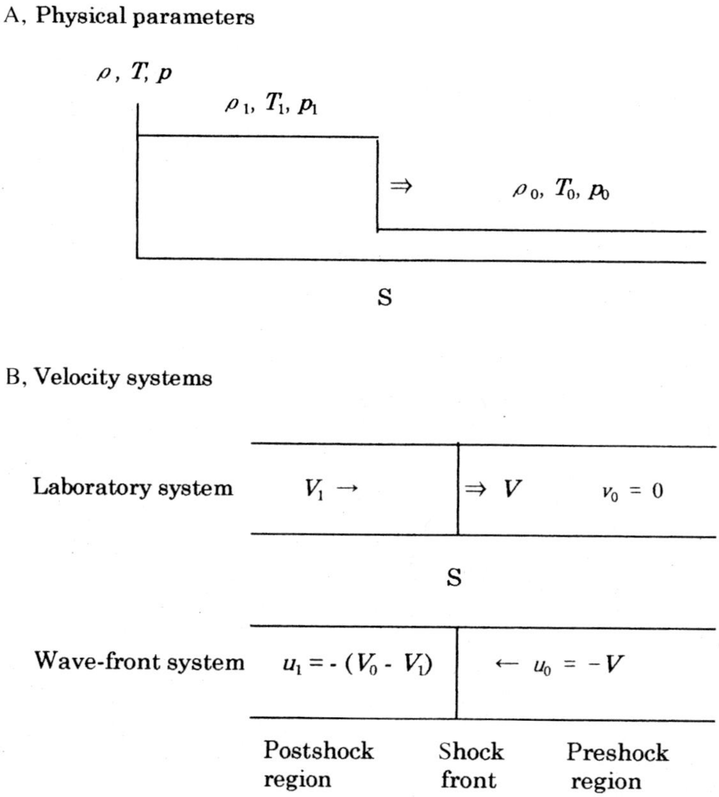

Figure 3.22: The state of gas and velocity system in the stationary shock wave. A, The jump of physical parameters in both sides of the front. B, the velocity systems. The suffixes 0, 1 denote the pre- and postshock states, v and u are the gas velocities in the laboratory and wave-front systems, respectivlely.

is sufficiently thin as compared to the hydrodynamical scale so that we can ignore the structure of the shock front and regard it as a simple discontinuous surface.

Consider a stationary shock wave propagating along x axis, and the uniform gas (ρ_0, p_0, T_0) has changed its state to (ρ_1, p_1, T_1) after passing the shock front S (Figure 3.22). The gas motion and shock waves can be described by two coordinate systems. One is the laboratory system in which the undisturbed gas is at rest and the shock front moves with the velocity V_S into this gas. The other is the wave-front system in which the shock front is at rest and the undisturbed gas flows into the front with velocity u_0 as shown in Figure 3.22B. The tangential component of the gas velocity is assumed zero in any case. These two velocity systems are connected by the relation

$$V_s = -u_0 \qquad v_1 = -(u_0 - u_1). \tag{3.4.2}$$

If there is no energy loss by the passage of shock front (adiabatic shock), the following conservation laws hold between two nearby planes in both sides of the front. In wave-front system we have

(a) mass flux

$$\rho_0 u_0 = \rho_1 u_1 = j, \tag{3.4.3}$$

where j denotes the mass flux.

(b) momentum flux

$$p_0 + \rho_0 u_0^2 = p_1 + \rho_1 u_1^2. \tag{3.4.4}$$

(c) energy flux

$$\frac{1}{2}u_0^2 + I_0 = \frac{1}{2}u_1^2 + I_1, \tag{3.4.5}$$

where enthalpy I is connected with the internal energy U of gas by the relation

$$I = U + \frac{p}{\rho}. \tag{3.4.6}$$

For ideal gas we have

$$U = C_V T, \qquad p = \frac{\rho R T}{\mu}, \qquad \frac{R}{\mu} = C_P - C_V, \tag{3.4.7}$$

or, if we use the specific heat ratio $\gamma = C_P / C_V$, enthalpy and internal energy are given as

$$U = \frac{1}{\gamma - 1}\frac{p}{\rho}, \qquad I = \frac{\gamma}{\gamma - 1}\frac{p}{\rho}. \tag{3.4.8}$$

By eliminating the velocity from the equations of mass and momentum conservations, we get

$$j^2 = \frac{p_1 - p_0}{V_0 - V_1}, \tag{3.4.9}$$

where $V = 1/\rho$ denotes the specific volume. Then, through some algebraic arrangement, we obtain

$$\frac{\rho_1}{\rho_0} = \frac{V_0}{V_1} = \frac{(\gamma - 1)p_0 + (\gamma + 1)p_1}{(\gamma + 1)p_0 + (\gamma - 1)p_1}. \tag{3.4.10}$$

This gives the relation of gas parameters in the jump of gas state for the adiabatic shock waves and is called the Rankine–Hugoniot relation (R-H relation). In Equation (3.4.10), p_1/p_0 is a parameter expressing the strength of the shock wave and is given as a function of the shock front velocity (see Equation (3.4.12)).

If we take p_1/p_0 and V_1/V_0 as the variables, Equation (3.4.10) yields a curve on the (p,V) plane which is called the Hugoniot curve and expressed as $H(p, V) = 0$. Figure 3.23 shows the relation between the change of the

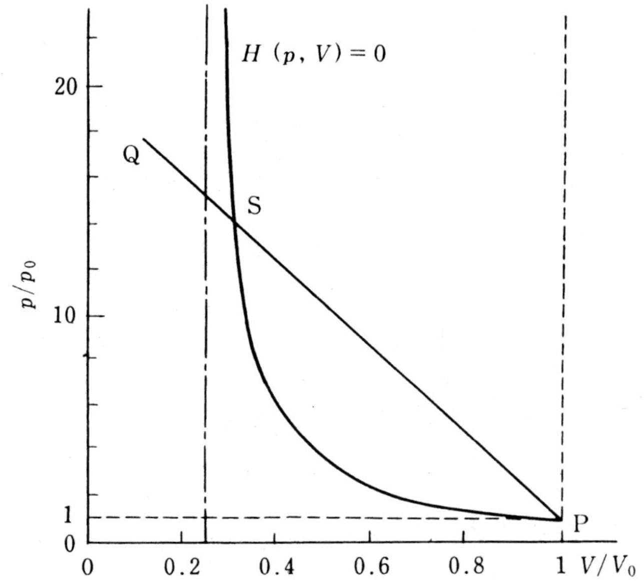

Figure 3.23: Hugoniot curve of adiabatic shock waves. Point P denotes the state of undisturbed gas, $H(p,V)$ the Hugoniot curve, and line PQ denotes the shock strength by its gradient. The vertical dotted-broken line at $V/V_0 = 0.25$ denotes the limiting gas compression rate in a strong adiabatic shock in monoatomic gas ($\gamma = 5/3$).

state of gas and the shock strength expressed by gradient j^2 on the $(p/p_0, V/V_o)$ plane. The solid curve gives the R-H relation and line PQ denotes the gradient j^2 on this plane which corresponds to the shock-front velocity by the relation $j^2 = \rho_0^2 u_0^2$ in Equation (3.4.3). Given this, the cross point S expresses the change of specific volume and gas pressure after the passage of shock wave.

As seen in Equation (3.4.10) or in Figure 3.23, there is an upper limit in the gas compression rate ρ_1/ρ_0 for $p_1/p_o \to \infty$ in the adiabatic shocks, and the limiting values for monoatomic gas ($\gamma = 5/3$) and diatomic molecular gas ($\gamma = 7/5$) are given as

$$\frac{\rho_1}{\rho_0} = \frac{\gamma + 1}{\gamma - 1} = \begin{cases} 4 & (\gamma = 5/3) \\ 6 & (\gamma = 7/5)\,. \end{cases} \tag{3.4.11}$$

The propagating velocity of shock front is often expressed by the Mach number $M = V_S/c_0$, where c_0 is the sound velocity of the undisturbed gas. In this case pressure and temperature jumps are expressed as

$$\frac{p_1}{p_0} = \frac{2\gamma}{\gamma + 1} M^2 - \frac{\gamma - 1}{\gamma + 1} \tag{3.4.12}$$

$$\frac{T_1}{T_0} = \left(\frac{\gamma - 1}{\gamma + 1}\right)^2 \left(\frac{2\gamma}{\gamma - 1} M^2 - 1\right) \left(\frac{2}{\gamma - 1}\frac{1}{M^2} + 1\right). \tag{3.4.13}$$

In strong shock waves ($M^2 \gg 1$), the above formulae are approximated by

$$\frac{p_1}{p_0} \approx \frac{2\gamma}{\gamma+1} M^2 = \begin{cases} 1.25\, M^2 & (\gamma = 5/3) \\ 1.16\, M^2 & (\gamma = 7/5)\,, \end{cases} \tag{3.4.14}$$

$$\frac{T_1}{T_0} \approx \frac{2\gamma(\gamma-1)}{(\gamma+1)^2} M^2 = \begin{cases} 0.312\, M^2 & (\gamma = 5/3) \\ 0.194\, M^2 & (\gamma = 7/5)\,. \end{cases} \tag{3.4.15}$$

Pressure and temperature both increase proportionally to the square of Mach numbers, and there are no upper limit. The remarkable feature of shock waves is the small compression ratio as compared to the large jumps in temperature and pressure.

3.4.2 Shock waves in stellar atmospheres

We have so far considered the basic properties of shock waves in the adiabatic condition. Shock waves in stellar atmospheres reveal some remarkable features due to the interaction between radiation and inhomogeneity of the medium. In this section we consider the jump condition and structure of the waves in stellar atmospheres.

Generalized R-H relation

The R-H relation (Equation (3.4.10)) yields the jump conditions in the adiabatic shocks. In case of strong shock waves, or shock waves in hot medium, the effects of ionization, dissociation, and the interaction with radiation become important on the jump condition.

Generally speaking, there are basically three different paths for changing gas state on the ($p/p_0 - V/V_0$) planes. One is the isothermal change which is given by pV = constant, the second is the adiabatic change given by pV^γ = constant, where γ is the specific heat ratio, and this is the so-called Poisson's law. The third is the R-H relation as given above. These paths are shown in Figure 3.24 by different symbols as T (isothermal), P (adiabatic), and H (R-H relation). Among these, R-H ralation gives the highest pressure jump with some compression limit, whereas P and T yield the pressure increase without comprsssion limit. In Figue 3.24, $\gamma = 5/3$ is assumed.

Now consider the effects of ionization in the R-H diagram. Ionization acts as a latent heat at the shock front absorbing the ionization energy from the shock kinetic energy so that the pressure jumps decrease and instead the density compression increase even beyond the limiting line at the R-H ralation. In Figure 3.24, the modified R-H curves, which take into account the effect of ionizatin, are shown in two simple cases for the initial state of gas: $T_0 = 6300$ K and $x_0 = 10^{-2}$(1), and $T_0 = 8400$ K and $x_0 = 0.1$(2), where x_0 stands for the ionization degree of the undisturbed gas. As seen in Figure 3.24, the R-H curves in these cases can enter beyond the compression limit with much low-pressure ratios. In weak shocks, ionization is not effective, thus the (1) and (2)

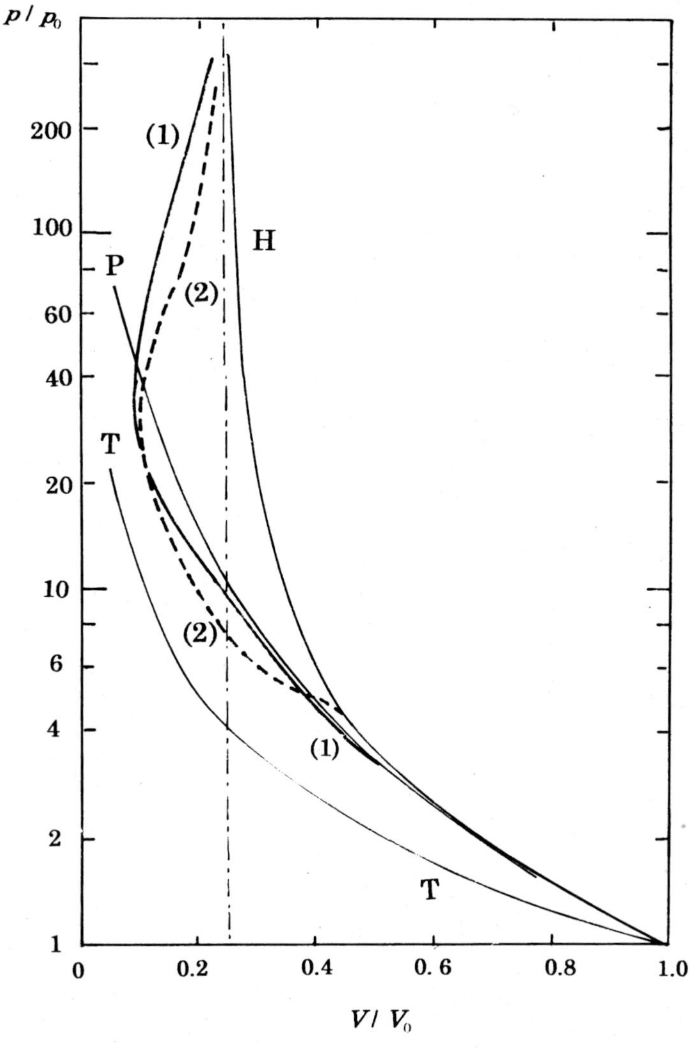

Figure 3.24: Changes of the state of gas in $(p/p_0 - V/V_0)$ plane. H is the Hugoniot curve, P the poisson curve (adiabatic compression), and T the isothremal curve. Curves (1) and (2) denote the R–H relatin including the effect of ionization in two cases as given in the text.

curves do not left from the R-H curve. In very stong shocks, ionizatin energy is negligibly smaller than the shock kinetic energy so that the curves (1) and (2) again approach the R-H curve.

A more sophisticated and generalized R-H relation containing the effects of ionization and dissociation of hydrogen gas without including the effect of radiation was derived by Nieuwenhuijzen et al. (1993). According to them, the generalized R-H relation can be written as

$$\frac{\rho_1}{\rho_0} = \frac{V_0}{V_1} = \frac{(\gamma_0 - 1)\,\gamma_1\,(\Gamma_0\,M^2 + 1) + X}{(\gamma_1 - 1)\,\{M^2\,(\gamma_0 - 1)\,\Gamma_0 + 2\gamma_0\}}, \tag{3.4.16}$$

where

$$X = \left\{M^4\,\Gamma_0^2\,(\gamma_0 - 1)^2 - 2M^2\,\Gamma_0\,(\gamma_0 - 1)\,(\gamma_1^2 - \gamma_0) + (\gamma_0 - 1)^2\,\gamma_1^2\right\}^{1/2} \tag{3.4.17}$$

and the shock strength is given as Mach number $M = V_S/c_0$. The specific heat ratio γ can be changed by the change of particle composition of gas in passing the shock front, and we define I_i, U_i and γ_i in the preshock ($i = 0$) and postshock ($i = 1$) regions as follows:

$$I_i = \frac{\gamma_i}{\gamma_i - 1}\frac{P_i}{\rho_i}, \qquad U_i = \frac{1}{\gamma_i - 1}\frac{P_i}{\rho_i} \tag{3.4.18}$$

$$\gamma_i = \frac{I_i}{U_i}, \qquad \Gamma_i = \left(\frac{d\ln P}{d\ln\rho}\right)_{\rm ad} = \left(\frac{dI_i}{dU_i}\right)_{\rm ad}, \tag{3.4.19}$$

where ad stands for the adiabatic change. When there is no change in ionization/dissociation, γ_i takes the common value of γ as

$$\gamma_i = \Gamma_i = \gamma,$$

and Equation (3.4.16) reduces to the R-H relation in adiabatic shock waves.

Equation (3.4.16) is called the generalized R-H relation and depends on the ionization and dissociation conditions for individual atoms or molecules. A sample of numerical calculations carried out by Nieuwenhuijzen et al. (1993) is illustrated in Figure 3.25, which shows the ratios of post- and preshock parameters (p, ρ, T) as functions of preshock temperature T_0 in the range (2000–40,000 K) for a shock strength $M = 2$. The preschock gas is assumed to be composed of 16 elements with the solar abundance and to have gas pressure 1 dyn cm^{-2} (= 0.1 Pascal). LTE is also assumed to calculate the ionization, but the effect of radiation is not included. For comparison, the jumps of parameters in adiabatic case are also shown by a horizontal line in each parameter. As seen in this figure the effect of ionization is small at low preshock temperatures, and it becomes remarkable in the temperature range $T_0 = 5000{-}20{,}000$ K where the deviation from the adiabatic case depends on T_0 due to the different ionization potentials of composing atoms. Roughly speaking, the postshock temperature deviates to lower values as the ionization absorbs a part of shock energy as latent heat. As a result, gas density deviates generally as an increase, but the pressure changes with more complexity by the different behaviors of tempeature and gas density. For the preshock temperature higher than around 30,000 K, the numerical solutions converge to the adiabatic case. This is for the same reasons as shown in Figure 3.24.

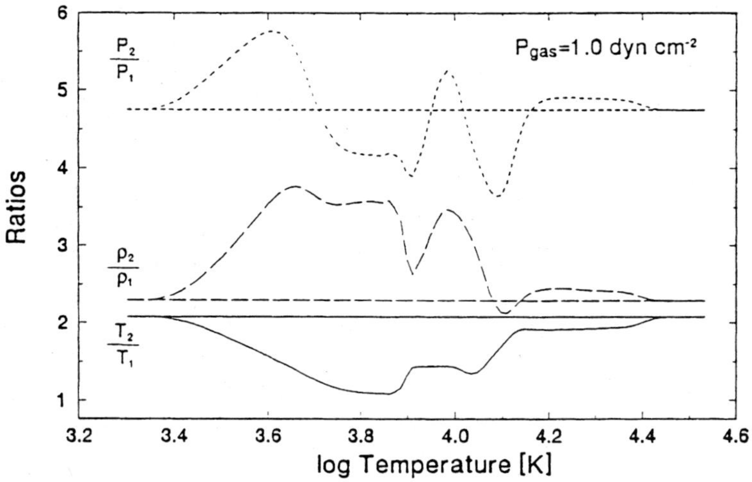

Figure 3.25: The changes of physical parameters in the generalized R–H relation. The ratios of post and preshock values of p, ρ, T for a shock with $M = 2$ are shown as functions of the preshock temperature. For comparison the results of adiabatic case are also shown by a horizontal line for each parameter. (From Nieuwenhuijzen et al. 1993)

The structure of radiative shock waves

Shock waves in stellar atmospheres exhibit some complicated multilayer structures due to their interaction with radiation. They are called radiative shock waves.

We now consider the structure of radiative shocks propagating in a medium such as found in protostars and cool star atmospheres. According to Gillet and Lafon (1983), the structure of one-dimensional stationary shock waves can be described by some distinguished zones as schematically illustrated in Figure 3.26. For the undisturbed layer in this figure, the following ranges of parameters are adopted:

$$N = 10^9\text{–}10^{15}\,\mathrm{cm}^{-3}, \qquad T = 500\text{–}5000\,\mathrm{K},$$

and the medium is assumed to consist of hydrogen (H_2, H, p, and e) only.

Each zone in Figure 3.26 is characterized as follows:

Undisturbed gas. The preshock region is not influenced by shock waves. Hydrogen is in state of H and H_2.

Zone 1, discontinuity, shock front. Shock front is defined as the thin zone where the physical parameters steeply change. The behaviors at the front differ for the heavy and light particles. The steep change occurs for heavy particles under the conditions of the classical R-H relation, since no effect is procured by radiation which is fully optically thin in the front. The electron gas, on the other hand, has much longer mean

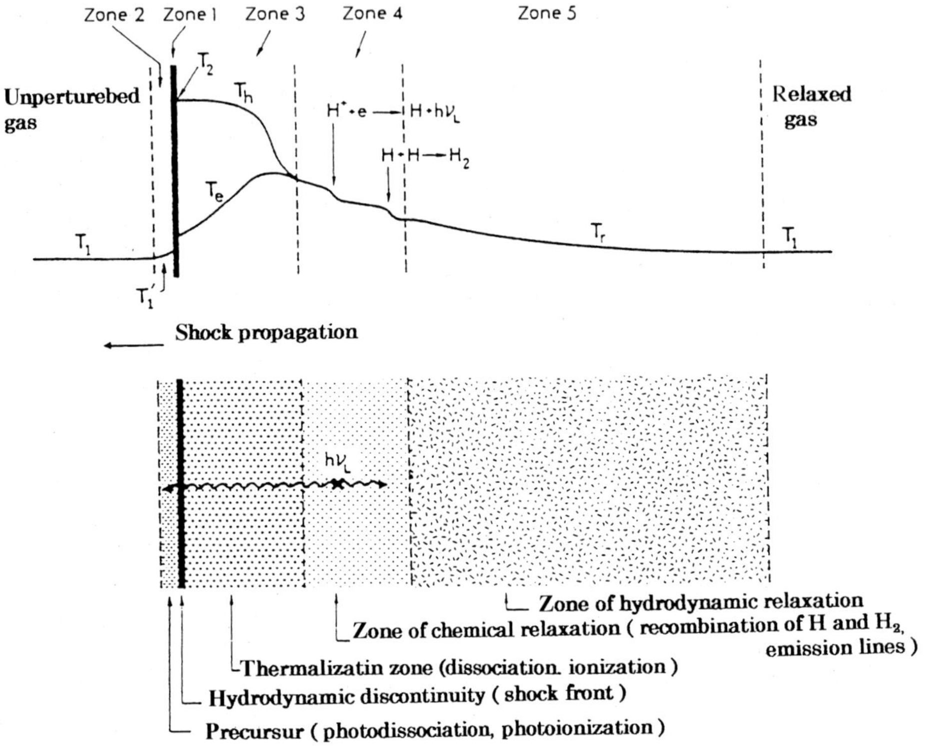

Figure 3.26: Schematic picture of the structure of radiative shock waves in cool stars. The waves are separated in some different zones by the dominant processes (see text). (From Gillet and Lafon 1983)

free path as compared to the thickness of the front, and then the density and velocity changes are much smaller as compared to heavy particles. The temperature is splitted into ion temperature T_{h} for heavy particles and electron temperature T_{e} for electrons. Ion temperature rises up to T_2 given in Figure 3.26, by the R.-H. relation, while the changes of temperature and density of electron gas are roughly approximated by the adiabatic compression so that the amplitudes of tempertraure change is much smaller than those of shock transitions.

Zone 2, precursur zone. In strong shocks, energetic photons produced in shocked region propagate much faster than the velocity of shock front and form a zone called precursur where hydrogen molecules are dissociated or atoms are ionized. The thickness of the precursur is of the order of the mean free path of the ionizing or dissociating photons.

Zone 3, thermalization zone. Behind the shock front, the energy stored in heavy particles is redistributed through collisions with electrons to give rise thermalization to the unified kinetic temperature T_{r}. Since the time scales of interaction is different for atoms (ionization, excitation, thermalization) and molecules (dissociation, excitation, thermalization), the thermalization zone may have some stratified structure.

Zone 4, chemical relaxation zone. The zone of relaxation can be divided into two parts—chemical and hydrodynamical ralaxation. In the former, ions and electrons recombine into atoms and then molecules, resulting in the emission of radiation in the Lyman and Balmer continua, along with the formation of emission lines. The Lyman continuum contributes to the formation of the precursor zone.

Zone 5, hydrodynamic relaxation zone. In the second relaxation zone, the gas is already stable and relaxes thermodynamically and hydrodynamically toward equilibrium.

Relaxed gas region, the temperature and density of gas approaches to the values of pre-shock undisturbed region.

Similar types of the shock waves can be seen in the early stage of protostar evolution as seen in Figure 3.13, where the spherically symmetric shock wave has a precursor zone, and is connected to the central hydrostatic core through the shock front and some relaxation zones. This type of radiative shocks, though with some modifications for individual cases, appear to prevail in a various kinds of active phenomena in stellar atmospheres and envelopes in many types of stars.

3.4.3 Stellar atmospheres and shock waves

In stellar atmospheres and envelopes, shock waves are generated by various mechanisms such as the growth of sound waves in convection layer, instabilities in stellar winds or accretion flows, and pulsational motions in variable stars. They play important roles in the dynamical structure (heat and momentum transfer) and radiation processes (such as X-ray emission, emission line formation). We now consider some typical shock phenomena in early-type and late-type stars.

Early-type stars

Remarkable shock phenomena in the envelopes of early-type stars are the instabilities of stellar winds and resultant X-ray emissions. In Section 3.2, we have considered the stellar winds as stationary flows. Actually, however, wind flows are generally very unstable. Carlberg (1980) has made a stability analysis of a supersonic stellar wind driven by radiation. By linearization analysis, Carlberg found two major instabilities to make the wind flows break up into slabs and clumps. One is the so-called sound-wave instability that acts in a vertical direction, and the other is the Rayleigh-Taylor instability that acts in a horizontal direction. The growth time scale is around 1000 s for the former and 10^5 s for the latter. The effects of wind clumping may be observable in line profiles and in X-ray spectra.

Now consider the X-ray emitting stellar winds of O type stars where the electron temperature is generally thought as of the order of 10^4 K. In such winds, X-ray emission needs the existence of some hot regions as high as 10^6–10^7 K. In order to generate such high temperature by the passage of shock waves, we need Mach number 20–60 in adiabatic shocks as seen in Equation (3.1.13) by putting $T_o = 10^4$ K and $\gamma = 5/3$. It can also be achieved if the shock propagation velocity reaches 400–1200 km s^{-1}. In the shocks with energy loss by radiation, the Mach number should be much higher as compared to the case of adiabatic shocks. Some instabilities inside the winds will give rise trains of strong shock waves, and shocked regions will contribute to the X-ray emission.

One example is the model calculation by Feldmeier et al. (1995). They assumed the undisturbed stationary wind from O type supergiants having $T_{\rm eff} =$ 42,000 K, $R_* = 19R_\odot$, $M_* = 42\,M_\odot$ with mass loss rate $\dot{M} = 3 \times 10^{-6}\,M_\odot$ per year and terminal velocity $V_\infty = 2000$ km s^{-1}. The perturbation of the wind is given by a sound-wave train originating in the photosphere with period 5000 s and density amplitude 1%. This wave train will grow to a train of shock waves in the outer envelope.

Figure 3.27 shows a snapshot of the wind structure 10 days after the start of perturbation. The gas density (g sm^{-3}), wind velocity (km s^{-1}), and

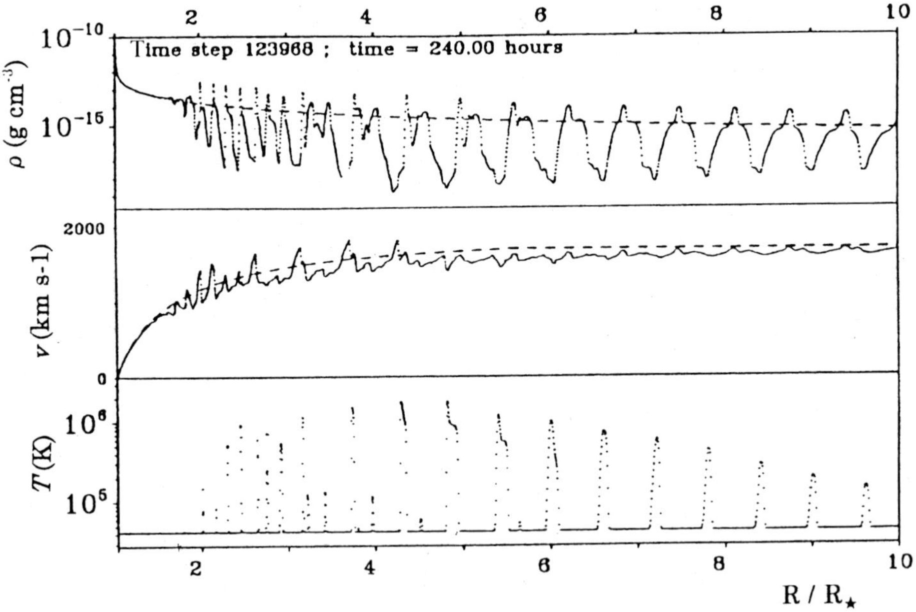

Figure 3.27: A sequence of narrow and dense shell formed in the radiatively driven winds. Each shell is surrounded by foregoing and reverse shocks in both sides, the central hot gas can contribute to the emission of X-rays. (From Feldmeier et al. 1995)

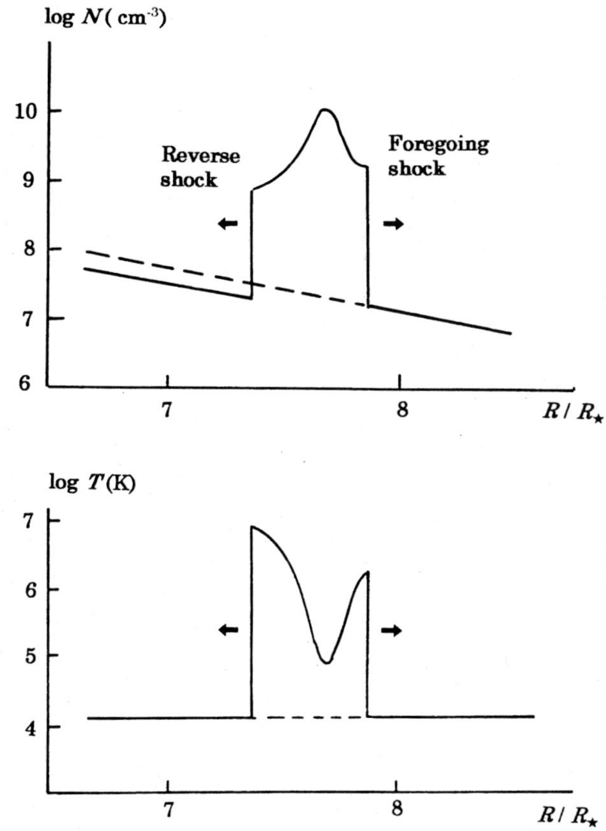

Figure 3.28: Schematic pictures of foregoing and reverse shock waves in a narrow and dense shells formed in the radiatively driven wind of O stars.

temperature (K) are shown in the range from stellar surface R_* to $10R_*$. The initial stationary wind shown by dashed lines is transformed into a sequence of a very narrow and hot shells. Each shell is enclosed by a strong compression wave of a forfacing shock wave on their outer/front side and by a reverse shock on their back side. Inside the shells, temperature can rize up to 10^6 K and emits the X-ray radiations. The hot region surrounded by two shock fronts is schematically illustrated in Figure 3.28. From each of hot shell, X-rays can be emitted.

Late-type stars

In late-type stars, shock waves play important roles in the heating of chromospheres and coronae, and in the generation of stellar winds that cause the formation of shock-emission line spectra. The structure and formation

Table 3.1: Sound-wave flux and frequency which yield maximum energy in K type stars (Bohn 1984)

	$\log g$	F_M (erg cm^{-2} s^{-1})	$P(\omega)$(s^{-1})
Giants	3	2.0×10^8	560
Dwarfs	5	2.5×10^7	5.6

of emission lines will be considered in Chapter 6. In this section, we consider the shock waves produced from the growth of sound waves according to Ulmschneider (1989).

Suppose giant ($g = 3$) and dwarf ($g = 5$) stars of effective temperature $T_{\rm eff} = 5000$ K (K0–K1). Stellar atmosphere is approximated as gray (absorption coefficient is independent of the wavelength) and a plane parallel layer in radiative equilibrium. In K type stars, the maximum energy flux $F_{\rm M}$ of the sound waves, generated in convection layer, and its frequency P are given as in Table 3.1. The energy flux $F_{\rm M}$ with the period P can be transported to the energy flux of piston motions with velocity amplitude u_0 by the relation

$$\frac{1}{2}\,\rho_0\, u_0^2 = F_{\rm M}, \qquad (3.4.20)$$

where ρ_0 denotes the gas density at the base of the atmosphere. If we put $u_0 = 0$, we get the solution for an equilibrium atmosphere. Time-dependent hydrodynamic calculations for $u_0 > 0$ yield the propagation of sound waves which grow up to a shock trains and disturb the upper atmosphere. A part of the results for the sound waves with the initial energy flux given in Table 3.1 is shown in Figure 3.29 for a dwarf star. This figure is a snapshot at 1.22 $\times 10^3$ s after the generation of sound waves at the surface. The temperature and density variations are shown as the functions of distance from the star's center. The equilibrium atmospheres are also illustrated by broken lines.

One may see that the sound waves grow to a shock wave train with increasing strength toward the upper atmosphere. This shock strength, however, attains a limiting value, where the increase of wave amplitude due to the energy conservation in the decreasing density is balanced by the amplitude decay due to shock dissipation. In case of dwarf stars, the limiting value rapidly declines near the transition layer due to prevailing dissipation, and thus shock waves partly contribute to the heating of corona.

In late-type supergiants, shock waves give raise the efficient mass loss (Cuntz 1990) and heat up coronae to emit the X-ray emission (Stepien and Ulmschneider 1989).

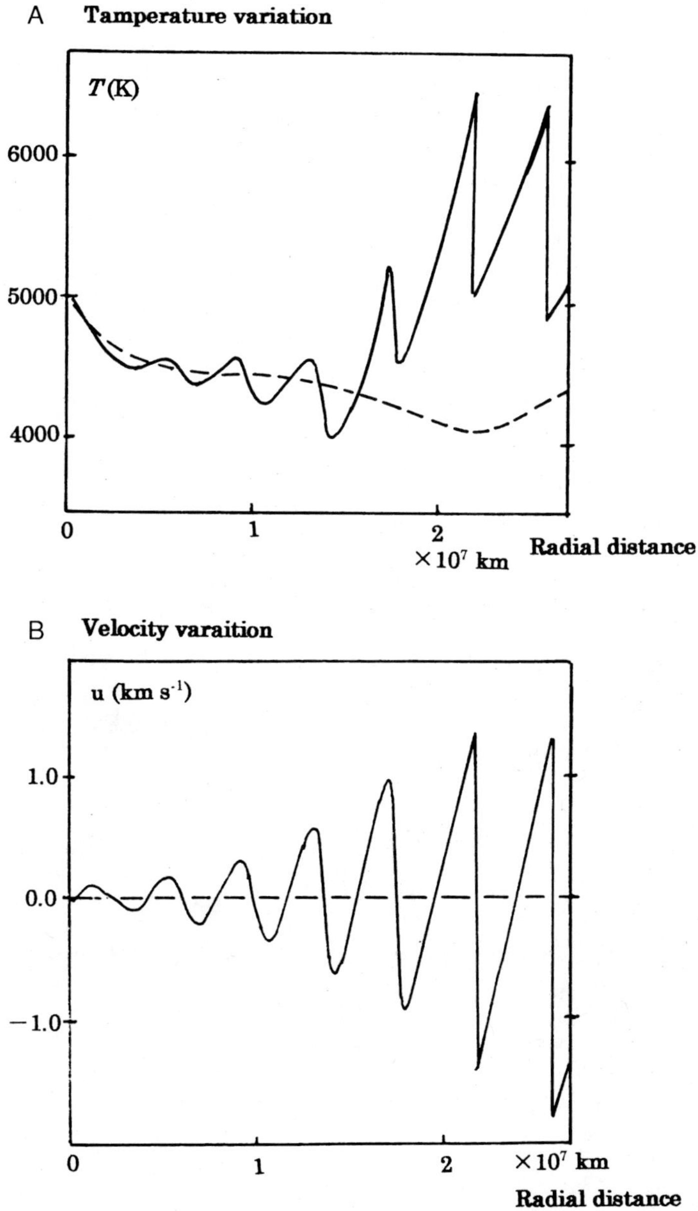

Figure 3.29: Acoustically heated theoretical model of a dwarf star with $T_{\rm eff} = 5012$ K and log $g = 5$ at time $t = 1.22 \times 10^3$ s. Panel A is the temperature variation, panel B the velocity variation, in both panels the equilibrium atmosphere is indicated by a broken line. (Partly reproduced from Ulmschneider 1989)

Further reading

Bianchi, L. and Gilmozzi, R. (eds.) (1988). *Mass Outflow from Stars and Galactic Nuclei.* Kluwer, Netherlands.

Hartmann, L. (1998). *Accretion Processes in Star Formation.* Cambridge University Press, Cambridge.

Lamers, H. J. G. L. M. and Cassinelli, J. P. (1999). *Introduction to Stellar Winds.* Cambridge University Press, Cambridge.

References

Abbott, D. C. (1980). The theory of radiatively driven stellar winds. I. A physical interpretation. *Ap. J.*, **242**, 1183–1207.

Abbott, D. C. (1982). The line acceleraton. *Ap. J.*, **259**, 282–301, ibid. II.

Armitage, P. J. and Livio, M. (1996). Accretion disks in interacting binaries: Simulations of the stream-disk impact. *Ap. J.*, **470,** 1024–1032.

Belcher, J. W. and MacGregor, K. B. (1976). Magnetic acceleration of winds from solar-type stars. *Ap. J.*, **210**, 498–507.

Bertschinger, E. and Chevalier, R. A. (1985). A periodic shock wave model for Mira variable atmospheres. *Ap. J.*, **299**, 167–190.

Bohn, H. U. (1984). Generation of acoustic energy from convection zones of late-type stars. *A. A.*, **136**, 338–350.

Brandt, J. C., Wolff, C., and Cassinelli, J. P. (1969). Interplanetary gas. XVI. A calculation of the angular momentum of the Solar wind. *Ap. J.*, **156**, 1117–1124.

Carlberg, R. G. (1980). The instability of radiation-deriven stellar winds. *Ap. J.*, **241**, 1131–1140.

Castor, J., Abbott, D., and Klein, P. (1975). Radiation-driven winds of Of stars. *Ap. J.*, **195,** 157–174.

Cuntz, M. (1990). On the generation of mass loss in cool giant stars due to propagating shock waves. *Ap. J.*, **353**, 255–264.

Dominik, C. (1990). Dust driven mass loss in HR diagram. *Rev. Modern Astron.*, **3**, 199–208.

Feldmeier, A., Puls, J., Reile, C., Pauldrach, A. W. A., Kudnitzki, R. P., and Owocki, S. P. (1995). Shocks and shells in hot star winds. *Ap. Sp. Sci.*, **233**, 293–299.

Gail, H. P. (1990). Winds of late-type Stars. *Rev. Modern Astron.*, **3**, 156–173.

Gail, H. P. and Sedlmayer, E. (1987). Dust formation in stellar winds. V. The minimum mass loss rate for dust-driven winds. *A.A.*, **177**, 186–192.

Garmany, C. D. and Conti, P. S. (1984). Mass loss in O-type stars: Parameters which affect it. *Ap. J.*, **284**, 705–711.

Gillet, D. and Lafon, J. P. J. (1983). On radiative shocks in atomic and molecular stellar atmospheres. I. *A.A.*, **128**, 53–63.

Harrop-Allin, M. K. and Warner, B. (1996). Accretion disc radii in eclisping cataclysmic variables. *M. N. R. A. S.*, **279**, 219–228.

Hartmann, L. (1998). *Accretion Processes in Star Formation.* Cambridge University Press, Cambridge.

Hartmann, L. and MacGregor, K. B. (1980). Momentum and energy deposition in late-type stellar atmospheres and winds. *Ap. J.*, **242**, 260–282.

Hearn, A. G. (1975). The energy balance and mass loss of stellar coronae. *A.A.*, **40**, 355–364.

Holzer, T. E. (1987). Theory of winds from cool stars, in *Circumstellar Matter*, Proc. IAU Symp. No. 122, I. Appenzeller and C. Jordan (eds.), D. Reidel Publ. Co., pp. 289–306.

Holzer, T. E., Flå, T., and Leer, E. (1983). Alfvén waves in stellar winds. *Ap. J.*, **275**, 808–835.

Iben Jr., I., Tutukov, A. V., and Yungelson, L. R. (1995). A model of the galactic X-ray binary population. I. High-mass X-ray binaries. *Ap. J. Suppl.*, **100**, 217–231.

Kippenhahn, R. (1973). Chromospheric activity and stellar evolution. *Stellar Chromospheres*, Proc. IAU Colloq., S. D. Jordan and E. H. Avrett (eds.), NASA, Washington, D.C., 265–278.

Knapp, G.R. and Morris, M. (1985). Mass loss from evolved stars. III. Mass loss rates for fifty stars from CO J = 1–0 observations. *Ap. J.*, **292**, 640–669.

Königl, A. (1991). Disk accretion onto magnetic T Tauri stars. *Ap. J.*, **370**, L39–L43.

Kwok, S. (1975). Radiation pressure on grains as a mechanism for mass loss in red giants. *Ap. J.*, **198**, 583–591.

Lamers, H. J. G. L. M. and Cassinelli, J. P. (1999). *Introduction to Stellar Winds.* Cambridge University Press, Cambridge.

Lamers, H. J. G. L. M. and Snow, T. P. Jr. (1978). Ionization conditions in the expanding envelopes of O and B stars. *Ap. J.*, **219**, 504–514.

Leitherer, C. (1988). Hα as a tracer of mass loss from OB stars. *Ap. J.*, **326**, 356–367.

Lim, J. and White, S. M. (1996). Limits to mass outflow from late type dwarf stars. *Ap. J.*, **462**, L91–L94.

Loup, C., Forveille, T., Omont, A., and Paul, J. F. (1993). CO and HCN observations of circumstellar envelopes. A catalogue—mass loss rates and distribution. *A. A. Supl.* **99**, 291–377.

Lucy, L. and Solomon, P. H. (1970). Mass loss by hot stars. *Ap. J.*, **159**, 879–893.

Mullan, D. J., Doyle, J. G., Rediman, R. O., and Mathioudakis, M. (1992). Limits of detectability of mass loss from cool dwarfs. *Ap. J.*, **397**, 225–231.

Nieuwenhuijzen, H., de Jager, C., Cuntz, M., Lobel, A., and Achmad, L. (1993). A generalized version of the Rankine-Hugoniot relations including ionization, dissociation, radiation and related phenomena. *A. A* , **280**, 195–200.

Olson, E. C. (1987). Photometry of long-period Algol binaries. III. The accretion disk and mass transfer in RZ Ophiuchi. *A. J.*, **94**, 1309–1317.

Ostriker, E. C. and Shu, F. H. (1995). Magnetocentrifugally driven flows from young stars and disks. IV. The accretion funnel and dead zones. *Ap. J.*, **447**, 813–828.

Paczynski, B. (1971). Evolutionary processes in close binary systems. *Ann. Rev. A. A.*, **9**, 183–208.

Proudman, I. (1952). The generation of noise by isotropic turbulence. *Proc. Roy. Soc. London A.* **214**, 119–132.

Richards, M. T., Albright, G. E., and Bowles, L. M. (1995). Doppler tomography of the gas stream in short-period Algol binaries. *Ap. J.*, **438**, L103–L106.

Stahler, S. W., Shu, F., and Taam, R. E. (1980). The evolution of protostars. I. Global formation and results. *Ap. J.*, **241**, 637–654.

Stein, R. F. (1968). Waves in the Solar atmosphere. I. The acoustic energy flux. *Ap. J.*, **154**, 297–306.

Stepien, K. and Ulmschneider, P. (1989). X-ray emission from acoustically heated coronae. *A. A.*, **216**, 139–142.

Theuns, T. and Jorissen, A. (1993). Wind accretion in binary stars. I. Intricacies of the flow structure. *M. N. R. A. S.*, 265, 946–967.

Ulmschneider, P. (1989). The chromospheric emission from acoustically heated stellar atmospheres. *A.A.*, **222**, 171–178.

Umschneider, P., Theurer, J., and Musielak, Z. E. (1996). Acoustic wave energy fluxes for late-type stars. *A.A.*, **315**, 212–221.

Weber, E. J. and Davis, L., Jr. (1967). The angular momentum of the Solar wind. *Ap. J.*, **148**, 217–228.

Weymann, R. (1963). Mass loss from stars. *Ann. Rev. A. A.*, **1**, 97–144.

Willson, L. A. and Hill, S. J. (1979). Shock wave interpretation of emission lines in long period variable stars. II. Periodicity and mass loss. *Ap. J.*, **228**, 854–869.

Winkler K. A. and Newman, M. J. (1980). Formation of solar-type stars in spherical symmetry. I. The key role of the accretion shock. *Ap. J.*, **236**, 201–211.

Wood, P. R. (1979). Pulsation and mass loss in Mira variables. *Ap. J.*, **227**, 220–231.

Chapter 4

Formation of Emission Lines

4.1 Theories of static envelopes

4.1.1 Dilution effect and the Rosseland cycle

Dilution effect

When we consider the radiation fields of stellar envelopes exposed to the photospheric UV radiation, the basic feature is their anisotropic nature. The degree of anisotropy is expressed by the geometrical dilution factor, W, defined as the ratio of the solid angle, ω, of the photosphere seen from a point of interest, P, relative to the total solid angle 4π. Thus we have

$$W = \frac{\varpi}{4\pi}. \tag{4.1.1}$$

If point P is located at a distance r from the star's center, and θ be the angle subtended by the stellar radius at point P, as seen in Figure 4.1, the solid angle ω is expressed as

$$\omega = 2\pi \int_0^\theta \sin\theta \; d\theta = 2\pi(1 - \cos\theta). \tag{4.1.2}$$

Then the dilution factor is given by

$$W = \frac{1}{2}(1 - \cos\theta). \tag{4.1.3}$$

Converting $\cos\theta$ to the ratio of stellar radius R and distance r, we have

$$W = \frac{1}{2}\left\{1 - \sqrt{1 - \left(\frac{R}{r}\right)^2}\right\}. \tag{4.1.4}$$

In case of $r \gg R$, we have

$$W \sim \frac{1}{4}\frac{R^2}{r^2}. \tag{4.1.5}$$

The value of W ranges from 0.5 at the stellar surface ($r = R$) to 10^{-15} in extended planetary nebulae. In stellar envelopes the typical value of W is in

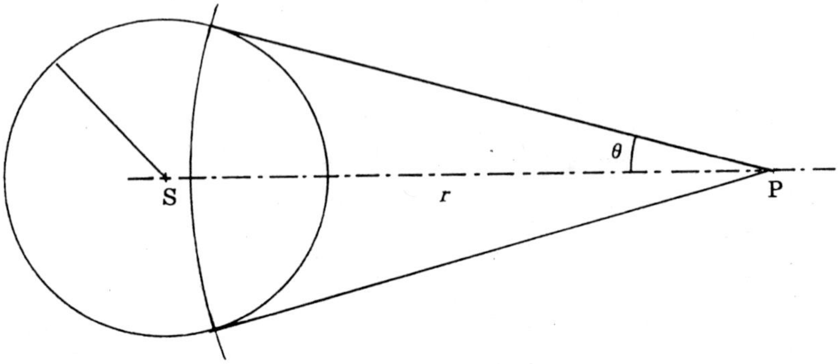

Figure 4.1: Dilution factor at point P is defined as the ratio of solid angle subtended to the photospheric disk, relative to the whole solid angel 4π.

between 10^{-1} and 10^{-5}. The relation between W and $x = r/R$ is partly given in Table 4.1.

From the definition of the dilution factor, it is clear that the radiation energy density $u(\nu)$ at the point P is simply reduced by a factor W from its value $U(\nu)$ at the surface of the photosphere. Thus we have

$$u(\nu) = WU(\nu). \tag{4.1.6}$$

This relationship is also used for the mean radiation intensity instead of energy density, i.e., the mean intensity $J_\text{v}(r)$ at point P is given by $J_\text{v}(r) = WI^*_\text{v}$, where I^*_v denotes the radiation intensity at the stellar surface. This approximation is valid when the envelope is optically thin for the ν radiation.

Table 4.1: The relation between the dilution factor W and $x = r/R$

W	0.2	0.1	0.05	0.02	0.01	10^{-3}	10^{-4}	10^{-5}
x	1.25	1.67	2.29	3.57	5.03	15.8	50.0	158

Rosseland cycle

Anisotropy of the diluted radiation field is the basic condition of emission-line formation in nebulae. The effect of dilution was first pointed out by Rosseland (1936) using a hypothetical three-level atom (levels 1, 2, and 3), as shown in Figure 4.2.

This atom emits or absorbs radiation through transitions between two energy levels. Now, we shall consider two transition cycles: cycle A (transitions $1 \to 3 \to 2 \to 1$) and cycle B (transitions $1 \to 2 \to 3 \to 1$). Cycle A denotes the absorption of short-wavelength radiation ($1 \to 3$), followed by two cascade downward transitions ($3 \to 2$, $2 \to 1$) to return to the ground level. Cycle B denotes the absorptions of two incident lights ($1 \to 2, 2 \to 3$) and emits one

Figure 4.2: Rosseland cycle in a three-level atom. Two transition cycles A and B are compared in diluted radiation field.

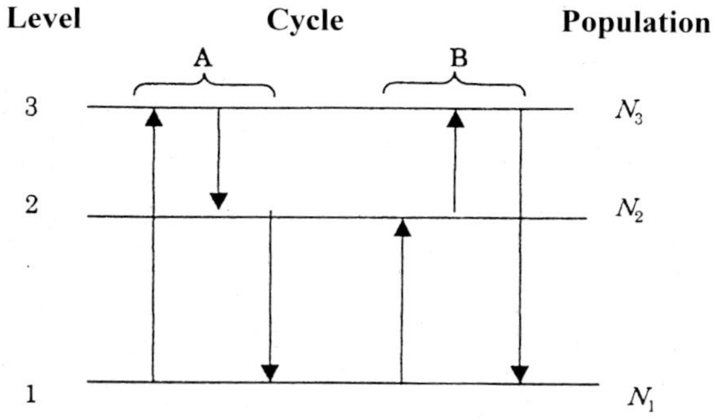

short-wavelength radiation (3 → 1). In the TE state, both cycles occur at the same rate so that the two cycles are in balance. In a diluted radiation field, the rates of cycles A and B significantly deviate from the balance as shown below.

For simplicity we consider the nebular gas in which the line radiations are optically thin. Now, let P_A and P_B be the occurrence rates of cycles A and B per unit volume, per unit time. These rates can be written by using the Einstein coefficients as follows:

$$P_A = N_1 B_{13} J_{13} (A_{32} + B_{32} J_{23})(A_{21} + B_{21} J_{12}) \tag{4.1.7}$$

$$P_B = N_1 B_{12} J_{12} B_{23} J_{23} (A_{31} + B_{31} J_{13}). \tag{4.1.8}$$

The Einstein coefficients between levels n, $n'(n > n')$, $A_{nn'}$, $B_{n'n}$, $B_{nn'}$, are connected by the following relations, as derived in Section 2.3.2.

$$\frac{A_{nn'}}{B_{n'n}} = \sigma_{n'n}, \; g_{n'} B_{n'n} = g_n B_{nn'}, \qquad \sigma_{n'n} = \frac{g_{n'}}{g_n} \frac{2h\nu^3_{n'n}}{c^2}, \tag{4.1.9}$$

where g_n denotes the statistical weight of level n. Inserting these relations into Equations (4.1.7) and (4.1.8), we can write the ratio of P_A and P_B as follows:

$$\frac{P_B}{P_A} = \frac{J_{12} J_{23} (\sigma_{13} + J_{13})}{J_{13} (\sigma_{23} + J_{23})(\sigma_{12} + J_{12})}. \tag{4.1.10}$$

In the right-hand side, the expressions $\sigma_{n'n} + J_{n'n}$ in brackets denote the sum of the spontaneous and induced transitions. Since $n'n$ radiation is assumed to be optically thin, we can write $J_{n'n} \cong I^*_{n'n} W$, where $I^*_{n'n}$ is the intensity at the stellar photosphere. Hence the induced transition term $J_{n'n}$ decreases with the decrease of W and can be ignored for the nebular field of $W \ll 1$. If we neglect this term, Equation (4.1.10) can be approximated as

$$\frac{P_B}{P_A} = \frac{J_{12} J_{23} \sigma_{13}}{J_{13} \sigma_{23} \sigma_{12}}, \tag{4.1.11}$$

where $\sigma_{n'n}$ are physical constants and $J_{n'n}$ can be taken as $I^*_{n'n}W$. Hence we have

$$\frac{P_{\mathrm{B}}}{P_{\mathrm{A}}} \propto W. \tag{4.1.12}$$

This implies that the rate of cycle B decreases proportionally with the dilution factor W. Though we have neglected the effect of induced transition, the Relationship (4.1.12) is generally valid for the region of $W \ll 1$, as in case of planetary nebulae where $W \sim 10^{-15}$. The prevalence of cycle A is apparent in such diluted nebular fields. When this cycle is prevailing, we say that the Rosseland cycle is operating in that radiation field. In the case of stellar envelopes where the dilution factor is in the order of 10^{-1}–10^{-5}, the effect of inverse cycle, cycle B, becomes nonnegligible. This point will be discussed later (Section 4.1.4).

4.1.2 Nebular approximation and recombination lines

The Rosseland cycle is based on a three-level atom, but the cyclic transition is applicable in nebular field for atoms having many energy levels. In this case, cycle A operates as an ionization followed by recombination and subsequent cascade transitions toward the ground level. If the nebular material is sufficiently transparent for the radiations in cascade transitions, we observe these radiations as emission lines. The emission lines formed in this way are called recombination lines, and the radiation fields where the Rosseland cycle is fully operating are called nebular type. This mechanism of emission-line formation was widely applied in 1930s, not only for planetary nebulae and diffuse emission nebulae but even for stellar envelopes.

The formation process of recombination lines of hydrogen and helium in the case of nebular-type field is given as follows. Let n, n' be two energy levels ($n > n'$), then the volume emissivity $\varepsilon_{nn'}$ for $\nu_{n'n}$ radiation can be written as

$$\varepsilon_{nn'} = \frac{h\nu_{n'n}}{4\pi} N_n A_{nn'}, \tag{4.1.13}$$

where N_n is the population of atoms in level n. If the nebular medium is transparent for this radiation, this volume emissivity yields the intensity of emission line, $I(\nu_{nn})$, when integrated over the entire nebular volume.

The volume emissivity $\varepsilon_{nn'}$ involves the level population N_n, which is given as a solution of the equations of statistical equilibrium for the whole energy levels of atoms in question. Energy levels, level populations, and involved transitions are shown in Figure 4.3. General forms of statistical equations can be derived as follows (Menzel and Baker 1937).

In Figure 4.3, the transitions leaving from level n are either downward to level n' (spontaneous and stimulated emission) or upward to n'' (absorption) or to ionized state κ (ionization). The transitions entering level n are two downward ones from ionized state κ (recombination) and from upper level

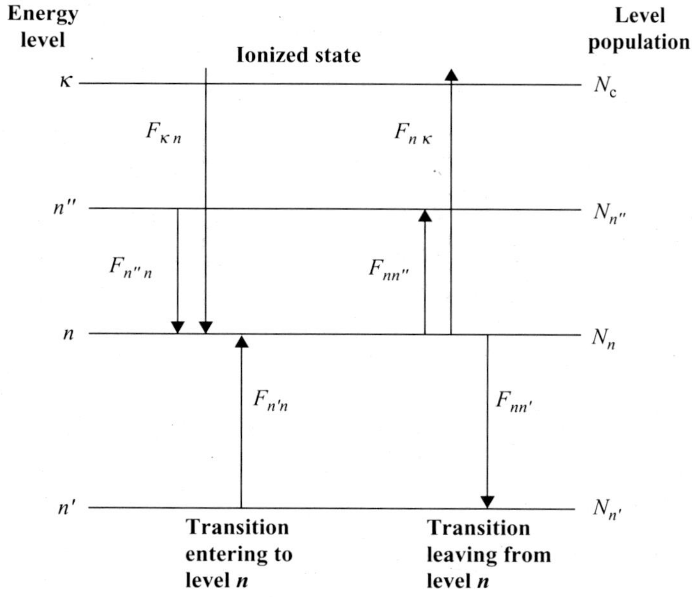

Figure 4.3: Energy level n and transitions entering to or leaving from level n. In the state of statistical equibrium both transitions should be balanced. Level κ denotes the ionization state.

n'' (cascade emission). Since time scales of these transitions are known to be sufficiently shorter than those of thermodynamical change of the nebular gas, transitions leaving from and entering into should be the same for every n level, i.e., electron transitions should be statistically in equilibrium state. Equations expressing this equilibrium are called the equations of statistical equilibrium.

If the rate of transition from n to n' is denoted as $F_{nn'}$, the equation of statistical equilibrium for level n generally takes the form

$$\sum_{n''=n+1}^{\infty} F_{n''n} + \int_{\nu_n}^{\infty} F_{\kappa n}\, d\nu + \sum_{n'=1}^{n-1} F_{n'n} = \sum_{n''=n+1}^{\infty} F_{nn''} + \int_{\nu_n}^{\infty} F_{n\kappa}\, d\nu + \sum_{n'=1}^{n-1} F_{nn'}, \tag{4.1.14}$$

where $F_{\kappa n}$ and $F_{n\kappa}$ denote the transitions between level n and the ionized state (ionization and recombination), and the lower boundary of each integral, ν_n, is the minimum frequency needed for ionization from level n.

If we can asume that ionization occurs only from the ground level as in the case of an actual nebulae, and the nebula is sufficiently transparent for every radiation due to transition $n \to n'$ between excited levels, then Equation (4.1.14) can be expressed as follows:

For $n > 1$, we have

$$\sum_{n''=n+1}^{\infty} F_{n''n} + \int_{\nu_n}^{\infty} F_{\kappa n}\, d\nu = \sum_{n'=1}^{n-1} F_{nn'}. \tag{4.1.15}$$

And for $n = 1$, we have

$$\sum_{n''=2}^{\infty} F_{n''1} + \int_{\nu_1}^{\infty} F_{\kappa 1}\, d\nu = F_{1\kappa} + \sum_{n''=2}^{\infty} F_{1n''}. \tag{4.1.16}$$

Now we consider atoms or ions with m energy levels, then we have m equations of statistical equilibrium, among which $(m-1)$ equations are mutually independent. Therefore we can get $(m-1)$ ratios of N_n/N_1 $(n = 2, \ldots, m)$, implying that we can only derive the relative intensities of emission lines.

Since the lines of Lyman series generally have large absorption coefficients, it is difficult to assume the transparency for these lines even in low-density nebulae. Taking this point into account, Baker and Menzel (1938) considered the following three cases, Cases A, B, and C, for applying the equations of statistical equilibrium in actual nebulae.

Case A: Nebula is transparent even for Lyman line radiations, and there is no reabsorption of Lyman line radiations, i.e., $F_{1n} = 0$ for $n \geqq 2$.

Case B: Nebula is very opaque for Lyman line radiations, and absorptions from level 1 to level n are exactly balanced by the inverse spontaneous transition, i.e., $F_{1n} = F_{n1}$ for $n \geqq 2$.

Case C: Nebula is transparent as in Case A, but with a central star radiating like a black body in the region of the Lyman lines (Baker et al. 1938).

When applied to hydrogen atoms in each of these cases, we can derive the relative intensities of emission lines in the Balmer and other series. In the Balmer series, relative intensities are usually expressed as the ratio to Hβ, and called the Balmer decrement, since the intensities are decreasing from Hα to Hβ, Hγ and higher members in ordinary nebulae.

In Case B, Equation (4.1.15) can be writen by using Einstein coeffcients as follows:

$$N_c N_e A_{cn} + \sum_{n''=n+1}^{n_m} N_{n''} A_{n''n} = N_n \sum_{n'=2}^{n-1} A_{nn'} \quad (n = 2, 3 \ldots, n_m), \tag{4.1.17}$$

where n_m denotes the highest energy level of hydrogen when approximated as a m-level atom. For Lyman line radiations, since we can write $N_1 B_{1n} = N_n A_{n1}$ for every level, Equation (4.1.16) takes the form

$$N_1 B_{1c} J_{1c} = N_c N_e A_{c1}. \tag{4.1.18}$$

This is a simple expression of the ionization in the nebular field. In this equation, B_{1c}, A_{c1} denote the transition probabilities integrated over the whole frequency range of the Lyman continuum.

Table 4.2: The Balmer decrements in nebular fields (Case B) (adapted from Pottasch 1984)

	Electron temperature (K)		
Balmer line	5000	10,000	20,000
Hα	3.00	2.85	2.74
Hβ	1.00	1.00	1.00
Hγ	0.460	0.469	0.476
Hδ	0.253	0.259	0.264
Hε = H7	0.155	0.159	0.163
Hζ = H8	0.102	0.105	0.107
H9	0.0714	0.0734	0.0746
H10	0.0162	0.0162	0.0161

Since the Balmer decrements of Case B provide good coincidence to observations in actual nebulae, this case has long been accepted as the standard decrement. Numerical decrements of this case are shown in Table 4.2 which was calculated by Pottasch (1984). As seen in this table the dependence of the decrements on electron temperature is not significant in Case B, i.e., the value of Hα/Hβ is from 3.00 for T_e = 5000 K to 2.74 for T_e = 20,000 K. On the other hand, Case A yields decrements that are generally flatter, such as Hα : Hβ : Hγ ~2.0 : 1.0 : 0.6, while Case C gives steeper decrements depending on the effect of UV radiation from stellar photosphere (Aller 1956). It is generally difficult to apply the decrements of Case C to stellar envelopes where we often see steeper decrements, since Case C is based, like Case A, on the asumption of transparency for the Lyman line radiations.

Although the decrements in nebular fields depend on electron temperature only slightly, close inspection of Table 4.2 shows that the values of decrements gradually flatter with the increase of electron temperature. This indicates the general tendency that the decrement becomes flatter when the excitation state is getting higher.

4.1.3 Generarization of nebular approximation and escape probability by scattering

Generalization of nebular approximation

The nebular approximation has been improved and extended mainly in an attempt to apply it to stellar envelopes characterized by high gas density and large-scale gas motions. The effects of gas motion will be considered in the next section.

We consider some points of improvement:

(i) Effects of sublevels (principal, azimuthal, magnetic, and spin quantum numbers). Energy levels of atoms and ions are generally designated by a set

Table 4.3: Balmer decrements when the effects of electron collisions are taken into account in nebular condition (adapted from Pottasch 1960)

$\tau(H_\alpha)$	0	1	3	10	20
Hα/Hβ	2.81	3.16	3.86	7.45	14.5
Hγ/Hβ	0.447	0.469	0.485	0.585	0.774
Hδ/Hβ	0.254	0.266	0.286	0.371	0.545

of quantum numbers (n, ℓ, m, s) as stated in Section 2.2. Brocklehurst (1971) solved the equations of statistical equilibrium for hydrogenic ions when the principal and azimuthal quantum numbers are taken into account. According to his numerical solution, the effects of azimuthal quantum numbers for the relative intensities of emission lines remain less than several percent.

(ii) Effects of collisional excitation and de-excitation. Effects of electronic collision were first considered by Pottasch (1960). He included the collision terms in the equations of statistical equilibrium in nebular condition like Case B for seven-level hydrogen atoms. He thus derived the Balmer decrements Hα : Hβ : Hγ : Hδ as a function of the optical depth τ (Hα) in the case of $T_e = 10,000$ K and $N_e \lesssim 10^5$ cm^{-3}. The results of his theoretical decrement are shown in Table 4.3. This is compared with the decrements in Case B given in Table 4.2. It is seen that the decrement is almost the same to Case B when τ (Hα) = 0, but deviates markedly with the increase of τ(Hα). That is, Hα/Hβ becomes steeper, while Hγ/Hβ and Hδ/Hβ are getting flatter.

(iii) Effects of strong incident radiation. When strong incident radiation exists, the induced transitions relative to the spontaneous ones are strengthened. Elitzur et al. (1983) examined the effects of such incident radiation for hydrogen atoms and showed that the decrement Hα/Hβ becomes flattened with increasing strong incident radiation. They applied this effect to explain the flat Balmer decrements of CVs (see Section 6.6.6).

Escape probability by scattering and the Balmer decrements

The next problem is concerned with the process of radiative transfer. Instead of solving the equations of radiative transfer for line radiations in cases of optically thick medium, the method of escape probability has been introduced. We can now define two kinds of escape probability. One is the escape probability by motion to be applied in moving envelopes. This will be discussed in the next section. The second is the escape probability by scattering in static envelopes. For line radiations in optically thick envelopes, any photons emitted or absorbed suffer frequency shift due to envelope motion or scattering and have a probability to escape from the envelope.

The escape probability by scattering in the case of a plane-parallel gas layer can be derived as follows. Let $\varphi(x)$ be a normalized profile of absorption coefficient in a line radiation, where the central frequency is ν_0 and x is the nondimensional frequency shift from the line center measured by the Doppler width $\Delta\nu_D$ as defined by

$$x = \frac{\nu - \nu_0}{\Delta\nu_D}. \tag{4.1.19}$$

The normalization of $\varphi(x)$ is given by

$$2\int_0^\infty \phi(x)\,dx = 1. \tag{4.1.20}$$

Now, let τ_0 be the optical thickness of the gas layer at the line center. Let this line radiation suffers isotropic scattering and frequency shift by a single scattering proportional to the line profile $\varphi(x)$. We define the critical frequency x_1 at which the optical thickness becomes unity in the line profile. Then, if the frequencey shift exceeds this critical value $|x_1|$, this line radiation can escape out of the layer. In Figure 4.4 we show the line profile $\varphi(x)$ and the critical frequency $x = x_1$, where the value $\varphi(x_1)$ corresponds to the point of $\tau = 1$. From this figure, the escape probabilty by scattering can be defined by

$$\beta = 2\int_{x_1}^\infty \phi(x)\,dx, \tag{4.1.21}$$

where x_1 denotes the critical frequency. Since optical thickness inside the line profile is assumed to be proportional to the profile, x_1 can be obtained by the proportionality relation

$$\frac{\tau_0\,\phi\,(x_1)}{\phi\,(0)} = 1. \tag{4.1.22}$$

The form of β in Equation (4.1.21) depends upon the form of $\varphi(x)$ taken from either Doppler, Voigt, or Stark-broadened profile. As an example, Hummer's

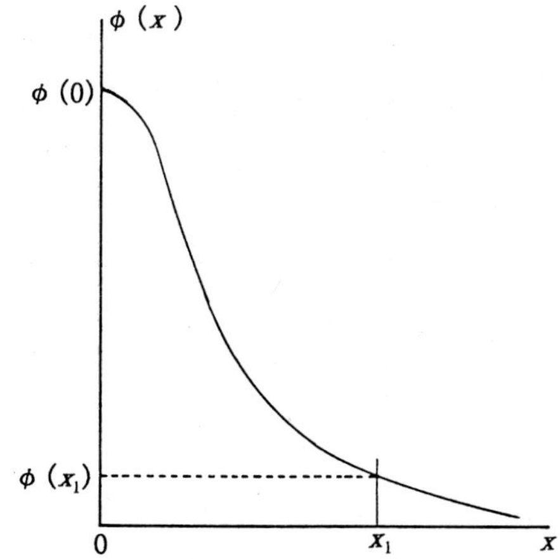

Figure 4.4: Profile of absorption coefficient and the critical frequency. The ordinate is the normalized profile $\phi(x)$, and the abscissa is the nondimensional frequency shift $x = (v - v_0)/\Delta v_D$, where Δv_D denotes the Doppler width. $x = x_1$ is the frequency at which the optical depth is equal to unity.

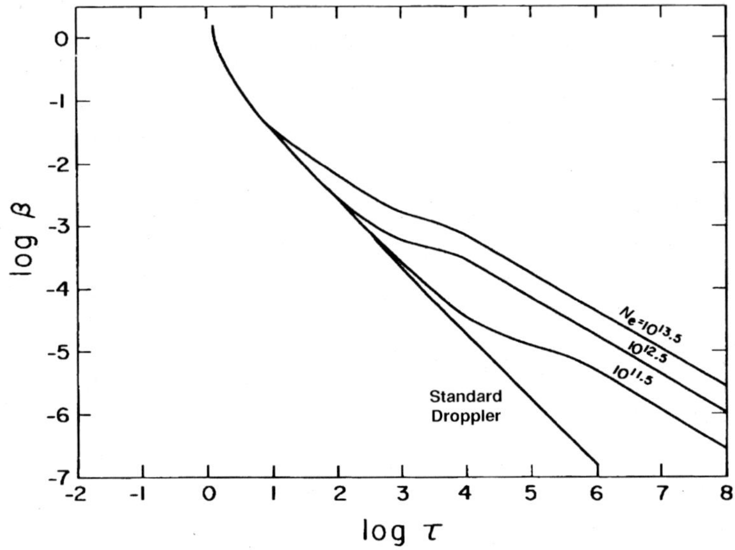

Figure 4.5: The escape probability by scattering as a function of optical depth of a gas layer. The escape probability curve calculated for the Voigt profile is labeled by Standard Doppler. The effect of the Stark broadening is shown for some values of electron density Ne. (From Drake and Ulrich 1980)

(1964) escape probability β based on the Voigt profile takes the form

$$\beta = \frac{1}{2\tau_0 \left(\pi \ln \tau_0\right)^{1/2}}. \tag{4.1.23}$$

When we consider layers of higher gas density, the effect of Stark broadening becomes more significant. Drake and Ulrich (1980) calculated the values of β by taking this effect into consideration. Their result is shown in Figure 4.5 as a function of optical thickness τ_0 at the line center. One may see that β decreases with increasing τ, while the Stark effect broadens the wing part of the line profile and then increases the value of β.

With the aid of escape probability $\beta_{nn'}$ for transition $n \to n'$, the equations of statistical equilibrium can be written as follows:

$$N_e N_c A_{cn} + \sum_{n''=n+1}^{n_m} N_{n''} \beta_{n''n} A_{n''n} = N_n \sum_{n'=1}^{n-1} \beta_{n'n} A_{nn'} \quad (n > 1) \tag{4.1.24}$$

$$N_1 R_{1c} = N_e N_c A_{c1} + \sum_{n''=2}^{n_m} N_{n''} \beta_{1n''} A_{n''1} \quad (n = 1). \tag{4.1.25}$$

In deriving these equations, it is assumed that the gas layer is transparent for the continuum radiation, and ionization occurs only from the ground level. Ionization is expressed in Equation (4.1.25) by the term $N_1 R_{1c}$, where R_{1c} is defined by

$$R_{1c} = 4\pi \int_{\nu_1}^{\infty} a_1(\nu) \frac{J_\nu}{h\nu} d\nu, \tag{4.1.26}$$

where ν_1 denotes the frequency of ionization limit and $a_1(\nu)$ is the bound-free absorption coefficient for the ground level. Thus R_{1c} (s^{-1}) expreses the number density of ionizing photons emitted in the gas layer per unit volume and per unit time. It is also possible to regard R_{1c} as the intensity of incident radiation from an outside light source. In this case, R_{1c} takes the value such as in between 3×10^{-4} s^{-1} (5000 K) and 3 s^{-1} (40,000 K). In the case of the stellar envelopes, R_{1c} should be a function of stellar temperature and the dilution factor of the gas layer. The terms of collisional transition and/or induced transition can also be added in the above equations.

By solving Equations (4.1.24) and (4.1.25), we can obtain the relative populations of $N_1 : N_2 : N_3 : \ldots : N_m$, and then the Balmer decrements can be calculated by

$$\frac{I(\mathrm{H}n)}{I(\mathrm{H}\beta)} = \frac{\beta_{2_n} N_n A_{n2} \nu_{2n}}{\beta_{2_4} N_2 A_{42} \nu_{24}}. \tag{4.1.27}$$

Usually, the gas layer is assumed as plane-parallel, homogenous and stationary. The parameters are electron temperature, electron density, chemical abundance, and intensity of incident radiation.

Based on this method, Drake and Ulrich (1980) calculated the Balmer decrements for the dense and optically thick gas layer, by taking a wide range of parameters. The parameters adopted by them are in the following range:

$$T_e = 5000\text{–}40000\ \mathrm{K}, \qquad N_e = 10^{11}\text{–}10^{15}\ \mathrm{cm}^{-3}$$
$$\tau(\mathrm{L}\alpha) = 10^3\text{–}10^6, \qquad R_{1c} = 3 \times 10^{-4}, 3 \times 10^{-2}.$$

Figure 4.6A and B illustrate a part of their results for the Balmer decrements Hα/Hβ in A and $\mathrm{H}n/\mathrm{H}\beta (n = 5\text{–}8)$ in B as a function of electron density. Adopted parameters are shown in the figures.

Some important features may be pointed out from these figures. First, the decrement Hα/Hβ gradually increases with the increase of electron density and reaches some maximum value at some point, thereafter rapidly decreasing. When $\tau(\mathrm{H}\alpha)$ is large, the decrement exceeds 10, while the decrement aproaches the nebular decrement when $\tau(\mathrm{H}\alpha)$ is small. On the contrary, the decrements in Hγ–H8 reveal minimum and maximun points at some values of electron density. These complicated behaviors are caused by combined effects of photoelectric ionization, electron collision, and self-absorption. In these figures, a large variatey of the Balmer decrements are seen. For this reason the Drake and Ulrich model has been widely applied to Be stars (Section 5.3.4), CV, symbiotic stars (Sections 6.6.6 and 6.7.1) and even to active galactic nuclei (Stirpe 1991).

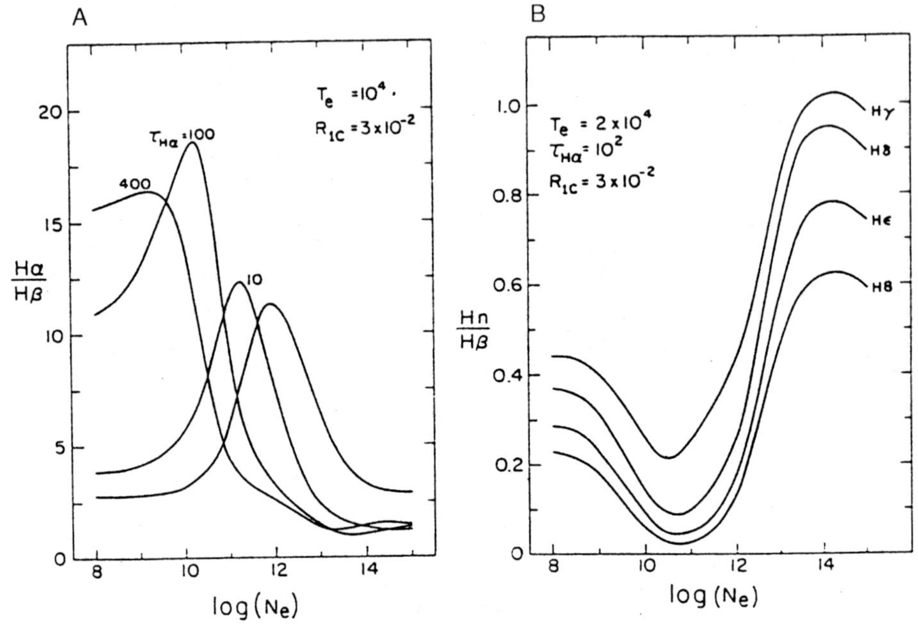

Figure 4.6: Balmer decrements calculated based on the escape probability by scattering (Drake and Ulrich 1980). (A) H_α/H_β, (B) H_n/H_β, the adopted physical parameters are given in the figures.

In this section the escape probability by scattering is calculated in the case of single scattering. An extension to multiple scattering were made by Bruevich et al. (1990) (Section 6.2.2).

4.1.4 Radiation field of the envelopes of early-type stars

In stellar envelopes, the dilution factor is usually in a range of 10^{-1}–10^{-5} (much larger than that of nebular fields) and electron density is as high as $N_e = 10^8$–10^{13} cm^{-3}. The radiation fields of stellar envelopes may have several characteristics:

(i) The inverse process for the Rosseland cycle can not be ignored due to large values of dilution factor.
(ii) The effects of radiative transfer is significant due to large optical depths for the lines in question.
(iii) The effects of gas motions inside the envelopes by rotation, winds, and others, are generally important.

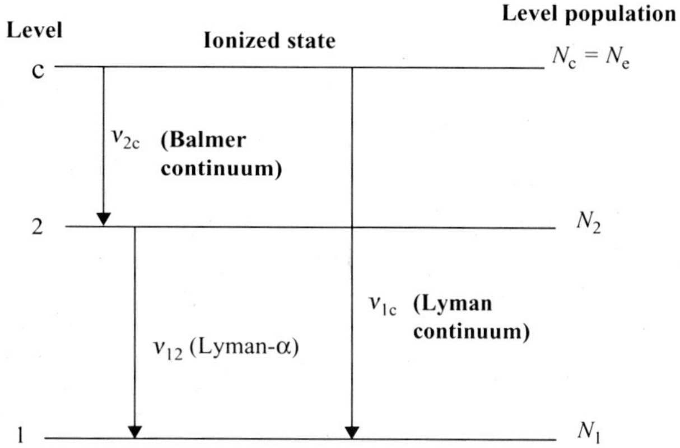

Figure 4.7: Fictitious hydrogen atom simplified to 3 levels (1, 2, c), where c denotes the continuum level.

In this section, we consider the case of static envelopes in the non-LTE. We first consider the basic characteristics of the radiation fields based on the static atmosphere theory proposed by Miyamoto (1949).

Three-level solution and Be-star type radiation field

We begin with the case of three-level atoms as in the case of the Rosseland cycle. We assume a hydrogen atom to have three levels (1, 2, c), where level c denotes the ionized state, as shown in Figure 4.7. For simplicity we regard level c as a discrete level averaged for all the ionizing radiation, and we define the transition coefficients A_{cj} (for $c \to j$ recombination), B_{jc} (for $j \to c$ ionization) as follows:

$$
\begin{aligned}
A_{cj} &= \frac{2^9 \pi^5}{(6\pi)^{3/2}} \frac{e^{10}}{m^2 c^3 h^3} \left(\frac{m}{k}\right)^{3/2} e^{h\nu_j/kT} \frac{1}{j^3 T^{3/2}} E\left(\frac{h\nu_j}{kT}\right) \\
E(x) &= \int_x^\infty \frac{e^{-x}}{x}\, dx \\
B_{jc} &= \frac{\displaystyle\int_{\nu_j}^\infty B_{jc}(\nu)\, I(\nu)\, d\nu}{\displaystyle\int_{\nu_j}^\infty I(\nu)\, d\nu} \\
B_{jc}(\nu) &= \frac{2^8 \pi^5}{3\sqrt{3}} \frac{m e^{10}}{c h^7 j^5} \frac{1}{\nu^4} \\
I(\nu) &= \frac{2h\nu^3}{c^2} \frac{1}{e^{h\nu/kT} - 1} \\
\nu_j &= \frac{2\pi^2 m e^4}{h^3} \frac{1}{j^2}.
\end{aligned}
\tag{4.1.28}
$$

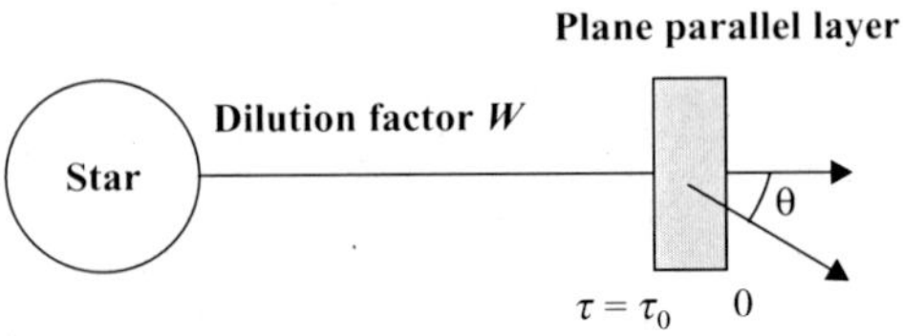

Figure 4.8: Plane-parallel gas layer placed at a dilution factor W. The angle θ gives the directon of observer relative to the radial direction.

The mean frequency ν_{jc} and the mean line width Δ_{jc} are defined by

$$\nu_{jc} = \frac{\int_{\nu_j}^{\infty} \nu B_{jc}(\nu) I(\nu)\, d\nu}{\int_{\nu_j}^{\infty} B_{jc}(\nu) I(\nu)\, d\nu}$$

$$\Delta_{jc} = \frac{\int_{\nu_j}^{\infty} I(\nu)\, d\nu}{I(\nu_{jc})}. \tag{4.1.29}$$

Now, we assume that the envelope is static, plane-parallel, composed only of hydrogen, and characterized by a dilution factor W as shown in Figure 4.8. In this case the equations of statistical equilibrium can be written as

$$\begin{aligned} N_1\,(B_{12}J_{12} + B_{1c}J_{1c}) &= N_2 A_{21} + N_e N_c A_{c1} \\ N_2\,(A_{21} + B_{2c}J_{2c}) &= N_1 B_{12} J_{12} + N_e N_c A_{c2} \\ N_e N_c\,(A_{c1} + A_{c2}) &= N_1 B_{1c} J_{1c} + N_2 B_{2c} J_{2c}. \end{aligned} \tag{4.1.30}$$

These equations are not mutually independent; the third equation is derived from the sum of the first and second equations. We then obtain the population ratios of N_2/N_1, $N_e N_c/N_1$ from Equation (4.1.30).

For the gas layer in Figure 4.8, the equations of radiative transfer for the respective radiation is written as follows:

$$\begin{aligned} \cos\theta\,\frac{dI_{1c}}{dz} &= -\frac{h\nu_{1c}}{4\pi}\left(N_1 B_{1c} I_{1c} - N_e^2 A_{c1}\right) \\ \cos\theta\,\frac{dI_{12}}{dz} &= -\frac{h\nu_{12}}{4\pi}\left(N_1 B_{12} - N_2 A_{21}\right) \\ \cos\theta\,\frac{dI_{2c}}{dz} &= -\frac{h\nu_{2c}}{4\pi}\left(N_2 B_{2c} I_{2c} - N_e^2 A_{c2}\right), \end{aligned} \tag{4.1.31}$$

where z sec θ is directed to the observer.

Since the gas layer can be supposed to be optically thin for Balmer continuum, we shall consider the equations of radiative transfer for Lyman continuum and Lyman-α radiation. Optical depths for these radiations are defined as

$$d\tau_{1c} = -\frac{h\nu_{1c}}{4\pi} N_1 B_{1c}\, dz,$$
$$d\tau_{12} = -\frac{h\nu_{12}}{4\pi} N_1 B_{12}\, dz = \varpi_{12}\, d\tau_{1c}, \tag{4.1.32}$$

where

$$\varpi_{12} = \frac{\nu_{12} B_{12}}{\nu_{1c} B_{1c}}. \tag{4.1.33}$$

The transfer equation for Lyman-α can be written using the optical depth as

$$\cos\theta \frac{dI_{12}}{\varpi_{12} d\tau_{1c}} = I_{12} - \frac{N_2}{N_1}\frac{A_{21}}{B_{12}}. \tag{4.1.34}$$

The second term of the right-hand side is the source function where N_2/N_1 can be derived by solving Equation (4.1.30) as follows

$$\frac{N_2}{N_1} = \frac{1}{A_{21}(A_{c1} + A_{c2}) + A_{c1} B_{2c} J_{2c}} [B_{12}(A_{c1} + A_{c2}) J_{12} + A_{c2} B_{1c} J_{1c}]. \tag{4.1.35}$$

We first insert this relative population into Equation (4.1.34) and use the definitions J_{2n}, K_{2n} for intensity I_{2n} (Section 2.5), then the equations of radiative transfer can be written under the Eddington approximation ($3K_{2n} = J_{2n}$) as

$$\frac{d^2 J_{12}}{d\tau_{1c}^2} = K\left(J_{12} - \frac{B_{1c} J_{1c} A_{c2} A_{21}}{B_{12} B_{2c} J_{2c} A_{c1}}\right), \tag{4.1.36}$$

where

$$K = \frac{3\varpi_{12}^2 \varsigma}{1+\varsigma}, \qquad \varsigma = \frac{B_{2c} J_{2c}}{A_{21}} \frac{A_{c1}}{A_{c1} + A_{c2}}. \tag{4.1.37}$$

The Balmer continuum is supposed as optically thin, and hence approximately written as

$$J_{2c} = W I_{2c}^*, \tag{4.1.38}$$

where I_{2c}^* denote the intensity of Balmer continuum at stellar surface.

In case of the radiation field of Be star envelopes, let us estimate the value of the constant K in Equation (4.1.37). Physical constants are taken from Kogure (1959b) for the case of $T = 19700$ K as

$$\varpi_{12} = 8.33 \times 10^3, \qquad A_{c1} = 1.32 \times 10^{-13}, \qquad A_{c2} = 5.33 \times 10^{-14},$$
$$B_{2c} = 5.83 \times 10^{-6}, \qquad A_{21} = 4.68 \times 10^8$$
$$J_{2c} = W I_{2c}^* = 2.22 \times 10^{12} W.$$

Inserting these values, we get

$$K = 4.10 \times 10^6\, W. \tag{4.1.39}$$

In case of Be star envelopes, since the dilution factor is in a range of $W = 10^{-1}$–10^{-5}, it leads to the value of K sufficiently exceeding unity. In addition, the optical thickness of the gas layer in the Lyman continuum lies in the order of unity, leading to the value of the left-hand side of Equation (4.1.36) sufficiently smaller than the first term, KJ_{12}, of the right-hand side. Therefore Equation (4.1.36) can be held only when we have

$$J_{12} \approx \frac{B_{1c}J_{1c}A_{c2}A_{21}}{B_{12}B_{2c}J_{2c}A_{c1}}.$$

Or, rewriting this equation, we get

$$B_{12}J_{12}B_{2c}J_{2c}A_{c1} \approx B_{1c}J_{1c}A_{c2}A_{21}. \tag{4.1.40}$$

This indicates that two cyclic transitions $1 \to 2 \to c \to 1$ and $1 \to c \to 2 \to 1$ in Figure 4.7 should be almost balanced. That is, the Rosseland cycle is nearly balanced with its inverse cycle, in contrast to the case of nebular field. Miyamoto (1949) called this type the Be-type radiation field, and this is limited to the region of $K \gg 1$, i.e., in the region of $W \gg 10^{-7}$.

In the radiation field of this type for hydrogen atoms, since the cyclic transition in three-level atom is balanced with its inverse cycle, the recombination lines are mainly formed by ionizations from the second energy level, and followed by recombination and cascade transitions. In this case, the ionization equation is approximately given in the form

$$N_2 B_{2c} J_{2c} = N_e^2 A_{c2}. \tag{4.1.41}$$

Or, we have

$$\frac{N_e^2}{N_2} = \frac{B_{2c} I_{2c}^* W}{A_{c2}}. \tag{4.1.42}$$

This indicates that the degree of ionization decreases proportionally to the dilution factor W in the Be-type radiation field. Equation (4.1.41) or (4.1.42) yields the simplified ionization equation corresponding to Equation (4.1.18) for nebular field.

Seven-level atoms and Balmer decrements

In order to derive the Balmer decrements, it is necessary to increase the numbers of energy levels of hydrogen atoms. Miyamoto (1952a,b) and Kogure (1959a,b) have extended the above three-level atoms to seven-levels and derived the decrement $\mathrm{H}\alpha/\mathrm{H}\beta$. The difference between these authors lies in the assumptions for the transparency of the Paschen and Bracket line radiations. Miyamoto has assumed complete opaqueness for these radiation, whereas Kogure assumed complete transparency. In both cases they assumed complete opaqueness for the Lyman line radiations and adopted Equation (4.1.41) for the ionization equation.

For hydrogen atomes simplified to seven levels (six disrecte levels, $n = 1, 2, \ldots, 6$ and one continuum level, c), the equations of statistical equilibrium for the levels higher than $n = 3$ are distinguished in both cases as follows:

(i) The envelope is optically opaque (Miyamoto case)

$$N_n \left(A_{n2} + B_{nc} J_{nc}\right) = N_2 B_{2n} J_{2n} + N_c N_e A_{cn} \quad (n > 3). \qquad (4.1.43)$$

It should be noted that the envelope is optically thin in the outer layer even in this case. Miyamoto has connected the equation of statistical eqluilibrium for opaque and transparent regions at some suitable depth. The latter equation is the same as given below.

(ii) The envelope is optically thin (Kogure case)

$$N_n \left(\sum_{n'=2}^{n-1} A_{nn'} + B_{nc} J_{nc} \right) = N_2 B_{2n} J_{2n} + N_c N_e A_{cn} + \sum_{n''=n+1}^{6} N_{n''} A_{n''n} \quad (n = 3, 4, 5, 6). \qquad (4.1.44)$$

The equations of radiative transfer for Balmer lines take the following form in both cases,

$$\cos\theta \, \frac{dI_{2n}}{dx} = -\frac{h\nu_{2n}}{4\pi} \left(N_2 B_{2n} I_{2n} - N_n A_{n2}\right). \qquad (4.1.45)$$

Now, we define the optical depths for $2n$ radiation measured from the surface as

$$d\tau_{2n} = -\frac{h\,\nu_{2n}}{4\pi} N_2 B_{2n} \, dx = \omega_n \, d\tau_\alpha, \qquad (4.1.46)$$

where we have

$$\omega_n = \frac{\nu_{2n} B_{2n}}{\nu_{23} B_{23}}. \qquad (4.1.47)$$

Equation (4.1.45) can be converted to the equations for the mean intensities as in case of three-level solution. Then we have

$$\frac{d^2 J_{2n}}{d\tau_{2n}^2} = 3 \left(J_{2n} - \frac{N_n}{N_2} \frac{A_{n2}}{B_{2n}} \right). \qquad (4.1.48)$$

Inserting the solutions of Equations (4.1.43) or (4.1.44) for N_n/N_2 into Equation (4.1.48), we have the equations of radiative transfer for J_{2n}. If we assume the source functions of these equations to be constant, we can solve them and finally derive the Balmer decrement $H\alpha/H\beta$ as a function of the dilution factor W and the stellar temperature T.

Miyamoto's and Kogure's decrements thus derived are shown in Figure 4.9 as a function of dilution factor W for early-type Be stars ($T = 19{,}700$ K), in

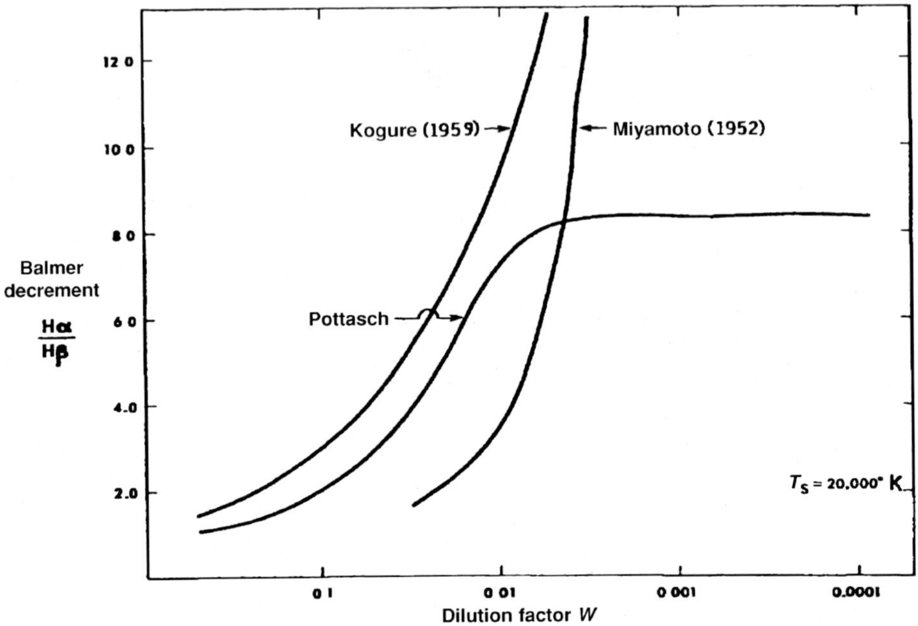

Figure 4.9: Balmer decrements H_α/H_β of Be-star type as a function of dilution factor in the casee of $T = 19700$ K. Miyamoto's solution (Paschen lines are opaque) and Kogure's solution (Paschen lines are transparent) are shown, along with Pottasch's decrement where the effects of electron collisions are taken into consideration in the case of $N_e = 10^{11}$ cm^{-1}. (Reproduced from Pottasch 1961)

which Kogure's solultion presents Case V as the typical one among his several solutions (Kogure 1959b).

In Figure 4.9 we see that the decrements in both cases increase with decreasing dilution factor W, and this increase can be explained by the decrease of ionization degree in the envelope given in Equation (4.1.42). Figure 4.9 also shows that Kogure's solution yields higher decrement for the same dilution factor. This difference gives different envelope dimensions for the observed decrements Hα/Hβ, ordinarily lying in a range of 2–5. The dilution factor giving this range of decrement is $W{-}10^{-3}{-}10^{-5}$ in Miyamoto's solution, while $W{-}10^{-1}{-}10^{-3}$ in Kogure's solution. Actually, however, both solutions are based on an assumed opaqueness for the Paschen lines, i.e., opaque or transparent, so that the real decrement will run between these two courves in Figure 4.9.

In Figure 4.9, the theoretical decrement by Pottasch (1961) is also shown. In this case the envelope is assumed to consist of ionized hydrogen at a constant electron density and temperature. By solving the non-LTE problem for four-level atom including the effects of electron collisions, he derived the decrement Hα/Hβ as a function of the dilution factor W. The theoretical curve is illustrated in case of $N_e = 10^{11}$ cm^{-3} and $T_e = 20,000$ K. In Miyamoto's and Kogure's case, the effects of electron collision are not included so that the decrements do not depend on electron density. The effects of collision will decrease with the decrease of electron density, and then with the decrease of dilution factor. By this reason, Kogure's case will be adopted for Be stars in Chapter 5.

The radiation field will change from Be type to nebular type for the envelopes with small dilution factor. Kogure (1961) showed that the decrement $H\alpha/H\beta$ increases up to some value at about W–10^{-5}, and then gradually decreases for smaller dilution factor and finally approaches the nebular-type (Case B) decrement.

The comparison with observed decrements will be given in Section 5.3.4.

4.1.5 Balmer decrements of emission-line stars

Balmer emission lines in stars usually appear superimposed on the strong continuous spectrum and photospheric absorption line profile, in contrast to the nebular spectra. The emission-line components are extracted from the stellar spectra and then their equivalent widths are measured to obtain the observed Balmer decrements. The observational procedure of these measurements is described below.

Measurement of emission equivalent widths

Profiles of spectral lines accompanied emission-line components are schematically shown in Figure 4.10. Let I_c be the intensity of a nearby continuum and line AA$'$ denote a position at wavelength λ inside the spectral line. AS and AO denote the depth of the photospheric absorption R_s and the observed depth R_o, respectively. Let W_S, W_{Ob} be the equivalent widths of photospheric absorption lines and of observed line profile, respectively. We then have

$$W_S = \int R_S \, d\lambda, \qquad W_{Ob} = \int R_{Ob} \, d\lambda, \tag{4.1.49}$$

where integration covers the whole spectral line. In the part of the emission line stronger than the continuum level we have $R_{ob} < 0$, and W_{OB} for the whole line can become negative. Let W_E be the equivalent width for the emission

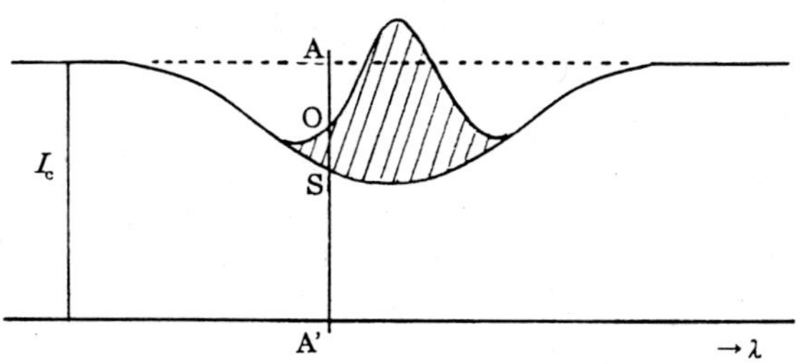

Figure 4.10: Profile of a spectral line with emission component. Line depths of observed (AO) and photospheric (AS) absorptions at a wavelength are indicated.

component shown by the shadow in Figure 4.10, we have

$$W_{\rm E} = \int (R_{\rm S} - R_{\rm Ob})\, d\lambda = W_{\rm S} - W_{\rm Ob}. \tag{4.1.50}$$

The equivalent width $W_{\rm S}$ of the photospheric absorption line can be obtained by the measurement of a standard star or by some model atmospheres for the same MK spectral type. By using the value of $W_{\rm S}$ thus obtained, we have the value of $W_{\rm E}(>0)$ for the emission-line component. Small-sized telescopes may measure emission-line equivalent widths for spectroscopic observations since no precise measurement of line profiles is needed.

Balmer decrements

The radiation flux $F(\rm Hn)$ of the Hn line emission can be expressed in terms of the equivalent width $W_{\rm E}(\rm Hn)$ in Equation (4.1.50) and the radiation flux $I_c(\rm Hn)$ for the nearby continuum as follows:

$$F(Hn) = W_{\rm E}(Hn)\, I_c(Hn). \tag{4.1.51}$$

Balmer decrement is defined as the flux ratio of emission components of Balmer lines relative to that of Hβ. Denoting $D_{n4} = F(Hn)/F(\rm H\beta)(n = 3, 5, 6, \ldots)$, we have

$$D_{34} = \frac{F(\rm H\alpha)}{F(\rm H\beta)}, \qquad D_{54} = \frac{F(\rm H\gamma)}{F(\rm H\beta)}, \qquad D_{n4} = \frac{F(Hn)}{F(\rm H\beta)}, \quad (n = 6, 7, \ldots). \tag{4.1.52}$$

Or, in terms of equivalent widths $W_E(Hn)$, we have

$$D_{34} = \frac{F(\rm H\alpha)}{F(\rm H\beta)} = G_\alpha \frac{W(\rm H\alpha)}{W(\rm H\beta)}, \qquad D_{n4} = \frac{F(Hn)}{F(\rm H\beta)} = G_n \frac{W(Hn)}{W(\rm H\beta)}, \tag{4.1.53}$$

where the conversion coefficients G_n are given by Equation (4.1.51) as

$$G_\alpha = \frac{I_c\,(\rm H\alpha)}{I_c\,(\rm H\beta)}, \qquad G_n = \frac{I_c\,(Hn)}{I_c\,(\rm H\beta)}. \tag{4.1.54}$$

When results of spectroscopic photometry for the standard stars are available, these coefficients are given by the ratios $I_c(Hn)/I_c(\rm H\beta)$. Some theoretical values may also be used. In earlier days before 1970s, black body radiation was usually used. In recent years, intensity ratios in some model atmospheres are often used. In actual cases, following corrections should be made when needed.

(i) Correction for interstellar reddening. Emission-line stars located in dusty regions or in remote regions suffer interstellar absorption and interstellar reddening by a stronger obscuration in shorter wavelength region. Balmer decrement can be corrected if the color excesses in multiband photometry of the stars are available. Actually, however, Balmer decrement is often used for the reddening correction assuming the nebular type decrement (Menzel-Baker's Case B). This is not the case for distant emission-line stars, for which we need to adopt some theoretical decrement instead of nebular Case B. For example, Gutierrez-Monero and Moreno (1996) derived the interstellar reddening using the Balmer decrements considering optical depth effects for the case of symbiotic stars.

(ii) Correction for the blending of other spectral lines. Blending of Balmer lines with other absorption lines is not significant in early-type stars (B, A types), where the spectra are mainly characterized by hydrogen and helium lines. In late-type stars, however, there are numerous lines and molecular bands so that the effects of line blending sometimes become crucial. For example, when Balmer emission lines are formed in a relatively deep layer of the atmosphere in Mira-type stars, absorption by TiO molecules decreases the strength of the Balmer lines in the order of $H\alpha > H\beta > H\gamma$. In order to correct this blending effect, one needs to compare the observed line profiles with those of standard stars or of some model atmospheres. This can only be done with high-resolution spectroscopy.

(iii) Correction for complex profiles of Balmer lines. When emission lines reveal some complex profiles, such as composed of a central narrow component and broad wing, as frequently observed in Of stars or active galactic nuclei, one should be careful of maintaining their separation, because both components may have been formed in different regions. Precise measurement of line profiles and/or model consideration on the formation of emission lines become crucial in these cases.

4.2 Theories of moving envelopes

4.2.1 Escape probability by motion

If there is a systematic motion with velocity gradient in an envelope, the Doppler effect will cause a shift of line center. Photons emitted inside the envelope can escape from its surface even in an optically thick envelope as the whole. That is, any photon in envelope has an escape probability due to the Dopple effect. This probability was first introduced by Sobolev (1947, 1960), and it has been widely applied not only to the spectral analysis of stellar envelopes but also to the analysis of molecular clouds. This type of probability is different from that of "the escape probability by scattering," and so we shall call it "the escape probability by motion," or the Sobolev-type probability. We shall derive this probability following Emerson (1996).

Now consider the escape probability for a photon which is emitted at an optical depth τ_ν from the surface at frequency ν. This photon can escape the medium with the probability $p_\nu = \exp(-\tau_\nu)$. The probability averaged over line frequency, β_e, weighted by the line emission profile ϕ_ν is given by

$$\beta_e = \int_0^\infty \phi_\nu p_\nu \, d\nu. \tag{4.2.1}$$

Furthermore, the probability averaged over angle, $\langle\beta_e\rangle$, at a point in a particular line but in all directions can be written

$$\langle\beta_e\rangle = \frac{1}{4\pi}\int \beta_e \, d\Omega. \tag{4.2.2}$$

Since the mean escape probability depends on the structure of the envelope, we first derive the probability β_e. For simplicity, consider an envelope moving toward an observer with a uniform velocity gradient dV/ds, and let a photon emitted at a point P with the central frequency ν_0 escape the envelope at distance L. This is defined by the distance in which the central frequency ν_0 shifts a Doppler width Δ_D due to the velocity gradient. This frequency shift is shown in Figure 4.11, where the velocity difference ΔV is given by

$$\Delta V = \frac{dV}{ds} L = \frac{\Delta_D}{\nu_0} c. \tag{4.2.3}$$

If the emission frequency is converted to a nondimensional variable $x = \Delta\nu/\Delta_D$, then the optical depth for this frequency, by taking the shift of frequency along the distance into account, is

$$\tau = \frac{\kappa'}{\Delta_D} \rho \int_0^\infty \phi_x\left(x - \frac{s}{L}\right) ds, \tag{4.2.4}$$

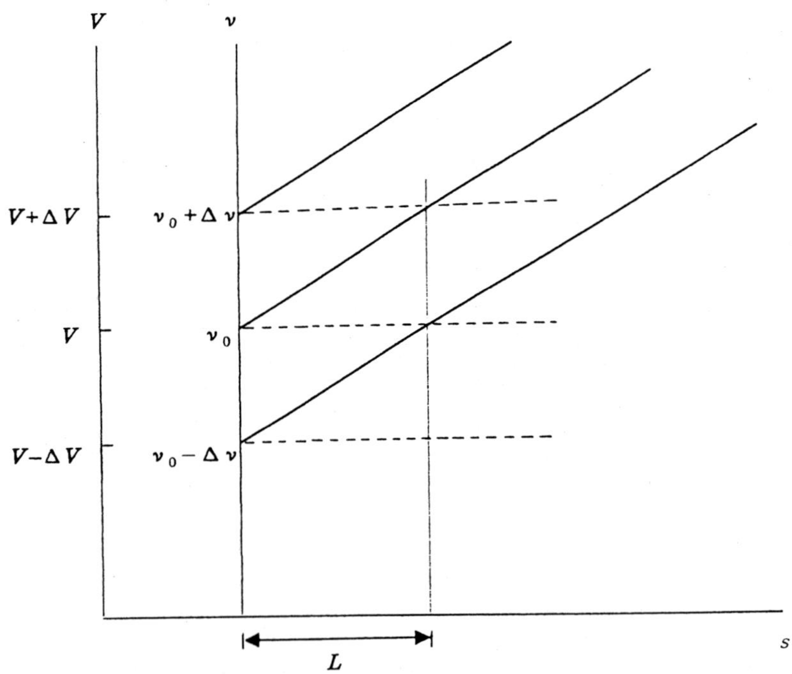

Figure 4.11: Frequency shift in line profile due to velocity gradient.

where $\kappa'(= \int \kappa_\nu \, d\nu)$ is the absorption coefficient integrated over the whole absorption line.

Frequency x at the surface changes to $x' = x - s/L$ at the geometrical depth s, then the integral in Equation (4.2.4) can be replaced by a frequency integral by using $ds = -L \, dx'$ as

$$\tau = \tau_S \int_{-\infty}^{x} \phi_x(x') \, dx', \tag{4.2.5}$$

where

$$\tau_S = \kappa' \rho \frac{L}{\Delta_D}. \tag{4.2.6}$$

The escape probability at frequency x is then given by

$$p_\nu = \exp\left[-\tau_S \int_{-\infty}^{x} \phi_x(x') \, dx'\right]. \tag{4.2.7}$$

Averaging over frequency, we have from Equation (4.2.1)

$$\beta_e = \int \phi_x \exp\left[-\tau_S \int_{-\infty}^{x} \phi_x(x') \, dx'\right] dx. \tag{4.2.8}$$

Changing the integration variable x to η defined by

$$\eta = \int_{-\infty}^{x} \phi_x(x') \, dx'. \tag{4.2.9}$$

Then we have for the escape probability β_e

$$\begin{aligned} \beta_e &= \int_0^1 e^{-\tau_S \eta} \, d\eta \\ &= \frac{1 - e^{-\tau_S}}{\tau_S}, \end{aligned} \tag{4.2.10}$$

where τ_S is given by inserting Equation (4.2.3) to Equation (4.2.6),

$$\tau_S = \kappa' \rho \frac{1}{\nu_0} \frac{c}{dV/ds}. \tag{4.2.11}$$

Physically, τ_S corresponds to the optical thickness of the characteristic length L and we identify it as the effective optical thickness. Equation (4.2.10) for β_e, combined with Equation (4.2.11), gives the general form of the escape probability by motion.

This result has still to be averaged over the direction to give $\langle \beta_e \rangle$. For this purpose we consider twe cases of spherically symmetric and plan-parallel envelopes.

(i) Spherically symmetric envelopes. Suppose we have a spherically symmetric envelope expressed in polar coordinates by r and θ as given in Figure 4.12. Let $V_r(r)$ be the radial outflow velocity, s the distance in

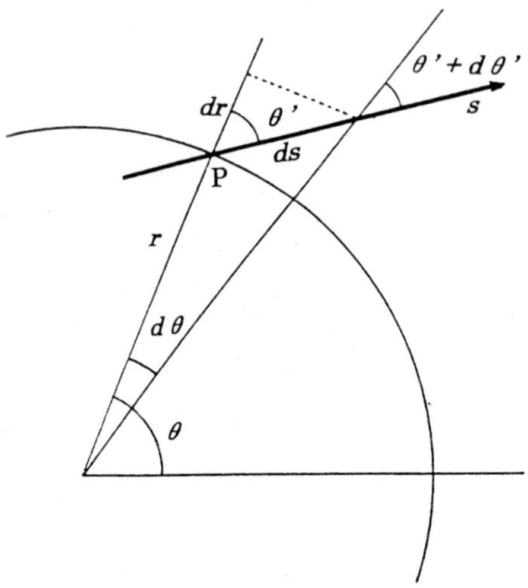

Figure 4.12: Spherically symmetric envelope and geometrical relation in the polar coordinates.

the direction toward the observer, and $V_s(s)$ the line-of-sight velocity. Measuring the angle between radial and line-of-sight directions by θ', we have

$$V_s = V_r \cos\theta'. \tag{4.2.12}$$

Consider a small displacement along the lins-of-sight ds. Since we have

$$dr = ds \cos\theta', \qquad d\theta = -d\theta', \qquad \sin\theta' = -r\, d\theta'/ds, \tag{4.2.13}$$

the velocity gradient along s direction can be derived by putting $\mu = \cos\theta'$

$$\frac{dV_s}{ds} = \frac{dV_r}{dr}\mu^2 + \frac{V_r}{r}\left(1-\mu^2\right). \tag{4.2.14}$$

If the velocity is linear, $V_r = kr$ (k is a constant), then $dV_s/ds = k$ and dV_s/ds and τ_S are independent of μ. Hence we have

$$\langle\beta_e\rangle = \frac{1}{2}\int_{-1}^{+1}\frac{1-e^{-\tau_S}}{\tau_S}\,d\mu = \frac{1-e^{-\tau_S}}{\tau_S}, \tag{4.2.15}$$

where

$$\tau_S = \frac{\kappa'\,\rho}{\nu_0}\,\frac{c}{k}. \tag{4.2.16}$$

(ii) Plane-parallel envelopes. Consider a plane-parallel envelope, taking z as the outward direction and $V_z(z)$ as the outflowing velocity (Figure 4.13). Observer's line-of-sight is in the angle θ' from z axis and

Figure 4.13: Velocity components in a plane-parallel envelope.

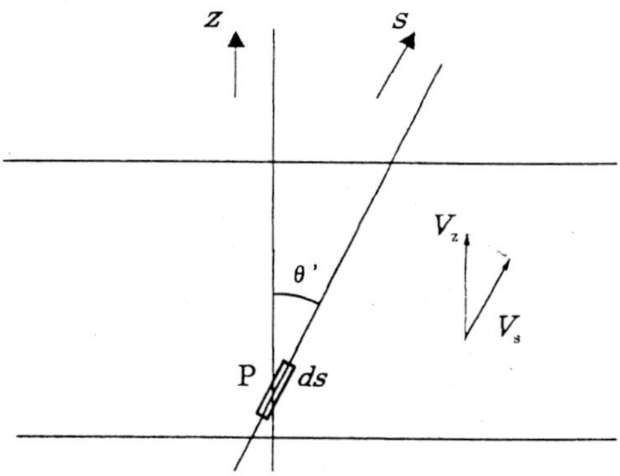

measures the distance s along this line. Then we have

$$V_S(s) = V_z(z)\cos\theta', \qquad \frac{dV_S}{ds} = \frac{dV_z}{dz}\mu^2. \tag{4.2.17}$$

In this case, τ_S in Equation (4.2.11) can be written

$$\tau_S = \kappa'\rho\frac{1}{\nu_0}\frac{c}{dV_S/ds}. \tag{4.2.18}$$

This is the effective optical depth in the line-of-sight. Similarly, the effective optical depth in z direction τ_S' can be defined as

$$\tau_S' = \frac{\kappa'\rho}{\nu_0}\frac{c}{(dV_z/dz)}. \tag{4.2.19}$$

Then we have the following relationship between two effective optical depths:

$$\tau_S = \tau_S' \cdot \frac{1}{\mu^2}. \tag{4.2.20}$$

Using these relations, the mean escape probability $\langle\beta_e\rangle$ can be written by using the effective optical depth in z direction,

$$\begin{aligned}\langle\beta_e\rangle &= \frac{1}{2}\int_{-1}^{+1}\frac{1-\exp\left[-\tau_S'/\mu^2\right]}{\tau_S'}\mu^2\,d\mu \\ &\cong \frac{1}{3\tau_S'}\left(1-e^{-3\tau_S'}\right).\end{aligned} \tag{4.2.21}$$

In both cases of spherical symmetric and plane-parallel, the escape probability by motion approaches to unity when optically thin ($\tau_S \ll 1$) and is inversely proportional to τ_S when optically thick ($\tau_S \gg 1$).

4.2.2 Escape probability and formation of emission lines

Sobolev's method

Sobolev (1947) has considered the formation of emission lines in terms of escape probability by motion. The outline is as follows.

Suppose an atom having energy levels n, n' ($n > n'$) as in Figure 4.3. Let $\nu_{n'n}$ be the frequency of n–n' transition, and $N_{n'}$, N_n the level populations in n', n levels, respectively. If $\nu_{n'n}$ radiation has an escape probability $\beta_{n'n}$, the volume emissivity $E_{n'n}$ can be written as

$$E_{n'n} = \frac{h\nu_{n'n}}{4\pi} N_n A_{nn'} \beta_{n'n}. \tag{4.2.22}$$

Hence the relative intensity of $n' \to n$ and $n' \to k (k > n')$ radiations is

$$\frac{E_{n'n}}{E_{n'k}} = \frac{N_n A_{nn'} \beta_{n'n} \nu_{n'n}}{N_k A_{kn'} \beta_{n'k} \nu_{n'k}}. \tag{4.2.23}$$

Radiation intensity $I_{n'n}$ emitted from the whole envelope can be given as a volume integral:

$$I_{n'n} = \frac{h\nu_{n'n}}{4\pi} \int N_n A_{nn'} \beta_{n'n} \, dV. \tag{4.2.24}$$

In these equations the level populations N_n, N_k are given as the solutions of the equations of statistical equilibrium, in which the net emission for n'–n transition can be expressed in terms of escape probability as

$$N_n A_{nn'} - N_{n'} J_{n'n} B_{n'n} = N_n A_{nn'} \beta_{n'n}. \tag{4.2.25}$$

With this approximation, we can avoid solving the equation of radiative transfer for $J_{n'n}$ and derive the relative intensities of emission lines directly by solving the equations of statistical equilibrium.

Sobolev (1947) originally derived the escape probability in the case of optically thick envelopes in Equation (4.2.10), and so the probability $\beta_{n'n}$ is inversely proportional to $\tau_{n'n}$. In this case the probability $\beta_{n'n}$ can be written by making use of the Einstein coefficients for κ' as

$$\beta_{n'n} = \frac{\nu_{n'n}}{N_{n'} B_{n'n} c} \frac{dV}{ds}. \tag{4.2.26}$$

This equation shows that factor $\beta_{n'n} N_{n'} B_{n'n} / \nu_{n'n}$ depends only on velocity gradient dV/ds so that $N_1 \beta_{12}$ for the resonance line can be taken as a basic parameter in the equations of statistical equilibrium. Sobolev assumed a homogenous medium where N_e is constant and N_1 is proportional to $1/W$, as infered from ionization Equation (4.1.18), and adopted parameter $x = \beta_{12}/\mathrm{W}$ to derive the relative intensities of emission lines. In this way he applied his method to WR stars, P Cygni stars, and novae in some reasonable agreement with observations.

Improvement of Sobolev's method

While Sobolev (1947) considered an infinite medium in uniform acceleration, Castor (1970) has developed the escape probability method for treating the transfer of line radiation in a stellar envelope, which is in rapid radial expansion and exposed to the radiation from the stellar core. Castor and van Blerkom (1970) applied it to the calculations of the emission lines of WR stars.

Castor (1970) rederived the escape probability by motion, starting from the formal solutions of the transfer equations and simplifying them under the conditions of a spherically symmetric and accelerating velocity field. The absorption line coefficients are sharply peaked at some point in the envelope and at a particular wavelength due to Doppler shift by expanding motion. He thus derived the mean intensity $J_{nn'}$ for $n'n$ radiation ($n > n'$) in the form

$$J_{nn'} = (1 - \beta_{nn'})\, S_{nn'} + \beta_c\, I_c, \tag{4.2.27}$$

where $S_{nn''}$, $\beta_{nn'}$ denote the source function and the escape probability for *nn'* radiation, respectively, I_c is the surface intensity for continuous radiation, and β_c is the probability of a photon escaping from the local region and striking the central star that is assumed to radiate continuously and practically constant over the line. Here the escape probabilities are given by

$$\begin{aligned} \beta_{nn'} &= \frac{1 - \exp(-\tau_{nn'})}{\tau_{nn'}} \\ \beta_c &= W\beta_{nn'}. \end{aligned} \tag{4.2.28}$$

In Equation (4.2.27), the first term of the right-hand side shows the contribution of radiation emitted and reabsorbed near the point in question with a probability $(1 - \beta_{nn'})$. The second term denotes the contribution from the diluted stellar radiation. Inserting the mean intensities from Equation (4.2.27) into the equations of statistical equilibrium, Castor obtained the relative populations as functions of their position inside the envelope and derived the relative intensities and profiles of emission lines in some simple cases. This method has been widely applied to the analysis of WR stars, P Cygni stars, and other early-type emission-line stars.

As an example of application, we show the analysis of HeII emission lines in WR stars carried out by Castor and van Blerkom (1970). The source function $S_{n'n}$ is written as

$$S_{n'n} = \frac{N_n A_{nn'}}{N_{n'} B_{n'n} - N_n B_{nn'}}, \tag{4.2.29}$$

and the net radiative rate between two levels n, n' ($n > n'$) takes the form

$$\begin{aligned} N_n A_{nn'} + N_n B_{nn'} J_{n'n} - N_{n'} B_{n'n} J_{n'n} &= N_n A_{nn'} \beta_{n'n} \\ &+ (N_n B_{nn'} - N_{n'} B_{n'n})\, \beta_c I_c. \end{aligned} \tag{4.2.30}$$

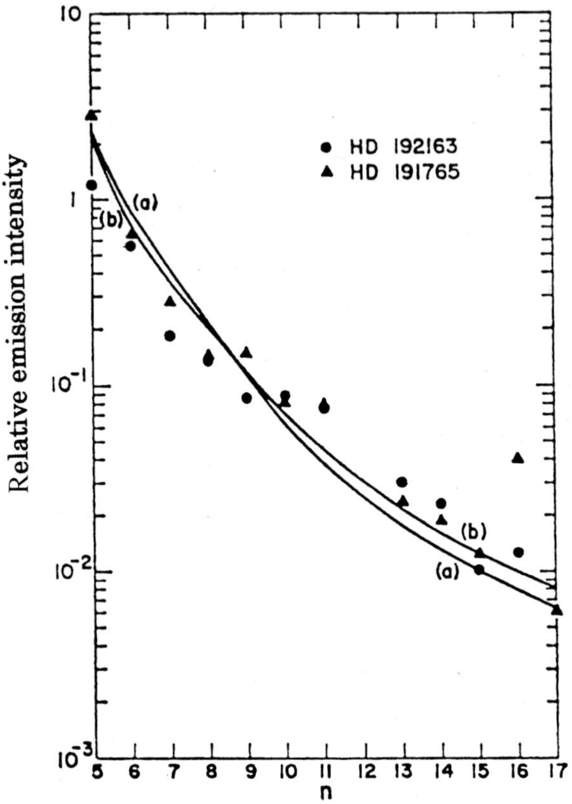

Figure 4.14: Relative emission intensities of HeII Pickering lines. Theoretical curves are shown for two models : (a) $T_e = 1 \times 10^5$ K and $N(\text{He}) = 2.5 \times 10^{11}$ cm^{-3}, and (b) $T_e = 2 \times 10^5$ K and $N(\text{He}) = 5 \times 10^{11}$ cm^{-3}. Observed data are for two stars of HD192163 (dots) and HD 191765 (triangles). (From Castor and van Blerkom 1970)

By using these relations, the equations of statistical equilibrium can be written as the algebraic equations for the level populations N_1, N_2, ..., $N_n, \ldots$, with escape probabilities $\beta_{nn'}$ which are obtainable through some suitable expanding envelope model. Castor and van Blerkom thus derived the relative intensities of emission lines in the Pickering series of ionized helium ($n = 5, 6, 7, \ldots, n' = 4$). The results of their theoretical intensities are shown in Figure 4.14, where the abscissa is the n number and the ordinates are the relative intensities, and compared with observed values. Theoretical curves are given for the two models with different electron temperatures and the number density of He ions. One may see a good agreememt between theory and observations in a wide range of n values. Thereafter, Castor and Lamers (1979) calculated the theoretical emission-line profiles for P Cygni-type stars based on this method and prepared an atlas of P Cyg line profiles.

Since 1980s, many attempts have been made to improve and extend the method of determining escape probability by motion, such as, including the extension to a wide velocity range from static to large gradient (Hummer and Rybicki 1982), the development of the Sobolev Exact Integration (SEI) Method (Lamers et al. 1987) and the development of Sobolev P method which is the escape probability method for polarized radiation (Jeffery 1990).

4.2.3 Method of velocity zones

A method to divide a moving envelope into zones of equal line-of-sight velocities and to solve the non-LTE problem for each of such zones was developed by Marlborough (1969) and Kogure (1969). Kogure applied the non-LTE solution in static envelope to each of the velocity zones divided by velocity interval of twice Doppler width. With this method, emission-line profiles (Kogure 1969) and formation of shell absorption lines in Be stars (Kogure et al. 1978) have been analyzed.

We herewith consider the formation of spectral lines in a flattened rotating envelope (disklike or ringlike) of Be stars with the aid of the method of velocity zones (Kogure et al. 1978).

Suppose a rotating disklike envelope having velocity zones divided by equal line-of-sight velocities v_k $(k = 0, \pm 1, \pm 2, \ldots)$ as shown in Figure 4.15. The velocitiy v_k is given by

$$v_k = 2kv_{\mathrm{D}}, \tag{4.2.31}$$

where v_{D} denotes the thermal Doppler velocity, and the factor 2 comes from the effective width of a line formed in a single zone. The kth velocity zone can absorb or emit the $2n$ radiation of Balmer series at the wavelength

$$\lambda_{2n}(k) = \lambda_{2n}(0)\left(1 + \frac{v_k}{c}\right), \tag{4.2.32}$$

where $\lambda_{2n}(0)$ stands for the wavelength at the respective line center.

For the kth velocity zone the screening factor $\beta(k)$ is defined as the fractional area of the stellar disk that is screened by the kth velocity zone lying in front of it. Furthermore, the following quantities are defined:

- $I_{2n}(k)$—the emergent intensity of the $2n$ radiation at the surface of the kth velocity zone directed toward the observer.
- $I^*_{2n}(k)$—the stellar emergent intensity at the stellar surface directed to the observer (Photospheric absorption line appears at the respective wavelength).
- $\pi F^*_{c,2n}$—the stellar emergent flux in the continuum adjacent to the $2n$ radiation.

The spectral width of these three quantities is assumed to be $2\Delta_{\mathrm{D}}$ in Å, where Δ_{D} corresponds to the mean thermal Doppler velocity v_{D} in Euqtion (4.2.31). In addition, we define

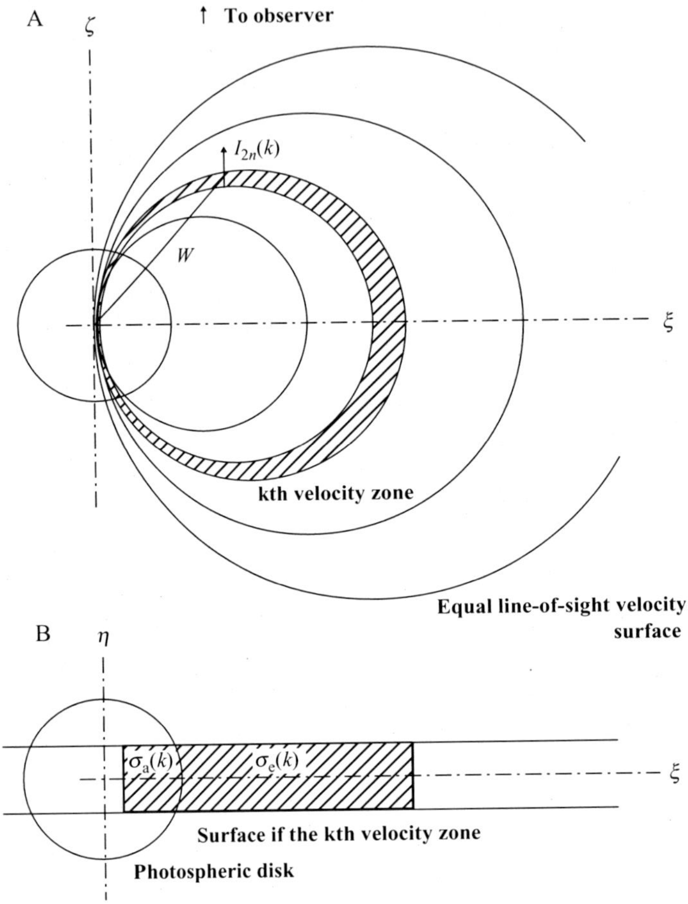

Figure 4.15: Rotationally flattened disklike envelope and velocity zones. (A) View from the rotation axis (ξ–ζ plane), and (B) View from the equatorial plane (ξ–η plane), ζ axis is toward the observer. Shaded areas denote the kth velocity zone.

A^*—the projected area of the stellar disk seen from the observer,
$\sigma\,(k)$—the projected area of the kth velocity zone,
$\sigma_a\,(k)$—the part of $\sigma(k)$ that screens the stellar disk,
$\sigma_e\,(k)$—the part of $\sigma(k)$ that extends outside the stellar disk.

For σs we have

$$\sigma\,(k) = \sigma_a\,(k) + \sigma_e\,(k)\,, \tag{4.2.33}$$

and

$$\sigma_a\,(k) = \beta\,(k)\,A^*. \tag{4.2.34}$$

With the aid of these symbols and assumptions for simplification, we can write the residual intensity $r_{2n}(k)$ at the wavelength $\lambda_{2n}(k)$ of the $2n$ radiation in the form

$$r_{2n}(k) = \frac{1}{\pi F^*_{c,2n} A^*} \left[\int_{\sigma_a(k)} \{ I_{2n}(k) + I^*_{2n}(k) \exp[-\tau(Hn,k)]\} \, d\sigma \right.$$
$$\left. + \int_{A^* - \sigma_a(k)} I^*_{2n}(k) \, d\sigma + \int_{\sigma_e(k)} I_{2n}(k) \, d\sigma \right], \tag{4.2.35}$$

where $\tau(\mathrm{Hn}, k)$ denotes the optical thickness of the kth velocity zone in the $2n$ radiation at the projected surface element $d\sigma$. Each term in Equation (4.2.35) has the following implications:

First term—the integral denotes the intensity of radiation emerging through the outer surface of the kth velocity zone lying in front of the stellar disk and composed of the diffused part $I_{2n}(k)$ and the direct part from the stellar surface $I^*_{2n}(k) \exp[-\tau(\mathrm{Hn},k)]$. The integration covers the area given by Equation (4.2.34). The intensity $I_{2n}(k)$ depends on the dilution factor at the surface element $d\sigma$ of the kth zone.

Second term—the intensity of $\lambda_{2n}(k)$ radiation emitted from the part of the stellar disk that is not screened by the kth zone.

Third term—the radiation intensity at the same wavelength emitted from the surface of the kth velocity zone outside of the stellar disk. This part of the radiation contributes mainly to the formation of the emission component.

For the purpose of simplifying Equation (4.2.35), we first define the average emergent flux of the kth zone by

$$\pi F_{2n}(k) = \frac{1}{A^*} \int_{A^*} I_{2n}(k) \, d\sigma, \tag{4.2.36}$$

and

$$\pi F^*_{2n}(k) = \frac{1}{A^*} \int_{A^*} I^*_{2n}(k) \, d\sigma = I^*_{2n}(k). \tag{4.2.37}$$

The last equality of the Equation (4.2.37) holds when the limb darkening, gravity darkening, and Doppler shift due to stellar rotation are ignored. With these emergent fluxes, one can rewrite Equation (4.2.35) as

$$r_{2n}(k) = \frac{F_{2n}(k)}{F^*_{c,2n}} \left[\beta(k) + \frac{\sigma_e}{A^*} \right]$$
$$+ r^*_{2n}(k) \{ \beta(k) \exp[-\tau(Hn,k)] + 1 - \beta(k) \}. \tag{4.2.38}$$

It is noticed that the kth zone has an uniform velocity field, and one can use the solution of a static envelope given in Section 4.1.4 for deriving the emergent fluxes from this zone.

Now, we consider some extreme cases for stars having suffiently developed envelopes. For the lower members of the Balmer series, Hα, Hβ, Hγ, ... in classical Be stars, the main part of the line profiles can be supposed as

$$\tau\left(Hn,k\right) \gg 1, \qquad F_{2n}\left(k\right) \geq F^*_{c,2n}. \tag{4.2.39}$$

In this case, Equation (4.2.38) can be written as

$$r_{2n}\left(k\right) \;\approx\; \frac{\sigma\left(k\right)}{A^*}\,\frac{F_{2n}\left(k\right)}{F^*_{c.2n}} + r^*_{2n}\left(k\right)\left\{1 - \beta\left(k\right)\right\}. \tag{4.2.40}$$

This expresses the main part of the profile, i.e., strong emission appears when $F_{2n}(k)$ is sufficiently larger than $F^*_{c,2n}$.

On the other side, one can put, for higher members,

$$\tau\left(Hn,k\right) \ll 1, \quad F_{2n}\left(k\right) \ll F^*_{c,2n}, \tag{4.2.41}$$

and then Equation (4.2.38) can be written as

$$r_{2n}\left(k\right) \;\approx\; r^*_{2n}\left(k\right)\left\{\beta\left(k\right)\exp\left[-\tau\left(Hn,k\right)\right] + 1 - \beta\left(k\right)\right\}. \tag{4.2.42}$$

This equation shows the formation of shell-absorption lines depending on the parameters τ(Hn,k) and $\beta(k)$. This point will be further discussed in Section 5.3.4. In this way, the method of velocity zones provides the uniform interpretation on the formation of emission lines and shell-absorption lines.

4.2.4 Other methods

Comoving frame method

Co-moving Frame method (CMF) has been developed by Mihalas et al. (1975) as one of the numerical methods to derive the relative intensities and line profiles of emission lines in moving envelopes (see Mihalas 1978, Chapter 14).

For simplicity we consider a monotonously expanding envelope with constant acceleration as shown in Figure 4.16. We also adopt the coordinate systems $(r,\, \mu = \cos\theta)$ and $(p,\, z)$, where, z axis is directed toward an observer and the p axis is rectangular to the z axis. The expanding velocity $V(r)$ is supposed to increase with radius r. Let R and $V_{\max}$ be the radius of the outer surface and the highest expansion velocity, respectively. The radiation intensity at a point $P(p,r)$ is expressed by $I(\nu,\mu,r)$, where frequency ν is measured by the system moving with point P. In this coordinate system, the CMF method is carried out through the following three steps (Rons et al. 1992):

First step—the source function $S(r,\, \nu_{\rm CMF})$ is calculated as a function of frequency and depth. In the CMF method, this is done by solving the equation of radiative transfer in a frame which instantaneously moves with the moving material.

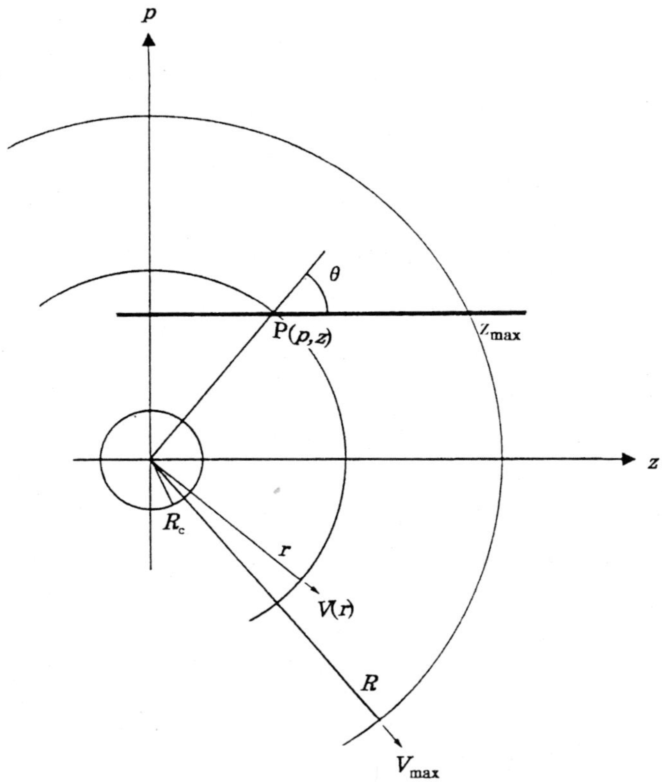

Figure 4.16: Coordinate system in the CMF method for a spherically symmetric moving envelope.

Second step—the emergent intensity at the outer surface is formally derived by the following integration parameterized by p

$$I(p, \nu_{\rm obs}) = \int S(r, \nu_{\rm CMF})\, e^{-\tau} d\tau + I^*(p, \nu_{\rm obs})\, e^{-\tau}, \qquad (4.2.43)$$

where $\nu_{\rm obs}$ is the frequency seen by the observer. The second term of the right-hand side contains the intensity emitted at the stellar surface and is added only when the value of p is smaller than the stellar radius R_c, i.e., $p < R_c$. The above integration is usually calculated numerically by converting it into the difference equations.

Third step—a spectral line profile is obtained after an integration of the emergent intensities over the disk surface of the stellar envelope (for all value of p).

Notice that, when p is fixed, the absorption line coefficient varies along the distance z, and takes the maximum value at some point. Before and after this point, the envelope is almost transparent for the line. The point of maximum absorption depends on the values of position p and the law of out-flowing motion. Then the model is constructed semi-empirically with prescribed velocity and opacity laws.

This method has been applied mainly to the analysis of WR stars (Mihalas and Kunasz 1978, Papkalla 1995), and a large variety of emission-line profiles have been derived by selecting suitable parameters.

Approximate lamda iteration method

An iterative method based on the lamda operation has been developed to solve the radiation field of moving envelopes. We begin with introducing the Λ-operator.

Suppose a plane-parallel atmosphere and write the equation of radiative transfer, in the form of Equation (2.5.9), as

$$\mu \frac{dI_\lambda}{d\tau_\lambda} = I_\lambda - S_\lambda, \tag{4.2.44}$$

where $\mu = \cos\theta$ and the source function S_λ is assumed to be isotropic.

Let $I_\lambda^+(\tau)$, $I_\lambda^-(\tau)$ be the outward and inward radiation intensity at the optical depth τ from the atmospheric surface. Then the formal solution of transfer equation can be written as

$$\begin{aligned} I_\lambda^+(\tau) &= \int_\tau^\infty S_\lambda\, e^{-(t-\tau)/\mu} \frac{dt}{\mu} \quad (0 < \mu < 1) \\ I_\lambda^-(\tau) &= -\int_0^\tau S_\lambda\, e^{-(t-\tau)/\mu} \frac{dt}{\mu} \quad (-1 < \mu < 0)\,. \end{aligned} \tag{4.2.45}$$

Inserting these expressions into the definition of mean radiation intensity in Equation (2.5.5), we have

$$\begin{aligned} J(\tau) &= \frac{1}{4\pi} \int I_\lambda\, d\omega = \frac{1}{4\pi} \left[\int_0^1 I_\lambda^+(\tau)\, d\omega + \int_{-1}^0 I_\lambda^-(\tau)\, d\omega \right] \\ &= \frac{1}{2} \left[\int_\tau^\infty S_\lambda(t)\, E(t-\tau)\, dt + \int_0^\tau S_\lambda(t)\, E(\tau - t)\, dt \right] \\ &= \frac{1}{2} \int_0^\infty S_\lambda(t)\, E(|t-\tau|)\, dt, \end{aligned} \tag{4.2.46}$$

where the E function is defined as

$$E(|t-z|) = \int e^{-(t-z)x} \frac{dx}{x}. \tag{4.2.47}$$

Now we define the Λ operator as

$$\Lambda(\cdot\cdot) = \frac{1}{2} \int_0^\infty (\cdot\cdot)\, E(t)\, dt, \tag{4.2.48}$$

where $(\cdot\cdot)$ denotes the operande, for which we insert the source function $S(t)$, then Equation (4.2.46) can be written as

$$J_\lambda(\tau) = \Lambda(\mathrm{S}_\lambda). \tag{4.2.49}$$

If the source funtion can be replaced by the mean intensity, as in case of the total radiation, $\mathrm{J}(\tau) = \int \mathrm{J}_\lambda(\tau)\, d\lambda$, in radiative equilibrium, or, in case of line radiation in a pure scattering atmosphere, we have

$$\mathrm{J}(\tau) = \Lambda(\mathrm{J}) \quad \text{for the total radiation,}$$

or,

$$\mathrm{J}_\lambda(\tau_\lambda) = \Lambda(\mathrm{J}_\lambda) \quad \text{for the line radiation,}$$

respectively. In these cases, the Λ operator directly provides a method of successive approximation by repeating the procedure of

$$J_{\mathrm{new}}(\tau) = \Lambda(J_{\mathrm{old}}) \tag{4.2.50}$$

toward a desired accuracy.

This is the principle of the so-called Hopf's Λ-operation method (Woolley and Stibbs 1953). However, this method is not practical for the line radiation, since there is some constraint for the selection of the initial function and also the convergency of successive approximation is very poor. Poor convergency is caused by treating the whole line profile with a single successive approximation, in spite of great difference in absorption coefficient in the line center and its wings.

This convergence difficulty in the Λ-operation method has been improved by Werner and Husfeld (1985) and Hamann (1985) by introducing the method of approximate lambda iteration (ALI method). This retains the simple structure of lambda iteration, but avoids the inherent convergence problems. This scheme is derived by introducing the approximate lambda operator Λ^* along with the ordinay operator Λ through the identity

$$J(\tau) = (\Lambda - \Lambda^*)(S) + \Lambda^*(S). \tag{4.2.51}$$

The iteration now may proceed as

$$J_{\mathrm{new}} - \Lambda^*(S_{\mathrm{new}}) = (\Lambda - \Lambda^*)(S_{\mathrm{old}}), \tag{4.2.52}$$

where the operator Λ^* is defined by adopting the idea of core saturation of Rybicki (1972) as

$$\Lambda^*(S) = \begin{cases} S & \text{in optically thick line cores} \\ 0 & \text{else where} \end{cases}. \tag{4.2.53}$$

By this operation the line cores remain unaffected as $J_{\mathrm{new}} = J_{\mathrm{old}}$, while the wing parts transform to the new form of J_{new}. This implies that the radiation field is locally determined for the optically thick radiation, whereas the optically thin radiation largely suffers the effects of radiative transfer inside the envelope. In moving envelopes, the shifts of absorption coefficient profiles and the boundary between line cores and wing parts are functions of place and velocity distribution in the envelope so that numerical iterations are conducted under some suitable envelope models. The ALI method has thus improved the convergency problem and been widely applied to the spectral analysis of WR stars (Hamann and Schmutz 1987) and other emission line stars, particularly by Hamann's group in Kiel University in 1985 through 1995. Useful references are given in Hamann et al. (1994).

Method of Doppler tomography

In close binary systems, emission line profiles can be obtained as a time sequence along the orbital motions. Marsh and Horne (1988) introduced the methods required to analyze such spectroscopic data applying the Doppler tomograph for emission-line profiles. This form of accretion-disk analysis was initially applied to CVs and then Algol-type systems and other types of binaries.

The principle of this method can be summarized in the case of CVs as follows (see Marsh 2001).

First step is to express the emission-line profiles in the velocity space. For simplicity we consider two-dimensional motions on the orbital plane viewed edge-on. Orbital phase $\phi = 0$ is defined to be the moment when the secondary (mass doner) is closest to observer, and the velocity axis V_x is directed to the secondary from the primary (mass gainer) at $\phi = 0$, V_y is pointed in the direction of motion of the doner. In this velocity space, the radial velocity $V_r(\phi)$ can be written as

$$V_r\,(\phi) = \gamma - V_x\,\cos 2\pi\phi\, + V_y\,\sin 2\pi\phi, \tag{4.2.54}$$

where γ is the systemic velocity of the star. With these definitions, the flux observed from the system that comes from the velocity element bounded by $(V_x, V_x + dV_x)$ and $(V_y, V_y + dV_y)$ is given by $I(V_x, V_y)\,dV_x\,dV_y$. This is called the velocity-space image.

The line profile $f(V, \phi)$ at phase φ as a function of velocity V can be expressed as

$$f\,(V, \phi) = \int_{-\infty}^{\infty}\int_{-\infty}^{\infty} I\,(V_x, V_y)\,g\,(V - V_r)\,dV_x\,dV_y, \tag{4.2.55}$$

where g is a function (of velocity) representing the line profile from any point in the image.

Second step is the inversion procedure. A series of line profiles at different orbital phases is therefore a set of projections of the image to the observer at different angles. The inversion of projections to reconstruct the image $I(V_x, V_y)$ is known as tomography, famous application is the medical X-ray imaging (CAT—computerized axial tomography).

Third step is the construction of the image $I(V_x, V_y)$ based on observed emission-line profiles. On the $(V_x - V_y)$ plane, we may have a map of intensity distribution reflecting the velocity structure of the disk. By analyzing these maps, disk structures such as the existence of bright spots, gas streams, and spiral shocks have been derived in many stars.

Doppler tomography is a useful tool of spectral analysis for close binary systems, but it is to be noticed that this method is applicable when the disk is sufficiently optically thin for emission lines in question. When the lines become optically thick, radiative transfer should be taken into consideration.

4.3 Formation of forbidden lines

4.3.1 Nebular-type forbidden lines

Emission lines are generally divided into two types: permitted and forbidden. Permitted lines are formed when electronic transitions obey the selection rules stated in Section 2.2.2. In two- or multielectron atoms, there often exist the metastable levels which are quasistable because the downward transitions are prohibited by the selection rules, as seen in CaII 3d level (Figure 2.6) or in HeI 2s ^{3}S1 level (Figure 2.7). However, the downward transitions from metastable levels are not strictly forbidden; there still exist some transition probabilities.

In classical electrodynamics, the permitted lines from ordinary energy levels correspond to the dipole radiations, whereas transitions from metastable levels correspond the quadrupole or magnetic-dipole radiations. Quantum mechanics give the transition probabilities in every transition. As compared to the dipole radiation, transition probabilities of quadrupole and magnetic dipole radiation are many orders of magnitude smaller than the former. They are almost unobservable in laboratories and hence called the forbidden lines.

In astrophysics, forbidden lines are observable in many types of interstellar medium and stellar envelopes. The most ubiquitous forbidden lines are those of atoms or ions which possess two quasistationary energy levels several electron volts higher than their ground level. This type of forbidden lines was first observed in planetary nebulae and identified by Bowen (1935) so that they are called nebular-type forbidden lines. These atoms or ions are characterized by two common properties: they are multielectron atoms (ions) and the configuration of valence electrons belongs to one of the three types shown in Figure 4.17. Table 4.4 gives the configuration of all electrons for some typical atoms (ions). As seen from Figure 4.17 and Table 4.4, the nebular-type forbidden lines are classified into three types of $2p^2$, $2p^3$, $2p^4$ (for the latter two, $3p^3$ and $3p^4$ for SI and SII) according to the configuration of outer valence electron(s).

Forbidden lines are formed by the transitions between ground and two metastable levels, and hence there are three different types as designated in Figure 4.17:

- N nebular line (transition between ground and first excited levels).
- A auroral line (transition between first and second excited levels).
- TA trans-auroral line (transition between ground and second excited levels).

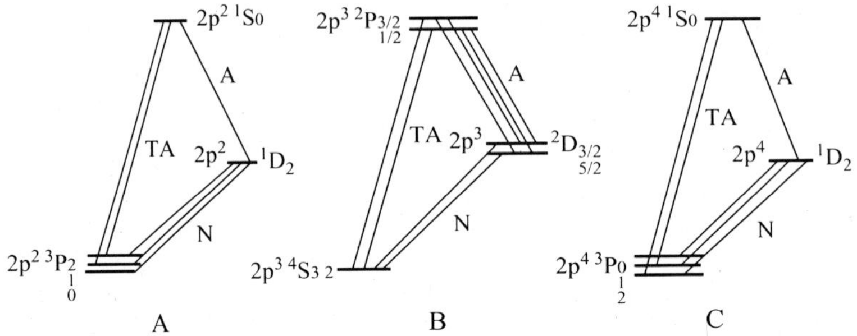

Figure 4.17: Energy-level diagram for three electron configurations. Transitions N, A, and TA denote nebular, auroral, and transauroral transitions, respectively. (A) $2p^2$ configuration [NII], [OIII], (B) $2p^3$ configuration [OII], [SII], and (C) $2p^4$ configuration [OI], [NeIII].

In spectroscopy, forbidden lines are usually designated by the names of atoms or ions with brackets, such as [OIII]λ 4959 Å and λ 5007 Å for the doublet nebular lines of twice ionized oxygen. Among three types of nebular-type forbidden lines, nebular lines usually appear in the optical region, whereas auroral lines are in the optical to infrared region, and transauroral lines are in the ultraviolet region.

4.3.2 Formation of forbidden lines and critical electron density

Consider a simplified two-level atom, composed of lower ground or metastable level (level 1) and one higher metastable level (level 2). The transition from level 2 to 1 is the sum of spontaneous emission (with Einstein coefficient A_{21}) and collisional de-excitation (coefficient α_{21}), and these downward transitions are supposed to be in statistical equilibrium with the upward collisional excitation (coefficeint α_{12}). Hence we have

$$N_2 (A_{21} + \alpha_{21} N_e) = N_1 N_e \alpha_{12}, \tag{4.3.1}$$

where N_1, N_2 are level populations and N_e the electron density.

Table 4.4: Electron configuration for typical atoms (ions)

Atom/Ion	Configuration	Atom/Ion	Configuration
OI	$\underline{1s^2}\ 2s^2\ 2p^4$	NI	$\underline{1s^2}\ 2s^2\ 2p^3$
OII	$\underline{1s^2}\ 2s^2\ 2p^3$	NII	$\underline{1s^2}\ 2s^2\ 2p^2$
OIII	$\underline{1s^2}\ 2s^2\ 2p^2$	SI	$\underline{1s^2}\ \underline{2s^2}\ \underline{2p^6}\ 3s^2\ 3p^4$
		SII	$\underline{1s^2}\ \underline{2s^2}\ \underline{2p^6}\ 3s^2\ 3p^3$

Note: Underline denotes closed-shell electrons.

In order to derive the relation between α_{12} and α_{21}, we start with the collisional equilibrium between two levels under the TE. In this case we have

$$N_1 N_e \alpha_{12} = N_2 N_e \alpha_{21}. \tag{4.3.2}$$

The relative population N_2/N_1 can be expressed by Boltzmann's law in Equation (2.3.6) as

$$\frac{N_2}{N_1} = \frac{g_2}{g_1} \exp\left(\frac{-\chi_{12}}{kT}\right), \tag{4.3.3}$$

where χ_{12} denotes the excitation potential of level 2. Combining Equations (4.3.2) and (4.3.3) we get

$$\alpha_{12} = \frac{g_2}{g_1}\alpha_{21} \exp\left(\frac{-\chi_{12}}{kT}\right). \tag{4.3.4}$$

The collisional de-excitation coefficient is given by

$$\alpha_{21} = 8.63 \times 10^{-6} \frac{\Omega\,(1,2)}{g_2 T_e^{1/2}} \quad (\mathrm{cm^3\,s^{-1}}), \tag{4.3.5}$$

where $\Omega(1,2)$ is a parameter called the collision strength.

Now we return to Equation (4.3.1). The photon number ($\mathrm{cm^{-3}\,s^{-1}}$) emitted by the transition $2 \rightarrow 1$ is given by

$$Q_{21} = N_2 A_{21}. \tag{4.3.6}$$

Solving N_2 from Equation (4.3.1) and inserting to Equation (4.3.6), we have

$$Q_{21} = \frac{N_1 N_e A_{21} \alpha_{12}}{A_{21} + \alpha_{21} N_e}. \tag{4.3.7}$$

We now define the critical electron density N_e^c

$$N_e^c = \frac{A_{21}}{\alpha_{21}} \tag{4.3.8}$$

and consider the two extreme cases of $N_e \ll N_e^c$ and $N_e \gg N_e^c$:

(1) $N_e \ll N_e^c$ (case of rarefied gas). Since $A_{21} \gg \alpha_{21} N_e$ in Equation (4.3.7), we have

$$Q_{21} \approx N_1 N_e \alpha_{12}. \tag{4.3.9}$$

In this case the volume emissivity $\varepsilon_{12} = h\nu_{12} Q_{21}$ is determined by the collisional excitation coefficient α_{12}, not depending on the value of A_{21}. All of atoms excited to level 2 transit to lower level by emitting radiation ν_{12}, since collisional de-excitation is ineffective by the small value of electron density. This radiation is just the forbidden line stated above, and the condition of the formation of forbidden lines is that the electron density is smaller than the critical density.

(2) $N_e \gg N_e^c$ (dense gas). Since $A_{21} \ll \alpha_{21} N_e$ in Equation (4.3.7), we have

$$Q_{21} \approx N_1 A_{21} \frac{\alpha_{12}}{\alpha_{21}}. \tag{4.3.10}$$

Combining with Equation (4.3.4) we get

$$Q_{21} = N_1 A_{21} \frac{g_2}{g_1} \exp\left(\frac{-\chi_{12}}{kT}\right) \tag{4.3.11}$$

The atoms excited to level 2 fall to the lower level mostly through collisional de-excitation. Therefore, though the emissivity is proportional to A_{21}, the radiation is not practically observable due to extremely low values of A_{21}.

The critical density defined by Equation (4.3.8) is applicable for any types of nebular (between 1 and 2 levels), transauroral (between 1 and 3) and auroral (between 2 and 3) transitions. Thus the nebular-type forbidden lines can be formed in rarefied gas where the electron density is lower than the respective critical density. Typical forbidden lines and their critical densitis are shown in Table 4.5. In the calculation of critical density, the electron temperature was taken to be 10^4 K, since the collision strength is weakly dependent on temperature in the range of 5000–20,000 K.

Table 4.5: Main forbidden lines and critical electron densities

Wavelength (A)	Atom/Ion	Type of transition	Transition probability A_{21} (s^{-1})	Collision strength Ω (1, 2)	Critical density N_e^c (cm^{-3})
3726.16	[OII]	N	1.64×10^{-4}	0.534	1.42×10^4
3728.9	[OII]	N	3.82×10^{-5}	0.801	3.32×10^3
3868.7	[NeIII]	N	1.71×10^{-1}	1.34	7.39×10^6
3967.5	[NeIII]	N	5.42×10^{-2}	1.34	2.34×10^6
4068.6	[SII]	TA	2.25×10^{-1}	1.52	6.86×10^6
4076.4	[SII]	TA	9.06×10^{-2}	0.759	5.53×10^6
4363.2	[OIII]	A	1.78	0.617	3.15×10^7
4958.9	[OIII]	N	6.74×10^{-3}	2.17	1.70×10^5
5006.8	[OIII]	N	1.96×10^{-2}	2.17	5.22×10^5
6300.2	[OI]	N	6.34×10^{-3}	0.266	1.38×10^6
6363.8	[OI]	N	2.11×10^{-3}	0.266	4.60×10^5
6548.1	[NII]	N	1.01×10^{-3}	2.68	2.18×10^4
6583.4	[NII]	N	2.99×10^{-3}	2.68	6.45×10^3
6716.4	[SII]	N	2.60×10^{-4}	4.19	4.31×10^4
6730.8	[SII]	N	8.82×10^{-4}	2.79	1.47×10^4
7318.6	[OII]	A	6.15×10^{-2}	0.295	4.83×10^6
7319.9	[OII]	A	1.17×10^{-1}	0.730	7.42×10^6
7329.9	[OII]	A	1.02×10^{-1}	0.275	8.58×10^6
7330.7	[OII]	A	6.14×10^{-2}	0.408	6.97×10^6

Type of transition: N (nebular), A (auroral), TA (transauroral). The values of A_{21} and $\Omega(1,2)$ are taken from Mendoza (1983), where suffix 1 = lower level and 2 = upper level.

Table 4.5 shows, for example, that the forbidden lines of [OIII] can be formed in nebular gas of electron density lower than around 10^4 cm^{-3}. Transitions with lower Einstein coefficient or higher collisional transition rate yield lower critical elecron density. Generally speaking, auroral lines have higher critical densities and the transauroral lines are intermediate. In case of stellar envelopes, electron density is ordinarily higher than around 10^8 cm^{-3} so that forbidden lines are not usuarlly observed in stellar emission-line spectra. In some types of stars, forbidden lines can be formed in the rarefied layer of the outer parts of the envelope.

Relative intensities of forbidden lines provide an useful tool for the spectroscopic diagnostics to estimate the physical parameters (T_e, N_e) of nebular gas. For example, the line ratio of [SII]λ6717 Å/λ6730 Å is not sensitive to the electron temperature, but it is sensitive to the electron density so that this ratio serves as a good probe of electron density. Similarly the line ratio of λ4363 Å/λ5007 Å of [OIII], i.e., auroral to nebular line ratio, is insensitive to the variation of N_e and provides a good probe of T_e. If we can select a suitable combination of forbidden lines or forbidden to permitted line ratios, we can estimate the physical state of the nebular gas with sufficient precision. Among the emission-line stars, B[e] stars, symbiotic stars, and T Tauri stars often show the presence of forbidden lines. The behaviors of forbidden lines in these stars will be shown in the respective chapters in this book.

4.3.3 Semiforbidden lines (intersystem lines)

In the ultraviolet spectral region, some stars exhibit the semiforbidden lines (or intersystem lines) in emission such as OIV] λ1661, 1666 Å, NIV]λ1483, 1488 Å (see Section 6.7.2). These lines are emitted as dipole radiations, but by transitions between different multiplicity systems. The transition probability A_{21} is of the order of $10^2\,\mathrm{s}^{-1}$, intermediate between permitted lines ($A_{21} \sim 10^5$–10^7 s^{-1}) and forbidden lines ($A_{21} \sim 10^{-2}$ s^{-1}). These lines are thus called semiforbidden lines or intersystem lines and are designated by half bracket such as OIV]. Some typical semiforbidden lines are listed in Table 4.6.

4.4 Nonthermal atmospheres

4.4.1 Late-type stars and basal atmospheres

Late-type stars generally possess developed convection zones, beneath the photospheres, where the kinetic energy is produced and then propagated outward to heat up the outer atmospheres. We call these outer layers the nonthermal atmospheres (or envelopes), since they are not in the radiative equilibrium. Nonthermal outer atmospheres are usually composed of two distinct layers: chromospheres and coronae. Chromospheres are characterized by the presence of emission lines in CaII H, K, and Hα, whereas coronae are observable by soft X-rays or by some coronal emission lines of highly ionized ions.

Table 4.6: List of typical semiforbidden lines

Ion	Transition	Excitation potential (eV)	Wavelength (Å)
CII]	2p $^3P^0$–2p^2 ^4P	5.33	2326, 2328
CIII]	2s^2 ^1S–2p $^3P^0$	6.46	1907, 1909
NII]	2p^2 ^3P–2p^3 $^5S^0$	5.79	2140, 2143
NIII]	2p $^2P^0$–2p^2 ^4P	7.08	1750, 1852
NIV]	2s^2 ^1S–2p $^3P^0$	8.31	1488
OIII]	2p^2 ^3P–2p^3 $^5S^0$	7.45	1661, 1666
OIV]	2p $^2P^0$–2p^2 ^4P	8.81	1406, 1410
OV]	2s^2 ^1S–2p $^3P^0$	10.21	1216
SiII]	3p $^2P^0$– 3p^2 ^4P	5.31	2335, 2345
SiIII]	3s^2 ^1S–3p $^3P^0$	6.56	1892, 1884
SiIV]	3p $^2P^0$–3p^2 ^4P	8.96	1393, 1404

Like the Sun, chromospheres and coronae often exhibit violent phenomena such as flares, prominences, and/or radio outbursts. In late-type stars, chromospheric activities are prevailing in more developed forms as compared to the Sun.

The main index of the chromospheric activity is the strength of emission lines in CaII H and K. If the level of activity is defined by the intensity of these lines, there exists large amplitude in the activity levels among stars and in the course of time variations. Schrijver (1987) has measured the emergent flux of CaII H + K emission and $B - V$ colors of stars. The results are shown in Figure 4.18, where one may see that there exists a lower boundary in the distribution of observed flux in both giants and dwarfs. This minimum flux density among stars for each color is called the basal flux, and atmospheres heated only through the basal flux are called the basal atmospheres. The outer atmospheres can be heated magnetically or nonmagnetically. Among some heating mechanisms, the basal atmospheres are supposed to be heated only by nonmagnetic sound waves propagated into the outer atmospheres (Mullan and Cheng 1993, Mauas et al. 1997).

In this section, we consider the structure of chromospheres and formation of emission lines in the basal atmospheres in order to understand the basic processes in the atmospheres of late-type stars. Chromospheric activities and characteristic features of main emission-line stars will be discussed in Chapter 6.

4.4.2 Models of chromosphere

Basic equations

We consider a basal atmosphere simplified as a chemically homogeneous plane-parallel layer. This layer is heated by the propagation of mechanical energy

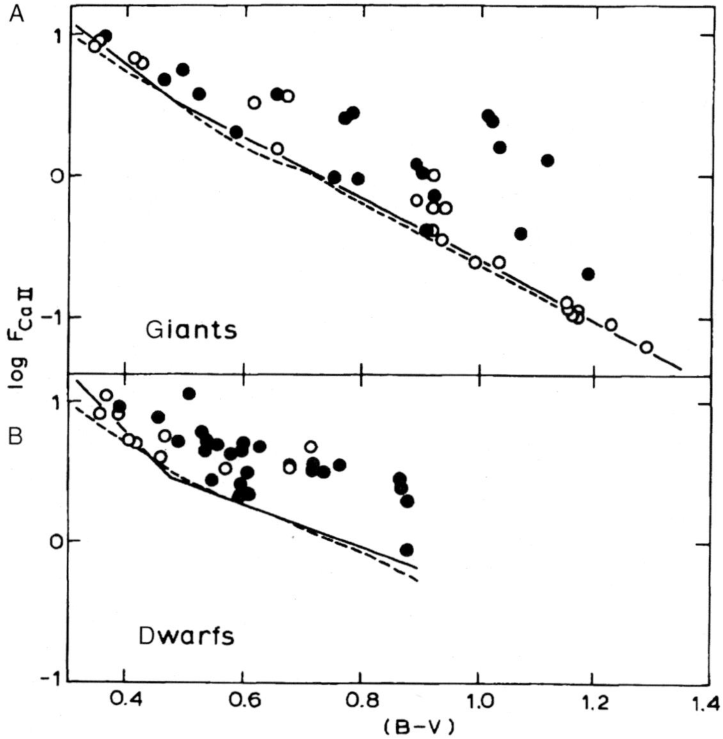

Figure 4.18: Emergent flux of chromospheric CaII H + K versus (B–V) color for (A) giants and (B) dwarfs in F, G, K types. Filled circles denote stars for which soft X-ray flux is known. The line segments represent the theoretical basal flux and the dashed curve represents an observational lower limit. (From Schrijver 1987)

generated in the deep convective layer. Hence this requires solving the equations of radiative transfer and hydrodynamic equations simultaneously. The basic equations in this case can be written as follows (Mullan and Cheng 1993):

(a) Hydrodynamical equations. Let us consider one dimensional flow in the plane parallel layer. Euler's equations are written as the conservation equations of mass, momentum, and energy in the form, respectively,

$$\frac{\partial \rho}{\partial t} + \frac{\partial}{\partial x} (\rho v) = 0, \tag{4.4.1}$$

$$\frac{\partial (\rho v)}{\partial t} + \frac{\partial}{\partial x} \left(\rho v^2 + p\right) = -\rho g, \tag{4.4.2}$$

$$\frac{\partial e}{\partial t} + \frac{\partial}{\partial x} \left[v\left(e + p\right)\right] = Q_{\text{mech}} + Q_{\text{rad}} + Q_{\text{cond}} - \rho v g, \tag{4.4.3}$$

where x is the outward geometrical distance, t the time, ρ the gas density, p the gas pressure, v the gas velocity, and g the gravitational acceleration. The energy density e per unit mass is given as, if we assume an ideal gas,

$$e = \frac{3}{2}nkT + \frac{1}{2}\rho v^2. \tag{4.4.4}$$

In the energy Equation (4.4.3), $Q_{\rm mech}$ denotes the rate of mechanical energy deposition per atom per second, $Q_{\rm rad}$ the rate of radiative cooling, and $Q_{\rm cond}$ the cooling rate by thermal conduction (unit is erg $\rm g^{-1}\,s^{-1}$).

In the basal atmospheres, mechanical energy deposit is due to dissipation of acoustic waves, hence $Q_{\rm mech}$ can be written from Equation (3.1.16) as

$$Q_{\rm mech} = -\frac{dF_A}{dx}. \tag{4.4.5}$$

The cooling rate $Q_{\rm rad}$ is connected with the net cooling function $\Phi_{\rm rad}$ by the relation

$$Q_{\rm rad} = -\frac{1}{\rho}\Phi_{\rm rad}, \tag{4.4.6}$$

where the function $\Phi_{\rm rad}$ takes different form for photosphere, chromosphere, and corona. For example, in optically thick layers like photosphere or lower chromosphere, main cooling source is the continuous radiation and the cooling function is given by

$$\Phi_{\rm rad} = -4\pi\,\kappa\,(J - B) \quad ({\rm erg\ cm^{-3}\ s^{-1}}), \tag{4.4.7}$$

where J, B are the mean intensity and Planck radiation integrated over the whole frequencey range, respectively, and κ the mean absorption coefficient. In the optically thin layers like upper chromospheres, cooling through spectral lines become dominant and contribute to the formation of emission lines. The conductive cooling rate $Q_{\rm cond}$ can be handled using the Spitzer conductivity of a fully ionized gas. This term is important in the transition region and in corona, where the state of full ionization can be assumed.

(b) Equation of radiative transfer. Consider the radiative transfer for an atom simplified to two levels. The equation of radiative transfer for the line at frequency ν can be written as

$$\frac{1}{3}\frac{d^2 J_\nu}{d\tau_\nu^2} = J_\nu - S_\nu, \tag{4.4.8}$$

where J_ν denotes the mean intensity at optical depth τ_ν, S_ν the source function which can be written as

$$S_\nu = \frac{J_\nu + \varepsilon B_\nu + \eta B_\nu^*}{1 + \varepsilon + \eta}. \tag{4.4.9}$$

This source function contains three terms of scattering, electron collision, and cascade processes in the numerator. The first term denotes the scattering process, i.e., absorption of ν radiation followed by reemission of the same ν radiation, and is proportional to the mean intensity J_ν. The second term represents the reemission of ν radiation from collisionally excited atoms with fraction ε, and is proportional to the Planck function $B_\nu(T_k)$ at temperature T_k in the chromosphere. The third term gives the contribution to ν radiation in photoelectrically controlled radiation field, i.e., photoionization followed by recombination and cascade transition to the upper energy level. Reemission of ν radiation takes place at a fraction η. The factor $1/(1 + \varepsilon + \eta)$ denotes the normalization of rates for each of the above three different processes.

According to the formulae of Cram and Giampapa (1987), the ratio $\eta B_\nu^*/(\varepsilon B)$ takes the value higher or lower than unity for the Hα line depending on the physical conditions of the atmospheres. In contrast, the third term is practically zero for the CaII line implying the prevalence of collisional excitation instead of photoionization. The difference in the controlling agency yields a different formation of emission lines as shown in Section 4.4.3.

Models of chromospheres heated by basal radiation flux

In order to derive the precise chromospheric model, one needs to solve simultaneously the above hydrodynamic equations and the radiative transfer equations. However, this direct approach is very complicated, and most of the present models are based on some simplification by focusing on either the hydrodynamic or the radiative equations as follows:

(a) Time-independent model (non-LTE model). This model simplifies the structure of the chromosphere by assuming a temperature structure (Cram and Giampapa 1987, Cuntz et al. 1994), or by assuming an input rate of mechanical energy of sound waves in some power law (Anderson and Athey 1989). Instead of ignoring hydrodynamics, non-LTE effects for line formation are treated in detail.

(b) Time-depending model (mechanical model). In this model, the hydrodynamics of acoustic wave are considered in some detail. Typically, a wave train is injected at the base of chromosphere and the hydrodynamic equations are integrated to determine the dynamical structure of the chromosphere. Several wave behaviors are averaged to construct a height profile. The radiative transfer is instead treated simply (Schmitz and Ulmschneider 1980, Ulmschneider 1991).

(c) Intermediate approach (hybrid model). Mullan and Cheng (1993) adopted an intermediate approach regarding both the hydrodynamics and the treatment of radiative losses.

4.4.3 Formation of emisision lines

Chromospheric emission lines

Main chromospheric lines and their ionization and excitation potentials are shown in Table 4.7 (see Grotrian diagrams in Section 2.2). Three types may be distinguished as

Hα line type	IP and EP_2 are both high (>10 eV)
CaII H, K type	IP is high but EP_2 is low
NI D type	IP and EP_2 are both low

Among these, CaII H and K and Hα are most ubiguitous in late-type stars, but they exhibit different behaviors in the formation of emission lines as seen below. NI D lines are often observed in low-excitation envelopes such as T Tauri stars.

Non-LTE model

We consider the formation of emission lines in the chromospheres heated by the basal flux following the stationary models of Cram and Giampapa (1987) based on the non-LTE treatments. Their numerical iteration proceeds as follows:

(a) A model photosphere is chosen appropriete for the given spectral type and luminosity.
(b) A suitable temperature distribution in the chromosphere is postulated and grafted to the top of the photospheric model.
(c) The equations of statistical equilibrium for multilevel atoms (hydrogen)/ions (calcium) are solved to obtain the ionization equilibrium in the model atmosphere.
(d) Theoretical line profiles are computed and compared with the available observations.
(e) The chromospheric temperature distribution is adjusted systematically to achieve the closest possible agreement between theory and observations.

Table 4.7: Main emission lines and excitation potentials

Line	λ(Å)	IP (eV)	EP_2(eV)	EP_1(eV)
H_α	6562.81	13.60	12.04	10.15
MgII h,k	2802.3, 2795.4	15.03	4.424, 4.435	0
CaII H,K	3968.5, 3933.7	11.86	3.152, 3.124	0
NaI D1,2	5895.9, 5889.9	5.14	2.103, 2.105	0

IP, ionization potential; EP_2 and EP_1, excitation potential for upper and lower energy levels, respectively.

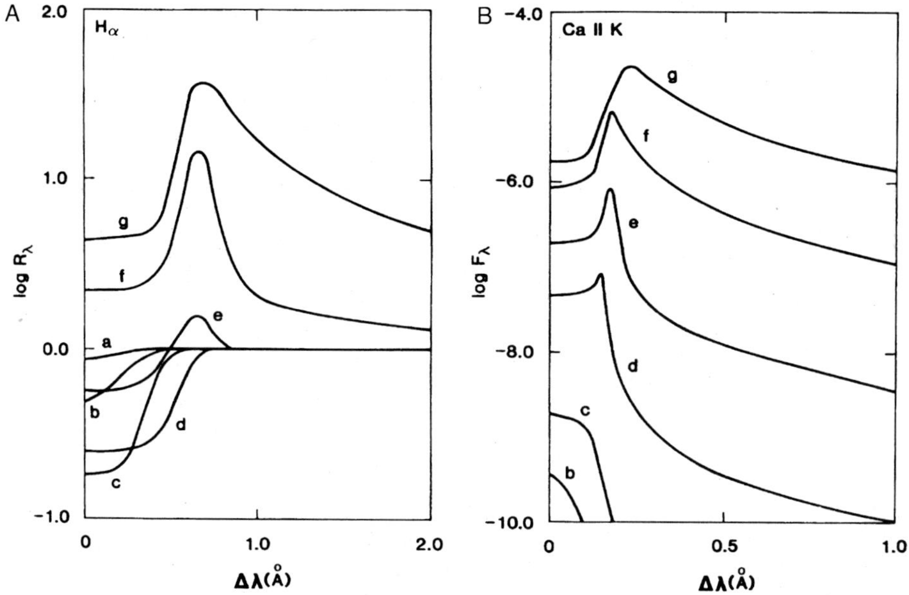

Figure 4.19: Theoretical profiles of the H_α and CaII K lines produced by a sequence of model chromospheres. The H_α profiles represent the logarithm of the relative intensity while the K-line profiles are the logarithm of the surface flux. The curves labeled a–g are corresponding to the chromospheric column densities $N(\mathrm{cm}^{-2})$ as given by log $N = -4.0$ (a), -3.0 (b), -2.5 (c), -1.75 (d), -1.5 (e), -1.25 (f), and -1.0 (g). Note that the profile (a) of CaII K is too faint to depict on the graph. (From Cram and Giampapa 1987)

This method was applied to the chromospheres of red dwarfs, keeping the effective temperature at 3000 K and the chromospheric temperature at 8000 K. They calculated the change of Hα and CaII K line profiles as a function of the chromospheric mass column density. The results are shown in Figures 4.19 (A) and (B).

Consider first the growth of Hα emission with the increase of the column density N. When N is small, giving small optical depth less than unity for the Hα, then the Hα line is in a photoelectrically controlled state and shows a weak absorption feature. As the column density increases, and the Hα optical depth becomes sufficiently large, the Hα line changes to a collisionally controlled state making the absorption progressively weaken and ultimately pass into emission. A self-reversal due to non-LTE scattering appears in the line core as the emission character becomes prominent. In conntrast to the Hα behavior, the CaII K line is always in a collisionally controlled state and its emision intensity monotonously increases with the increasing column density. When the chromospheric optical thickness in the K line exceeds a value of the order of 100, a self-reversal appears in the core of the line profile. In this way Cram and Giampapa showed that the formation of Hα and CaII K lines in emission or absorption can be explained by the amount of the column density in the chromosphere.

In this section, we have considered the formation of emission lines in the inactive basal atmospheres. In most late-type stars with developed chromospheres and coronae, chromospheric activity is high and heating by shock waves and/or magnetohydrodynamic waves may have important roles. Such cases will be considered in Chapter 6.

4.4.4 Chromospheric activities of A-type stars

Late-type stars keep their chromospheric activities through kinetic energies generated in the deep convection layers as considered in Chapter 3. Hence it has long been supposed that the stars showing chromospheric activities are confined to the stars later than F type, which have developed convection layers. Recently, however, some evidence is reported showing the existence of chromospheric activity in A type stars as shown below.

(a) Observations of X-ray emission. Simon et al. (1995) have measured the soft X-ray flux, or flux upper limit, for 74 A-type stars based on the ROSAT archive data.

For late-type A stars ($0.20 < B\text{–}V < 0.35$), nine supposedly single stars are found to coincide with X-ray sources. The X-ray luminosity ranges from levels comparable to the active Sun (log Lx $\sim$ 27.6 erg s^{-1}) to the level for active late-type binaries (log Lx $\sim$ 30.1).

For early-type A stars ($0.0 < B\text{–}V < 0.20$), X-ray sources are identified in 10 stars, among which 5 are confirmed double stars, the rest are ostensibly single. The maximum luminosity is log $Lx = 30.1$ which is 3.5 orders of magnitude brighter than the X-ray upper limits for undetected stars. Although there remains a possibility that the X-ray emissions from single A-type stars originate from unseen binary companion, Simon's group suggests the possibility of their chromospheric origin.

(b) Detection of UV emission lines caused by chromospheric activity. Simon and Landsman (1991), based on the low-dispersion IUE spectra, inspected the chromospheric activities for 176 stars (A6–G0, V–III) by measuring the emission-line strength of CIIλ1335 Å. Stars later than F5 type showed chromospheric activity as expected. Among 13 stars of late-A type, emission lines are detected in 5 stars suggesting the existence of chromospheric activity. Later on, Simon and Landsman (1997) detected emission lines in CIII (λ1176 Å), SiIII (λ1206 Å), and NV(λ1239 Å) for two A7V stars, α Aql and α Cep, in the Hubble Space Telescope low-dispersion spectra. The existence of these lines reveals that the chromospheric activity can be traced up to A7 (effective temperature 8200 K). In Figure 4.20 are shown the UV spectra of these stars observed by Simon and Landsman.

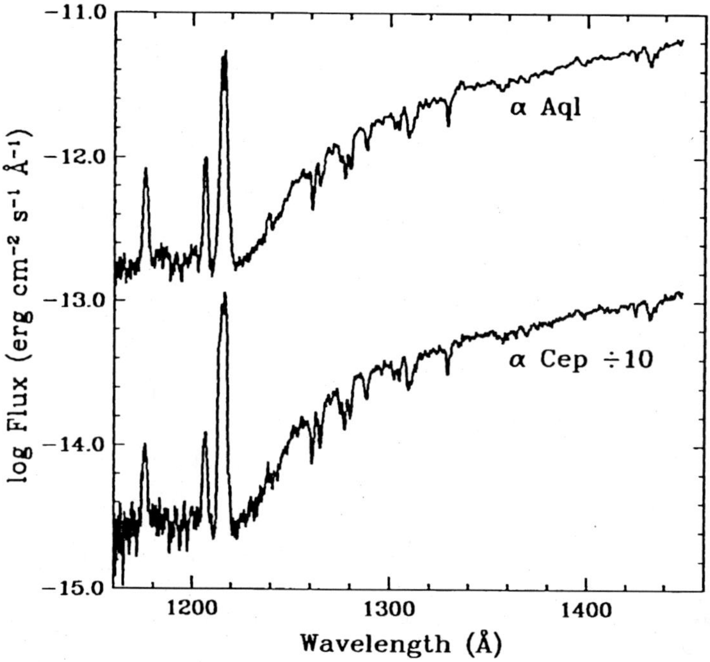

Figure 4.20: Low-resolution HST spectra of α Aql and α Cep, showing emission lines in CIII (λ 1176 Å), SiIII (λ 1206 Å), and NV (λ 1239 Å). (From Simon and Landsman 1997)

(c) Chromospheric emission line in Herbig Ae star. According to observations by Böhm and Catala (1995), nine Herbig Ae/Be stars exhibited emission line in HeIλ5876 Å, while six stars did not show it, and some others were unknown. The emission-line equivalent widths of HeI emission in detected stars were in a range of 25–100 mÅ. Exceptionally strong emission of 230 mÅ was found in HD 100546 (Ae). Most of these emission lines shifted toward the violet side, for which Böhm and Catala suggested the existence of expanding chromospheres.

The observations cited above seem to show the possibility of existence of chromospheric activities for A and late B type stars. These stars are usually supposed not to have convection zones according to the traditional theories. Hence the existence of chromospheres in these stars offers a new problem on the generation of mechanical energies and formation of chromospheres.

Further reading

Kitchin, C.R. (1982). *Early Emission Line Stars.* Adam Hilger Ltd., Bristol.

Williams, R. and Livoi, M. (eds.) (1995). *The Analysis of Emission Lines.* Cambridge University Press, Cambridge.

References

Aller, L. H. (1956). Chapter IV, Section 2. The Balmer decrement. *Gaseous Nebulae.* Chapman & Hall, London.

Anderson, L. S. and Athay, R. G. (1989). Model solar chromosphere with prescibed heating. *Ap. J.*, **346**, 1010–1018.

Baker, J. G. and Menzel, D. (1938). Physical processes in gaseous nebulae. III. The Balmer decrements. *Ap. J.*, **88**, 52–64.

Baker, J. G. Menzel, D., and Aller, L. H. (1938). Physical processes in gaseous nebulae. V. Electron temperature. *Ap. J.*, **88**, 422–428.

Böhm, T. and Catala, C. (1995). Rotation, winds and active phenomena in Herbig Ae/Be stars. *A. A.*, **301**, 155–169.

Bowen, I. S. (1935). The spectrum and composition of the gaseous nebulae. *Ap. J.*, **81**, 1–16.

Brocklehurst, M. (1971). Calculations of the level populations for the low levels of hydrogenic ions in gaseous nebulae. *M. N. R. A. S.*, **153**, 471–490.

Bruevich, E. A., Katsova, M. M., and Livshits, M. A. (1990) Kinetics of hydrogen in the chromospheres of red dwarfs. *Sov. Ast.*, **34**, 60–65.

Castor, J. I. 1970. Spectral line formation in WR envelopes. *M. N. R. A. S.*, **149**, 111–127.

Castor, J. I. & van Blerkom, D. (1970). Excitation of HeII in WR envelopes. *Ap. J.*, **161**, 485–502.

Castor, J. I. and Lamers, H. J. G. L. M. (1979). An atlas of theoretical P Cyg profiles. *Ap. J. Suppl.*, **39**, 481–511.

Cram, L. E. and Giampapa, M. S. (1987). Formation of chromospheric lines in cool dwarf stars. *Ap. J.*, **323**, 316–324.

Cuntz, M., Rammacher, W., and Ulmschneider, P. (1994). Chromospheric heating and metal deficiency in cool giants: Theoretical results versus observations. *Ap. J.*, **432**, 690–700.

Drake, S. A. and Ulrich, R. K. (1980). The emission-line spectrum from a slab of hydrogen at moderate to high densities. *Ap. J. Suppl.*, **42**, 351–383.

Elitzur, M., Ferland, G. J., Mathews, S. G., and Shields, G. A. (1983). Stimulated emission and flat Balmer decrements in cataclysmic variables. *Ap. J.*, **272**, L55–L59.

Emerson, D. (1996). *Interpreting Astronomical Spectra.* John Wiley & Sons, West Sussex, UK.

Gutierrez-Moreno, A. and Moreno, H. (1996). Spectroscopic observations of some D-type symbiotic stars. *P. A. S. Pacific*, **108**, 972–979.

Hamann, W. R. (1985). Line formation in expanding atmospheres: Accurate solution using approximate lambda operators. *A. A.*, **148**, 364–368.

Hamann, W. R. and Schmutz, W. (1987). Computed HeII spectra for WR stars: a grid of models. *A. A.*, **174**, 173–182.

Hamann, W. R., Wessolowski, U., and Koesterke, L. (1994). Non-LTE spectral analysis of WR stars.: the nitrogen spectrum of the WN6 prototype HD 192163 (WR136). *A. A.*, **281**, 184–198.

Hummer, D. G. (1964). The mean number of scatterings by a resonance-line photon. *Ap. J.*, **140**, 276–281.

Hummer, D. G. and Rybicki, G. B. (1982). A unified treatment of escape probabilities in static and moving media. I. Plane geometry. *Ap. J.*, **254**, 767–779.

Jeffery, D. (1990). The Sobolev-P method.III. The Sobolev-P method generalized for three-dimensional systems. *Ap. J.*, **352**, 267–278.

Kogure, T. (1959a). The radiation field and theoretical Balmer decrements of Be stars. I. *P. A. S Japan*, **11**, 127–137.

Kogure, T. (1959b). The radiation field and theoretical Balmer decrements of Be stars. II. *P. A. S Japan,* **11**, 278–291.

Kogure, T. (1961). The radiation field and theoretical Balmer decrements of Be stars. III. *P. A. S Japan*, **13**, 335–360.

Kogure, T. (1969). Contribution à l'étude des profils d'émission d'étoiles Be dites "Pole-on." *A. A.*, **1**, 253–269.

Kogure, T., Hirata, R., and Asada, Y. (1978). On the formation of hydrogen shell spectrum and the envelopes of some shell stars. *P. A. S Japan*, **30**, 385–407.

Lamers, H. J. G. L. M., Cerruti-Sola, M., and Perinotto, M. (1987). The "SEI" method for accurate and efficient calculations of line profiles in spherically symmetric stellar winds. *Ap. J.*, **314,** 726–738.

Marlborough, J. M. (1969). Models for the envelopes of Be stars. *Ap. J.*, **156**, 135–155.

Marsh, T. R. (2001). Doppler tomography. *Astrotomography, Indirect Imaging Methods in Observational Astronomy*, H. M. J. Boffin, D. Steeghs, and J. Cruypers (eds.), *Lecture Notes in Physics*, 573, 1–27.

Marsh, T. R. and Horne, K. (1988). Images of accretion discs – II. Doppler tomography. *M. N. R. A. S.*, **235**, 269–286.

Mauas, P. J. D., Falchi, A., Pasquini, L., and Pallavicini, R. (1997). Chromospheric models of dwarf M stars. *A. A.*, **326**, 249–256.

Mendoza, C. (1983). Recent advances in atomic calculation and experiments of interest in the study of planetary nebulae. *Planetary Nebulae*, IAU Symp., No. 103, D. R. Flower (ed.), D. Reidel Publ. Co., Dordrecht, 143–172.

Menzel, D. H. and Baker, J. G. (1937). Physical processes in gaseous nebulae. II. Thoery of the Balmer decrement. *Ap. J.*, **86**, 70–77.

Mihalas, D. (1978). *Stellar Atmospherres*, 2nd edition. Freeman & Comp. San Francisco.

Mihalas, D., Kunasz, P. B., and Hummer, D. G. (1975). Solution of the co-moving frame equation of tranfer in spherically symmetric flows. I. Computational method for equivalent-two-level-atom source functions. *Ap. J.*, **202**, 465–489.

Mihalas, D. and Kunasz, P. B. (1978). Solution of the comoving-frame equation of transfer in spherically symmetric flows. V. Multilevel atoms. *Ap. J.*, 219, 635–653.

Miyamoto, S. (1949). On the radiation field of Be stars. I. *Jap. J. Ast.*, **1**, 17–25.

Miyamoto, S. (1952a). On the radiation field of Be stars. II. *P. A. S. Japan*, **4**, 1–10.

Miyamoto, S. (1952b). On the radiation filed of Be stars. III. Theoretical Balmer decrements. *P. A. S. Japan*, **4**, 28–36.

Mullan, D. J. and Cheng, Q. Q. (1993). MgII and Lyα fluxes in M dwarfs: Evaluation of an acoustic model. *Ap. J.*, **412**, 312–323.

Papkalla, R. (1995). Line formation in accretion disks. 3D comoving frame calculations. *A.A.* **295**, 551–564.

Pottasch, S. R. (1960). Balmer decrements: The diffuse nebulae. *Ap. J.*, **131**, 202–214.

Pottasch, S. R. (1961). Balmer decrements: II. The Be stars. *Annales d'Astrophys.*, **24**, 159–167.

Pottasch, S. R. (1984). *Planetary Nebulae.* D. Reidel Dordrecht.

Rons, N., Runacres, M., and Blomme, R. (1992). Comoving frame calculations for λ-Cephei. *Nonisotropic and variable outflows from stars, ASP Conf. Ser.* **22** L. Drissen, C. Leitherer, and A. Nota (eds.), 199–202.

Rosseland, S. (1936). Theory of radiative transformations in nebulae. Chapter 22 *Theoretical Astrophysics.* Clarendon Press. Oxford.

Rybicki, G. B. (1972). A novel approach to the solution of multilevel transfer problems. *Line Formation in the Presence of Magnetic Fields*, R. G. Athay, L. L. House, and G. Newkirk, Jr. (eds.). High altitude observatory, Boulder, CO., p. 145.

Schmitz, F. and Ulmschneider, P. (1980). Theoretical stellar chromospheres of late type stars. *A.A.*, **84**, 191–199.

Schrijver, C. J. (1983). Coronal activity in F, G, K type stars. *A. A.*, **127**, 289–296.

Schrijver, C. J. (1987). Magnetic structure in cool stars. XI. Relation between radiative fluxes measuring stellar activity, and the evidence for two components in stellar chromospheres. *A.A.*, **172**, 111–123.

Simon, T., Drake, S. A., and Kim, P. (1995). The X ray emission of A type stars. *P.A. S.P.*, **107**, 1034–1041.

Simon, T. and Landsman, W. B. (1991). The onset of chromospheric activity among A and F stars. *Ap. J.*, **380**, 200–207.

Simon, T. and Landsman, W. B. (1997). High chromospheres of late A stars. *Ap. J.*, **483**, 435–438.

Sobolev, V. V. (1947). *Moving Envelopes of Stars* (in Russian). English translation, 1960. Harvard University Press, MA.

Stirpe, G. M. (1991). Broad emission lines in active galactic nuclei. *A. A.*, **247**, 3–10.

Ulmschneider, P. (1991). Acoustic heating. *Mechanisms of Chromospheric and Coronal Heating.* P. Ulmschneider, E. R. Priest, and R. Rosner (eds.), Springer-Verlag, Berlin, pp. 328–343.

Werner, K. and Husfeld, D. (1985). Multi-level non-LTE line formation calculation using approximate Λ-operators. *A. A.*, **148**, 417–422.

Woolley, R.v.d.R. and Stibbs, D. W. N. (1953). Chapter III, The integral equations of radiative equilibrium: Exact solutions. *The Outer Layers of a Star.* Clarendon Press, Oxford, pp. 30–51.

Part II

Emission-Line Stars

Chapter 5

Early-type Emission-line Stars

5.1 Wolf–Rayet stars

5.1.1 Spectral classification and basic stellar parameters

What are Wolf–Rayet stars?

Wolf–Rayet (WR) stars owe its name to the discoverers, C. Wolf and G. Rayet. They are one of the most conspicuous stellar objects in spectral features, activities, and in the processes of stellar evolution. They are spectroscopically characterized by the following features.

(i) The spectrum consists almost wholly of emission lines on the O-type continuum. Sometimes absorption lines appear at the violet edges of emission lines.

(ii) The emission lines are very broad. Interpreted as Doppler broadening, the widths correspond to some hundreds to thousands of km s^{-1} and not necessarily the same for all ions.

(iii) The lines represent a wide range of excitation and ionization even in any one star. The excitation level is generally much higher than that estimated from its color temperature of the continuous spectrum.

(iv) The spectrum falls into three groups due to the appearance of C, N, O lines:

WN stars—Emission lines of N and He ions are strong and predominant.

WC stars—Lines of He, C, O ions are strong.

WO stars—Lines of O ions are predominant; C ions are also seen. The number of stars of this group is small as compared to the above two groups.

There also exist stars showing both characteristics of WN and WC group.

In the category of WR stars, there exist two major groups: One is the classical WR stars, which are young population I stars with high masses and high luminosities. The other is the central stars of planetary nebulae with WR-type spectra, which are of considerably lower masses and lower luminosities, belonging to population II. The latter group is now labelled as [WR], according to the notation of van der Hucht et al (1981). In this section we consider the classical WR stars of population I, and [WR] will be discussed in Section 5.2.3.

WR stars are often observed in binary systems, in which WR stars are usually low-mass components in combination with massive O stars. Single WR stars are in many times located as the central stars of ring nebulae.

Spectral classification

WR stars ordinarily do not show Fraunhoffer lines of photospheric origin. This is because the photospheres corresponding to O stars are embedded inside optically thick stellar winds. By this reason, classification of WR spectra is based on emission lines, and, accordingly, the classification system is not closely coupled to the stellar parameters of effective temperature and luminosity. The main emission lines usually used for classification in the optical region are as follows:

NIII—λ4634–4641 Å (blend), λ5314 Å
NIV—λ3479–3484 Å (blend), λ4058 Å
NV—λ4603 Å, λ4619 Å, and λ4933–4944 Å (blend)
CIII—λ5272 Å, λ5696 Å
CIV—λ4650 Å, λ5805 Å, λ5470 Å, λ4684 Å
OV—λ4159 Å, λ5592 Å
OVI—λ3834 Å, λ3811 Å, λ5290 Å
HeII—λ4686 Å, λ5412 Å
HeI—λ5876 Å

Table 5.1 shows the classification criteria based on these emission lines for WN, WC, and WO, separately (van der Hucht 1992, Kingsburg et al. 1995). In this table inequality denotes the relative strength of emission lines. For WO group, the revised classification scheme by Kingsburg et al. (1995) is adopted.

Catalogues and selected WR stars

WR stars in the Galaxy are usually named by the HD catalogue number or by the number of van der Hucht et al. (1981) catalogue. This catalogue, which contains 158 WR stars along with their identification charts and main characteristics, is revised and extended by van der Hucht (2001, VIIth catalogue) to 227 WR stars, comprising 127 WN star, 87 WC stars, 10 WN/WC stars, and 3 WO stars.

There are several spectral atlases in the optical region. Conti et al. (1990) published an atlas of 83 stars (47 WN and 36 WC stars) with the spectral resolution of 4.6 Å mm^{-1} in the region λ6000–10,150 Å, and Hamann et al. (1995) published an atlas of 62 WN stars with the resolution $\lambda/\Delta\lambda \sim 2000$ in the wavelength region $\lambda 3800 - 7000$ Å.

WR stars in the Galaxy are widely distributed along the galactic plane, and mostly associated with star-forming regions containing OB-associations

Table 5.1: Spectral classification of WR stars (van der Hucht 1992, Kingsburg et al. 1995)

(1) WN sequence	Nitrogen emission lines	Other emission lines
WN2	NV weak/absent, NIV absent	HeII strong
WN3	NIV $\ll$ NV, NIII weak/absent	
WN4	NIV $\sim$ NV, NIII weak/absent	
WN5	NIII $\sim$ NIV $\sim$ NV	
WN6	NIII $\sim$ NIV, NV weak	
WN7	NIII $>$ NIV, NIII $<$ HeII 4686 Å	HeI weak (P Cyg profile)
WN8	NIII $\gg$ NIV, NIII $\sim$ HeII 4686 Å	HeI strong (P Cyg profile)
WN9	NIII strong, NIV weak NIIweal/abs.	HeI (P Cyg profile)
WN10	NII $\sim$ NIII, NIV absent	Balmer line, HeI (P Cyg profile)
WN11	NII strong, NIII absent, NIV absent	Balmer lines, HeI (P Cyg profile)
(2) WC sequence	**Carbon emission lines**	**Other emission lines**
WC4	CIV strong, CII weak/absent	OV intermediate
WC5	CIII $\ll$ CIV	CIII $<$ OV
WC6	CIII $\ll$ CIV	CIII $>$ OV
WC7	CIII $<$ CIV	CIII $\gg$ OV
WC8	CIII $>$ CIV	CII absent, OV weak/absent
WC9	CIII $>$ CIV	CII weak, OV weak/absent
(3) WO Sequence	**Oxygen emission lines**	**Other emission lines**
WO1	OVI $\gg$ OV, no OIV	OV $\gtrsim$ CIV
WO2	OVI $>$ OV, no OIV	OV $<$ CIV
WO3	OVI $>$ OV, OVI $>$ OIV	*OV* $\ll$ CIV
WO4	OVI $>$ OV, OVI $\approx$ OIV	*OV* $\ll$ CIV
WO5	OVI $>$ OV, OVI $\approx$ OIV	OV $\ll$ CIV

and/or HII regions. A list of selected WR stars in the northern sky is given in Table 5.2.

Distance and luminosity of WR stars

Distance to a WR star which belongs to a star cluster or an association is usually estimated from the distance to the parent cluster or association, by using its color-magnitude diagram, or kinematic properties for the Galactic rotation. When necessary, the distance is estimated by measuring the equivalent width of interstellar lines such as NaD, or by using the photometric properties of the stars. The distances in Table 5.2 are those derived from the spectroscopic parallaxes by Conti and Vacca (1990), which are derived from the measurements of visual magnitude V, absolute magnitude M_V, and

Table 5.2: Selected WR stars in northern hemisphere

WR No[a]	HD[b]	Star[c]	V[d]	Spectral[e]	Distance[f]	Binarity/P[g]	Ass./HII[h]
			mag		Kpc	P(day)	
1	4004		10.54	WN5	1.7	—	Cas OB7
2	6327		11.43	WN2	5.4	—	Cas OB7
3	9974		10.79	WN3	8.6	SB	
4	16523		10.61	WC5	3.9	—	
5	17638		11.12	WC6	2.5	—	
6	50896	EZ CMa	6.94	WN5	1.0	SB1/3.76d	Cr121?, S308
111	165763		8.25	WC5	1.9	—	
113	168206	CV Ser	9.43	WC8+O8–9	2.5	SB2e/29.71d	SerOB2,S54
127	186943		10.36	WN4+O9	4.9	SB2/9.55	Vul OB2, S92
128	187282		10.56	WN4	5.0	—	
134	191765		8.31	WN6	2.2	—	S109
135	192103	V1042Cyg	8.51	WC8	3.1	—	S109
136	192163		7.73	WN6	1.5	SB1/ 4.5d	CygOB1,S105
137	192641		8.18	WC7+abs.	1.6	SB	CygOB1,S109
138	193077		8.21	WN5+abs.	3.4	SB	CygOB1,S109
139	193576	V444Cyg	8.27	WN5+O6	2.4	SB2e/4.21	CygOB1,S109
140	193793		7.19	WC7+abs.	0.80	SB	S109
141	193928		10.15	WN6	2.1	SB1/21.64	S109
148	197406		10.50	WN7	5.2	SB1/4.32	
153	211853	GP Cep	9.20	WN6+O	3.0	SB2e/6.69	CepOB2?.S132
155	214419	CQ Cep	8.94	WN7+O	3.8	SB2e/1.6	CepOB1
157	219460		10.03	WN4.5+BOIII	2.3	VB	Ma50, S157

[a]WR number is taken from van der Hucht et al.(1981).
[b,c]HD and name of star.
[d]Visual magnitude.
[e]Spectral type is mainly taken from van der Hucht et al. (1981).
[f]Distance to the star is taken from the spectroscopic parallax of Conti and Vacca (1990).
[g]Binarity: VB (visual binary), SB (spectroscopic binary), SB1, 2 (single and double lined binary, respectively). "e" denotes the eclipsing binary; P denotes the orbital period.
[h] Association/HII regions associated. HII regions are shown by Sharpless catalogue number.

interstellar absorption $A_{\rm v}$. The seventh catalogue of van der Hucht (2001) also provides the photometric distances for 227 galactic WR stars.

Recently it becomes possible to measure the precise distance of nearby WR stars by using the Hipparcos parallax with the errors of 1 milliarc second. For example, van der Hucht et al. (1997) measured the distance of the binary WR star γ^2 Vel (WR11, WC8 + O) to be $d = 258 + 41$, -31 pc, and thus determined the absolute magnitudes of the both components as $M_{\rm v}\,({\rm WC8}) = -3.7, M_{\rm v}({\rm O}) = -5.0$. The Hipparcos census of nearby OB associations by de Zeeuw et al. (1999) containe two WR stars γ^2 Vel (WR 11) and EZ CMa (WR 6).

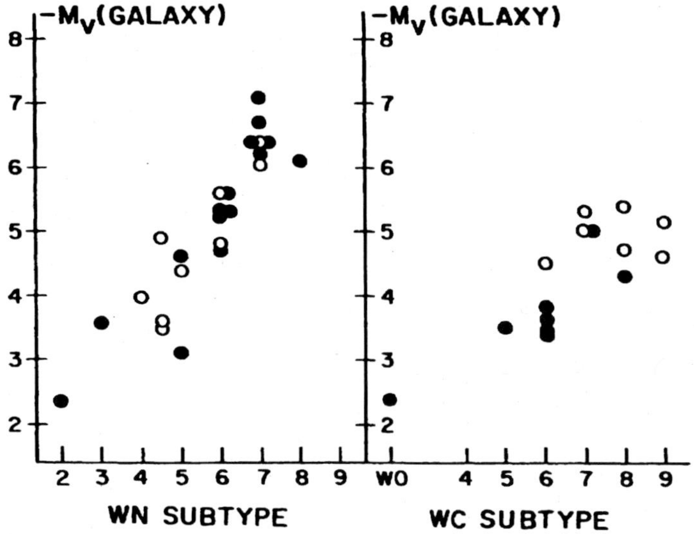

Figure 5.1: Absolute visual magnitudes versus spectral subtypes in the galactic WR stars. The left and right panels show the WN and WC series, respectively. Filled symbol indicates WR single star, and open symbol is WR binary corrected for companion. (From Conti 1986)

According to Conti (1986), the absolute magnitude M_v of WR stars shows a remarkable dependence on the spectral subclass, particularly for WN class, as shown in Figure 5.1. It varies from $M_v = -2$ to -3 in earlier types to -7 to -8 in later types in WN stars, while WC classes have no stars brighter than $M_v \sim -5$ in later subtypes. Similar dependence is also seen in the WR stars in the Large Megellanic Cloud (LMC).

Stellar colors, defined by UBV bands, and spectral types are not in parallel in WR stars. This is because WR stars are generally located in dusty star-forming regions and also surrounded by optically thick stellar winds. By making use of IUE data for the correction of interstellar absorption, Vacca and Torres-Dodgen (1990) have derived the color and absolute magnitude of WR stars that are supposed to be single stars in the Galaxy and LMC. The results showed that the stars are getting brighter toward the late type as seen in Figure 5.1, whereas the intrinsic colors do not show appreciable difference for early- and late-type stars. The difference is not significant in these features for WR stars in the Galaxy and in LMC on average.

5.1.2 Spectral features of WR stars

Spectral energy distribution

A wide range of spectral energy distribution (SED) of WR stars from X-ray to radio region is now available from ground based and space observations.

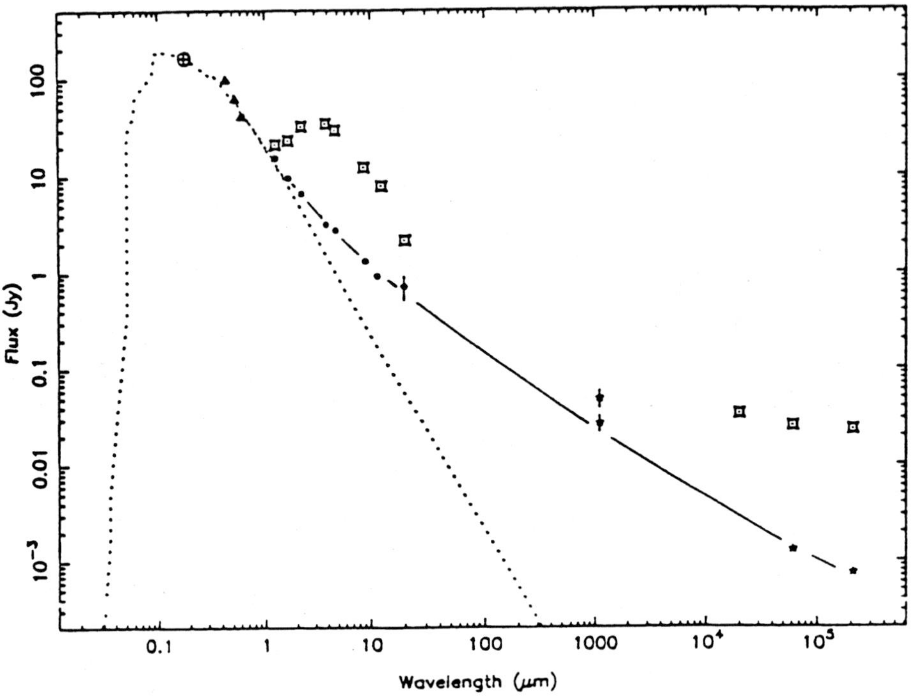

Figure 5.2: Spectral energy distribution of WR 140. The broken line represents the spectrum of a hot-star model, and the line through the lower sets of infrared and radio data points defines the system at quiescence. Square symbols give the values at active phase. (From Williams and van der Hucht 1994)

As a typical case we show in Figure 5.2 the SED of WR 140 (HD 193793, WC7 + abs) observed by Williams and van der Hucht (1994), where observed points in the infrared and radio regions are shown. In the figure, the broken line represents the spectrum of a hot-star model and the thin line defines the SED at quiescent phase.

Since WR stars are embedded in optically thick envelopes, the observed SEDs are conspicuously different from those of photospheric origins. The SED in quiencent stage exhibits a slower gradient from ultraviolet to radio regions as compared to that of stellar origin, and it can be approximated by a power law function with the index α in the form $S_\nu \sim \nu^\alpha$. Morris et al. (1993) measured the index α of WR stars by combining the spectra of the IUE and ground-based spectrophotometry. The average index thus measured for Galactic (69 stars), LMC (55 stars), and SMC (5 stars) is given by

$$\alpha = 0.85 \pm 0.40 \; (0.15 - 1.0\,\mu\text{m}). \tag{5.1.1}$$

There is no significant difference in the value of α for Galactic and Magellanic WR stars.

Similar spectral index is also obtained in the longer spectral region extended from infrared to millimeter wavelengths as given by Williams et al. (1990), who measured the value of α for γ^2 Vel as

$$\alpha = 0.69 \pm 0.02 \; (1.25\,\mu\mathrm{m} - 11\,\mathrm{mm}). \tag{5.1.2}$$

In this way, the SEDs of WR stars can be represented by a single power low function with the index $\alpha = 0.7$–0.8. This shows a remarkably slower gradient as compared to the value of $\alpha \sim 2$ for ordinary stars with the same effective temperature.

The SED in infrared and radio region of WR stars can generally be explained by thermal emissions originated from stellar winds, for which we shall consider in the next subsection. In radio wave region, however, there are stars that show nonthermal characteristics which are distinguished from thermal source by the form of radio spectra and the comparison of observed radio flux with theoretically predicted flux.

Abbott et al. (1986) thus examined 40 Galactic WR stars by comparing the very large array (VLA) 6-cm data and optical data, and found that 77% (33 stars) are thermal, 12% (5 stars) are nonthermal, and the remaining are unknown. For the origin of nonthermal emission, White (1985) suggested a possibility of synchrotron radiation by shock-accelerated electrons under a weak magnetic field. Dougherty and Williams (2000) detected nonthermal spectra in 9 WR stars among the 23 WR stars they examined. Of these nine stars, seven are known spectroscopic or visual binaries with O or early B-type companions. This suggests a possible connection between the appearance of nonthermal emission in WR stars and the presence of a massive companion.

In Figure 5.2 one may see that WR 140 showed remarkable IR excesses in its active stage. Such excess has been observed three times in 20 years since 1975, suggesting temporary formation of dust shell around the star. Time variation of WR 140 is illustrated in Figure 5.8 in Section 5.1.3.

Thermal radio spectrum of stellar envelopes

The thermal raio emission from an isothermal homogeneous gas cloud takes the spectral form as given by (see Section 2.4.2)

$$I_\nu = S_\nu (1 - e^{-\tau_\nu}), \tag{5.1.3}$$

where S_ν is the source function and can be expressed by Rayleigh–Jeans approximation in the radio spectral region. The optical depth τ_ν is proportional to the absorption coefficient κ_ν.

In the limiting case we have

$$I_\nu \rightarrow \begin{cases} S_\nu \propto \nu^2 & (\tau_\nu \gg 1) \\ \tau_\nu S_\nu \propto \nu^{-0.1} & (\tau_\nu \ll 1)\,. \end{cases} \tag{5.1.4}$$

In this equation, since κ_ν is known to be proportional to $\nu^{-2.1}$, the optical depth τ_ν is high for the lower frequency region and low for the higher

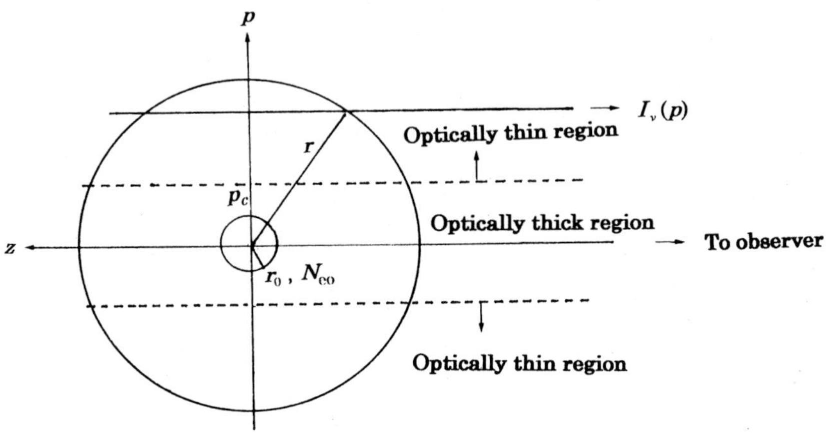

Figure 5.3: A model of stellar envelope emitting thermal radio waves. The electron density in spherical and isothermal envelope is assumed to decrease with radius as $N_e = N_{e0}(r/r_0)^{-2}$. The envelope is divided into two regions optically thin and thick as seen from an observer. (From Panagia and Felli 1975)

frequency region. Hence the radio spectrum has the spectral index 2 in lower frequency region and –0.1 in higher frequency region. At a certain frequency the spectrum takes a maximum value.

In the case of stellar envelopes, Panagia and Felli (1975) derived the radio spectrum based on the envelope model as shown in Figure 5.3. The envelope is assumed spherically symmetric and in the state of LTE. They divided the envelope viewed from an observer into inner optically thick region ($\tau_\nu > 1$, $p < p_c$) and outer optically thin region ($\tau_\nu < 1, p > p_c$) at a boundary distance p_c, where p_c is defined as the distance from the star by the condition $\tau_\nu(p_c) = 1$. By using the approximate Formulae (5.3.9) and (5.3.10), the total emergent intensity I_ν is obtained by integrating $I\ (p)$ for the whole envelope surface as

$$I_\nu = 2\pi \left[\int_0^{p_c} I_\nu\,(p)\; p^2\,dp + \int_{p_c}^{\infty} I_\nu\,(p)\,p^2\,dp \right] . \tag{5.1.5}$$

The critical distance is a function of frequency and the distribution of electron density inside the envelope. The latter can be given by a power law as

$$N_{\rm e} = N_{\rm e0}\,(r/r_0)^{-s} . \tag{5.1.6}$$

Panagia and Felli numerically integrated Equation (5.1.5) in the case of $s = 2$ and derived the spectral form

$$I_\nu = K\,\nu^{0.6}\,T^{0.1}, \tag{5.1.7}$$

where K denotes a constant determined by the boundary conditions including $N_{\rm eo}$ and $r_{\rm o}$. This equation yields the spectral index of 0.6, which is close to the observed value of 0.86 in (5.1.1) and 0.67 in (5.1.2).

Table 5.3: Spectral index of radio emission versus density distribution index s in the envelope (adapted from Panagia and Felli 1975)

Density distribution index	s	1.5	2.0	2.5	3.0
Spectral index	α	-0.1	0.60	0.95	1.16

It is noticed that the spectral index largely depends on the assumed distribution of electron density in Equation (5.1.6). The spectral index as a function of power law index s is shown in Table 5.3. One may see that the spectrum approaches to the transparent case ($\alpha = -0.1$) when density distribution is slower (s smaller), while it approached the opaque case ($\alpha = 2$) when density distribution is steeper (s larger). Spectral index $\alpha = 0.6$ in Equation (5.1.7) just corresponds to the case of $s = 2$.

In addition, Panagia and Felli considered the thermal radio spectrum emitted from an expanding envelope with mass loss. If the mass-loss rate $\dot{M}$ and expanding velocity $V_{\rm exp}$ are constant, then the radio spectrum takes the following form

$$I_\nu \propto \nu^{0.6} T_e^{0.1} \left(\frac{\dot{M}}{V_{\rm exp}} \right)^{4/3} . \tag{5.1.8}$$

It is again noticed that the spectral index 0.6 is deduced by the density distribution index $s = 2$, which is the consequence of mass conservation law in the stationary expanding envelope.

X-ray emission

Since the launching of the Einstein Observatory in 1978, it has been known that WR stars are generally strong X-ray emitters. Pollock (1987) has made Einstein observations for 48 WR stars in the energy range 0.2–4 K eV, and detected significant X-ray emission from 60% of the stars. As the results, Pollock showed that (a) the X-ray luminosity $L_{\rm X}$ is in a range of 10^{32}–$10^{33.5}$ erg s^{-1}, (b) the binary WR stars (single or double lines) are usually stronger X-ray emitters than the single WR stars, (c) X-ray luminosities of WN stars are about three times higher than those of WC stars, and (d) WR stars with nonthermal radio emission yields higher X-ray luminosities among WN and WC stars.

Since X-ray emission is not expected from the stellar winds in radiative equilibrium, some heating mechanisms are required to emit X-ray radiation. In single WR stars, X-rays are supposed to be emitted from shock-heated hot region caused by some wind instabilities. In contrast, WR + O binary systems produce excess X-ray emission by wind interactions from both stellar components (cf. Corcoran 1996).

Baum et al. (1992) have proposed a semi-empirical model of X-ray emission from single WR stars by dividing the stellar wind into normal and hot components. The *normal components* are in radiative equilibrium, and the *hot components* are the shock-heated gas, which are assumed to have electron temperature T_e and the filling factor f as free parameters. Each *hot component* occupies a small volume and the whole of *hot components* is assumed to be homogeneously distributed throughout the WR envelope. By adjusting the parameters T_e (about 4×10^6 K) and f (a few percent in mass), they obtained the X-ray intensity and spectrum comparable with observed values. Ignace et al. (2000) carried out more realistic model calculations for single WR stars and found a higher gas temperature of around 10^8 K and filling factors of $\sim 10^{-4}$ for the X-ray emitting regions. They also showed that the mean X-ray luminosity is higher for WN stars than for WC stars as

$$L_X = \begin{cases} 4.3 \times 10^{32}\ \mathrm{erg\,s^{-1}} & \text{for WN stars} \\ 1.5 \times 10^{32}\ \mathrm{erg\,s^{-1}} & \text{for WC stars} \end{cases}$$

in the ROSAT energy band of 0.2–2.4 keV.

Emission-line spectrum

WR stars exhibit strong emission lines in a wide spectral region from ultraviolet to infrared. Figure 5.4 illustrates low-dispersion spectra (30 $\mathrm{\AA\,mm^{-1}}$) of three WR stars taken from the atlas of Lundström and Stenholm (1984). One may see strong and wide emission lines without appreciable absorption components.

Figure 5.4 also shows a wide range of ionization state, such as NIII to NV and CII to CIV. In order to explain these multiple ionization degrees, two major mechanisms may be considered. One is the stratification of emitting layers and the other is the shock-wave heating.

The stratification appears when emission lines are formed through the ionization–recombination process, and the ionization degree depends on the distance from the stellar surface, as originally suggested by Beals (1929). If some velocity gradient exists as in accelerating stellar winds, the emission-line widths should be correlated with the ionization degree. On the other side, if the emission lines are formed by some shock-heated region, such correlation will not be expected, since shocked regions may be distributed throughout the envelope and emission lines are formed in the relaxation layers with some stratification behind the shock fronts (Section 3.4.2). In Figure 5.5, we show the correlation between ionization potential E (eV) and the Doppler width (km $\mathrm{s^{-1}}$) of emission lines, based on the data of Smith and Kuhi (1981). It is noticed that WR 3 (WN3 type) shows no particular correlation, whereas WR 136 (WN6 type) reveals the decreasing line width for higher ionization potential. Such correlation similar to WR 136 is also seen in WR 78 (WN7), and in WR 111 (WC5) (Schulte-Ladbeck et al. 1995).

In the photoionization model, emission lines of higher ionization potentials are formed in the region nearer to the star so that the correlation as seen in WR 136 may be explained by the stratification in an accelerative expanding

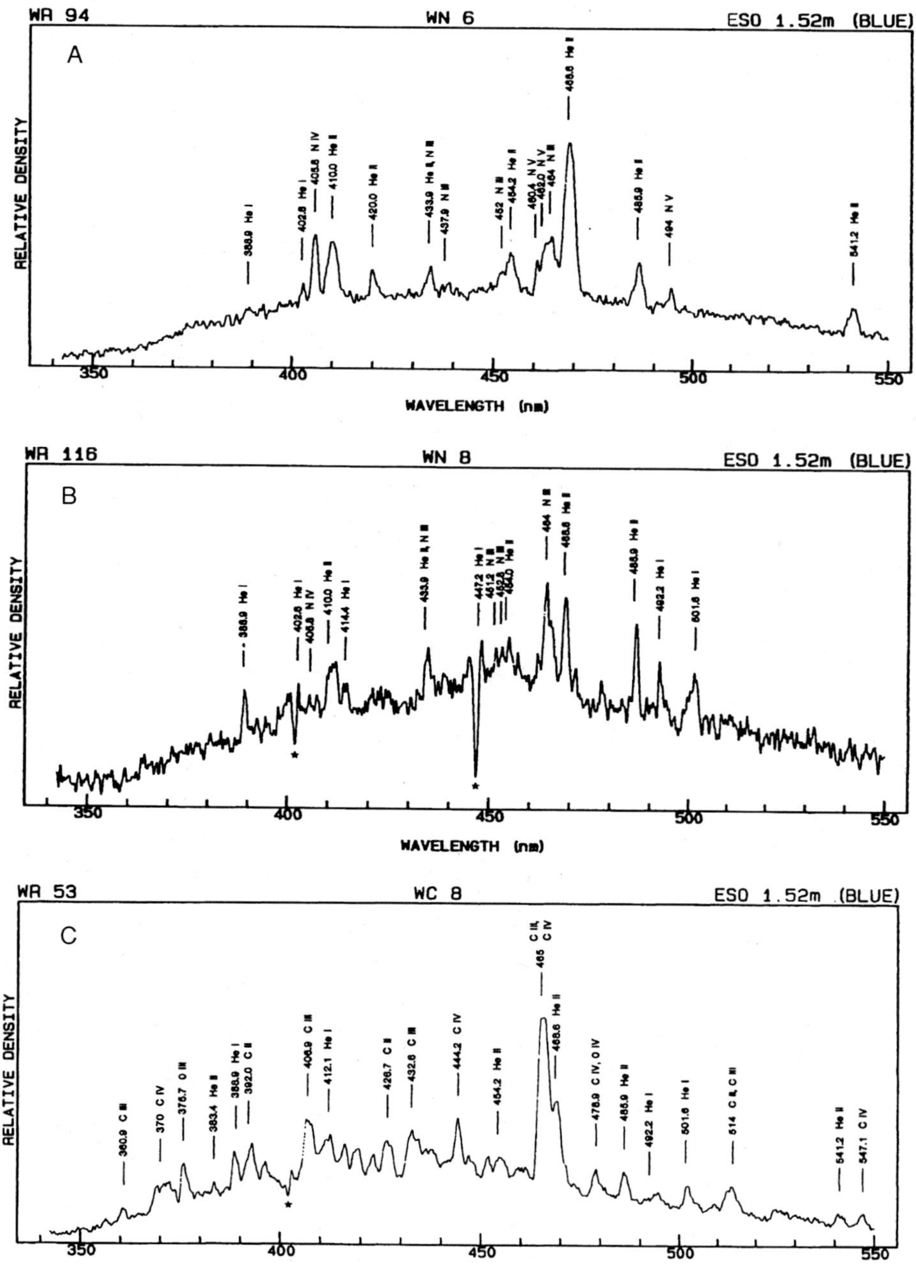

Figure 5.4: Samples of optical spectra of WR stars in the wavelength range 3500–5500 Å obtained by the ESO 1.52-m reflector. Stars are (A) WR 94 (WN 5), (B) WR 116 (WN 8), and (C) WR 53 (WC 8). (From Lundström and Stenholm 1984)

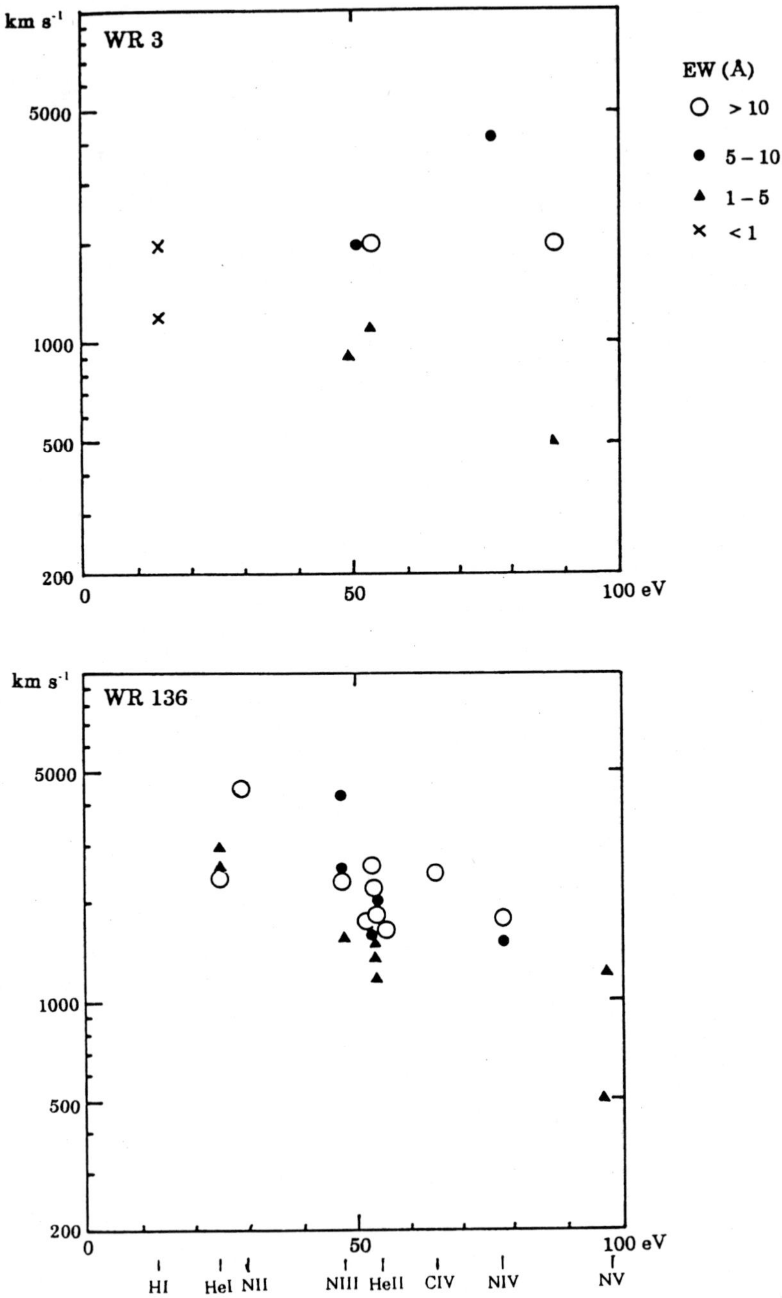

Figure 5.5: Correlation between emission-line width and ionization potential in two WR spectra: WR 3 (WN3) and WR 136 (WN6). Ions are distinguished by the emission equivalent width, and the names of ions are indicated in the bottom. (Data are taken from Smith and Kuhi 1981)

envelope. In contrast, WR 3, which shows no correlation, suggests the existence of some nonthermal ionization mechanisms like shock-wave heating.

Based on the measurements of emission line widths (FWHM) along the series numbers in HeII Fowler ($n-3$) and Pickering ($n-4$), Niedzielski (1994) examined the stratification in WR envelopes. That is, FWHM has shown larger value in lower members (n: smaller) and smaller value in higher members (n: larger), for all observed 4 WR stars (WR 1, WR 134, WR 136, and WR 141). Since lower members generally have larger optical depths, and then are formed in the outer parts of the envelope, the above correlation suggests the existence of stratification in accelerating stellar winds. It seems that the stratification structure is prevailing among the WR stars.

In the ultraviolet region there are a number of emission lines; their line intensities exhibit a loose correlation with the spectral type determined by optical emission lines. Niedzielski and Rochowicz (1994) have made low-dispersion spectroscopic observations with the IUE satellite for 94 single WR stars. The observed emission lines are helium (HeI, HeII), nitrogen (NIII–NV), oxygen (OIII–OVI), carbon (CII–CIV), silicon (SiIII, SiIV) and others. Figure 5.6 illustrates the relation between spectral subtypes and relative emission intensities of NIIIλ 1750/NVλ 1241 for WN stars and of CIIλ 2837/CIIIλ 1909 for WC stars. It is seen that the degree of ionization declines along the spectral subtype in optical classifications, though WN stars are somewhat largely scattered in distribution. Niedzielski and Rochowicz explained this scattering by the combination of two regions with different electron temperatures.

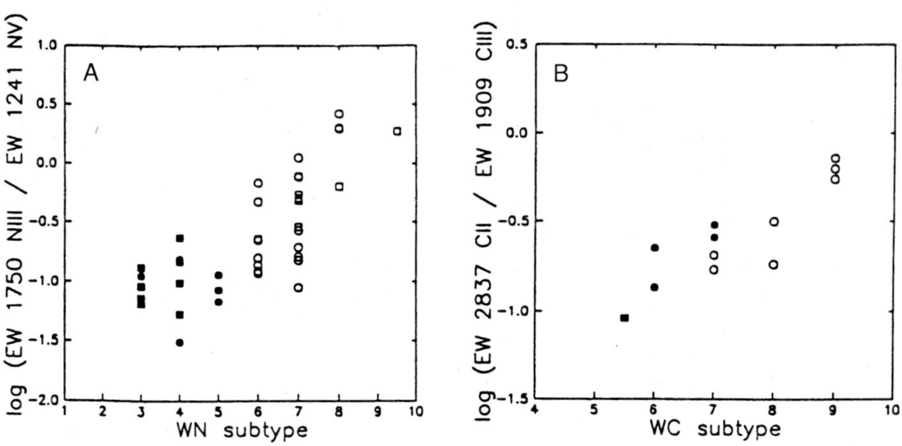

Figure 5.6: Correlation between spectral subtype and the equivalent widths of different UV emission lines of (A) nitrogen in WN stars and of (B) carbon in WC stars. Galactic WR stars are denoted by filled (early type) and open (late type) circles, WR star in LMC are shown by filled (early type) and open (late type) squares. (From Niedzielski and Rochowicz 1994)

Near infrared spectrum (7000–10,000 Å) exhibits the characteristic emission lines in elements similar to those given in Table 5.1 as a continuation from optical to infrared in both WN and WC stars. There are several spectral atlases in this region, by Vreux et al. (1989) (12 stars, 6150–10,350 Å), Conti et al. (1990) (47 stars, 7000–10,000 Å), and by Howarth and Schmutz (1992) (24 stars, 9700–11,200 Å). In these atlases the line identifications and estimated physical parameters are given. Conti et al. (1990) measured the relative intensities of main emission lines and found that the state of excitation is in parallel with the spectral subtypes.

Mid-infrared spectroscopic observations were carried out by van der Hucht et al. (1996) with the Infrared Space Observatory (ISO). Numerous emission lines were detected in the region 2.3–29.5 μm. Figure 5.7 illustrates a sample of infrared spectra for WR 11 (γ^2 Velorum, WC8 + O9I), where we can see rich permitted lines of HeI, HeII, CIII, and CIV, along with some forbidden lines such as [CaIV]λ 3.207 μm, [SIV]λ 10.510 μm, [NeII]λ 12.815 μm, and [NeIII]λ 15.554 μm. Since forbidden lines are supposed to be formed in the outermost region of the wind, it is suitable to measure the terminal velocity from the widths of these lines. Thus van der Hucht et al. obtained the following terminal velocities for some forbidden lines:

$$
\begin{aligned}
V_\infty &= 1540 \pm 30\,\mathrm{km\ s^{-1}} && \text{for [NeII], [NeIII],}\\
V_\infty &= 1560 \pm 25\,\mathrm{km\ s^{-1}} && \text{for [SIV]}\\
V_\infty &= 940 \pm 100\,\mathrm{km\ s^{-1}} && \text{for [CaIV]}
\end{aligned}
$$

This result indicates that the [CaIV] line is formed where the wind is still under acceleration.

5.1.3 Time variations

Long-term variations

The long-term variations of WR stars are not significant in the optical region, but remarkable in the infrared and radio regions. In Figure 5.8, the light curves in the infrared L band are shown for three WR stars observed by Williams and ven der Hucht (1994). In these stars, WR 140 and WR 137 are the spectroscopic binaries.

WR 140 (O4 – 5 + WC7) has the orbital period 2900 days and high orbital eccentricity ($e \sim 0.7$–0.8). The UBV colors show light variation around 0.6 magnitude in amplitude (Seggewiss et al. 2002), whereas the IR light curve exhibits remarkable outbursts in conjunction with the orbital period. These outbursts appear markedly in the range 3–10 μm region as seen in Figure 5.2. Williams and van der Hucht explained this IR excess as the temporary formation of a dust shell, which occurs during periastron passage when interaction of WC7 and O4–5 stellar winds is at its greatest. They supposed that the wind interaction region is sufficiently compressed and cooled down to allow the dust condensation. They also assumed that this dust shell is in the

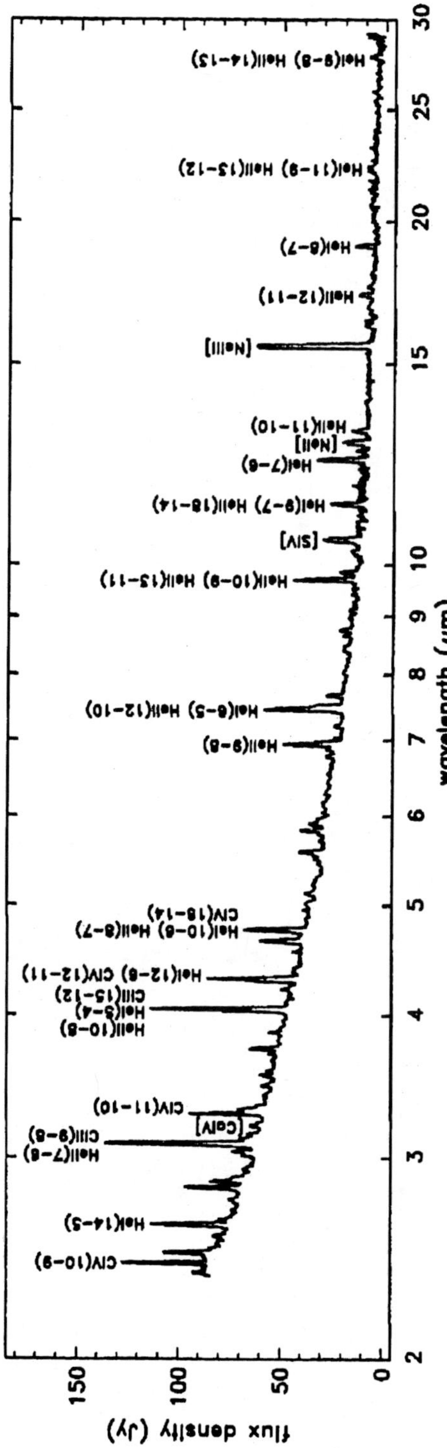

Figure 5.7: Mid-infrared spectrum of WR 11 (γ^2 Velorum, WC8 + O9I) obtained with the ISO satellite. Line identification (name of ion/atom, energy level in transition) is given for main emission lines. (From van der Hucht et al. 1996)

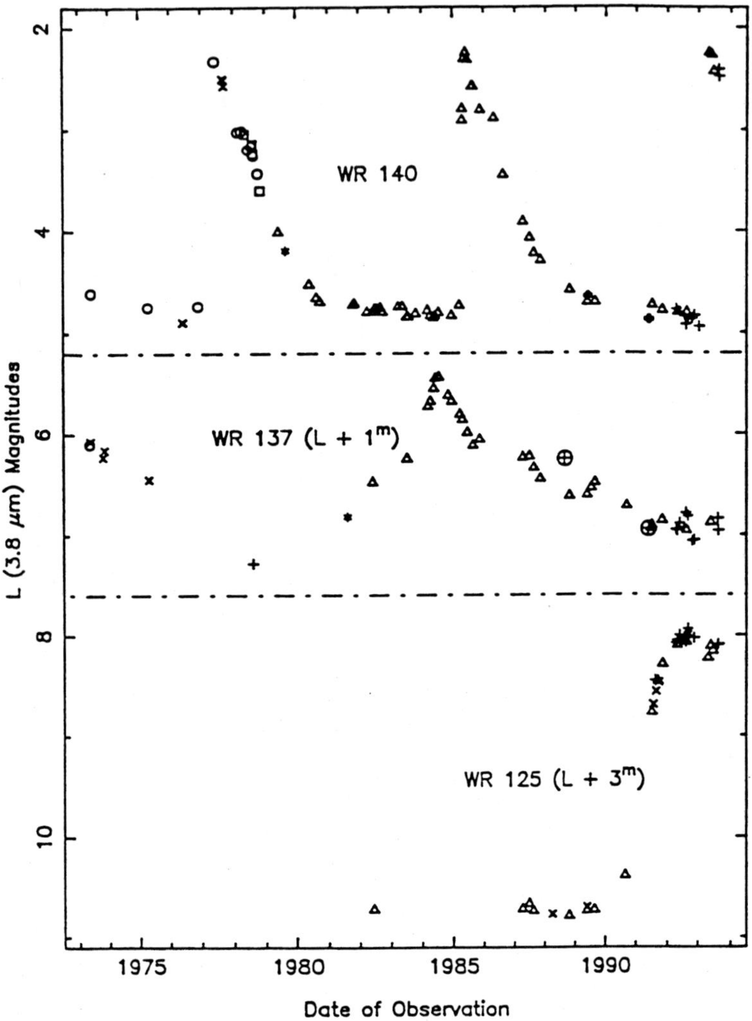

Figure 5.8: Infrared light curves of WR 140, WR137, and WR 125 at 3.8 μm (L band). The abscissa is the date, the ordinate is the L band magnitude. For WR 137 and WR 125, additional magnitude in order to arrange the magnitude scale is given in each bracket. Different symbols mark observations at different observatories. (From Williams and van der Hucht 1994)

radiative equilibrium in the diluted stellar radiation. Under these assumptions they derived the amount of dust and estimated the dust production rate to be 0.13% of the total mass-loss rate.

The colliding processes of the winds are also observed at X-ray and radio wavelengths, making WR 140 one of the brightest noncompact X-ray sources and a nonthermal radio source. Dougherty et al. (2003) carried out a model calculation under the assumption of equipartition of relativistic energy density and magnetic energy density and applied the model to WR 140 for explaining the dramatic variation of radio emission due to eccentric orbital motion. Zhekow and Skinner (2000) also applied the colliding wind model to WR 140 and reproduced the X-ray spectrum along with its variation at different orbital phases.

WR 137 (WC 7 + OB) is also a long-period (∼13 years) binary with random light variations with small amplitude. This star also exhibit infrared variations, producing the dust shell similar to WR 140 (Williams et al. 2001).

Short-term variations

WR stars often show short-term variability in brightness, emission-line intensity, polarization, etc., in a scale of less than several days. According to Moffat and Shara (1986), light variations in B band larger than 0.02 magnitude are observed in more than 50% of 18 WR northern bright stars and its rate reaches 70% if we allow the variability more than 0.01 magnitude. Although some periodic variations appear in coincidence with their orbital motions in binary WR stars, irregular variations are mostly predominant due to unstable stellar winds for both of singles and binaries (Moffat and Robert 1991).

Stars showing periodic short-term variations are summarized by van der Hucht (1992) as given in Table 5.4. The cause of the periodic variability has been supposed as binary interaction and/or nonradial pulsation.

As an example of short-term variability we show the line variations of WR 46 (HD 104994, WN3p) observed by van Genderen et al. (1991). This star has a period of 0.2824 days (6.77 h) and the amplitude of light variation $\Delta V \sim 0.1$, color variation $\Delta(V - B) \sim 0.01$, along with marked emission-line variations. Figure 5.9 shows the variation of the emission equivalent widths in a phase diagram for some prominent lines. The variations are generally in phase with the light curve, in the sense that the emission lines take the

Table 5.4: WR stars showing periodic short-term variability

WR	Name/HD	Spectral type	Period (d)
6	EZ CMa	WN5 + ?	$P \sim 1$
11	γ^2 Vel	WC8 + O9I	$P = 1.26$
16	86161	WN8	$P_A = 9.09$, $P_B = 15.15$, $P_C = 5.46$
40	96548	WN8	$P_A = 6.25$, $P_B = 2.5$
46	104994	WN3p	$P = 0.28$
50	LSS 3013	WC6 + abs.	$P = 1.06$
123	177230	WN8	$P = 1.94$
136	192163	WN6 + c?	$P_A = 0.45$, $P_B = 0.31$

Note: the suffix for the period A, B, C, denotes the multiple components (adapted from van der Hucht 1992).

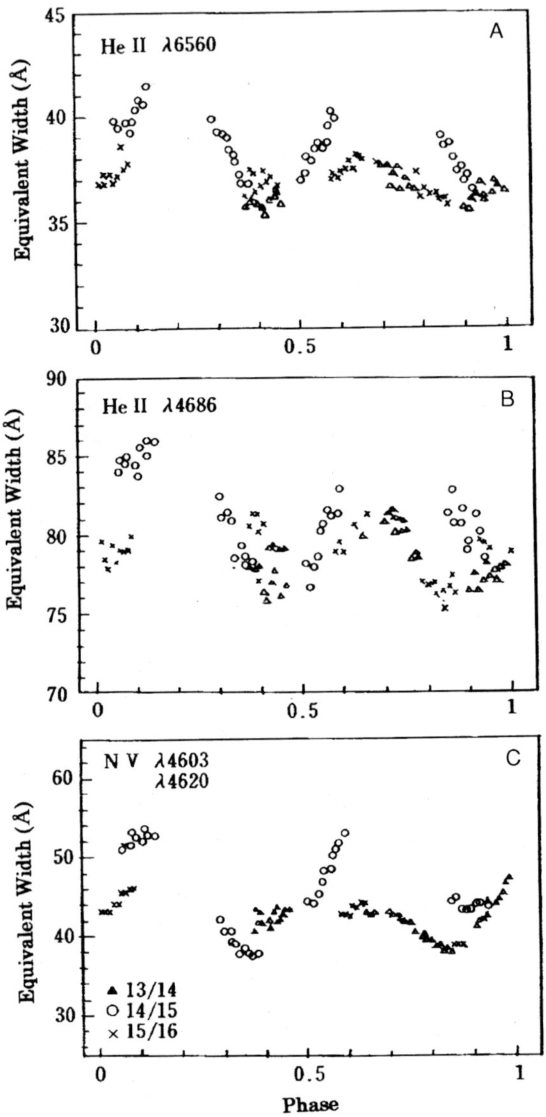

Figure 5.9: The variation of the equivalent widths for some emission lines of WR 46 in the phase diagram for the period $P = 0.2824$ (d). Different symbols denote the different cycles given by days in March 1989, as shown in the figure. (From van Genderen et al. 1991)

maximum equivalent widths near the brightness maxima (phases around 0.2 and 0.7), but showing large changes from cycle to cycle. Van Genderen et al. argued that WR 46 is a binary system, possibly composed of a WR primary (with mass of 4–9 $M_\odot$) and a white dwarf companion (with mass of 0.4–0.6 $M_\odot$). Its emission-line emitting region shows a stronger distortion than the region which emits the continuum light (the pseudo-photosphere). They thus suggested that the companion revolves within an outer envelope, and the emission-line emitting region closely surrounds both Roche lobes. Large

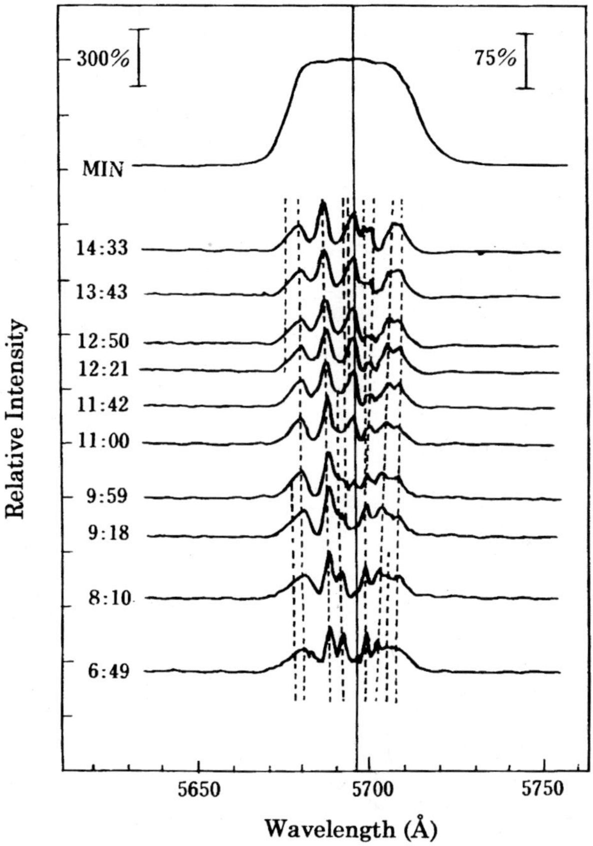

Figure 5.10: Short-term variation of line profile of CIIIλ 5696 Å observed in WR 135 (HD 192103, WC8). The top spectrum is the slightly smoothed minimum spectrum of four night's data. Below this, a series of spectra show the variation of profiles obtained by subtracting the minimum spectrum from each individual spectrum and displayed on a four times enlarged scale. UT times are indicated. Dashed straight lines are shown as a guide to trace substructures on successive spectra. Vertical bars indicate the scale in continuum unit. (From Moffat 1994)

scatter in Figure 5.9 seems to reflect geometrical changes of these regions in every cycle of binary motion.

We now consider the short-term nonperiodic irregular variations in the case of WR 135 (HD 192103, WC8) that was monitored spectroscopically by Moffat (1994), using the 3.6-m CFHT (Canada–France–Hawaii Telescope) at Mauna Kea. The rapid change of line profile for CIIIλ 5696 Å is shown in Figure 5.10, where the top profile designated as MIN is the minimum line profile of 4 nights' data. Below it, the series of spectra are those obtained by subtracting this minimum spectrum from each individual spectrum and displayed on a four times enlarged scale. Each spectrum is labeled by the observed UT time. Dashed straight lines show the variation of the same substructure on successive spectra.

These substructures are observed in some other emission lines, and the following properties are pointed out by Moffat:

(a) The number of substructures for a given emission line is proportional to the wind terminal velocity of the star.

(b) Line variability at different Doppler velocities across an emission line is directly proportional to the underlying line flux.
(c) Lines of high-ionization species (formed in hotter regions close to the central core) tends to be less variable.
(d) The substructure tends to move away from the line center with time.

These substructures imply that some kind of density enhancements or clumps in the winds form, grow, and decay as they propagate outward superposing on some steady stellar wind. Lépine and Moffat (1999) observed similar profile variations in HeIIλ 5411 emission lines of four WN stars and in CIII λ 5696 line of five WC stars. They introduced a phenomenological model that depicts WR winds as being made up of a large number of randomly distributed, radially propagating, discrete wind emission components.

Large variety is seen in the appearance of short-term variations of WR stars, in brightness, line intensity, and/or in line profile. The cause of variation may not be simple. Various mechanisms have so far been proposed such as binary interaction (van Genderen et al. 1991), radial pulsation (Maeder 1985), nonradial pulsation (Vreux 1985), variation in magnetic fields (Underhill 1983, Poe et al. 1989), and wind instability (Owocki 1990). Stellar rotation and orbital motion will cause the periodic variations as in case of Table 5.4, while wind instability or magnetic activity will cause irregular variations. There remains a possibility for complex causes in some stars.

5.1.4 Spectroscopic binaries and mass of WR stars

Spectroscopic binaries

There are two main types in the WR binaries: WR + O and WR + c, where O and c denote O-type stars and compact stars, respectively (Cherepashchuk 1991).

1. WR + O star system. Smith and Maeder (1989) have compiled the basic data for 27 galactic WR stars in binaries with O or B stars. It contains orbital period, eccentricity, spectral type, mass ratio, mass function, mass, and references. Among these properties, we show the relation between mass ratio $q = M(\mathrm{WR})/M(\mathrm{O})$ and spectral type in Figure 5.11. One may see some different trend in WN and WC stars. In WN stars, mass ratio markedly increases with spectral subtype up to the maximum ratio of $q = 2.78$ for WR 22 (WN7). In this binary system, while the O star component has 24 $M_\odot$, mass of the WR component reaches up to 72 $M_\odot$, the largest mass ever found in WR stars (Rauw et al. 1996). Generally the value of q is less than unity and shows a trend that smaller the value of q, smaller the mass of WR star. In contrast, WC stars have the values of q smaller than around 0.5 regardless the spectral type.

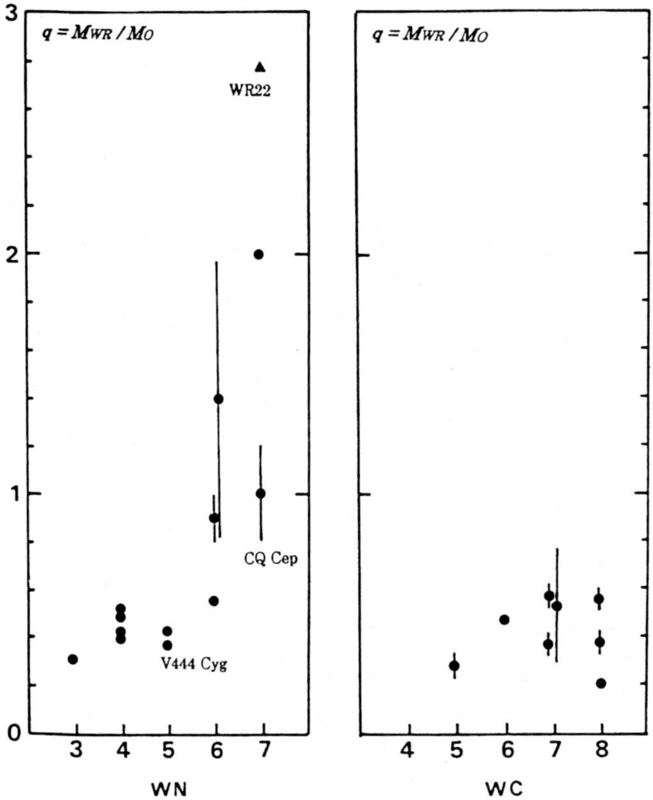

Figure 5.11: Correlation between mass ratio and the spectral subtype of WR binaries. Data points are plotted based on Smith and Maeder (1989) and Rauw et al. (1996).

Mass of WR stars widely ranges from 1.8 to 72 $M_{\odot}$ and no appreciable difference is seen for WN and WC stars, with a trend that late type stars have larger mass. The largest mass WR star is WR 22, as stated above, and this star is supposed to be at an early stage of the WR evolution, by the reason that it is rich in hydrogen and has large mass ratio.

Using ground-based and ISO IR observations, van der Hucht et al. (2001) found active dust formation in 26 WC stars but non in WN stars. Since dust formation is supposed to occur in the wake of the colliding stellar winds in WC + OB binaries, they argued that all dusty WC stars might be WC + O type binaries.

2. WR + c system.
   Moffat (1983) listed 12 galactic WR stars with suspected compact companions. These candidate stars were selected by the short-term

variations of emission-line profiles and photometric variability. For these stars, however, Vreux (1985) suggested the possibility of nonradial pulsation, based on some regularity in the distribution of periodicity among listed stars. Such regularity can also be seen in other WR stars showing nonradial pulsation.

According to Cherepashchuk (1991), WR binaries combined with compact stars (neutron stars or black holes) are known for the following three systems:

WR 6	HD 50896 (EZ CMa)	WR + c (M (c) $\sim 1 M_\odot$)	$P = 3.77$ days
WR148	HD197406	WR + c (M(c) $> 4\ M_\odot$)	$P = 4.32$ days
Cyg X-3		WR + c	$P = 4.8$ hours,

where M(c) is the mass of compact star and P denotes the orbital period.

Among these stars, WR 6 (EZ CMz) is supposed to have a neutron-star companion and WR 148 a black-hole companion judged from the mass of its companion. In contrast, nature of the companion of Cyg X-3 is not certain, though its binarity was confirmed by van Kerkwijk et al. (1992, 1996). Cherepashchuk and Moffat (1994) suggested, based on their orbital analysis of Cyg X-3, that the primary of this system is a peculiar WR star with small mass ($< 3 - 6\ R_\odot$) and of high temperature ($T_{\rm eff} = 70000 - 90000$ K). Ergma and Yungelson (1998) examined a possibility of black-hole companion of this system by means of population analysis for the expected Galactic number of black holes, and of orbital analysis of this system, but without reaching clear conclusion.

V444 Cyg and CQ Cep

Among WR + O type binaries, V444 Cyg and CQ Cep, which were discovered in 1939 and 1941, respectively, are typical short-period eclipsing binaries and have been sufficiently observed. The orbital and stellar parameters are summarized by Underhill (1991) as shown in Table 5.5.

The binary system V444 Cyg is composed of O and WR stars, both surrounded by developed stellar winds. This system has been intensively investigated since 1940s to determine the dimension and masses of stars. Based on the radial velocity curves for O stars (primary) and WR stars (secondary), Marchenko et al. (1994) derived a spectroscopic model of this system as shown in Figure 5.12, where phase $\phi = 0$ is defined as the mid-eclipse time when WR star is in the front of the O star. This figure illustrates the relative size of the both stars and envelope, along with the emitting regions of HeII and HeI lines. In the successive spectroscopic observations, Marchenko et al. (1997) analyzed the wind structure close to the WR component and derived the significant stratification in the order NV, HeII, and HeI in outward direction. They also

Table 5.5: Binary parameters of WR stars V444 Cyg and CQ Cep (adapted from Underhill 1991)

Star Spectral type Magnitude	V444 Cyg WN5 + O6 V = 8.27	CQ Cep WN7 + O $V = 8.94$
(1) Orbital parameters		
Orbital period (d)	4.212424	1.6412436
Orbital inclination (deg)	78 ± 4	74 ± 6
Orbital radius a(WR)/$R_\odot$	27.4	9.7
Orbital radius a(O)/R	9.1	8.2
Mass $M(\mathrm{WR})/M_\odot$	9.8	(13.6)
Mass $M(\mathrm{O})/M_\odot$	29.6	(16.0)
(2) WR parameters		
Effective temperature (K)	27,200	29,000
Luminosity log $(L/L_\odot)$	3.91	4.27
Effective radius $(R_\odot)$	4.0	5.5

Note: The masses of WR and O star components of CQ Cep are those derived by assuming the mass ratio $M(\mathrm{WR})/M(\mathrm{O}) = 0.85$.

estimated the mass-loss rate from the WR star as $(1.3 - 0.5) \times 10^{-5}\, M_\odot$ per year depending on the adopted wind-velocity law.

V444 Cyg is also known as an X-ray source. Corcoran et al. (1996) observed this star with ROSAT and detected the phase variation of X-ray intensity, where X-ray intensity is lower around the primary eclipse and higher out of eclipse. From the X-ray light curve and energy spectral distribution, they suggested that the X-ray be emitted in the region of wind interaction from both stars, in addition to the wind regions of both individual stars. Based on the ASCA (Advanced Satellite for Cosmology and Astrophysics, launched in 1993) observations, Maeda et al. (1999) found that the X-ray spectrum is composed of at least two components, with a soft component at $kT_1 \approx 0.6$ keV and a hard component at $kT_2 \approx 2$ keV. The X-ray luminosity of the soft component is $(6 - 11) \times 10^{32}$ ergs s^{-1}, with a maximum value at phase 0.5 and minimum at 0.25. Since this X-ray luminosity is of the same order as those from either single O6 or WN5 stars, the soft component can be attributed to the individual O6 and WN5 stars. The hard component, on the other hand, has a temperature two to three times higher than that usually found in single massive stars. Such high temperature can be naturally explained by a wind–wind collision with wind speed of a few thousand km per seconsd. It is also found that the highest temperature plasma produced by the collision of the winds is located nearly on the line connecting the O6 and WN5 stars, and then the X-rays are subject to heavy absorption mainly by the dense stellar wind of the WN5 star.

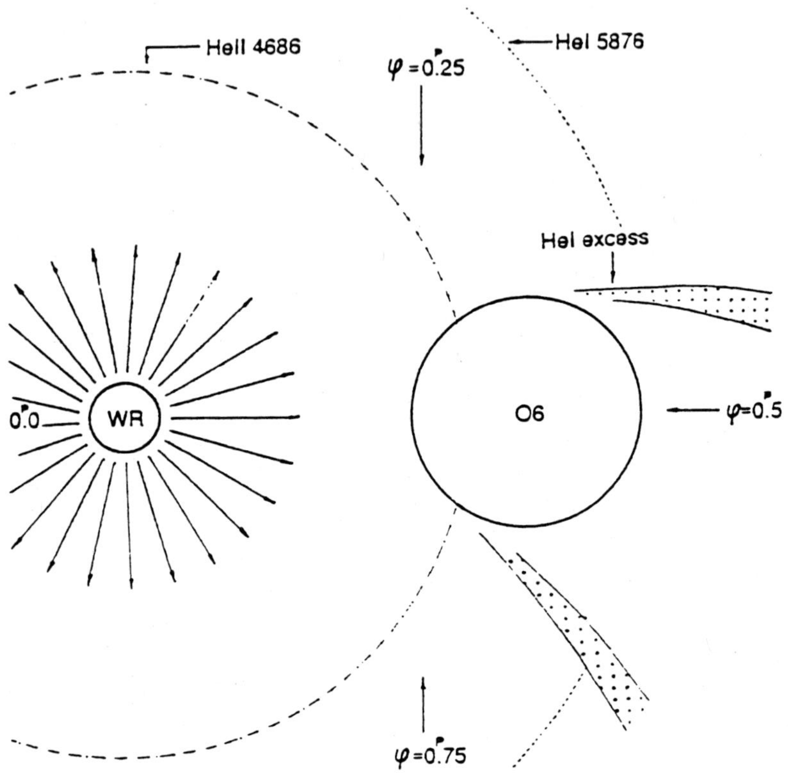

Figure 5.12: Spectroscopic model of the V444 Cyg system. Scale of the stars and envelopes in the orbital plane is shown, along with schematic ionization structure of the WR wind. (From Marchenko et al. 1994)

5.1.5 Spectroscopic models and chemical composition

A model of WR stars is called *standard model*, when it starts with the following basic assumptions for the stellar envelopes: (1) spherical symmetry, (2) monotonous velocity field, (3) homogeneous chemical composition, (4) radiative equilibrium, and (5) steady state.

This standard model has been introduced by Castor and Van Blerkom (1970) and Castor and Lamers (1979), who adopted the escape probability by motion (Sobolev type) and solved the non-LTE radiative transfer. Since 1980s, standard model has been developed by W. R. Hamann's group in Germany and by D. J. Hillier's group in America. To treat the non-LTE processes, Hamann's group has made use of the approximate lamda-iteration method (Section 4.2.4), while Hiller's group has developed the so-called complete linearization method (Hillier 1987, Hillier and Miller 1998). Both groups have carried out model calculations for a wide grid of stellar parameters and calculated emission-line profiles, ionization structure of stellar winds, and SED. By comparing with observations, they derived the stellar parameters, mass-loss rates, and chemical composition. Both groups started their model calculations from WN stars, since nitrogen abundance is low in WN stars (1–2% in mass) and the model for pure helium envelope yields sufficient approximations. Contrarily, carbon abundance is high (20–50% in mass) in WC stars, hence the simultaneous treatment of helium and carbon abundance becomes important. Koesterke and Hamann (1995) solved a series of model envelopes of WR stars composed of helium and carbon and derived the physical parameters and abundance ratio C/He.

Physical properties deduced from these model calculations are summarized in Table 5.6, for WN stars (Crowther et al. 1995) and for WC stars (Koesterke and Hamann 1995). Although the models used for WN and WC stars are different, one may see the characteristic difference for both series of stars, that is, WN stars are generally larger in radius, lower in temperature, and smaller in wind velocity as compared to WC stars. Particular difference can be seen in the chemical composition. In Koesterke and Hamann's model, though the envelopes of WC stars are assumed to be composed only of helium and carbon, the abundance of carbon relative to helium reaches as high as 30–50% in mass. In contrast, WN stars disclose that the abundance of carbon and nitrogen is less than around 1%. Hydrogen abundance of WN stars is ordinarily 10–30% as seen in Table 5.7, though there are some peculiar WN stars with extremely low hydrogen abundance less than 2% as in case of WR123 (WN8). In Table 5.7, we show the results of another model calculations by Cohen (1991) for the abundance of He, C, N, and CNO (hydrogen is not included) for WN, WC, and WO stars. It is seen that WO stars are characterized by the deficient helium and abundant oxygen. Average abundance of interstellar medium is also shown in Table 5.7 for comparison. High abundance of CNO elements in WR stars is remarkable, suggesting the appreciable contribution of CNO elements from WR winds to the general interstellar medium. Large difference of chemical composition among WR stars reflects the difference in the evolutionary state of these stars (for the evolution of WR stars, see Section 5.5).

As seen above, the standard models have so far been widely calculated and successfully applied to the interpretation of many spectroscopic behaviors of the WR stars. There remain, however, several problems in the standard models restricted by the basic assumptions, particularly in spherical symmetry, monotonous velocity field, and/or stationary state of gas motion. Hence several nonstandard models have been proposed to explain the variability and dynamical structure of the WR envelopes. Among others, Vreux et al. (1992) have analyzed line profile variations of WR 134, based on the axis-symmetric bipolar wind model; Poe et al. (1989) proposed a rotating, magnetic, radiatively driven wind model; and Underhill et al. (1990) also considered the rotating wheel model with variable nature. Since the stellar winds from the WR stars are unstable, time-depending models such as by Owocki et al. (1990) should be valuable in the analysis of some WR spectra.

Table 5.6: Physical parameters of WR stars deduced from model calculations

WR No.	Sp. type	R_* $R_\odot$	T_* 10^3 K	$T_{\rm eff}$ 10^3 K	$\log L_*$ $L_\odot$	$\log\dot{M}$ $M_\odot$ per year	V_∞ km s^{-1}	$X_{\rm H}$ (%)	$X_{\rm He}$ (%)	$X_{\rm C}$ (%)	$X_{\rm N}$ (%)
WN stars (Crowther et al. 1995)											
16	WN8	15.5	34.5	26.3	5.49	−4.22	630	20	79	0.01	1.6
40	WN8	14.4	35.9	24.7	5.49	−4.01	840	15	83	0.01	1.7
78	WN7	23.3	33.6	30.4	5.80	−4.07	1385	11	87	0.03	1.7
123	WN8	15.3	33.9	26.3	5.45	−4.04	970	<2	96	0.05	1.7
124	WN8	15.2	33.5	25.4	5.41	−4.11	710	13	86	0.03	1.2
156	WN8	17.6	31.8	26.7	5.45	−4.36	660	27	71	0.20	1.7
WC stars (Koesterke and Hamann 1995)											
4	WC5	1.8	74.1	—	5.0	−4.2	1900	—	50	50	—
111	WC5	2.7	62.5	—	5.0	−4.3	2000	—	50	50	—
15	WC6	1.7	75.7	—	4.9	−4.0	2700	—	70	30	—
56	WC7	3.9	60.3	—	5.3	−3.9	1800	—	50	50	—

Note: Among column items, stellar radius R_* is defined as the radius at the mean optical depth 20, and T_* is the temperature at that radius. X_A (A : H, He, C, N) indicates the chemical abundance expressed in mass fraction. In WC stars the chemical abundance is assumed only composed of He and C.

Table 5.7: Average chemical composition of WR stars and interstellar (adapted from Cohen 1991)

WR	X_{He} (%)	X_{C} (%)	X_{N} (%)	X_{O}%	X_{CNO}%
WN	97	0.038	2.2	0.11	2.4
WC	32	39	0.0	25	64
WO	9	21	0.0	66	87
Interstellar medium	22	0.34	0.12	0.82	1.3

Note: $X_{\mathrm{CNO}} = X_{\mathrm{C}} + X_{\mathrm{N}} + X_{\mathrm{O}}$; hydrogen is not included in Cohen's estimation.

5.2 O-type emission-line stars

5.2.1 Of stars

Classification

In O-type stars there are two groups of emission-line stars, i.e., Oe and Of stars. According to Harvard classification, Oe stars are O-type stars showing hydrogen emission lines superimposed on photospheric absorption lines, whereas Of stars are the O-type stars that exhibit strong emissions in NIII ($\lambda\lambda$ 4634, 4640, 4642 Å) and HeII (λ 4686 Å) lines. Later on, Walborn (1971) distinguished three groups of Of stars by the appearance of NIII and He II lines and designated as Of, O(f), and O((f)). Spectral features of these groups are shown in Table 5.8.

Three groups of Of stars also reveal the difference in luminosity (Conti and Leep 1974). Of and O(f) stars are more luminous than the main sequence O stars, while O((f)) stars lie among the main sequence. Thus HeII emission intensity seems somehow related to the luminosity as seen in Table 5.8.

Though all of these Of stars are similar to WR stars in excitation levels of emission lines, they exhibit narrower emission lines and higher or similar mass-loss rates as compared to WR stars. In addition, Of stars do not show the series of different chemical composition by N, C, O ions. In Table 5.9 we show a list of selected bright Of stars.

Table 5.8: Spectral features of Of stars (adapted from Walborn 1971)

Types of stars/lines	NIII $\lambda\lambda$ 4634, 4640, 4642 Å	HeII λ 4686 Å	Number of stars
Of	Strong emission	Strong emission	16
O(f)	Emission	Absence or weak absorption	13
O((f))	Emission	Strong absorption	20

Note: The number of stars is taken from Conti and Leep (1974).

Table 5.9: List of selected bright Of stars

HR	Name	V mag.	Spectral type	
			CL	W
1879	λ Ori A	3.66	O8III((f))	O8III((f))
2422		6.06	O7.5III(f)	O8p
2456	15 Mon	4.66	O8III((f))	O7V(f)
2781	29 CMa	4.98	O8.5If	O7Ia:fp
3165	ζ Pup	2.25	O4ef	O4I(n)f
6245		5.22	O8If	O8Iaf
6272		5.77	O8If	O8:Iafpe
6672		6.20	O8.5((f))	O7.5II((f))
6736	9 Sgr	5.97	O4((f))	O4V((f))
7574	9 Sge	6.23	O8If	O7.5Iaf
8469	λ Cep	5.04	O6ef	O6I(n)fp

Note: Reference of spectral type—CL, Conti and Leep (1974); W, Walborn (1971).

Emission-line spectrum

Typical line profiles of Of stars are schematically illustrated in Figure 5.13 based on the observations of Conti and Frost (1977) at Kitt Peak and some other Observatories. It contains four stars of Of, O(f), O((f)) types and three profiles of NIIIλ 4641 Å region ($\lambda\lambda$ 4634, 4640, 44642 Å), HeIIλ 4686 Å, and Hα lines. Here, one may see characteristic features of line profiles given in Table 5.8 for NIII and HeII lines. The Hα line shows behaviors somewhat different from the former two lines.

As seen in Figure 5.13, emission lines of Of stars are generally narrower than those of WR stars; their half-widths are usually less than several hundred km s^{-1} in velocity scale. For the line width, however, Underhill et al. (1989) claimed that Of stars often reveal broad wings in both sides of strong narrow emission lines. An example is seen in HeII and Hα lines of HD15570 in Figure 5.13. For these stars Underhill et al. proposed a double-structure model of Of envelopes, that is, sharp emission lines are formed in the envelope relatively stable and close to the star, whereas broad emission features are formed in tenuous high-velocity winds of the star.

Rotational velocity

The rotational velocity $V \sin i$ has been measured by Penny (1996) for 177 O and Of stars, based on the IUE high-dispersion spectra of the UV photospheric lines. By making use of this data set, let us compare the rotational velocities of O and Of stars. In Table 5.10, the average values of $V \sin i$ for both types of stars are shown.

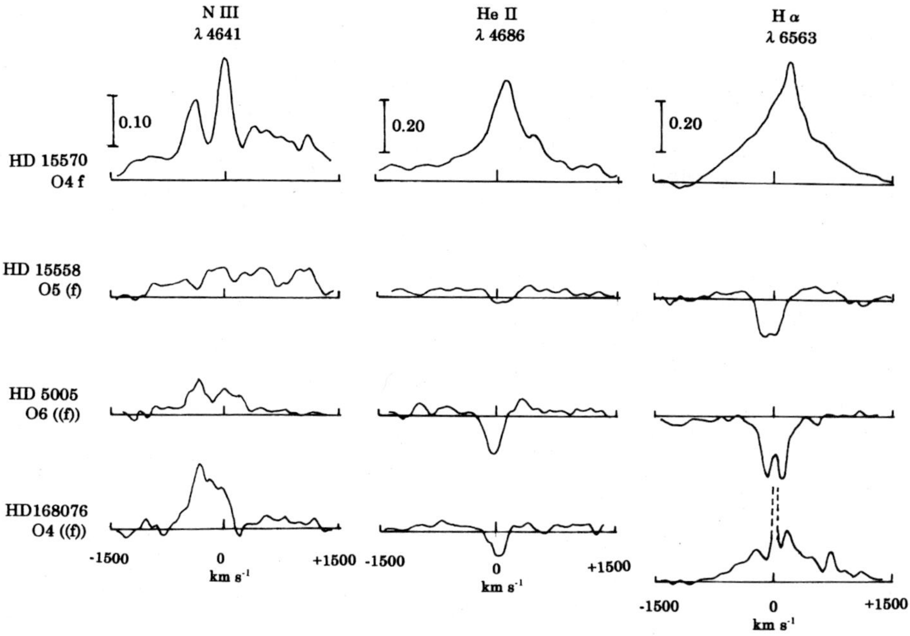

Figure 5.13: Comparison of three typical lines of Of spectra: Blended NIII (λ 4634, 4640, 4642 Å), HeIIλ 4686 Å, and Hα, are shown for four stars in different types of Of, O(f) and O(f)). Intensity scale is given in upper part. Profiles are schematically drawn based on the observations by Conti and Frost (1977).

It seems that there is no essential difference between O and Of stars in the average rotational velocities. This infers that stellar rotation may not be the important agency for the formation of emission lines in Of stars.

Binary system

Since WR stars contain binary systems in a high fraction, it has been expected that Of stars may be also in a similar situation. Actually, however, the number of binaries so far confirmed by radial-velocity observations is not

Table 5.10: Comparison of average rotational velocity V sin i for Of and O stars (based on the data of Penny 1996)

Luminosity class	V–III			II–I		
	$\langle V \sin i \rangle$ (km s^{-1})	σ (km s^{-1})	N	$\langle V \sin i \rangle$ (km s^{-1})	σ (km s^{-1})	N
Of, O(f), O((f))	118	70	38	127	37	23
O	138	99	60	107	50	50

Note: N is the number of stars used.

high. According to Underhill and Gilroy (1990) and Underhill (1995), only two stars have been confirmed the binarity (single-lined) out of 17 Of stars examined. Since many stars show radial-velocity variations, Underhill (1995) suggested that higher fraction of spectroscopic binaries could be found in the future.

Among the binaries in Of stars, HD 188001 (9 Sagittae, O7.5Iaf, SB1 = single-lined binary) has been frequently observed since 1950s. Underhill has first found its binarity (SB1), and estimated its orbital period as 78.74 days, though the orbital parameters are not yet fully determined. This star has shown a long-term variation in Hα emission profile, in the way that the asymmetric double peaks of emission components have changed from $V/R > 1$ (violet peak is stronger than red peak) in 1950s to $V/R < 1$ (red peak is stronger) in 1990s. This V/R variation is often observed in Be stars (see next Section) and HD 188001 is the first star that shows V/R variation among Of stars (Underhill 1995).

Stellar winds and mass-loss rate

Of stars are known to have developed stellar winds and belong to the stellar group of highest mass-loss rate together with WR stars. In Table 5.11 we show the mass-loss rates and terminal velocities for some Of stars taken from Lamers and Leitherer (1993), who derived the empirical mass-loss rates from thermal radio emission and Hα recombination radiation. Both radiations are supposed to be formed in the outer part of stellar winds, which are optically thin for the Hα radiation. The stars in Table 5.11 possess rather large terminal velocities of the order of 2000–3000 km s^{-1}, which resemble to the stars noticed by Undehill (1995) to have high-velocity stellar winds in Figure 5.13. If these

Table 5.11: Stellar winds and mass-loss rates of Of stars (adapted from Lamers and Leitherer 1993)

Star HD	Name	Spectral type	Mass-loss Rates $\log\dot{M}(M_\odot \mathrm{yr}^{-1})$ (Hα)	(radio)	Terminal velocity V_∞ (km s^{-1})	V_∞/V_{esc}
14947		O5If+	−5.32	<−4.76	2300 ± 70	2.61
15558		O5III(f)	−5.61	—	3350 ± 200	3.60
15570		O4If+	−5.02	−5.33	2600	3.92
15629		O5V((f))	−5.77	—	2900 ± 70	2.79
24912	ξ Per	O7.5IIIn(f)	−5.89	—	2400 ± 100	2.70
36861	λ Ori	O8III((f))	−6.20	<−6.04	2400 ± 150	2.79
66811	ζ Pup	O4I(n)f	−5.45	−5.62	2200 ± 60	2.27
151804		O8Iaf	−5.00	−5.00	1600 ± 70	2.50
188001	9 Sge	O7.5Oaf	−5.38	—	2950 ± 150	2.86
190429A		O4 If	−5.16	—	2300 ± 70	2.50
210839	λ Cep	O6I(n)fp	−5.46	−5.68	2100 ± 60	2.53

stars represent the general samples of Of stars, the mass-loss rates are as high as 10^{-5}–10^{-6} $M_{\odot}$ per year, independent of the spectral subtype in Of stars.

In their evolutionary state, Of stars are supposed to be located on the way from O stars to WR stars (see Section 5.5.1). They have enough strong stellar winds, though not yet fully developed as compared to the WR stars (Conti 1976).

5.2.2 Oe stars

Oe stars are the O-type stars that show emission lines in Balmer series. In contrast to Of stars, they do not show emission lines in highly excited ions such as NIII and HeII. The spectral features are rather similar to Be stars (see next Section), and supposed they are the extension of Be stars toward O-type stars (Frost and Conti 1976). For example, Negueruela et al. (2004) observed HD 39680 (O8.5 Ve) in the wavelength range λ 3900–4550 Å and found double-peaked emission in Hγ–H8 and in HeI λ 4474 Å, showing the similarity with early-type Be stars. In Table 5.12 we show a list of selected Oe stars observed by Conti and Leep (1974), Andrillat et al. (1982), and Negueruela et al. (2004).

Though the number of Oe stars with measured rotational velocity is small, they have larger values in average than those of O and Of stars (Penny 1996), and this is consistent to regard Oe stars as the extension of rapidly rotating Be stars. However, the frequency of Oe stars tends to become very rare among O stars (Negueruela et al. 2004). This can be attributed to the competition of strong stellar winds and rapid rotation. Existence of strong stellar winds in O stars, usually prevent the formation of disks, while rapid rotation causes the cooling of equatorial regions of stars, which gives rise to dense and slowly expanding rotational envelopes, and forms lower excitation emission lines as compared to Of stars.

Table 5.12: A list of selected Oe stars

Star	Spectral type		$V \sin i$ (km s^{-1})			
HD	(1)	(2)	(1)	(2)	H$_\alpha$	Var. of Hα
39680	O6Vnpe	O8.5 Ve	170	197	Strong emission	(Few var.)
45314	OBe, Ope	B0IVe	—	243	Week emission	
46056	O8V(e), O9V	—	235	337	Week emission	(Intermit)
60848	O8V:pe	O9.5 IVe	275	231	var. emission	
149757	ζ Oph O8V(e),O7.5III:n((f))	O9.5 IV	385	348	Week emission	(Intermit)
155806	O7.5 Vnpe,O7.5IIIe	O7.5 IIIe	150	92	Strong emission	

Note: Spectral type—(1) Conti and Leep (974) and Andrillat et al. (1982), (2) Nugueruela et al. (2004). Rotational velocity: (1) Uesugi and Fukuda (1982), (2) Penny (1996) Variability of Hα emission: few variation or intermittent.

5.2.3 Central stars of planetary nebulae (PNCSs)

Spectral types and spectroscopic characteristics

The central stars of planetary nebulae (PNCSs), as the sources of ionizing radiation are hot, dwarf stars with diversity in spectral types. The main spectral types of PNCSs and some properties of the stars and nebulae are summarized by Pottasch (1984). Each type is characterized as follow:

[WR] stars—These stars have very broad emission lines of H, He, C, N, and O. As compared to WR stars (Population I), [WR] stars are several magnitude fainter, and the predominance of WC stars is apparent. That is, six [WR] stars in Table 5.13 are all of the WC sequence, and only two stars in Table 5.14 show possible WN characteristics among [WR] stars.

Of stars—Very similar to Population I Of stars having emission lines of H, HeII, NIII, and usually CIII. As in case of [WR] stars, these stars are also several magnitudes fainter than Population I Of stars.

Of + [WR]—The spectra show a combination of the above two types and do not have a counterpart in normal Of and WR stars.

OVI stars—The spectra are very similar to [WR] or Of stars except for the fact that they show very prominent emission lines of OVI at $\lambda\lambda$3811, 3834 Å. These stars are supposed to be a group of [WR] (WC sequence) stars extended to hotter side.

Continuous spectra (c stars in Table 5.13)—This group of stars have emission lines of H, HeII, C, N, O on the continuum without showing absorption features, though, in high-dispersion spectra some stars show weak underlying absorption lines. In the far UV region some stars exhibit strong P Cygni profiles in CIV and NV.

Table 5.13: Spectral categories of the central stars of planetary nebulae (PNCSs) (adapted from Pottasch 1984)

	Central Star		Nebula	
Type of spectra	Average gravity (log g)	Surface temperature (10^4 K)	Excitation class	Number of stars/nebulae
[WR]	4.5	3.5–5	1–4	6
Of	4.4	3–4	3–5	8
Of + WR	5.6	4–7	5–6	5
OVI	5.5	5–10	6–9	12
c (continuous)	5.9	5–12	6–9	20
O type	4.9	3.5–6	—	9
Subdwarf O	7.1	8–12	—	8
Peculiar	—	—	—	6
Faint	7.8	12–14	7–10	8

Notes: Excitation class is the index of nebular spectrum from low excitation (Class 1) through high excitation (class 10), introduced by Aller (1956).

Table 5.14: A list of typical PNCSs

Nebula NGC,IC	Central star Name	V-mag.	Spectral type P-K	vdH	Pott	$T_{\rm eff}$ ($\times 10^3$ K)
I 351		15	W	c	c	40
I 418	HD 35914	9.57	OW, W, etc.	WC7	Of	32
N 5315		14.9	—	WC6	WR	70
N 6543	HD 164963	10.4	OW6,W,WN6	WN6	Of-WR	45
N 6210	HD 151121	11.3	OW,WC,O7	WC7	Of	45
N 6826	HD 186924	10.2	c,OW,O5 etc.	WN6	Of	42
N 6891	HD 192563	11.09	c,O7,Of-WR	Of-WR	Of	38
N 5189	HD 117622	14.1	—	WC7-8	OVI	100
N 246		11.95	O6,sdOfe,etc.	OVI	OVI	100
N 7009		>10.9	C,WN	—	c	100
N 7662		13.2	c	—	c	80

Notes: Nebula—N, NGC, I, IC; Spectral type—P-K (Perek and Kohoutek 1967), vdH (van der Hucht et al. 1981), Pott (Pottasch 1984). The designation of WR should be read [WR] by modern nomenclature. Effective temperature $T_{\rm eff}$ are derived from the fitting of continuous spectra with model-atmosphere calculations.

O-type stars—The absorption lines of O stars, especially H and HeII, are observable. HeIIλ 4542 Å is often broader than that in Of stars, indicating high surface gravity.

Subdwarf O stars—The absorption lines are much broader than the above O-type stars. Balmer series are often disappearing by Stark broadening in around H12, indicating subdwarf nature (see Section 2.6.2 for the Stark broadening).

Peculiar—The spectra indicate the temperature much too cold to cause the ionization of the nebulae, i.e., spectral type later than A0. These stars are supposed as close binaries of which only bright components are observed.

Faint stars—In about 10% of the nebulae, no exciting central star is detectable, or it is very faint, even though the nebula itself is quite bright. For example, NGC 7027, one of the brightest nebulae, shows no central star. This type of stars usually occurs in small nebulae with high surface brightness, which makes them difficult to find the central faint stars. There are two possibilities for their faintness one is the very small, but very hot stars, and the other is the stars obscured by dust.

This spectral category of Pottasch (1984) has been amended later for some types:

(i) Tylanda et al. (1993) compiled observational data of [WR] type stars for 77 PNCSs and showed that the stars are classified into two types of ordinary [WR] stars (39 stars) and weak emission-line stars (WEL)

(38 stars). Ordinary [WR] stars show strong and wide emission lines similar to the case of Population I WR stars. They are all of the WC type with one exception of suspected WN type. On the other hand emission lines of WELs are significantly weaker and narrower. According to Peña et al. (2001), WELs are mainly distributed in later subtype ([WC5]–[WC9]) and showing relatively low expanding velocities of nebular gas. There is no substantial difference between ordinary [WC] stars and WELs in the excitation level and chemical abundance.

(ii) Type of OVI stars are extended to higher excitation of OVII and even OVIII stars. Feibelman (1999) detected OVIIλ 1522 Å emission or absorption in 14 PNCSs, while Feibelman (1996) found OVIII emission lines such as $\lambda\lambda$ 1930, 1932, 2977, 6068 Å in 10 PNCSs.

The common feature of these central stars is their low luminosity in spite of high surface temperature. In Table 5.14 a list of selected PNCSs is given, where the three spectral types by different sources show some diversity of classification.

Binary systems in PNCSs

It has become widely accepted that many of PNCSs are binary systems. Based on the photometric surveys Bond (2000) showed that around 10% of all PNCSs are close binaries with the orbital periods ranging from 2.7 h to 16 days. Radial velocity survey was carried out by De Marco et al. (2004) for selected 11 stars having low mass-loss rates and optical spectra dominated by photospheric absorption lines (i.e., non-[WR] nuclei). They thus detected radial velocity variations in 10 stars out of 11 stars. The periods of variation, though could not determined meaningfully, are supposed to be from days to a few hundred days (intermediate periods). On the other side, wide binaries with period longer than ~1000 days could be detected as resolved visual binaries. Ciardullo et al. (1999), using the wide field cameras of the Hubble Space Tellescope (HST) searched the fields of 113 planetary nebulae, and found 19 double stars, among which 10 binary nuclei are very likely to be physically associated and another six are possible binaries, and the remaining three being doubtful.

Planetary nebulae with binary nuclei exhibit the morphologies showing axisymmetric, either elliptical or more pronounced butterfly or bipolar shapes. It may be certain that the binary fraction of PNCSs is very high (if not all), and the binary nuclei are closely related to the axisymmetrical morphologies of planetary nebulae.

Emission-line profiles and mass loss rates

Reflecting the diversity of spectral types, the central stars of planetary nebulae exhibit various profiles of emission lines that are often variable. In many stars, P Cygni type profiles are observed in emission lines, suggesting that the mass-loss process is prevailing. Figure 5.14 illustrates some emission-line profiles of the central star He2-131 (O8(f)p type), observed by Mendez et al. (1988) with

Figure 5.14: The spectrum of He2-131, central star of planetary nebula, and its time variation in two epochs in May 1985 (upper) and in January 1986 (lower). Prominent P Cygni profile is seen in January 1986. (From Méndez et al. 1988)

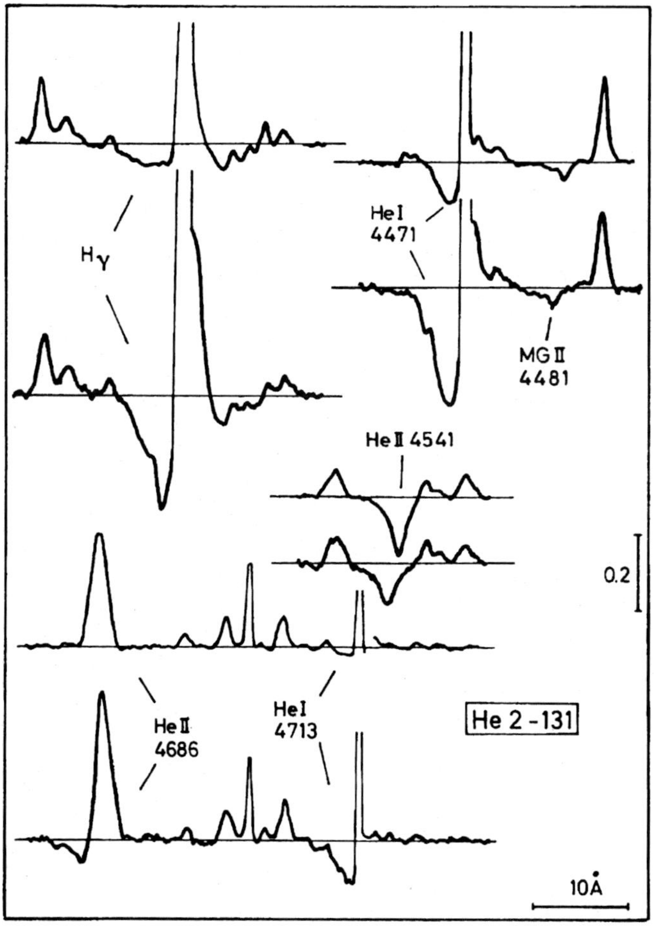

the ESO 3.6-m telescope. Two line profiles are traced for each of emission lines in different epochs of observations, i.e., upper one is taken in May 1985, lower is in January 1986. Clear P Cygni profile is conspicuous in the spectra of 1986.

There appear rich emission lines in the ultraviolet region in the lines of NV, NIV, OIV, OU, CIV, etc, and, they exhibit P Cygni profiles in many cases. Cerruti-Sola and Perinotto (1989) have observed with high-dispersion IUE spectrograph for central stars of some planetary nebulae. All of these stars presented P Cygni profiles in the ultraviolet lines. Based on the Sobolev exact integration (SEI) method, they analyzed the line profiles and found the mass-loss rates to be in the orders of $\dot{M} = 10^{-9} - 10^{-7}\, M_{\odot}$ per year. The stellar luminosity versus $\dot{M}$ diagram is shown in Figure 5.15, where the same data for the population I OB stars are also plotted for comparison. One may see that the central stars are located in the area expected from the extrapolation to lower luminosity of OB stars. From this result Cerruti-Sola and Perinotto

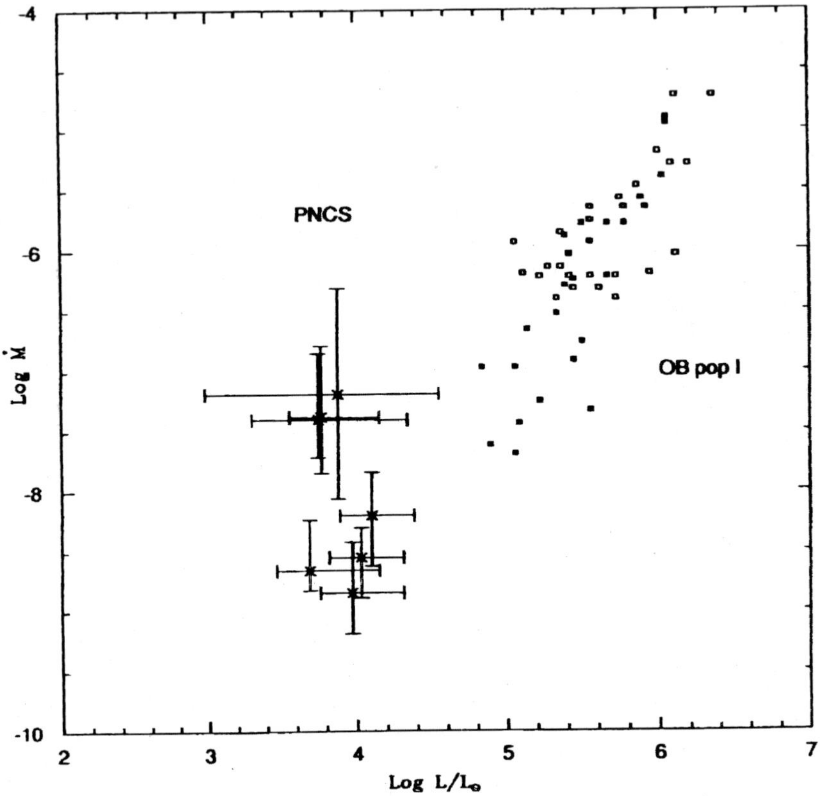

Figure 5.15: log $\dot{M}$ versus log $L/L_{\odot}$ for the central stars of planetary nebulae (PNCS) and for Population I OB stars. (From Cerruti-Sola and Perinotto 1989)

(1989) suggested the possibility that the driving mechanism of mass loss in the PNCSs is the same as in population I OB stars, i.e., the radiation pressure on heavy element ions.

5.3 B-type emission-line stars (Be stars)

5.3.1 What are Be stars?

Be stars are in its broad sense B-type stars that show emission lines in the Balmer lines (Hα, Hβ, ...), singly-ionized metals (FeII, TiII, ...), and sometimes neutral helium in their optical spectra. Bidelman (1976) classified Be stars in this broad sense, including supergiants and quasi-planetary nebulae. Later on, Jaschek et al. (1981) gave a more precise definition confining them to the stars of luminosity class V–III. Be stars in this definition are usually called the classical Be stars. Oe and Ae stars are often included in this category. Among nonclassical Be stars are supergiant Be stars (Section 5.4), peculiar Be stars (Sections 5.3.7 and 5.4.3), and Herbig Be/Ae stars (Chapter 7). These stars will be considered in the respective sections. In this section we focus the problems of classical Be, or simply, Be stars.

It is now widely accepted that Be stars are the rapidly rotating stars surrounded by disks or ring-like envelopes in the low-latitude regions, where the emission lines are formed. In addition, with the advent of ultraviolet observations in 1970s, it became clear that the hot and high-velocity winds flow out from the high-latitude regions of the stars.

Porter and Rivinius (2003) concisely review the recent development of our knowledge on the classical Be stars, and Yudin (2001) compiled the basic data of intrinsic polarization, IR excess, and projected rotational velocity ($V \sin i$) of a sample of 627 classical Be stars.

5.3.2 Basic types and catalogues

Classification by emission-line profiles

In Be stars, emission lines usually appear on the broad underlying photospheric absorption lines. Figure 5.16 illustrate the schematic line profiles of

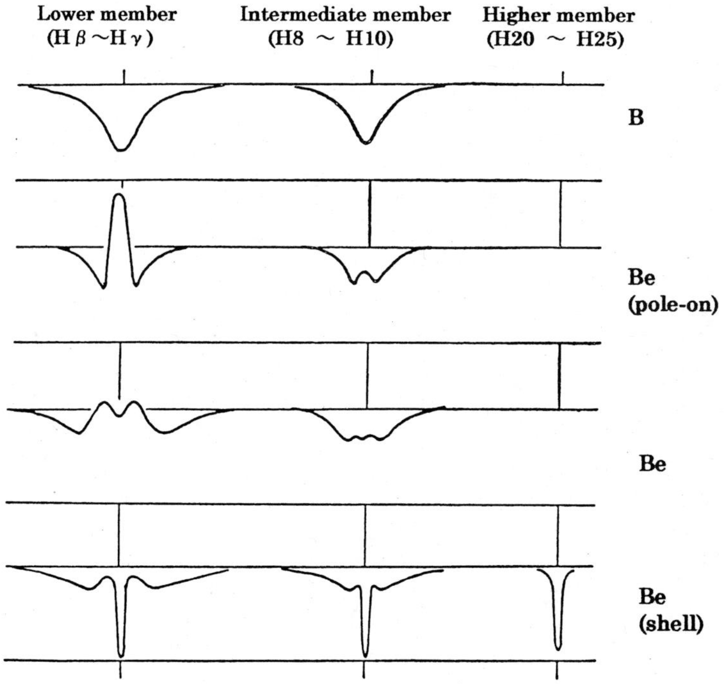

Figure 5.16: Typical schematic profiles of B and Be stars.

lower and higher Balmer lines in B and Be stars. Be stars are classified into three types according to their line profiles as follows.

(1) Pole-on stars: stars that are characterized by single-peaked narrow emission lines superimposed on the photospheric absorption lines.
(2) Be star (ordinary): stars showing double-peaked emission-line profiles. Since this type is most prevalent among Be stars, they are often called the ordinary Be stars.
(3) Be-shell stars: stars with sharp and deep absorption components in the centers of double-peaked emission lines are called Be-shell stars. Equator-on Be stars are usually included in this type.

The classical explanation for the origin of these line profiles has been brought by Struve (1931), who argued that the emission lines are formed in the disks or rings surrounding the equator of rapidly rotating B stars. According to this picture, the difference in the above different types of profiles can be explained by the different angles of line-of-sight with respect to the rotational axis of the star as schematically shown in Figure 5.17.

The emission-line profiles of Be stars are often strongly variable: sometimes emission lines disappear (normal B stars) and reappear (Be stars); sometimes the type changes such as from pole-on to ordinary Be, from ordinary Be to Be-shell; and sometimes emission-line profiles markedly changes. By this variable nature, Be stars are often regarded as Be phenomena rather than the special type of stars.

Catalogues and spectroscopic atlases

The first comprehensive catalogue of Be stars is the Mt. Wilson Catalogue (MWC), which has been successively published by Merrill and Burwell (1933, 1943, 1949, 1950) and contains 1600 Be, Ae stars and related objects, together with detailed references. Later on, several spectral atlases in optical, ultraviolet, and infrared regions have been published including the followings:

(i) Photographic spectral atlas by Hubert-Delplace and Hubert (1979). This atlas contains spectrum of 148 Be stars in the wavelength region around 3800–6600 Å, along with some physical data (spectral type, V-magnitude, color index $U-V$, $U-B$, $V \sin i$). Short descriptions on long-term variability are given for 35 stars, for each of which a sequence of 20 to 50 spectra is arranged to facilitate the tracing of variations.
(ii) Atlas of high-dispersion line profiles of Be stars by Hanuschuk et al. (1996). This atlas is a collection of high signal-noise ratio, high-resolution line profiles of Hα and FeII (mostly λ 5317 Å) for 77 Be and shell stars. Short-term variability (time scale between 5 days and a few minutes) and a classification scheme of line profiles are also overviewed.
(iii) Atlas of Far UV and Optical High-resolution Spectra by Doazan et al. (1991). Spectral traces of selected regions of far UV ($\lambda\lambda$ 1200–3000 Å) and optical (Hα, Hβ) spectra for 166 Be stars are collected. Particular

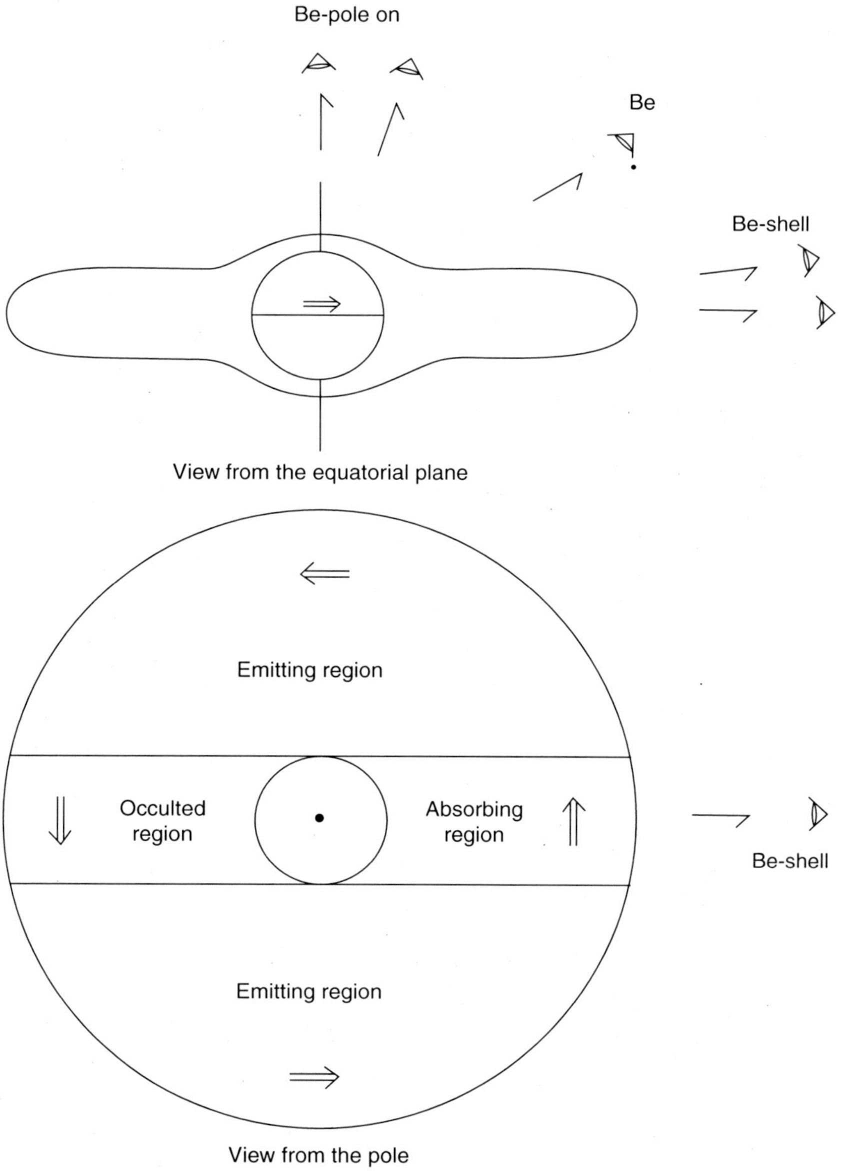

Figure 5.17: Classical model of rotating envelope of Be stars and the types of Be star emission-line profiles originally proposed by Struve (1931).

attention is given for the time variation of Pleione (28 Tau) in the UV spectrum during Be and Be-shell phase transitions.

(iv) In the near infrared, spectral surveys have been carried out by Andrillat et al. (1988, 1994) in $\lambda\lambda$ 7500–8800 Å region for 97 Be stars, and in $\lambda\lambda$ 9840–10,200 Å region centered upon Paschen 7 line for 74 Be stars. Spectral tracings are given in both papers.

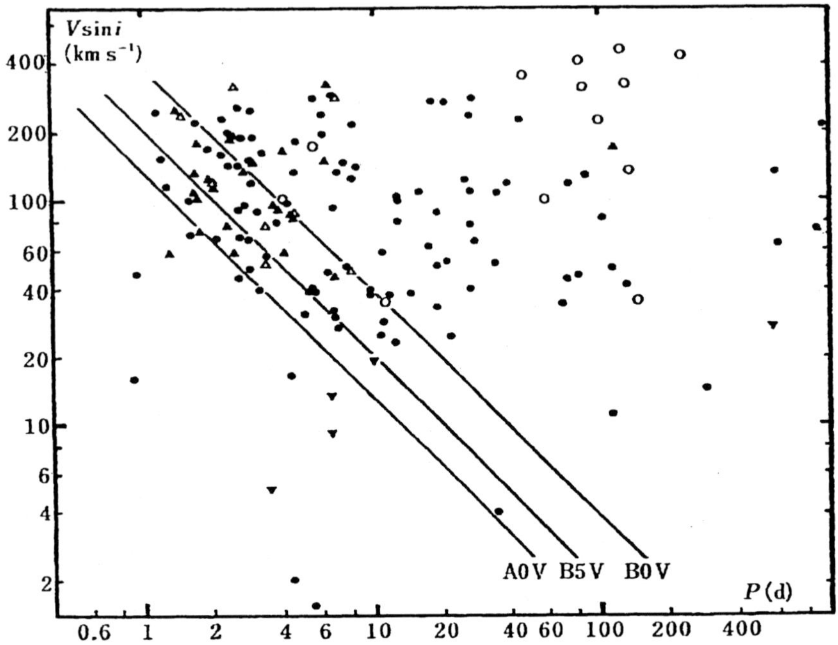

Figure 5.18: The period-rotational velocity relation for the spectroscopic binaries of B0–B9, V–III. Circles and triangles denote the noneclipsing and eclipsing binaries, respectively. The inverted triangles denote Bp stars. The emission-line stars are designated by open marks in each case. The lines of synchronization between axial rotation and orbital motion are also shown (Kogure 1981).

and $P_{\rm ob}$ the orbital period, then we have

$$V_* = \omega_* R_* , \qquad V_{\rm ob} = \omega_{\rm ob}\, a , \qquad \omega_{\rm ob} = \frac{2\pi}{P_{\rm ob}} \tag{5.3.1}$$

and the condition of synchronization is given by $\omega_* = \omega_{ob.}$ This gives

$$V_* \sin i = \frac{2\pi R_* \sin i}{P_{\rm ob}.} \tag{5.3.2}$$

The line of synchronization is given by inserting the value of R_* for given spectral type and assuming the inclination angle $i = 90°$. In Figure 5.18 three lines for B0V, B5V, and A0V thus obtained are indicated. One may see that the Be stars can be statistically separated into two groups of short-period ($P < 30$ day) and long-period ($P > 30$ day) binaries. The short-period group is well mixed in distribution with nonemission, line stars, and they are closely related to the Algol binaries. In contrast, the long-period group occupies the highest part of $V \sin i$ in Figure 5.18, where the rapid rotation should play an important role in the formation of envelopes.

According to Tarasov (2000), massive interacting binaries with Be components can be separated into two groups, taking into account the value of the orbital period as a primary parameter. The first group contains short period (2–7 days) binaries or "classical Algols." Most of these stars are dominated by the spectrum of a late B or early A star just like Algol systems. The second group contains more massive binaries with a wide range of periods from 6 to 7 days to a few hundred days, forming the so-called W Serpentis type binaries. The radii of gainers of this group are small relative to the binary separations and mass flows from the losers can form classical accretion disks observed in Be stars. The period–rotational velocity relation shown in Figure 5.18 may support this grouping of Be binaries.

Rotational velocities

The average rotational velocities $V \sin i$ take the maximum values in B-type stars along the spectral type as shown in Figure 2.18. Particularly, Be stars belong to the highest velocity class of all. Individually, however, a large variety is seen in $V \sin i$ among Be stars, from a few 10 km s^{-1} up to more than 400 km s^{-1}. Since inclination angle i can be supposed as randomly distributed, small values of V sin i may reflect the effect of small inclination angle i, and there is a possibility that the true value of rotational velocity is sufficiently large for every Be star.

Let us consider the real distribution of equatorial rotational velocity V in Be stars following the simple method of Fukuda (1982). For the distribution function $F(u)$ of the nondimensional rotational velocity $u = V/\langle V \rangle$, he assumed a simple step function such as the variable u is distributed uniformly over a range between $(1-p)$ and $(1+p)$, where p is an adjustable constant $(0 \leqq p \leqq 1)$, called the width parameter. Then the form of distribution function can be derived from the condition of random distribution of the inclination angle i, and the value of p is estimated by comparing with the histogram of observed rotational velocity u. The result is illustrated in Figure 5.19, where the vertical stripe in each spectral subtype corresponds to the value of $2p$ around the mean velocity $\langle V \rangle$. The hatched and open stripes denote the normal main sequence stars (luminosity class V) and the emission-line stars, respectively. The figure also shows the curve of break-up velocity, V_c, which is defined as the critical rotational velocity to make the gravity zero by the centrifugal force. The value of V_c is obtained by equating the centrifugal force, V_c^2/R, and surface gravity, GM/R^2, as

$$V_c = \sqrt{\frac{GM}{R}}, \tag{5.3.3}$$

where the mass, M, and radius, R, of the star are given by some model atmosphere calculation.

Figure 5.19 shows that B and Be stars are distributed well below the curve of break-up velocity as a whole, except for the late-type Be stars, whose

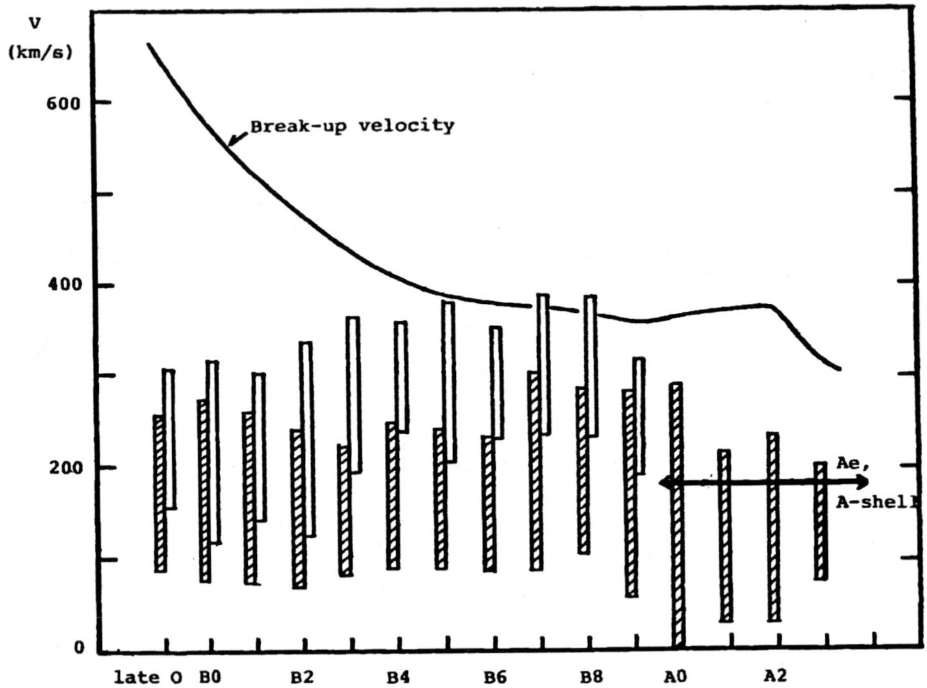

Figure 5.19: The distribution of rotational velocity V in km s^{-1} is shown along the spectral sequence. The hatched and open stripes denote the distribution areas of main sequence and emission-line stars, respectively. In A type stars, the average rotational velocity for Ae and A-shell stars is presented. The curve of break-up velocity by Collins model (1974) for the main sequence stars is also illustrated (Kogure and Hirata 1982).

rotational velocities apparently overtake the break-up velocity at their highest part. It is also noticed that the stripes of B and Be stars are mostly overlapping for stars earlier than B2, whereas the stripes of Be stars are definitively higher than those of B stars in the spectral range B3–B9. This implies that the late-type Be stars are essentially rapid rotators as compared to B stars.

The averaged values of $\langle V \sin i \rangle$ and $\langle V \rangle / V_c$ for different spectral subtypes and different luminosity classes of Be stars are compiled by Yudin (2001) for 418 stars and given in Table 5.16.

It is seen that the average ratios $\langle V \rangle / V_c$ are 0.67 for III, III–IV stars and 0.63 for IV–V stars, not significantly different each other. For all of Be stars, Yudin (2001) obtained the values of

$$\langle V \sin i / V_c \rangle = 0.50 \quad \text{and} \quad \langle V \rangle / V_c = 0.64.$$

Similarly, based on the model fitting for the line profile of He Iλ 4471 Å, Chauville et al. (2001) derived the values of

$$\langle V \sin i \rangle / V_c = 0.65 \quad \text{and} \quad \langle V \rangle / V_c = 0.83.$$

for around 100 Be stars.

Table 5.16: Average values of V sin i, number of stars, and ratio of $\langle V \rangle / V_c$ for different spectral subtype and different luminosity class of classical Be stars (adapted from Yudin 2001)

Luminosity class	III, III–IV				
Spectral subtype	O–B1.5	B2–B2.5	B3–B5.5	B6–B9.5	O–B9.5
Number of stars	37	30	28	40	135
$\langle V \sin i \rangle$	203 ± 11	211 ± 10	192 ± 14	207 ± 10	201 ± 6
$\langle V \rangle / Vc$	0.57	0.66	0.64	0.83	0.67
Luminosity class	IV–V, V				
Spectral subtype	O–B1.5	B2–B2.5	B3–B5.5	B6–B9.5	O–B9.5
Number of stars	66	78	78	61	283
$\langle V \sin i \rangle$	202 ± 9	207 ± 8	236 ± 7	243 ± 7	229 ± 4
$\langle V \rangle / Vc$	0.49	0.55	0.67	0.78	0.63

Note: Vc is the critical velocity defined by Equation (5.3.3).

Much higher value of $\langle V \rangle / V_b = 0.95$ has been obtained by Townsend et al. (2004) by taking into account the effects of the von Zeipel gravity darkening law ($T_{\rm eff}^4 \sim g_{\rm eff}$) in rapidly rotating stars. Thus they claimed that Be stars may be rotating much closer to their critical velocities than is generally supposed. If Be stars are actually rotating at or near the critical velocities, the dynamical processes responsible for the formation and development of Be envelopes should be affected seriously.

5.3.4 Balmer line spectrum

Emission-line intensities and Balmer decrements

Be stars are characterized by the existence of emission lines in the Balmer series, though their intensities are scattered and often variable. In Figure 5.20 the equivalent widths of the Hα emission, W(Hα), along the spectral subtype are shown based on the data of Sletteback et al. (1992) and Mennickent et al. (1994). It is apparent that there is an upper limit in the distribution of observed values, and it reaches the maximum at B2 and markedly decreases toward the late-type stars.

This distribution seems to show that early-type Be stars have envelopes more developed than late-type Be stars, not reflecting the effects of rotation, since rotational velocities are far from the corresponding break-up velocity in early types as seen in Figure 5.19. If the rapid rotation is the primary cause of envelope formation, the strongest Hα emission would appear in late-type stars. In addition, no correlation is seen between W(Hα) and $V \sin i$ in the above data set, also suggesting that the inclination angle may not be the primary parameter for the distribution of Hα intensity in Figure 5.20.

Next we consider the emission intensities of the Hα and Hβ lines and the decrement D_{34} = Hα/Hβ, which can be derived either by direct measurement

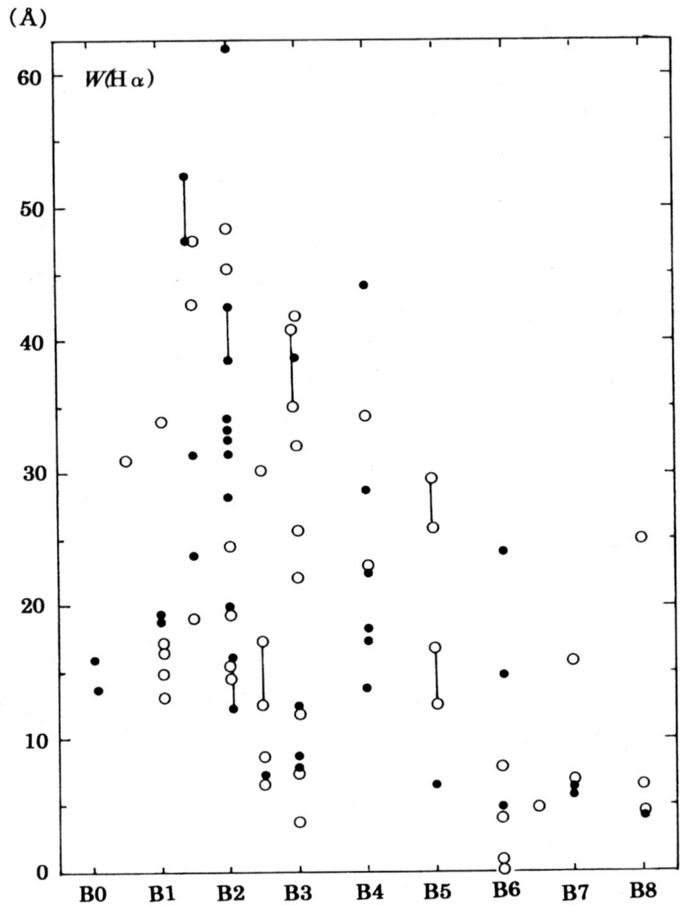

Figure 5.20: Equivalent width of the Hα emission along the spectral type of Be stars. Open circles denote the values of Sletteback et al. (1992), and filled circles those of Mennickent et al. (1994). The same star at different epoch of observations is connected by a line.

of the line fluxes, or by measurement of the equivalent widths of emission lines by using the conversion factor $G\alpha$ (Section 4.1.5). In both cases, the correction for the photospheric absorption profiles should be made. The observed data D_{34} of Briot (1971), Dachs et al. (1990), and Slettebak et al. (1992) are unified by adopting the common values of $G\alpha$ based on Collins model (Collins et al. 1991) and plotted in Figure 5.21. It is seen that the decrements D_{34} largely scatter from 1 to around 10, with a gradual increase toward later type Be stars.

The decrement D_{34} has been theoretically interpreted, as stated in Chapter 4, by two different approaches: nebular approach and envelope approach.

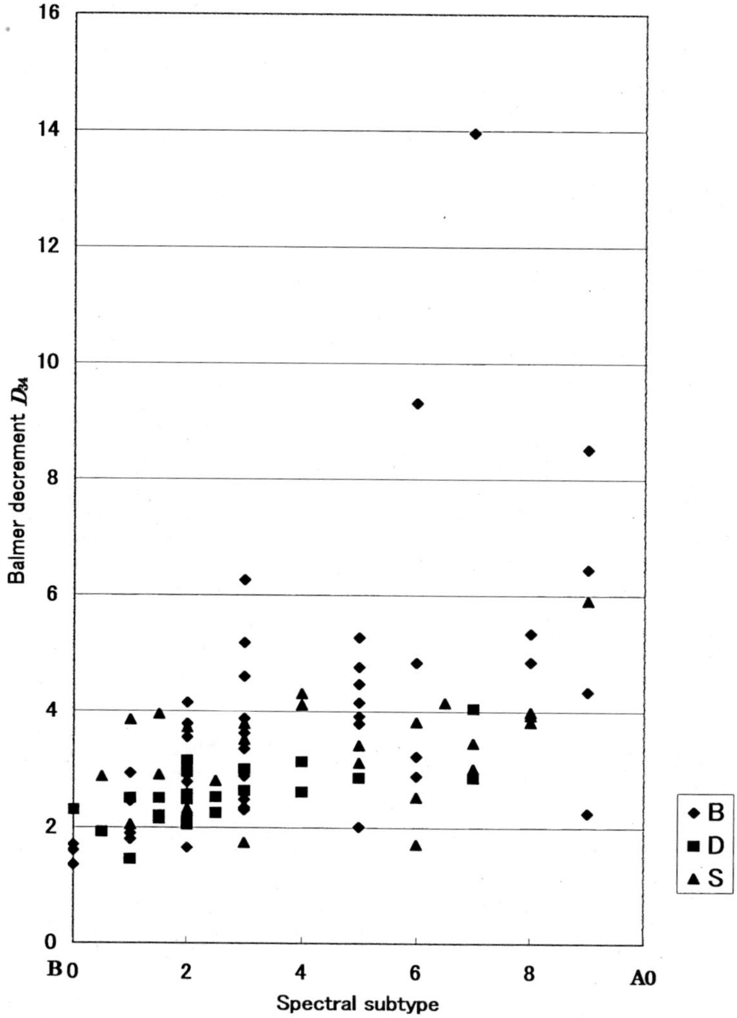

Figure 5.21: Balmer decrement D_{34} versus spectral subtype of Be stars. Different symbols indicate different observers: B (Briot 1971), D (Dachs et al. 1990), and S (Slettebak et al. 1992). When the decrement showed variability, average value is plotted. All of these observed decrements were derived from the emission equivalent widths so that the conversion factor G_α was unified by adopting Collins model (Collins et al. 1991).

Nebular approach, which has been developed by Drake and Ulrich (1980) based on the escape probability by scattering (see Figure 4.6), has been applied by Dachs et al. (1990) and Slettebak et al. (1992) for the decrements of Be stars. Both of them assumed the plane-parallel gas layer for the envelopes of Be stars, and applied the decrements D_{34} and D_{54}, of Drake and Ulrich by taking the electron density N_e as a free parameter among their three parameters of

N_e, T_e, and R_{1c}. The ionization parameter R_{1c} is introduced to adjust the amount of ionizing radiation. Dachs's and Slettebak's groups explained the variety of D_{34} by the variation of electron density under the conditions of constant R_{1c} and T_e. On this application some remarks are needed.

(1) Since constant value of R_{1c} fixes the spectral type of stars, it is difficult to explain the distribution of D_{34} in Figure 5.21. The theoretical decrement needs to be adjusted at least by two parameters of N_e and R_{1c} in the way to fit with the structure of the envelope.

(2) By comparing the decrements of Be and Be-shell stars, Slattebak et al. (1992) concluded that the envelopes of Be-shell stars are more compact and have lower electron density than those of Be stars. If so, the emission measure of the envelopes of Be-shell stars must be smaller than that of Be stars and this makes the Hα emission weaker. Actually, however, the Hα emissions of Be-shell stars are almost the same with Be stars. In addition, existence of shell-absorption lines indicates that the envelopes of Be-shell stars should be much optically thicker than those of Be stars (see next section). In order to promote the nebular approach following Drake and Ulrich (1980), more extensive grid calculations taking the N_e and R_{1c} as free parameters would be needed.

The envelope approach has been developed by Miyamoto (1949, 1952a,b), Kogure (1969, 2000), and Pottasch (1961) (Section 4.1.4). According to Kogure's model, the values of D_{34} are determined as the function of stellar effective temperature ($T_{\rm eff}$), dilution factor (W), and the optical depth for the Lyman continuum radiation (τ(Lc)).

The results of calculation in Kogure's case are shown in Figure 5.22, which was derived under the assumption of plane-parallel layer exposed to a diluted stellar radiation and of the constant optical depth for Lyman continuum throughout the envelope. It is shown in Figure 5.22 that the decrement D_{34} is getting steeper as the stellar temperature declines and the dilution effect increases. For comparison, the average curves of observed decrements are also delineated. One may see that the scattering of observed decrements is understandable in terms of the two parameters—stellar temperature and dilution factor—and the average decrements take values of $W = 0.1$–0.2, not depending on the spectral type, indicating that the average size of envelopes relative to the stellar radius remains nearly constant throughout the spectral type.

Emission-line profiles

As seen in Section 5.3.2, Be stars are classified into three types of pole-on, ordinary Be, and Be-shell stars according to the profiles of emission lines. In this section we briefly consider the formation and variation of profiles.

(a) Pole-on stars. There are two types of pole-on stars: one is the intrinsic pole-on stars, which exhibit always single-peaked emission lines

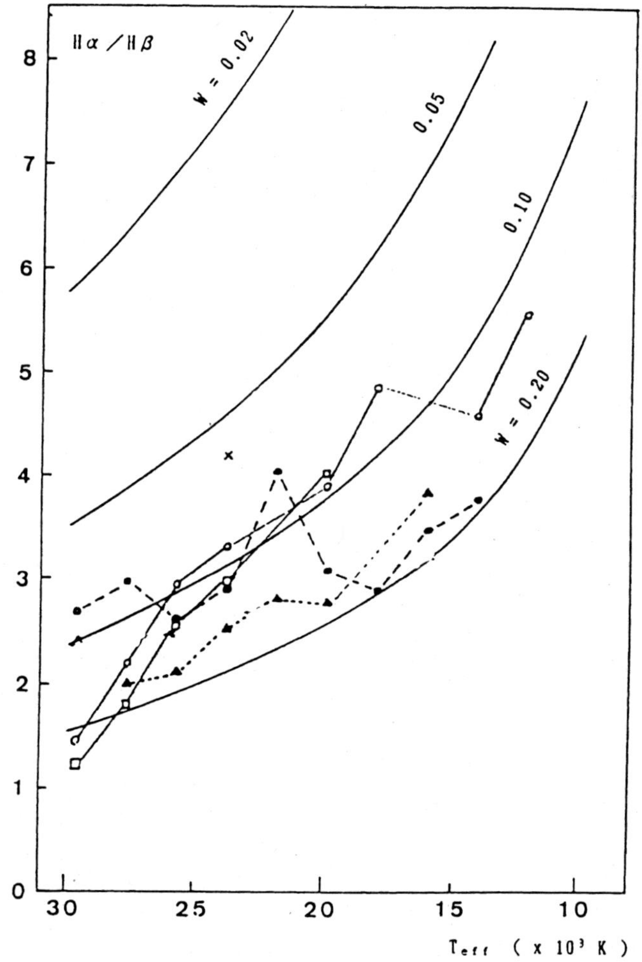

Figure 5.22: The Balmer decrement D_{34} of Be stars in Kogure's solution. Average decrements by several observers are shown by different symbols and lines. Theoretical curves denote the decrement with constant dilution factors W (Kogure 2000).

without changing to double-peaks, whereas the other is the temporary pole-on stars which sometimes show the exchange of double- and single-peaked emission in long-term variations.

It is usually supposed that the pole-on stars are the stars observed nearly from their rotational axes. Kogure (1969) has explained the profile of Hβ emission and its variations in terms of the development of the disklike envelopes seen from small inclination angles. He calculated the emergent flux of the $H\beta$ based on the method of velocity zones (see Section 4.2.3). A part of his model calculations in case of $V \sin i = 40$ and 80 km s^{-1} is shown in Figure 5.23 (a) and (b). Both figures show

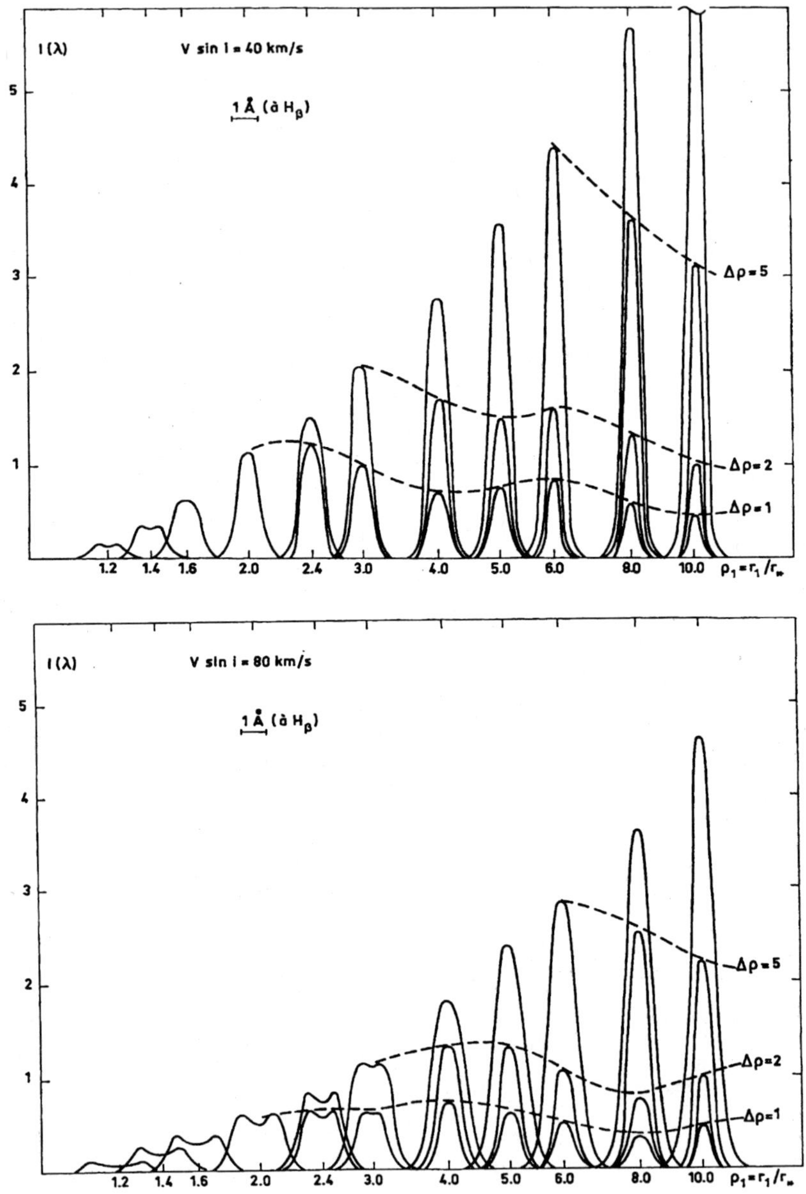

Figure 5.23: The growth of Hβ emission-line profile in nearly pole-on stars. The ordinate indicates the emission intensity in unit of the nearby continuum and the abscissa the outer radius of envelope, ρ_1, corresponding to the formation of emission-line profiles at the respective radius. The dashed lines connecting the emission peaks denote the level of emission-line peaks emitted from the detached ring with constant ring width in three cases of $\Delta\rho = \rho_1 - \rho_0 = 1$, 2, and 5, where ρ_0 is the inner radius of the ring. (a) Case of $V \sin i = 40$ km s^{-1} ($i = 6°$) (b) Case of $V \sin i = 80$ km s^{-1} ($i = 30°$), where the inclination angle given in the bracket corresponds to the rotational velocity of $V = 400$ km s^{-1} (Kogure 1969).

the variation of the Hβ profile when the outer radius $\rho_1 = r_1/r_*$ is gradually increased. If we assume $V = 400\,\mathrm{km\ s^{-1}}$, these figures correspond to the inclination angle $i = 6°$ and $30°$, respectively.

One may see that the profile is almost always single-peaked in case of $i = 6°$, whereas the profile changes from double to single peak as the envelope develops to larger size in case of $i = 30°$. Thus these figures may be the samples of intrinsic pole-on ($i = 6°$) and temporary pole-on ($30°$) stars, caused by the difference of inclination angle. In Figure 5.23, the case of ring structure with constant ring width ($\Delta\rho = \rho_1 - \rho_0 =$ constant) is also shown. The peak intensity remains almost constant or gradually decreases after detachment of the ring.

(b) Ordinary Be stars. Double-peaked emission lines are formed when the inclination angle i is sufficiently large. The violet and red components of the profile are formed in the approaching and receding parts of the rotating envelope, respectively. If the rotational velocity is decreasing monotonously with the radius, the peak separation enables us to estimate the outer radius by a simple formula as follows.

Let 2 Δv (km s^{-1}) be the peak separation measured by velocity, then the Δv is expressed as the function of the outer radius r_1 and rotational velocity $V \sin i$, as given by (Huang 1972, Hirata and Kogure 1984)

$$\Delta v = V \sin i(r_1/r^*)^{-j}, \tag{5.3.4}$$

where j is a parameter expressing the velocity law inside the envelope, i.e., the case of the angular momentum conservation ($j = 1$) and the Keplerian motion of the envelope ($j = 1/2$). Thus the peak separation is a good measure of the outer radius r_1 provided that we know the circular-velocity law, i.e., the value of j, and that the radial motion in the envelope is small. In Be stars, many attempts have been made to estimate the outer radii of Be star envelope based on this formula. Jaschek and Jaschek (1993) for 119 Be stars and Mennickent et al. (1994) for 42 Be stars have measured the peak separation and estimated the outer radii based on Equation (5.3.4), and assuming the Keplerian motion ($j = 1/2$).

According to Jaschek and Jaschek, outer radii of Be stars is almost independent on the spectral subtype and on the values of $V \sin i$, and the values of r_1 derived from the Hα line lie in a range of several to several ten times of the photospheric radii. Furthermore, the outer radii derived from the Hα, Hβ, and Hγ are decreasing in this order, implying that the Hα emission lines which are optically thick are formed in wider regions of the envelopes, whereas Hβ and Hγ lines are formed in the inner region with higher gas density.

On the value of j, Hummel and Vrancken (2000) presented two distinct methods to constrain its value: one is the line profile modeling and the other is the comparison of interferometric and spectroscopic disk radii. Applying these methods to several Be stars they concluded that Keplerian rotation is a valid approximation.

(c) Be-shell star. The sharp and deep absorption profiles which characterize the Be-shell stars are formed in the optically thick parts of envelopes lying in front of the stellar disk, Be-shell stars are then thought to be the stars with higher inclination angles (nearly equator-on). Cohen and Taylor (1994) argued that Be-shell stars are distinguished from Be stars by larger equivalent widths of the Hα emission and higher values of $V \sin i/V_c$, where V_c denotes the critical rotational velocity against the surface gravity. They thus supported the view that Be-shell stars have higher inclination angle i in general. The requirements for the formation of shell lines are considered in some detail in the next section.

Shell-absorption lines

Shell stars are mostly observed among the stars with large values of $V \sin i$. In Figure 5.24, we show the spectrum of 59 Cyg observed by Barker (1982) as an example. In this spectrum, the shell-absorption lines of the Balmer series can be traced up to H25, and, in addition, there appear many shell-absorption lines of HeI and ionized metals such as FeII, CrII, TiII, MgII, and CaII.

We now consider the hydrogen Balmer line spectrum of shell stars. The broad photospheric absorption lines can be seen up to around H14–H16, while emission components usually disappear in lower members less than Hε or so. In contrast, shell-absorption lines remain visible often for higher members higher than H20 or even H30. Let n_1 be the highest quantum number of the visible shell-absorption lines, then n_1 is a good measure of the development of shell-forming envelopes. We call the strong shell stars when n_1 reach the value of 20 or more, as seen in 59 Cyg ($n_1 = 25$) in Figure 5.24, or in EW Lac which has shown the highest record of $n_1 = 42$. On the other hand, weak shell stars reveal lower values of $n_1 = 10$–12, as in ψ Per ($n_1 \sim 12$) (see Figure 5.25).

The central depth of the shell-absorption line decreases with increasing quantum number. From this relation we can estimate the optical depth of the absorbing layer for Hα line, τ (Hα), as follows. Let r_{2n} be the residual intensity of the Hn line, and consider higher members of shell lines for which $\tau(\mathrm{H}n) < 1$, then we can make use of the Equation (4.2.42) in the case of $k = 0$, which denotes the zero line-of-sight velocity as expected in the gas layer lying in front of the photosphere under the assumption that the envelope has no radial motion. Then Equation (4.2.42) can be written as

$$1 - \frac{r_{2n}}{r_{2n}^*} = \beta\,\{1 - \exp\left[-\omega_n\, \tau\,(H\alpha)\right]\}\,, \tag{5.3.5}$$

where

$$\omega_n = \frac{\nu_{2n} B_{2n}}{\nu_{23} B_{23}} = \frac{\tau\,(\mathrm{H}n)}{\tau\,(\mathrm{H}\alpha)} \tag{5.3.6}$$

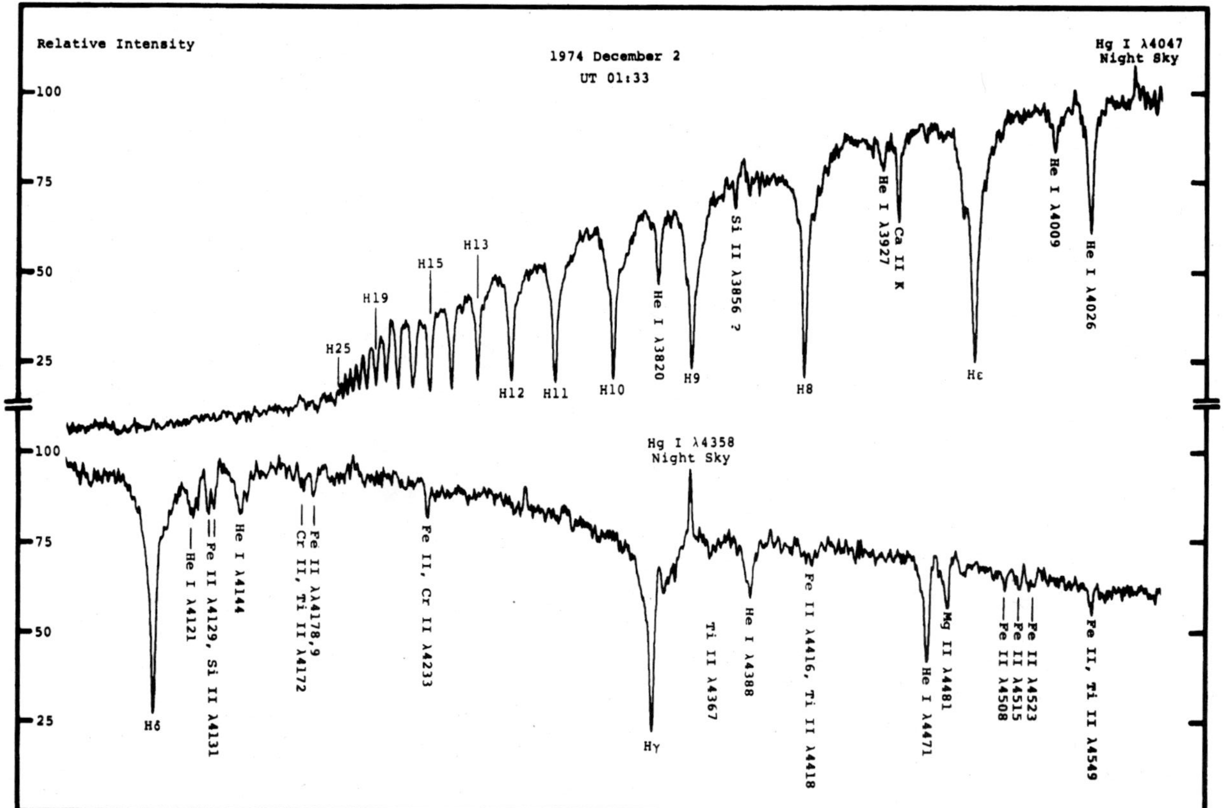

Figure 5.24: Spectrum of 59 Cygni, observed by Barker (1982) with the 61-cm reflector at the Colorado University in December 1974. The shell absorption lines in Balmer series are seen up to H25, along with many metallic and helium shell lines. These shell lines were disappeared next year in October 1975.

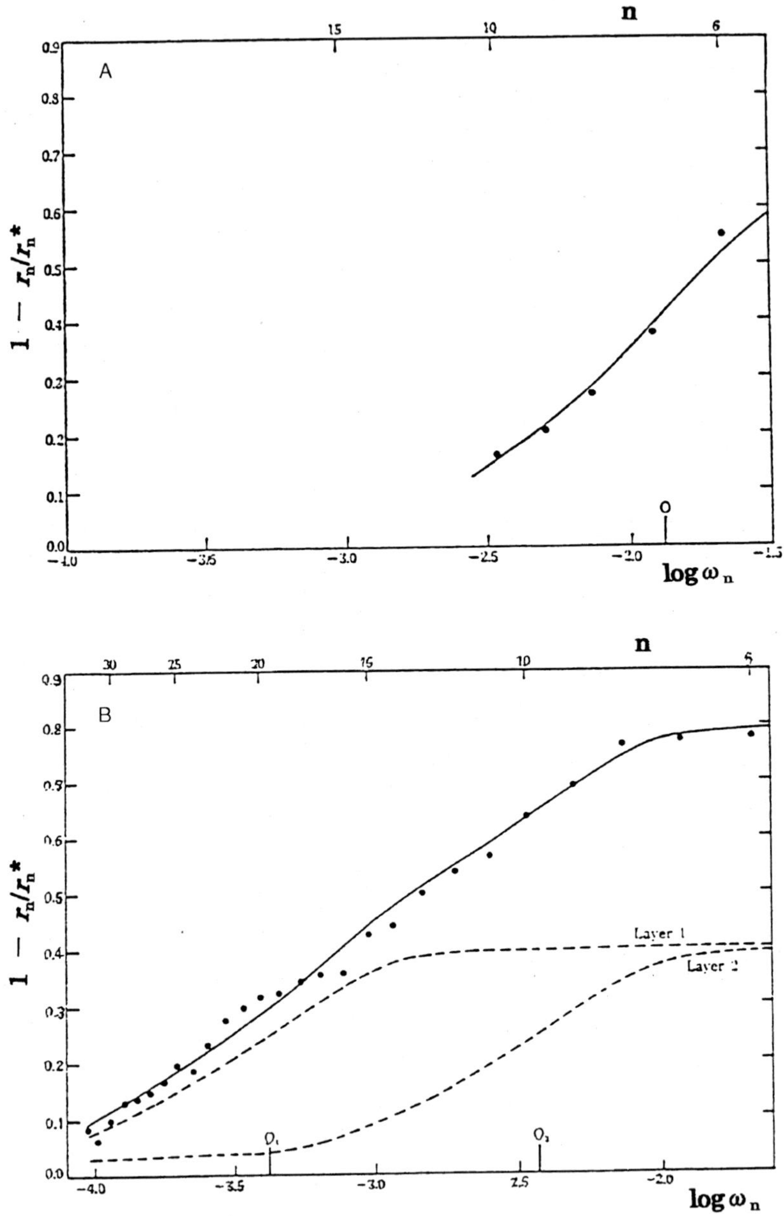

Figure 5.25: The central depths and relative optical thickness of the shell absorption lines. The abscissa gives the value of log ω_n for the line Hn, and the position of Hn line is shown in the upper side. A. The relation for the weak shell star ψ Per. Theoretical curve by a single-layer approximation is shown, the position O denotes the point at which theoretical curve take the value of log $\omega_n = 0$. B. The relation for the strong shell star ζTau. The observed points can be fitted by a combination of two layers, optically thick (Layer 1) and optically thin (Layer 2). The higher members correspond to Layer 1 and the lower members to Layer 2 (Kogure et al. 1978).

and β denotes the fractional area of the photospheric disk that is screened by the $(k = 0)$-th velocity zones (see Figure 4.15). The residual intensity of the broad photospheric absorption lines of H_n, r^*_{2n} can be taken as unity for n higher than around 20.

Let us consider a simple function

$$y = \beta\left(1 - e^{-x}\right) \tag{5.3.7}$$

with a free parameter β. When x takes a discrete value of $x_n = \omega_n \tau(H\alpha)$ (n = 3, 4, ...), the corresponding value of y_n gives the observable value of $1 - r_{2n}/r_{2n}{}^*$ by adjusting the value of β. For a lower member of the Balmer lines, since x_n is generally large ($x_3 = \tau$ (Hα)), we get $y_n \approx \beta$. On the $(\log \omega, y)$ plane we can plot the observed values of y_n, using the values of ω_n given by (5.3.6). On this plane, we search for the best fitted position by horizontal sliding of the theoretical curves. With this fitting we obtain both the value of β and the position of $\log \omega = 0$ or $x = 1$. By reading the value of ω at this point, we can derive the value of $\tau(\alpha)$ by the relation $x = \omega\tau(\mathrm{H}\alpha) = 1$. An example of the fitting is shown in Figure 5.25 A for the weak shell star ψ Per.

In many cases, however, we need two groups of Equation (5.3.7) in order to get the sufficient fitting with observations. In this case we have two points O_1 and O_2 on the abscissa by the two fitting procedures as shown in Figure 5.25 B for ζ Tau. Thus we get two pairs of $(\beta_1, \tau_1(\mathrm{H}\alpha))(\beta_2, \tau_2(H\alpha))(\tau_1 > \tau_2)$ from the fitting. The appearance of two pairs can be explained by the double structure of the envelope as shown in Figure 5.26. If we assume that the shell stars are almost equator-on, then the derived two pairs of $(\beta_1, \tau_1(\mathrm{H}\alpha))$ and $(\beta_2, \tau_2(\mathrm{H}\alpha))$ indicate the vertical structure composed of two layers of optically thick and thin layers. The thick layer has the optical thickness $\tau_1(\mathrm{H}\alpha)$ for Hα radiation and some vertical height to be calculated from the fractional area of the photospheric disk given by β_1. Similarly, optically thin layer has the thickness $\tau_2(\mathrm{H}\alpha)$ and some vertical thickness in both sides of the optically thick layer. Figure 5.26 illustrates a schematic vertical structure of the shell-star envelope. The optical depths thus derived for some shell stars are shown in Figure 5.27. It is seen that the optical depth for weak-shell stars (ψ Per or a phase of ϕ Per) takes the value around $\tau(\mathrm{H}\alpha) \sim 100$, whereas the strong sell stars usually take two values of $\tau_1(\mathrm{H}\alpha) \sim 2000$–$5000$ and $\tau_2(\mathrm{H}\alpha) \sim 200$–$300$, regardless the values of $V \sin i$. Strong shell stars usually exhibit two-layer structure as seen in Figure 5.27.

Disks to rings?

Concerning the evolution of circumstellar envelopes of Be stars, Rivinius et al. (2001) suggested that the disk structure can be changed to the ringlike structure in the course of envelope formation around Be stars. They argued that the quasi-static disk can be formed after an outburst event, and, after some weeks or months, a low-density region seems to develop above the star growing up to a ringlike structure. The formation of ring structure can be traced

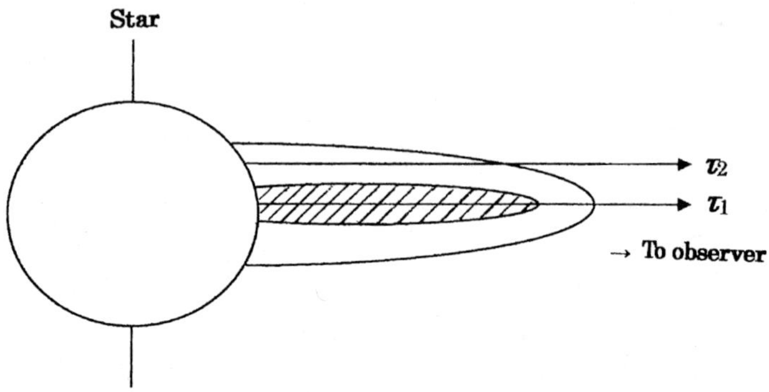

Figure 5.26: Schematic picture of the two-layer structure of the disk-envelopes of shell stars seen equator-on. The optically thick layer concentrates to the stellar equator, and the optically thin layer lie in both sides of the equator. The vertical thickness and optical thickness of these layers are obtainable by the shell-line analysis, if the equator-on view can be assumed (see text).

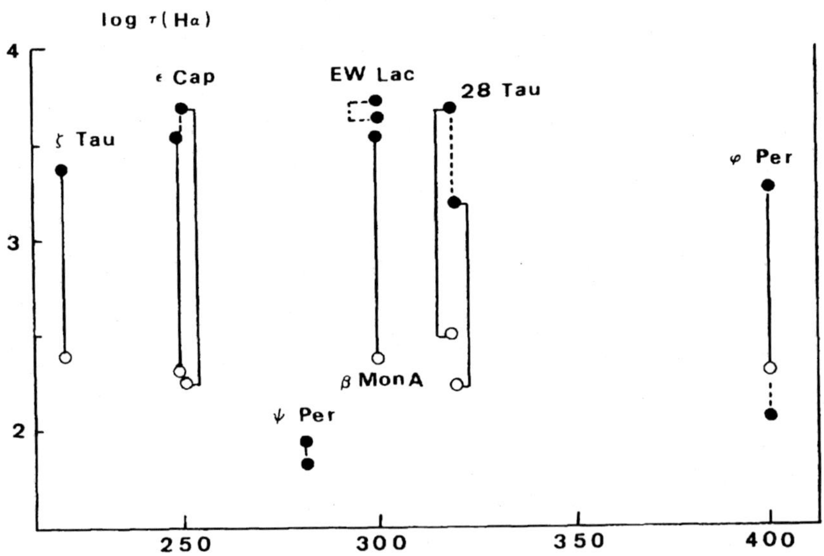

Figure 5.27: The optical depths for the Hα line for some shell stars. Abscissa gives $V \sin i$. When a two-layer approximation is adopted, the optically thick and thin layers are designated by filled and open circles, respectively. Two layers for each star are connected by a thick line, and the same star at different époque are linked by dotted lines Kogure (1990).

spectroscopically by the following features:

(1) Emission-line width decreases, particularly in the wings of lines, by the disappearing of fastest portion of the rotating disk.
(2) Equivalent widths of emission lines remain constant or even decrease with the expansion of the ring.
(3) Balmer decrement D_{34} steepens by the decrease of dilution factor.

These changes can be understood in Figure 5.23 in case of near pole-on stars and in Figure 5.22 for the dilution effect on D_{34}. Rivinius et al. furthermore argued that a subsequent outburst may replenish the inner cavity and new disk structure will develop. They observed the possible formation of rings in two Be stars of μ Cen (B2 IV-Ve) and ω CMa (B2.5 Ve).

Spatially resolved structure

First attempts to apply optical interferometry to Be stars have been made by Thom et al. (1986), based on a long-baseline Michelson interferometry at Côte d'Azur Observatory. Since then, disk-like structure of Be star envelopes has become directly observable. Recently, Tycner et al (2005) carried out interferometric observations with the Navy Prototype Optical Interferometer to derive the properties of the Hα-emitting envelopes of Be stars η Tau and β CMi. By combining with existing data, they derived the relationship between the net Hα-emission and the physical extent of the Hα-emitting regions. A part of their results is shown in Table 5.17, where the stars are arranged in the order of spectral subtype. Though the number of sample stars is not fully

Table 5.17: Physical parameters of the Hα-emitting region of Be stars, derived from optical interferometer (adapted from Tycner et al. 2005)

Star	Spectral type	d (pc)[a]	$\theta_{\rm mj}$ (mas)[b]	r[a]	$D_{\rm mj}$ (10^9 cm)[d]	$D_{\rm mj}/R_*^e$	$E(H\alpha)$ Å[f]	L(Hα) (10^{25} W)[g]
γ Cas	B0 IVe	188	3.67	0.79	103	18.6	−25.1	287
ϕ Per	B2 Vpe	220	2.67	0.46	87.8	21.0	−38.7	165
48 Per	B3 Ve	170	2.77	0.89	70.4	20.2	−26.8	77
ζ Tau	B4 IIIpe	128	3.14	0.31	60.1	14.4	−25.5	72.2
ψ Per	B5 Ve	215	3.26	0.47	105	32.1	−38.9	126
η Tau	B7 IIIe	113	2.08	0.75	35.1	10.1	−10.2	25.6
β CMi	B8 Ve	52.2	2.13	0.69	16.6	6.62	−9.42	4.66

[a] d, distance (based on Hipparcos parallax).
[b] $\theta_{\rm mj}$, angular diameter of the major axis of the elliptical $H\alpha$-emitting envelope.
[c] r, axial ratio of elliptical Gaussian model.
[d] $D_{\rm mj}$, $\theta_{\rm mj}$ in physical units.
[e] $D_{\rm mj}/R_*$, $D_{\rm mj}$ in unit of stellar radius.
[f] $E(H\alpha)$, emission equivalent width of the $H\alpha$,
[g] $L(H\alpha)$, $H\alpha$ luminosity.

large, we can see some behaviors characterizing the Be star envelopes from this table as

(1) Size of the Hα-emitting region generally decreases toward later spectral type. This reflects a tendency of hotter stars to have larger envelopes, and this tendency can be attributed to the amount of ionizing radiation from the stars. In early-type stars such as γ Cas the circumstellar envelope is almost completely ionized, in contrast to the case of late-type Be stars (like β CMi) whose envelopes may have smaller Hα-emitting regions due to ionization-bounded conditions in their circumstellar envelopes.
(2) Size of the Hα-emitting region relative to stellar radius, $D_{\rm mj}/R*$, in a range around 10 to 20 do not show particular tendency with spectral type. This is consistent with the trend seen in Figure 5.22, which indicates that the mean dilution factor of Be star envelopes is nearly constant along the spectral subtype.
(3) The Hα luminosity exhibits a clear dependence on the linear size of the envelope, implying that the Hα emission is formed in an optically thick region, since in that case, the emission intensity is directly proportional to the effective area of the emitting disk.

5.3.5 Other spectroscopic properties

Linear polarization

Linear polarization of Be stars is usually attributed to the electron scattering of light in a highly flattened envelopes extending surrounding the star's equators. First detection of intrinsic polarization in Be stars was made by Behr (1959) and, since then, a number of observations has been carried out both in the continuum and in emission lines (Coyne 1976). Since polarization degree of Be stars is mostly less than 2%, and, moreover, Be stars are distributed along the Galactic plane, the subtraction of intrinsic polarization from interstellar polarization is essential in the polarimetry of Be stars. Removing of interstellar polarization can be done in principle by using the properties of interstellar polarization (wavelength dependence and polarization position angle for nearby field) (see Quirrenbach et al. 1997). We here consider some characteristic properties of the intrinsic polarization in Be stars after corrected for the interstellar polarization.

(a) Wavelength dependence
The observations of linear polarization in Be stars cover a wide wavelength range from ultraviolet to infrared regions. If the origin of linear polarization is attributed to the pure electron scattering, the polarization degree will not depend on wavelength. Actually, however, self-absorption of starlight by hydrogen atoms in free-free and free-bound transitions acts as a source of depolarization and modifies the wavelength dependence, as first noticed by Coyne and Kruszewski (1969).

As an example, we consider the disk-model fitting by Wood et al. (1997) made for ζ Tau (B1 IVe-sh, $V \sin i = 220$ km s^{-1}). They carried out numerical simulations for the transfer problem that determine the polarization due to multiple photon scattering, including the effects of continuous hydrogen absorption and emission from a disklike circumstellar envelope. For an isothermal envelope with a constant opening angle, they calculated the spectrum over a wavelength range from ultraviolet (λ 1500 Å) to infrared (60 μm), so as to match the size of the Balmer jump and the slope of the Paschen and IRAS continua with observational data. Calculation was made in two cases of geometrically thin envelope (opening angle 2.5°) and thick envelope (opening angle 52°), among which they preferred the thin envelope as a realistic model. Figure 5.28 illustrates the result of calculation in the geometrically thin case along with observed spectrum. The upper panel shows the model (thick line) and observed (thin line) spectra, and the lower panel gives the corresponding linear polarization. The depression of polarization in the short wavelength sides of Balmer and Paschen series limits (λ 3646 Å, λ 8203 Å,

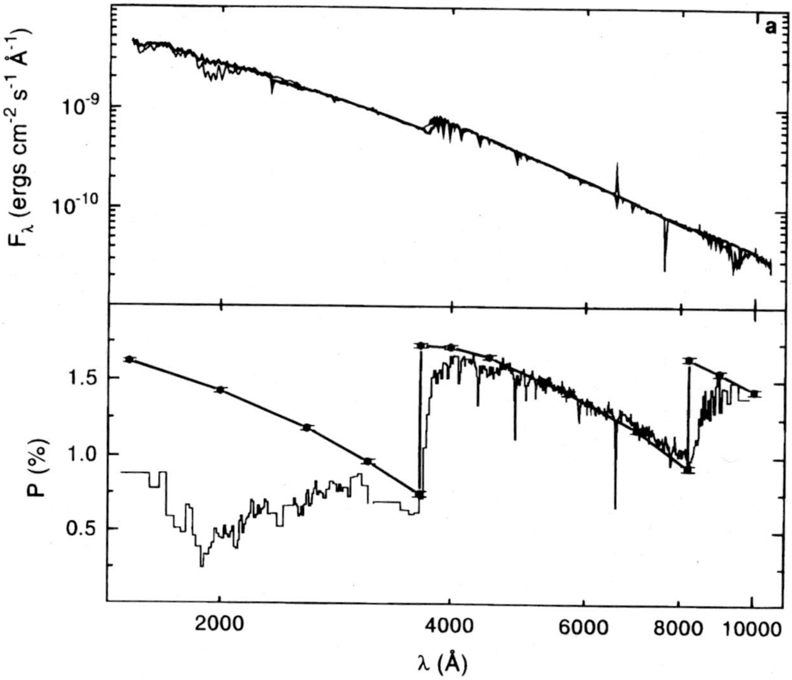

Figure 5.28: Model fitting for the Be star ζ Tau. Upper panel: thin disk model (thick line) and observed (thin line) spectra. Lower panel: linear polarization for the respective case. Polarization is corrected for interstellar reddening and polarization. (From Wood et al. 1997)

respectively) correspond to the increase of bound-free absorption coefficients in these regions. A large discrepancy of linear polarization in the UV spectral region between model and observation seems to require some additional depolarization agencies, for which Wood et al. suggested the existence of many metallic lines in the UV spectral region.

It is also noticed that the Balmer emission lines are formed by isotropic scattering of light emitted from ionized hydrogen so that the emission lines give rise to the depolarization effect similar to continuous radiation. Since stronger emission shows stronger depolarization effect, the depression of linear polarization will be higher in the central part of the emission lines. Poeckert and Marlborough (1977) showed the depression of around 0.2% inside the Hα emission of γ Cas, and similar depression more than 0.5% in the Hα of ζ Tau can be seen in Figure 5.28.

Relationship between linear polarization in B band and Hα emission intensity has been examined by Ghosh et al. (1999) for 29 Be stars. They showed that the linear polarization is low for stars with weak Hα emission, and gradually increases with increasing Hα emission. This relationship is consistent with the model that the circumstellar envelopes can be sufficiently optically thick in electron scattering.

(b) Time variations

A remarkable polarization change was observed in the Be star π Aqr (B1 III-IVe). This star is known as one of the largest polarization among Be stars. In the monitoring multicolor polarimetry for some Be stars from 1985 to 1998, McDavid (1999) found that π Aqr showed a gradual declining of polarization, particularly in B color, from 2% in 1986 to almost disappearing in late 1990s. During this period, Bjorkman (2000) observed large polarimetric "outbursts," three times in 1989, 1991, and 1994. The polarization variation in V-band is shown in Figure 5.29. The origin of gradual declining and occasional polarization outbursts may be related with the dynamic change of the envelope structure of this star, though not well understood yet.

Another example is X Per (O9.5 IIIe). Kunjaya and Hirata (1995) observed time variatioins of linear polarization $P(\%)$ in the optical continuum and equivalent width of the Hα emission in the period from 1989 through 1992. They found that the polarization decreases with the darkening in V magnitude; they explained these variations in terms of the variation of electron density in a simple rotating disk with a constant opening angle and a constant radial distribution law for the electron density. Since polarization is originated by the electron scattering, decreasing of electron density causes the decrease of optical depth for electron scattering and then the decrease both of the polarization and the Hα intensity, simultaneously. In this way they explained the variable nature observed in X Per.

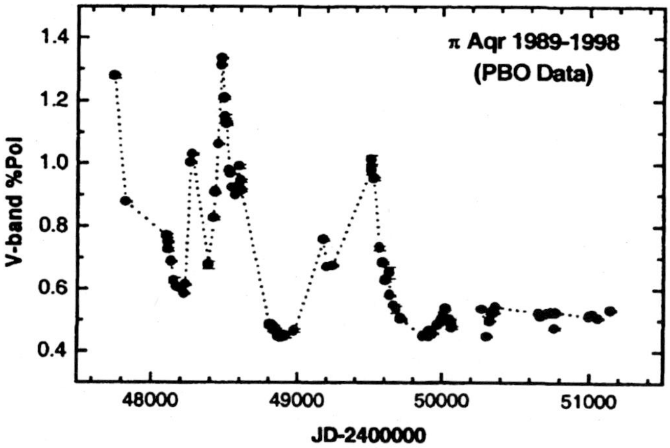

Figure 5.29: The variation of linear polarization in V band in the Be star π Aqr, observed from 1989 to 1998. Note the relatively large polarimetric "outbursts" which occurred in 1989, 1991, and 1994. (From Bjorkman 2000)

Ultraviolet spectrum

In the ultraviolet spectral region of Be stars we observe absorption lines both of photospheric and stellar-wind origins. The latter absorption lines are formed in outflowing gas, and distinguished from photospheric origins by their high excitation and large terminal velocities. Spectral classification of B0 to B8 type stars has been made based on the IUE spectral observations by Rountree and Sonneborn (1991) (in the spectral range λ 1200–1900 Å) and by Slettebak (1994) (in the spectral range λ 1200–2600 Å). Slettebak compared the spectral type derived from the photospheric absorption lines in IUE spectra with those in MK classification in optical region, and shown that the equivalent widths and relative intensities of UV absorption lines are nearly in parallel with those of MK classification as seen in Figure 5.30. No significant difference is seen between B and Be stars. Figures 5.30A and B exhibit those of photospheric absorption lines and C, the equivalent width of wind-origin absorption lines. Open and filled marks denote the B and Be stars, respectively. Parallel relation of UV lines with the MK classification is apparent.

Ultraviolet spectra of wind origin often show the strong resonance lines of CIV, SiIV, and NV, apparently in a higher degree of ionization than expected from the photospheric lines of corresponding MK type. The formation of these lines is called superionization phenomena (Hubeny et al. 1985, Marlborough and Peters 1986), and characterized as follows.

(i) The lines usually show broad and asymmetric profiles extended toward violet side. While O type stars exhibit P Cygni profiles in the UV spectra, B type stars are lacking of emission components due to low gas density of the stellar winds and hence only show some asymmetric absorption profiles.

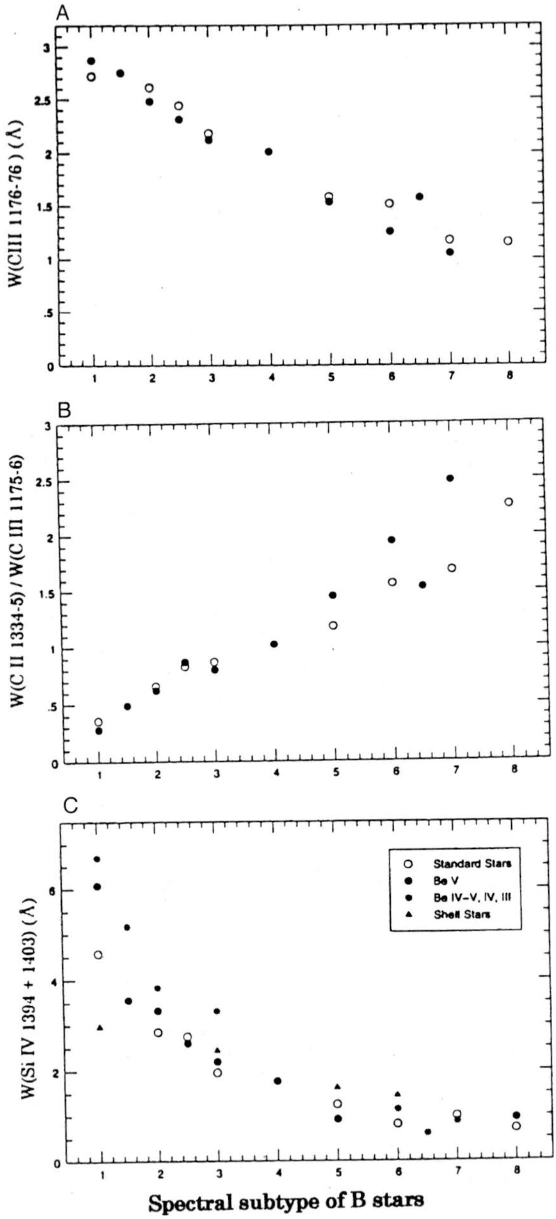

Figure 5.30: Relation of UV absorption lines with MK spectral type. A. Equivalent widths of CIIIλ1175,1176 (photospheric origin). B. Relative intensity of CIIλ1334-1335/CIIIλ1175-1176 (photospheric origin). C. Equivalent widths of SiIVλ1394 + 1403 (wind origin). Open circle denotes B type standard stars and filled circles denote Be stars for which luminosity class and shell stars are distinguished. (From Slettebak 1994)

(ii) The lines sometimes have discrete absorption components shifted toward the violet side. It particularly occurs in CIV, SiIV lines. The amount of shift is around 1000 km s^{-1} in velocity unit (Smith et al. 1997).

These line features indicate the existence of hot stellar winds with low gas densities, sometimes with multiple structures, which are supposed to be different from the disklike envelope around the equatorial plane. It is now widely accepted that the envelopes of Be stars are composed of two different components: one is the cool, disklike envelope around the equator with higher gas density and very small expanding velocity, and the other is the hot stellar wind with lower gas density and high flowing-out velocity. The latter is supposed to be originated from higher latitudes of the star's surface.

Infrared spectrum and mass-loss rates

Be stars show a large variety of infrared excess from almost lacking to significantly high excess. At one time, Allen (1973) classified Be stars, based on the H (1.6 μ), K(2.2 μ), L(3.5 μ) band photometry, into four types in the IR excess from X (no IR excess) to R, F, and D type (large IR excess). He attributed the large IR excess in D type to reradiation from circumstellar dust clouds. However, Gehrz et al. (1974) have denied the existence of dust shells, based on a wide infrared photometry from 2.3 to 19.5 μm, and argued that the infrared spectral feature can be explained by free–free emission from hot (Te $\geqq$ 10,000 K) stellar envelopes. This explanation was supported by the far-infrared observations by the IRAS satellite (Coté and Waters 1987).

In Figure 5.31, the observed IR spectra for two stars are illustrated and compared with the model calculation by Waters (1986). The resultant energy distribution of the disk emission superimposed on the photospheric radiation is shown by dotted line. For comparison, the energy distribution of stellar photosphere by Kurucz (1979) model is also shown by solid line. According to Waters (1986), the slope of energy distribution curve depends on the density distribution inside the disk. He approximated the density distribution by a power law with an index s as follows

$$\rho(r) = \rho_0 \, (r/R_*)^{-s} \, , \tag{5.3.8}$$

where ρ_0 denotes the gas density at the base of the disk R_*. By the best-fit curve shown in Figure 5.31 for the energy distribution, he adopted the value of $s = 2.4$. Waters et al. (1987) applied the same disk model to 101 Be stars observed by Coté and Waters (1987), and obtained the values of $s = 2.0$–3.5. In addition, based on this disk model, they derived the mass-loss rates from Be stars as follows.

If the power-law index s is given, the condition of mass continuity in the outflowing disk yields the velocity distribution as

$$V(r) = V_0 \left(\frac{r}{R_*}\right)^{s-2} , \tag{5.3.9}$$

where V_0 is the radial velocity at the base of the disk. Then, by making use of the formula (3.2.52) in the case of wind with a given opening angle $\theta_{\rm w}$, we can obtain the mass loss rate $\dot{M}$. Assuming the values of $V_0 = 15$ km s^{-1} and $\theta_{\rm w} = 15°$, and substituting the derived values of ρ_0 and s from the above

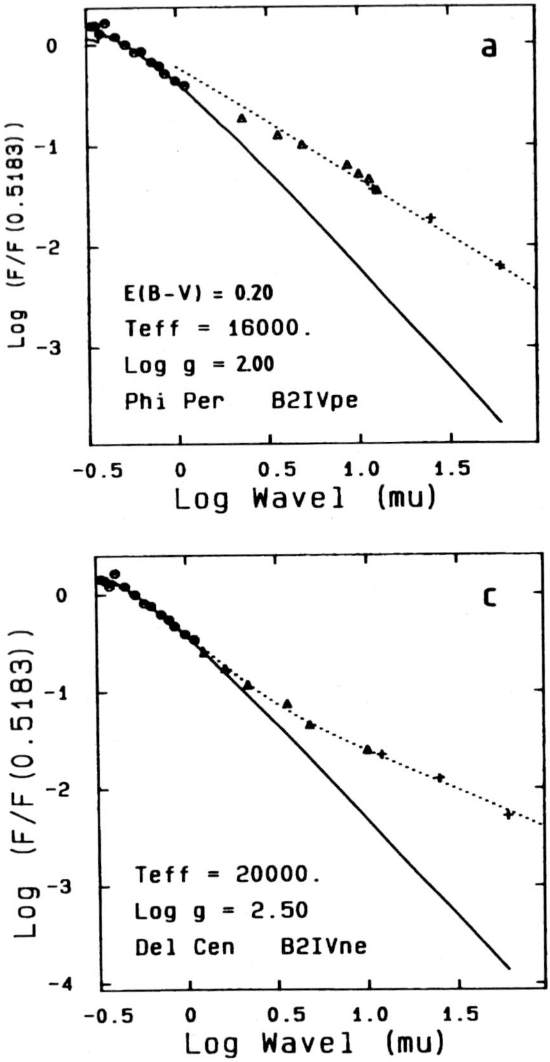

Figure 5.31: Infrared energy distributions of φ Per (upper panel) and δ Cen (lower panel), compared with Kurucz (1979) models (solid lines, stellar parameters are given in the figures). The total energy distribution combining the contributions from the photospheric and disk radiation is shown by dotted lines. Observed points by different observers are indicated by different symbols. (From Waters 1986)

disk model, Waters et al. (1987) estimated the mass-loss rates for individual stars. The results are shown in Figure 5.32 in the form of mass-loss rate versus stellar luminosity. Here, the mass-loss rates derived from the profiles of resonance lines in the UV region are also plotted.

Figure 5.32 reveals some interesting features on the mass loss–luminosity relation for Be stars as follows.

(a) The mass-loss rates of the Be stars derived from UV lines, $\dot{M}_{\rm UV}$, are on the line approximately extrapolated from the relation for O stars.

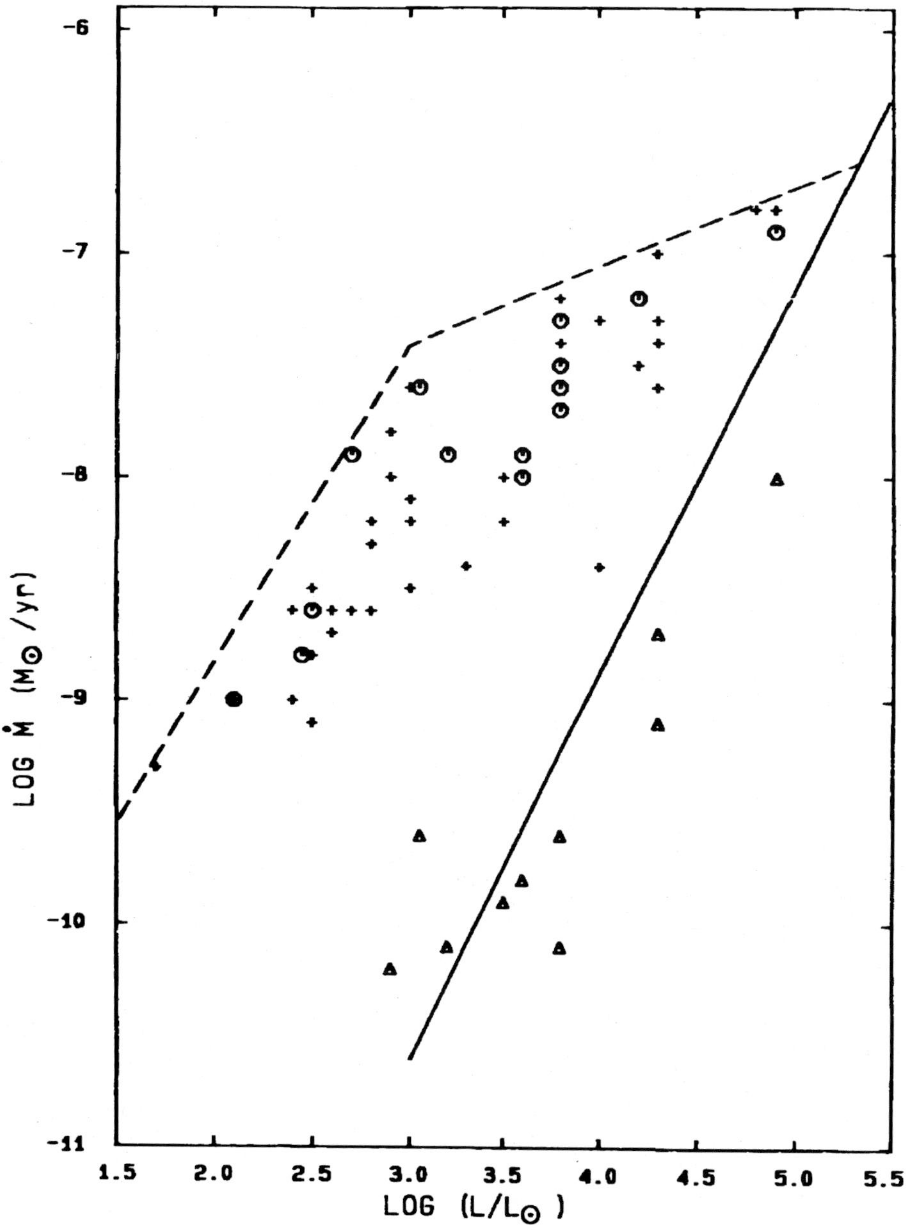

Figure 5.32: The mass loss-luminosity diagram for Be stars. Circles are the stars observed with 12, 25, and 60 μm IR bands, pluses the ones with 12 and 25 μm data. The dashed line is the upper limit for $\dot{M}_{\rm IR}$. Triangles are the mass-loss rates derived by Snow (1981) based on the profiles of UV resonance lines. The solid line gives extrapolation of the $(\dot{M}, L)$ relation for luminous O type stars. (From Waters et al. 1987)

This infers that the mass-loss from the polar region (i.e., outside the disk) is due to the same mechanism as the mass-loss from O stars. High terminal velocities of $V_\infty > 1000$ km s^{-1} and superionization in the polar winds also support the close relation with O star winds.

(b) The mass-loss rates of the Be stars derived from the IR excess, $\dot{M}_{\rm IR}$, occupy a wide band in the $\dot{M}-L$ diagram with a spread of $\Delta \log \dot{M} \approx 1$ at any luminosity. The values of $\dot{M}_{\rm IR}$ are exceeding the extrapolated relation of the O stars by a factor 10 to 10^3! Though $\dot{M}_{\rm IR}$ also shows good correlation with stellar luminosity, the mechanism of mass-loss process may be different from that of $\dot{M}_{\rm UV}$.

(c) There is an upper limit in the mass-loss rates $\dot{M}_{\rm IR}$, which is approximated by two straight lines with a bend at $\log (L/L_\odot) \approx 3.0$. This limit approaches the line of $\dot{M}_{\rm UV}$ near $\log (L/L_\odot) \approx 5.0$, approximately corresponding to late O-type stars. In O-type stars Be characteristics are observed as Oe stars (see Section 5.2.2), but only in O types later than O7.5. This infers that there exists no winds from disk envelope for stars with higher luminosity than this limit, and radiation-driven winds predominate from the whole stellar surface. Oe stars are just the intermediate stars between Be stars and luminous O stars.

In Be stars, the hydrogen recombination lines and infrared excess emission are supposed to originate in the same disk envelopes. It is therefore interesting to see whether any correlation exists between infrared continuum radiation and the emission lines. According to the observations of Ashok et al. (1984), there is a positive correlation between them as shown in Figure 5.33, which illustrates the relation between infrared luminosity $L_{\rm IR}$ and the Hα emission intensity $L_{\rm H\alpha}$. One may see the parallel relation as approximately written as $L_{\rm IR}/L_{\rm H\alpha} \sim 10^2$. Under the assumptions that the both emission is formed in the same region of the disk envelope and the envelope is optically thin for the Hα radiation, Ashok et al. derived the following theoretical relation

$$\frac{L_{\rm IR}}{L_{\rm H\alpha}} = 3\frac{\displaystyle\int_{V_{\rm IR}} N_{\rm e}N_{\rm i}\, dV_{\rm IR}}{\displaystyle\int_{V_{\rm H\alpha}} N_{\rm e}N_{\rm i}\, dV_{\rm H\alpha}}, \qquad (5.3.10)$$

where $V_{\rm IR}$, $V_{\rm H\alpha}$ denote the volumes of IR and Hα-emitting regions, $N_{\rm e}$, $N_{\rm i}$ the electron and ion number densities, respectively. If we assume that the both emission is emitted from the same region, i.e., $V_{\rm IR} = V_{\rm H\alpha}$, then Equation (5.3.10) yields $L_{\rm IR}/L_{\rm H\alpha} \sim 3$. This value is too small to compare with the observed relation shown in Figure 5.33. On the cause of this discrepancy, Ashok et al. suggested two possibilities: (1) there is a great difference in the volumes of emitting regions and (2) the envelope is opaque for the Hα radiation so that the actual value of Hα intensity is severely reduced from that of optically thin case. Since the envelopes of Be stars are generally sufficiently opaque for the Hα radiation, the second point should be reconsidered by taking into account

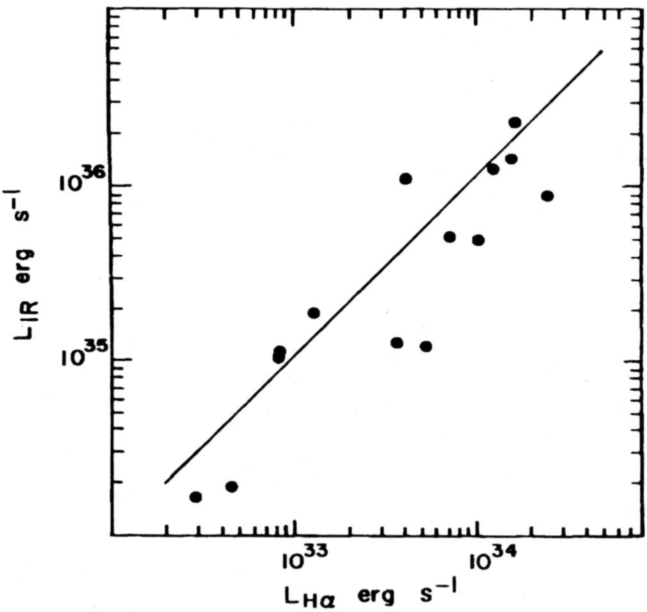

Figure 5.33: Correlation of luminosity (erg s^{-1}) between infrared (H, J, K bands) and Hα emission in Be stars. A straight line is added to show the relation $L_{IR}/L_{H\alpha} = 10^2$. (From Ashok et al. 1984)

the transfer process as high as accurately for line radiaitons. Similar observations have been carried out by van Kerkwijk et al. (1995), and the equivalent width of Hα emission (EW) versus (J–L) color diagram for 69 B and Be stars was drawn, showing similar correlation with that of Ashok et al. (1984). Large scatter in EW(Hα) was explained as a genuine star-to-star variation in physical condition. Van Kerkwijk et al. have tried to interpret their diagram by more realistic models of Poeckert and Marlborough (1978) and Waters (1986), but failed to reproduce the observed correlation. Some refinements of models for the cool envelopes of Be stars seem required.

In the near infrared region there are numerous emission lines. Andrillat et al. (1988, 1990a,b, 1994) have made the spectroscopic observations for the wavelength range λ 7500–10,500 Å for several tens of Be stars at the Haute Provence Observatory and detected emission lines such as listed below.

HI: P7 (λ 10049 Å)–P23 (λ 8345 Å)
OI: λ 7772, 7774-5, 8446 Å
CaII: λ 8542, 8662, 8498 Å
NI: λ 8629, 8686, 8680, 8683, 8712, 8719 Å

Near infrared spectra for some selected stars are exhibited in Figure 5.34, where one may see that the occurrence of emission lines shows a large variety.

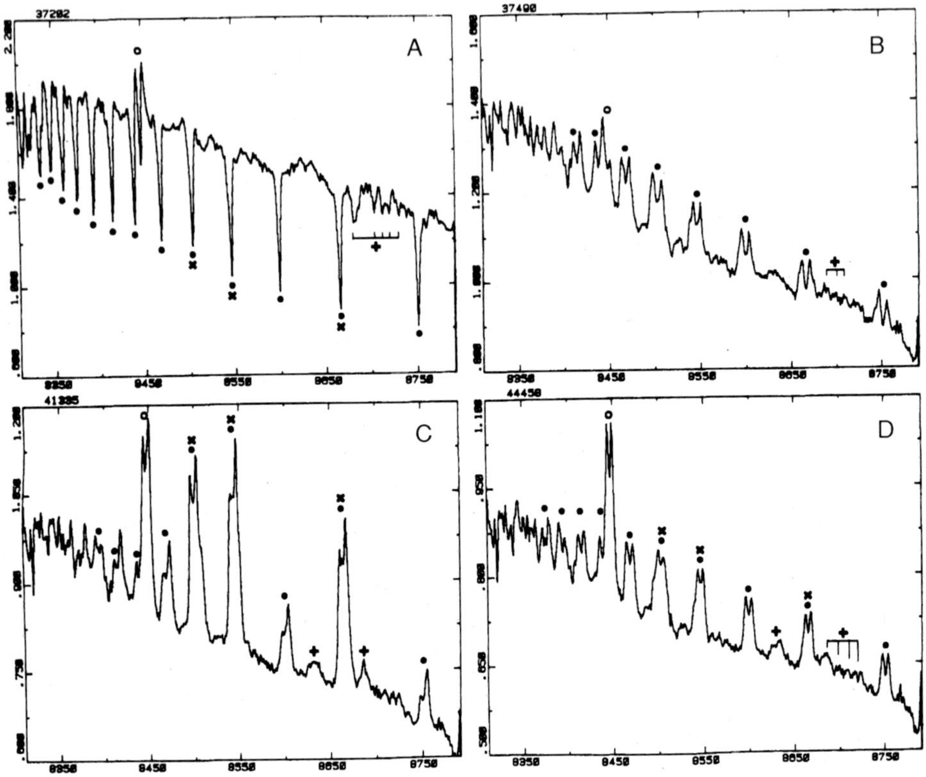

Figure 5.34: Samples of near IR spectrum of Be stars. (**A**) HD 37202 (B1 IVe-shell). Strong shell absorption lines are remarkable in P7–P24 and NI. (**B**) HD 37490 (B2IIIe). Double-peaked emission lines are seen in the Paschen series up to P19 and in OI line. NI is weak emission. (**C**) HD 41335 (B2IVe). Asymmetric doubled-peaked emission in the Paschen series up to P20 ($R > V$), CaII, OI also show strong double peak emission. (**D**) HD 44458 (B1.5 IVe). Paschen lines show symmetric double peaks. CaII is weak emission and OI is strong emission. Symbols of lines; ●: Paschen lines (from P12 at λ 8750.46 Å to the left); ○: OI; **x**: CaII; +: NI. (From Andrillat et al. 1990a)

Some stars show strong emission in the Paschen series, while others show sharp and deep absorption lines in the same series. Emission lines often reveal the double-peaked profiles as shown in Figure 5.34. Andrillat et al. (1990a) measured the peak separation of P7 and FeII λ 9997 Å and estimated the average size of the envelopes. As the results, they found that the outer radius is around four times of stellar radius in case of Keplerian motions ($j = 1/2$). These results indicate that the envelopes emitting IR emission lines are significantly smaller in size than those emitting Balmer emission lines. This infers that the emission lines in the Paschen series are formed relatively inner region of the envelope than those in the Balmer series. Double-peaked emission indicates more or less axisymmetric structure of the emitting envelope, whereas sharp and strong shell-absorption lines observed in ζ Tau (HD 37202) may be formed in an optically thick and sufficiently compact envelope due to the absence of emission components.

In the mid-infrared spectral region (λ 8–13.3 μm), Rinehart et al. (1999) carried out spectrophotometric observations for 11 Be stars and found a large number of hydrogen recombination lines H$m \rightarrow n$, where the upper level m ranges from 7 to 31, and the lower level n ranges from 6 to 10. In Table 5.18, the wavelength, emission line width, and line flux in the $m \rightarrow 9$ series of hydrogen atoms are shown for ζ Tau and γ Cas, based on the data of Rinehart et al. The mean errors are omitted in the table for simplicity. The emission line fluxes show some scattered values along the series number, and this makes it difficult to explain the formation of these lines by optically thin envelopes. Instead, Rinehart et al. argued that they must originate at optical depths of around unity. Emission line width, $\Delta\,\lambda$, is also scattered. It is interesting to see that F(Hm) tends to take larger value for larger line width in ζ Tau, but rather opposite in γ Cas. If we assume that the envelope is rotating with Keplerian velocity low, broader lines may originate in regions nearer to the stellar surface, where the emission lines are stronger in ζ Tau, but weaker in γ Cas. ψ Per shows the trend similar to ζ Tau. This picture, however, needs to be examined by comparing with other spectroscopic behaviors.

Table 5.18: Hydrogen $m \rightarrow 9$ series in ζ Tau and γ Cas, the flux is based on the data measured by Rinehart et al. (1999)

Line		ζ Tau			γ Cas		
m	λ	$\Delta\lambda$	F(Hm)	F(Hm)/F(H14)	$\Delta\lambda$	F(Hm)	F(Hm)/F(H14)
14	12.587	6.88	4.23	1.00	6.37	1.05	1.00
15	11.540	10.00	6.20	1.46	8.03	1.11	1.05
16	10.804	9.71	10.54	2.49	6.39	1.41	1.34
17	10.261	7.96	8.71	2.06	7.62	1.28	1.22
18	9.847	7.23	9.15	2.16	6.38	1.74	1.65
19	9.522	—	—	—	2.70	0.74	0.70
20	9.261	5.88	7.34	1.73	9.17	0.88	0.84
21	9.047	6.44	5.61	1.32	2.27	2.17	2.06
22	8.870	5.78	4.66	1.10	6.64	1.34	1.27
23	8.721	6.64	7.86	1.86	—	—	—
24	8.594	5.84	6.67	1.57	2.10	2.02	1.92
25	8.485	2.75	3.02	0.71	6.74	1.43	1.36
26	8.391	4.99	4.60	1.08	4.28	1.61	1.53
27	8.309	4.37	3.55	0.84	—	—	—
28	8.236	—	—	—	3.65	3.58	3.40
29	8.173	—	—	—	—	—	—
30	8.116	6.33	8.12	1.92	6.80	1.71	1.62

Note: λ and $\Delta\lambda$ (μm), F(Hm) (10^{-19} W cm^{-2} μm^{-1}).

X-ray observations

There are two types in the X-ray sources related to Be stars. One is the Be/X-ray binaries which are composed of a compact star and a Be star, in which X-ray emission originates from the gas flow from Be to compact companion. The second is the stellar winds of Be stars. We consider these two categories in the order.

(i) X-ray binaries. Around 20 Be/X-ray binaries have so far been known (van den Heuvel and Rappaport 1987, Apparao 1994). In the catalogue of high-mass X-ray binaries (HMXBs) presented by Liu et al. (2000) which contains 130 stars, more than 40 are designated as Be/X-ray binaries.

In a Be/X-ray binary, the compact component is supposed to be a neutron star when a definitive X-ray pulse is observable. When pulse is not detected, the compact star should be either a neutron star or a white dwarf. Some compact stars occasionally exhibit X-ray bursts, and some others show the relationship with the variation of Hα emission intensity.

A list of selected Be/X-ray binary systems is given in Table 5.19, where each column except last one may be self-explanatory. The last column gives the three burst types given by Stella et al. (1986) as characterized,

Type I—Periodic transient activity. Transient outbursts that recur periodically and have $L_X(\text{max})/L_X(\text{min}) > 100$. The maximum X-ray luminosities $L_X(\text{max})$ take the values of $10^{34} \sim 10^{39}$ erg s^{-1}.

Table 5.19: Selected Be/X-ray binaries (adapted from Stella et al. 1986)

	X-ray source	Spectral type	Orbital period (d)	Eccentricity e	Pulse period (s)	L_x(max) (erg s^{-1})	Burst type
1	A0538–66	B2 IIIe	16.7	>0.4	0.069	1.2 (39)	I
2	4U0115+634	Be	24.3	0.34	3.6	8 (36)	II
3	V0332+53	Be?	34.2	0.31	4.4	2 (37)	I + II
5	A0535+262	O9.7 IIIe	111	0.2–0.4	104	2 (37)	II+I(?)
6	GX 304–1	B2Ve	133	?	272	3 (35)	—
7	4U 114–62	B1Ve	188	?	292	6 (36)	II+I(?)
8	0352+309 (X Per)	O9.5(III-V)e	580?	?	835	1 (34)	—
10	A1118−62	O9.5(III-V)e	?	?	405	5 (36)	II(?)
18	A014+65	B0.5IIIe	11.6	?	2	(34)	—
19	0053+604 (γCas)	B0.5(III-V)e	?	?	?	2 (33)	—

Note: L_x (max) denotes the X-ray luminosity at maximum phase; the numbers in parenthesis give the power indices. For burst type, see text.

The X-ray binaries of this type generally have moderate orbital eccentricity (e $\lesssim$ 0.5), and the outbursts recur periodically when neutron star is near the passage of periastron. This suggests that the Be star companion has a developed envelope in which the orbit of the companion penetrates to cause enhanced accretion onto the neutron star component giving rise to the X-ray outburst.

Type II—Irregular transient activity. Transient outbursts that typically last several tens of days and exhibit luminosity increase in the orders of 100–1000. The stars of this type show high X-ray luminosity not directly related with their orbital periods. Negueruela et al. (1998) carried out monitoring spectroscopic observations of 4U0115 + 634 (V635 Cas) and A0535 + 262 among the stars in Table 5.19 for 7 years. They found the relation between Hα intensities and Type II X-ray outbursts, that is, the Hα emission is strengthened during the outburst and closely associated with V/R variations. By the existence of such correlation, they suggested a possibility of interaction between the compact companion and the disk.

In addition, there is a third group of X-ray binaries that have lower X-ray luminosity ($L_X \lesssim 10^{36}$ erg s^{-1}), and do not show large transient outbursts. The X-ray binaries with no sign in the burst type in Table 5.19 mostly belong to this group. X Per and γ Cas are among the typical case of this type, even though the X-ray behaviors of both stars are rather different while their optical spectra are similar (White et al. 1982).

Let us compare the X-ray characteristics of X Per and γ Cas. The compact companion of X Per is definitively a neutron star with clear pulse of 835 seconds. In contrast, γ Cas has no clear pulse so that it is not certain whether its companion be a neutron star or a white dwarf. This star has long remained as something enigmatic. Its variability of X-ray intensity has been monitored since the first observations by UHURU in 1970. Horaguchi et al. (1994) compared the long-term variations in optical observations (V magnitude, Hα emission intensity, and V/R ratio) with that of X-ray intensity, and found no particular correlation among them (see Figure 5.39 in Section 5.3.6). Recently binary nature of γ Cas has been found by Harmanec et al. (2000) and Miroshnichenko et al. (2002). Orbital period is around 205 days, and orbital eccentricity is nearer to zero according to Miroshnichenko et al. This favors a nondegenerate nature of the secondary and seems to support a hypothesis that the companion is not directly connected with the X-ray generation.

Concerning the origin of X-ray emission in γ Cas, Smith (1995, 1998) suggested the flare activities at the surface of Be stars, based on the similarity in short-term variations (within one hour or so) between X-ray and FUV continuum radiation. Robinson et al. (2002) monitored cyclic flux variations of X-ray flux and photometric variations (in B and V bands) with cycle lengths of 55–93 days, which are much shorter than the orbital period. Thus they supported the view of surface activity of the Be star component. They suggested a mechanism of cyclic magneto dynamo excited in the inner part of the circumstellar disk to explain the relationship between X-ray flux and optical variability.

(ii) X-ray emission from stellar winds. It is known that the Be stars have also stellar winds as a continuation from O type stars, and then X-ray emission of wind origin is expected. However, the winds of Be stars are not so strong as O stars due to their smaller terminal velocities and lower gas densities therefore, X-ray emission should be much weaker than O stars. Cassinelli and Cohen (1994) and Cassinelli et al. (1994) carried out X-ray observations with ROSAT (energy range 0.1–2.4 keV) for the main sequence B stars and obtained the following X-ray characteristics.

(a) X-ray luminosity L_X of B, Be, and β Cep stars ranges from $\sim 10^{32}$ erg s^{-1} in early-type B down to around 10^{28} erg s^{-1} in late-type B stars. No significant difference is seen among B, Be and β Cep stars. When compared with O-type stars for which $L_X = 10^{32}$–$10^{33.5}$ (erg s^{-1}), B type stars are markedly weak X-ray emitters.

(b) The correlation between the ratios of X-ray versus. bolometric luminosity, log (L_X/L_{bol}), and spectral types from O9.5–B7 is shown in Figure 5.35. For very early B stars, the ratio is not much different from the value 10^{-7} typical in the O-type stars. However, in B2–B3 spectral range there is a large drop in the L_X/L_{bol} ratio to values near a few times of 10^{-9}. For late-type B stars this ratio remains in a range of 10^{-8}–10^{-9}.

(c) X-ray spectra are generally soft, and attributable to the thermal emission from gas at a temperature range 1–3 $\times$ 10^6 K. This range is about a factor of 2 below the temperature found in O-type stars.

(d) No evidence of variability is found above the level of around 25% of X-ray intensity on time scales up to a few thousand seconds for all types of observed stars.

As for the mechanisms of X-ray emission, shock heating by some form of instability in winds is widely accepted, while some other mechanisms, such as coronal heating, inverse-Compton emission, are also suggested (Cassinelli et al. 1994).

5.3.6 Time variations

Large diversity in time variations

Remarkable behaviors of the Be star phenomena are the time variations of various forms and time scales. The spectroscopic variability, especially variations of emission-line features, has been observed since the middle of the nineteenth

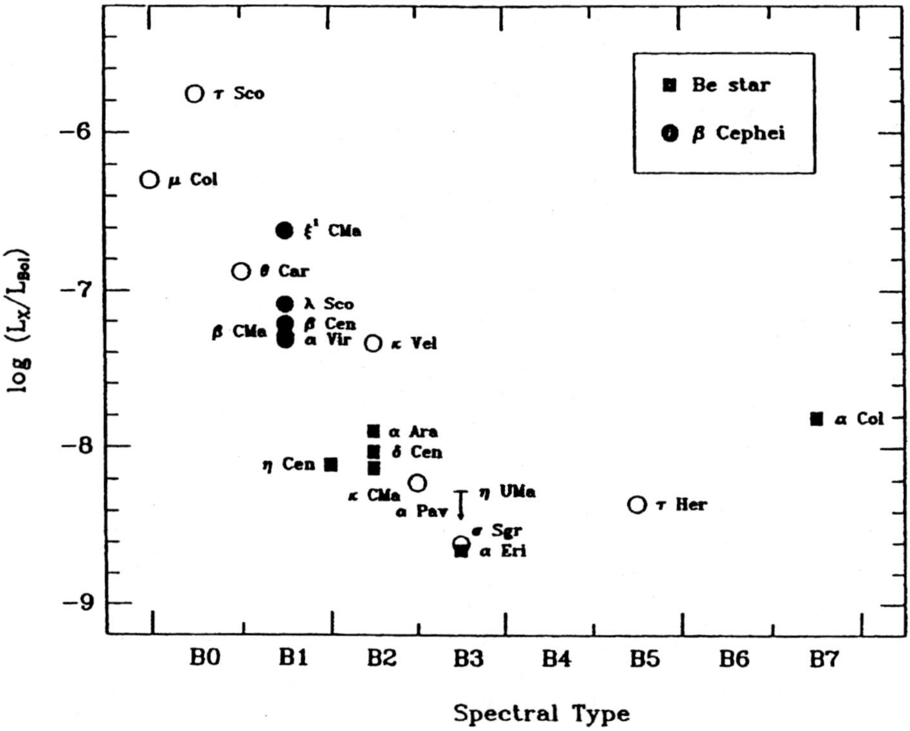

Figure 5.35: The relation between the ratio of X-ray luminosity to bolometric luminosity and the spectral type of B stars. Stars are designated by open circles for B stars, filled circles for β Cephei and filled squares for Be stars. (From Cassinelli and Cohen 1994)

century. In contrast, it is rather recently that the photometric observations for brightness and color variations have widely been carried out. Observations of variability are now extended to wide spectral regions from ultraviolet to infrared and even to radio regions. Variability is also found in a wide range of time scales from short-term variations, less than a few days or even in hours and minutes, up to long-term variations more than several months or years. In intermediate-term variations, we often meet with variations associated with binary interaction.

In the variation of emission-line spectrum, the following three types are well known since more than one century ago.

1. Phase change. The most remarkable variations in the Be star phenomena are the phase changes of stellar types between B$\longleftrightarrow$ Be, Be $\longleftrightarrow$ Be-shell, and B $\longleftrightarrow$ Be-shell, in time scales from a few years up to several decades. Spectroscopic models of the phase change will be discussed in the next paragraph.

2. E/C variations. Time variation of emission-line intensities (E) relative to the adjacent continuum (C). There are two possible cases of E/C variation: one is the true change of emission-line radiation, and the other is the variation of continuum level (i.e., the brightness variation). The so-called veiling effect, which occasionally appears in early-type Be stars, is a direct spectroscopic consequence of the latter case. Hubert-Delplace et al. (1982) examined the E/C variations of 140 northern Be stars in a time span of 24 years, and found that 81% of Be stars showed E/C variation for B0–B5 stars, while it was 48% for B6–A0 stars.
3. V/R variations. Time variation of relative intensities of violet (V) and red (R) components in double-peaked emission-line profiles. V/R variation is usually expressed as the ratio of respective emission-peak heights above the underlying photospheric absorption profile. We shall discuss V/R variations more in detail later in this section.

The photometric observations of Be stars have practically started by Mendoza (1958), who obtained V magnitude and B–V color for 187 northern Be stars, while the variability has been first confirmed by Feinstein (1968). He detected V variation larger than 0.06 mag. for 33 stars out of 72 Be stars.

Since 1980s, observing campaigns for photometric observations of Be stars have been organized in Ondrejov, Hvar, Dominion, and other Observatories, and the basic data for the photometric behaviors on variability have widely been accumulated. Percy and Bakos (2001) extended the monitoring observations up to 1999, for 15 Be stars, mainly to find the trace of long-term variations.

Phase change and long-term variations

In Figure 5.36 the phase-changing variations for five stars in a period of 1900–1980 are schematically shown. These stars experienced the phase change in time scales of several years to several decades. The phase change between normal B stars and Be stars are readily understandable by the formation or destruction of envelopes, whereas the transformation between Be and Be-shell stars should be related to the structural changes of stellar envelopes. In the classical view, the difference of Be and shell stars is usually attributed to the difference of inclination angle of rotational axis to the observer, i.e., shell stars are near equator-on while Be stars are those seen from some definite angle of inclination (Figure 5.17). However, this picture can not be applied to the transformation between Be and shell, since the inclination angle should be fixed for any particular stars.

We now consider the phase change between Be and shell stars from the viewpoint of the formation of shell-absorption lines (Kogure 1990). As seen from Equation (5.3.5), the condition of shell-line formation is the sufficiently large optical depths for the lower members of Balmer lines. Strong shell stars require the optical depth in the Hα radiation to be as high as 2000 or even 5000 as seen in Figure 5.27. In contrast, the envelopes of ordinary Be stars should be

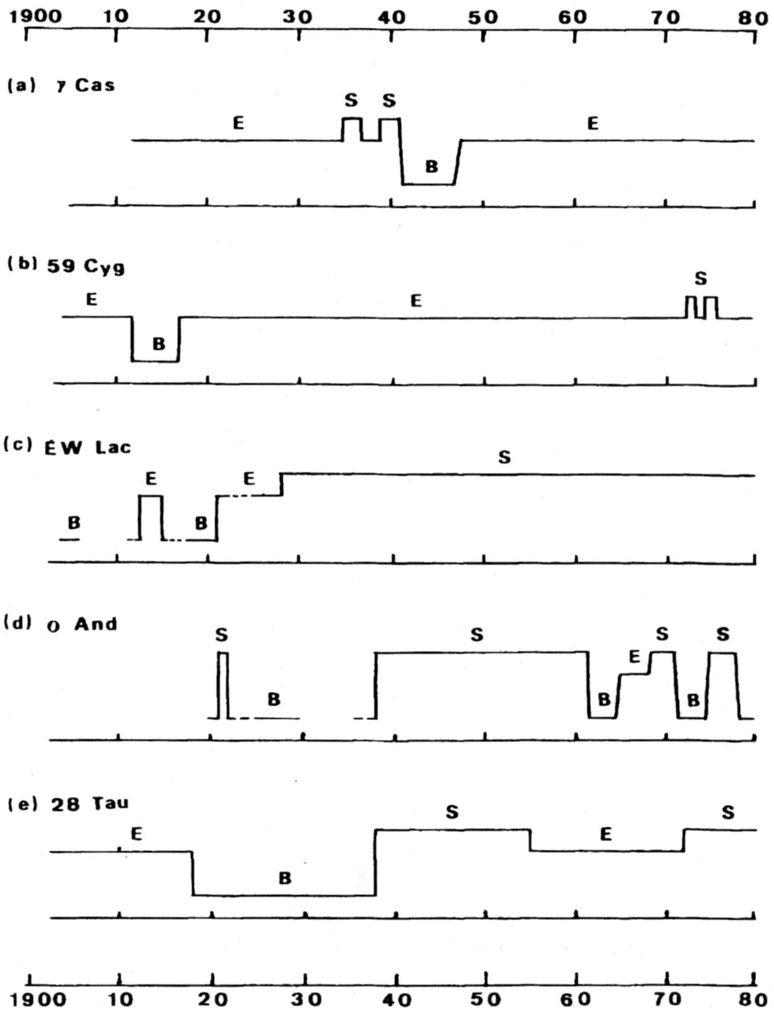

Figure 5.36: Schematic traces of the histories of phase-changing variations among B, Be, and Be-shell phases, which are designated by B, E, and S, respectively (Kogure 1990).

optically thin for the Hα line in order not to show shell absorption lines, i.e., τ (Hα) must be smaller than around 100. Since the optical thickness for the Hα is related to the square of electron density, the electron density of the envelopes of shell stars must be sufficiently higher than those of Be stars, if the extension of the envelopes is not much different in both type stars. According to the shell-line analysis based on Equation (5.3.5), the values of β, the fractional area of the stellar photospheric disk covered by the envelope in front, are typically less than unity. This implies that only a part of the photospheric

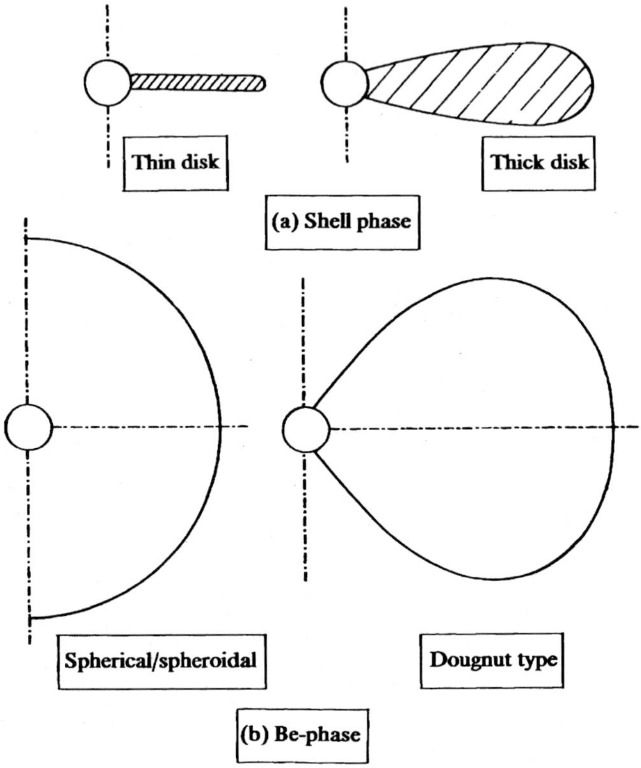

Figure 5.37: Possible structures of axisymmetric envelopes in the shell and Be phases. A. Shell stars have vertically thin disks, optically sufficiently opaque in the Balmer lines ($\tau(\text{H}\alpha) > 100$). B. Be stars with vertically expanded envelopes, optically thin in the Balmer lines ($(\text{H}\alpha) < 100$) (Kogure 1990).

disk is nakedly seen to the observer, i.e., the geometrical thickness in vertical direction should be smaller than the photospheric radius. This is in contrast to the case of Be stars where the gas density is low and then the volume of envelopes must be sufficiently large as compared to that of shell stars. In addition, one notices that (1) the outer radii of the envelopes are almost similar in Be and shell stars as seen by the similar peak separation of emission double peaks, and (2) the emission intensity of the Hα line has no apparent difference in both stars, i.e., the emission measure of the envelopes is similar with each other. From these facts of evidence one is led to the conclusion that the envelopes of Be stars must be extended vertically. In this way, Kogure (1990) argued that the difference of Be and shell stars lies in the vertical development of the envelopes as schematically shown in Figure 5.37, though the actual forms of the envelopes remain uncertain.

As an example of phase change, we consider the spectral variation of Pleione (28 Tau, B8 Ve-s). As seen in Figure 5.36, this star entered the second shell phase in 1972 with the formation of weak shell-absorption lines, after a long Be star phase. The shell spectrum has developed in 1974 to 1979 by increasing the central depths of absorption and the highest quantum number of visible shell lines, n_1, in the Balmer series (see Shell-Absorption lines

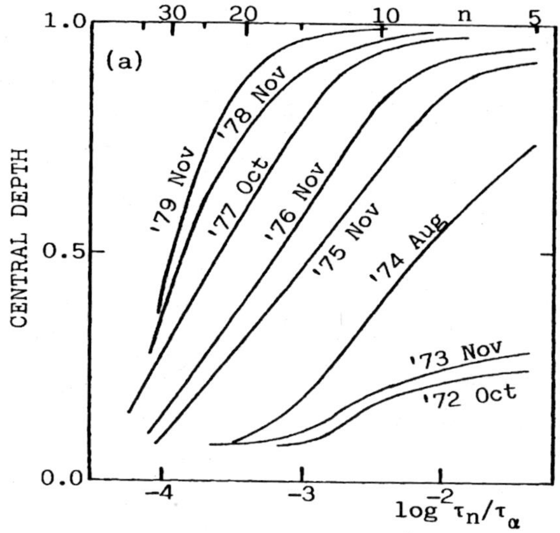

Figure 5.38: Development of the shell absorption system in Pleione. The observed points for the central depths of shell lines are expressed by smoothed curves. Observations were carried out at the Okayama Astrophysical Observatory for 8 epochs in 1972–1979. (From Hirata et al. 1982)

in Section 5.3.4). The maximum development has been reached in around 1980, then the shell spectrum has gradually declined and returned to the Be phase in 1988 (Hirata 1995). We can trace the development of shell absorption system in the Balmer series by plotting the central depths versus relative optical thickness (or the position of Balmer series number) of the shell-absorption lines, as has been made in Figure 5.25. The case of Pleione is schematically shown in Figure 5.38 for the early developing period from 1972 up to 1979, where the smoothed curves of observed points are drawn for eight epochs observed at the Okayama Astrophysical Observatory. Based on the data of these curves we can estimate the pairs of β and τ (Hα) for each epoch of observations. These curves have been fitted by two theoretical curves as in case of ζ Tau in Figure 5.25. If we assume the equator-on view of Pleione due to its large rotational velocity of $V \sin i \sim 320$ km s^{-1}, the existence of two curves implies a vertical stratification of the envelope such as shown in Figure 5.26. At its maximum phase of shell development, the optical depth of the Hα line reached the value of $\tau\,(\mathrm{H}\alpha) \sim 5000$ in the inner optically thick layer and the value of $\tau\,(\mathrm{H}\alpha) \sim 200$–$300$ in the outer optically thin layer. The vertical thickness of the whole envelope has increased from 0.2 R_* in 1972 to more than stellar diameter in 1978. In addition, the envelope mass has also increased from $1 \times 10^{-10}\,M_\odot$ up to $5 \times 10^{-10}\,M_\odot$ during the developing period.

Phase change between Be and shell star usually occurs in stars with large projected rotational velocities larger than, say 300 km s^{-1}, as in case of Pleione. However, it is notable to see that γ Cas which has smaller rotational velocity of $V \sin i = 230$ km s^{-1} and its inclination angle is supposed as around 70°, has also experienced twice the phase-changing variations in the middle and late 1930s as seen in Figure 5.36. This infers that the optically thick envelope having high electron density sufficient to produce the shell absorption has been expanded up to high-latitude region in front of the stellar surface.

In the occasion of the phase change from Be to shell phase, conspicuous variations in magnitude, colors, and/or spectral features are usually accompanied to the formation of shell lines. In case of Pleione, remarkable decrease of light in U band and broadening of photospheric absorption line of CaII K line have been observed just before the transition to the shell phase. If the line broadening is explained by the rotational Doppler motion, this implies that the rotation is accelerated from deeper layers to the stellar surface in the equatorial zone. Since the effective gravity is lower in the equatorial zone in highly rotating stars, the surface temperature in this zone will decrease by von Zeipel's theorem to allow the formation of the low-excitation CaII K line.

Similar phenomena have also been observed in 88 Her (B6 Ve) and some other late-type Be stars in their phase-changing variations. Based on these phenomena, Hirata (1995) interpreted these broadening as the result of angular momentum transfer from the stellar interior to the surface to cause the accompanied changes in colors and spectral features. In addition, Hirata suggested that this momentum transfer activity is the cause of phase-changing variations in late-type Be stars.

The rotating disklike envelopes of Be stars are usually assumed to be vertically thin. In contrast, the possibility of vertically expanded envelope has been first suggested by Doazan (1987), who has developed the Be star theory based on the spheroidal/ellipsoidal, variable mass-loss, decelerated models. According to Doazan, the transition from Be to Be-shell phase occurs when the mass-flow is enhanced to produce the shell absorption lines.

V/R variations

In Be stars with double-peaked emission, we often observe the V/R variations. Figure 5.39 illustrates an example of remarkable V/R variations observed in γ Cas (Horaguchi et al. 1994). The variation in the equivalent width of the Hα line is also shown in the same period from 1970 to 1990. In between these variations a weak correlation is seen but not parallel with each other.

The V/R variations have so far been spectroscopically interpreted by the two models: the symmetric rotating–pulsating envelope model and the elliptical disk or ring model (Ballereau and Chauville 1989).

(i) Symmetric rotating–pulsating envelope. If the disk envelope is purely rotating, emission lines will show symmetric $V = R$ profiles, while if some expanding motion is added, asymmetric $V < R$ profiles will appear just like P Cyg stars. In contrast, rotating–contracting motion will make asymmetric $V > R$ profiles. By adding suitable expanding–contracting motions to the rotational motion, we can obtain the V/R profile variations though phenomenological. In actual Be stars, however, expanding motion due to stellar wind is predominant so that we

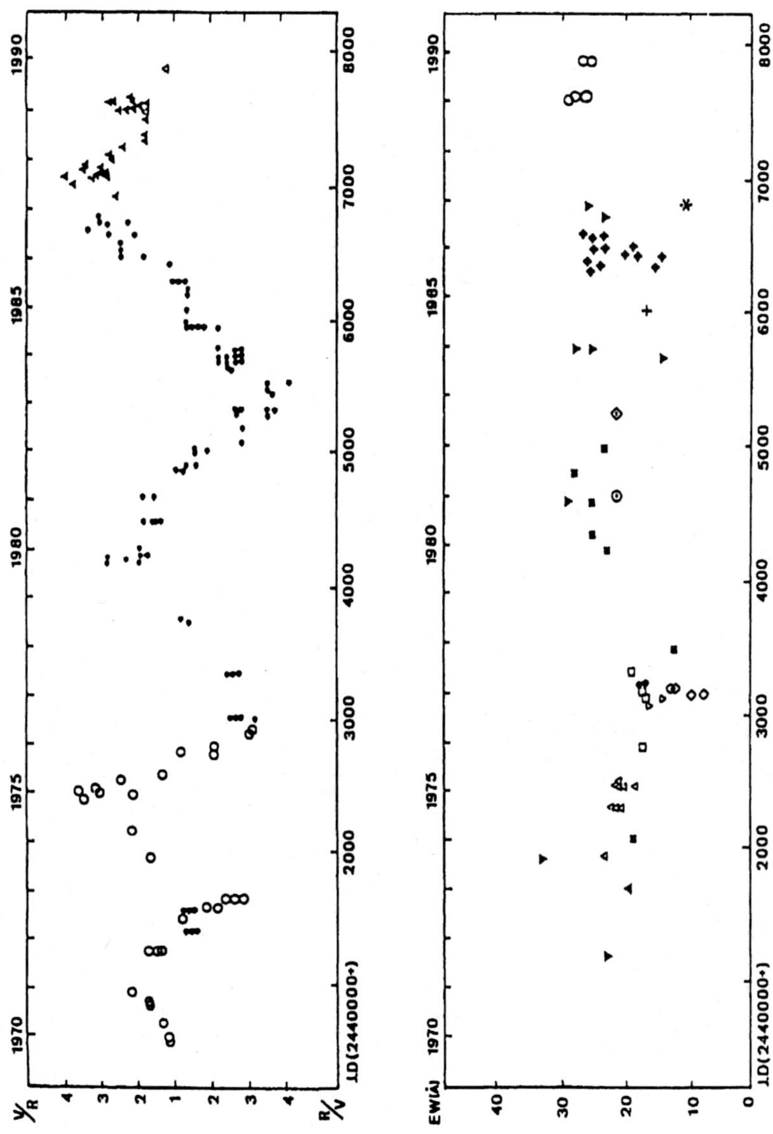

Figure 5.39: Long-term variations in γ Cas observed in the period 1970–1990. The upper panel denotes the V/R (the upper and lower parts of the ordinate give the values of V/R and R/V, respectively) and the lower panel shows the emission-equivalent width of Hα, EW(Å). Different symbols denote the values by different observers. (From Horaguchi et al. 1994)

should mostly observe the $V < R$ profiles. This is in contrast with observations. Furthermore, nonaxisymmetrical structure of the Be star envelopes have been already resolved by the optical interferometeric observations (Table 5.17). By these reasons the expanding–contracting model is not supported.

(ii) Elliptical disk or ring model. Consider a rotating elliptical disk or ring in which gas undergoes the Keplerian motion. Gas particles move faster near the periastron passage than near the apoastron passage. The gas then accumulates more near the apoastron passage and emits stronger emission lines, than near the periastron. If this elliptical disk or ring rotates around the star by some long-term apsidal motion, observers will see the V/R variation along the rotation of this envelope. This spectroscopic model was first suggested by Struve (1931) and theoretically examined by Huang (1972, 1973) and Kriz (1976, 1979).

The elliptical disk or ring model has been reformed dynamically as a one-arm oscillation of rotating disk. Kato (1983) first pointed out that one-armed, low-frequency waves can exist stably in the geometrically thin, nearly Keplerian disks. He suggested the applicability of this oscillation to Be stars, and Okazaki (1991, 1996) has developed this theory as the dynamical explanation of the V/R variations.

As an undisturbed disk Okazaki assumed an axisymmetric, isothermal Keplerian disk with constant vertical height and hydrostatic equilibrium in vertical direction. On this disk, he considered the oscillation in the form of $\exp[i(\omega t - m\varphi)]$, where $\omega = 2\pi\,\nu$ denotes the circular frequency, φ the phase, and m the mode of oscillation. One-arm oscillation is given by $m = 1$. The enhancement of the surface density is expressed by $\sigma_0 + \sigma_1$, where σ_0 denotes the undisturbed density and σ_1 the disturbed part.

Let the inner radius of the rotating disk be the stellar radius $r_{\rm s}$ and the outer radius is $r_{\rm out}$. Assume the distribution of gas density in the undisturbed disk is expressed by a power law of the radius $r^{-\alpha}$. In this case the periods of oscillation depend on the parameter α and the ratio $r_{\rm out}/r_{\rm s}$, but only weakly on the spectral type. Okazaki (1996) calculated the profiles of the Balmer lines in some phases of the rotating disk and showed the well-defined V/R variability under the suitable conditions of Be envelopes. The oscillation period falls in a range from several years to several decades, which are just the observed periods of V/R variations.

The global oscillation model of Okazaki (1991) has been applied by Hanuschik et al. (1995) to the slow quasi-cyclic V/R variation observed in δ Cen (HR 4621, B2 IVe, $V \sin i = 220$ km s^{-1}). Figure 5.40 illustrates how the V/R variability arises from a global density wave. The central black-and-white picture shows the density distribution of a perturbed disk. For clarity only the perturbed part, σ_1, of the z-integrated density is shown, which has to be superimposed onto a unperturbed density law $\sigma_0 \propto \mathrm{r}^{-2}$. Four profiles

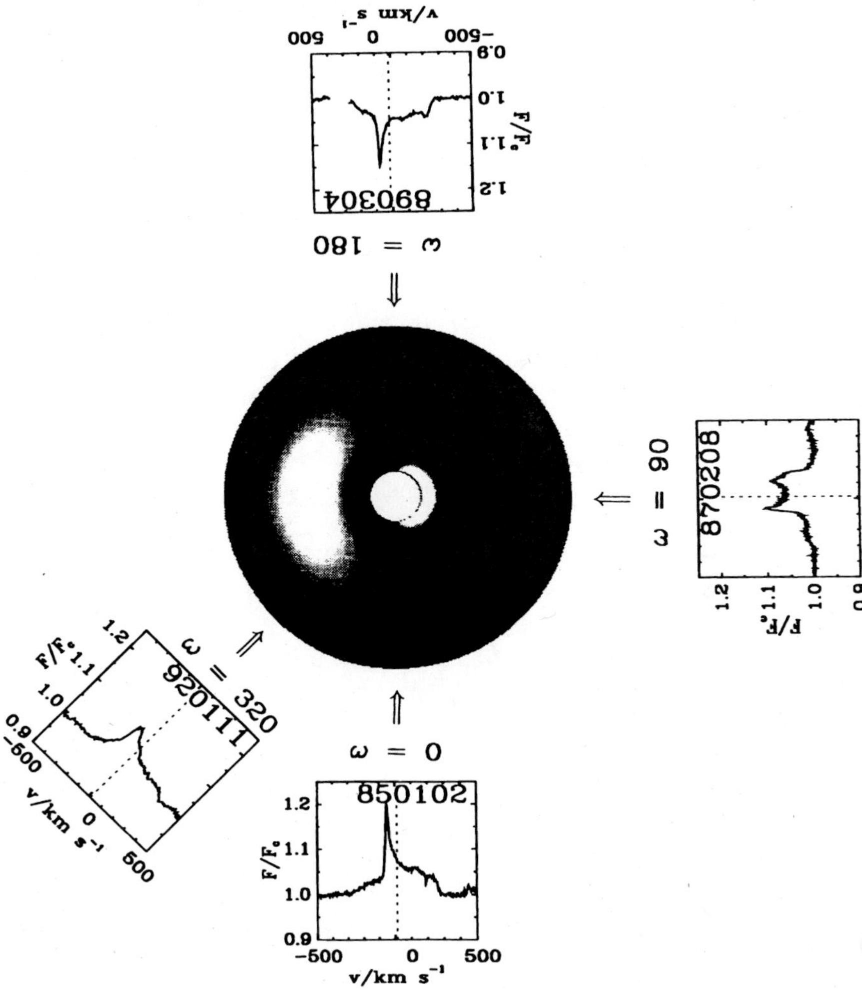

Figure 5.40: Perturbed part of the z-integrated density, σ_1, in the disk, and observed FeII λ 5317 Å profiles of δ Cen, plotted at approximate phase angles $\omega = 0°$, $90°$, $180°$, and $320°$. Dark areas denote, $\sigma_1 < 0$, bright ones $\sigma_1 > 0$. (From Hanuschik et al. 1995)

of FeII λ 5317 Å of δ Cen observed at the respective approximate phase angle ω are plotted. The numbers above the profiles denote the date of observations such as 850102 at $\omega = 0$ meaning January 2, 1985. It is clearly visible how this anisotropic density distribution produces symmetric line profiles at $\omega \approx 90°, 270°$ and steeple-type asymmetric profiles at $\omega \approx 0°$ and $180°$ for FeII lines. For the Hα line, the V/R variation is less remarkable in the central part of the profile, since the contribution from the undisturbed region should be large by its optically thick nature.

Short-term variations and nonradial pulsation

The short-term variability shorter than around 1 day was first discovered by photoelectric observations of Walker (1953) for the Be star EW Lac (HD 217050). Since 1970s, photometric and spectroscopic observations have been carried out widely to elucidate the various forms of short-term variabilities in brightness, colors, and spectral line profiles. Percy (1987) has listed 40 Be stars showing rapid variability and 38 probable or possible rapid variables in Be stars. Variable nature of these stars can be briefly summarized as follows.

(a) Short-term variability occurs in time scales ranging from around 0.3 to 2.0 days, with small amplitudes of typically 0.01–0.1 magnitude.
(b) Light curves are occasionally sinusoidal, but are more usually nonsinusoidal. Double-wave light curves, or light curves with two unequal minima are rather common.
(c) There is no firm evidence for a substantial period change.
(d) The spectroscopic period is the same as the photometric period, when both periods are known.
(e) Rapid spectroscopic variability is seen in virtually all Be stars in changing asymmetries and sometimes in bumps and wiggles in line profiles. They often appear as the transient events.

The amplitudes of rapid variations in line profiles are generally of the order of around 1% relative to the continuum. This means that high signal-noise ratio is required to the spectroscopic observations of rapid variations. The advent of high sensitive CCD detectors in the late 1980s made possible such precise spectroscopic observations.

The short-term variations of Be stars have been interpreted mainly by two different points of view: nonradial pulsation (NRP) and the corotating circumstellar material. We shall now consider these problems in the order.

Nonradial pulsation

The excitation of NRP in massive stars was theoretically predicted in early 1970s (Osaki 1974), but it was not until the1980s that a wide observational evidence for the NRP has been accumulated for Be stars (Baade 1987).

Typical behavior of the NRP in B and Be stars is the rapid variation of line profiles in the form of subfeatures (bumps or dips) traveling through the profiles. Variations of subfeatures suggest the occurring and traveling of hot or cold regions on the photospheric surface. Figure 5.41 illustrates the profile variations of HeI λ 6678 Å of the Be star λ Eri, observed by Kambe et al. (1993). Observations at the Okayama Astrophysical Observatory were made in two different nights in a quiescent phase of 1991. Remarkable rapid variation in line profile was seen each night as the traveling of subfeatures (dips and bumps).

The behaviors of the nonradial oscillations in stellar bodies are generally ascribed as follows.

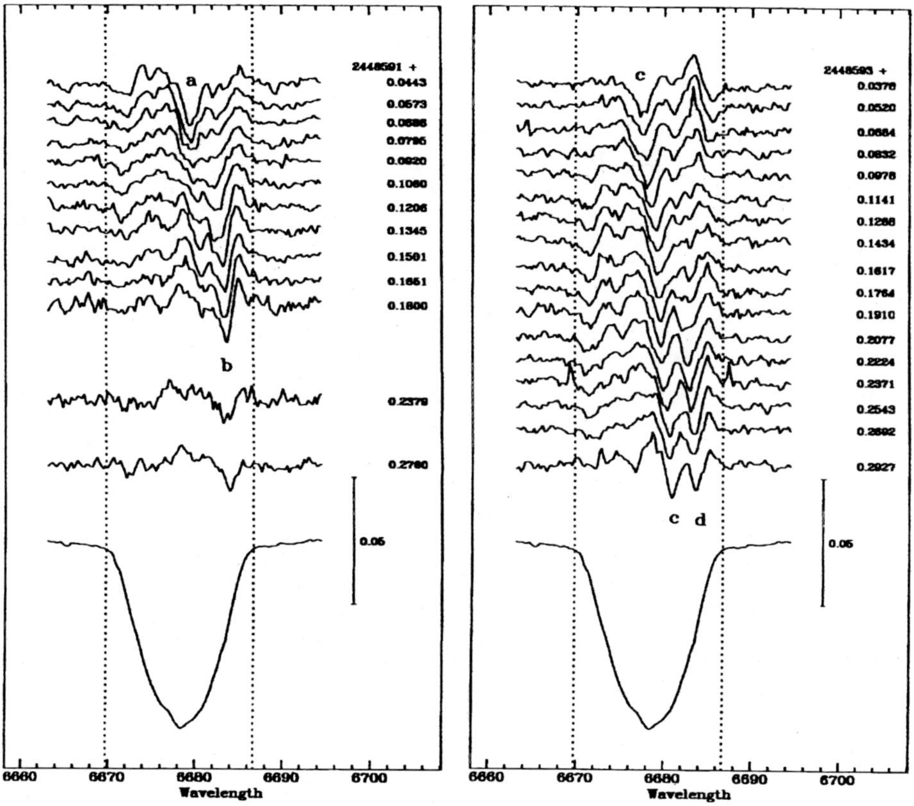

Figure 5.41: The line-profile variations of HeI λ 6678 Å of the Be star λ Eri in a quiescent phase. Observations were carried out for 2 days in November 1991. The average profiles are shown in the bottoms, and the residuals from the average profile are arranged in the order of observed time (heliocentric Julian day). The marks a, b, c, d denote the main traveling subfeatures. The abscissa is the wavelength and the dotted lines correspond to the line-of-sight velocity ± 380 km s^{-1} in both sides of the line center. (From Kambe et al. 1993)

Perturbations give rise to proper oscillations. Let us express the oscillations on the stellar surface by polar coordinates $(r,\ \theta,\ \varphi)$ and time t, then the proper oscillations can be expressed as the deformation $\Delta R(r,\ \theta,\ \varphi,\ t)$ of the surface by the three parameters (quantum numbers) $(n,\ l,\ m)$ as follows.

$$\Delta R\,(r, \theta, \phi, t) = a_{nl}\,(r)\ P_l^m\,(\cos\theta)\ \exp\left[i\,(\omega t + m\phi)\right], \qquad (5.3.11)$$

where $a_{nl}\,(r)$ denotes the amplitude, $P_l^m\,(\cos\theta)$ the Legendre polynomial, and ω the circular frequency. Among the three parameters, n denotes the order of the oscillations giving the fundamental oscillation $(n = 1)$ and the higher order oscillations ($n > 1$). The mode of oscillation is determined by the two parameters l, m as follows.

- $l = 0$: radial oscillation. Stars oscillate keeping the spherical symmetry.
- $l \neq 0$: nonradial oscillation. The deformation of the surface increases its complexity as the parameter increases as $l = 1, 2, \ldots$.
- $m = 0$: stationary waves. Surface deformation stays the same place.
- $m < 0$: prograde mode. Deformation waves propagate in the same direction of the stellar rotation.
- $m > 0$: retrograde mode. Waves propagate in the opposite direction to the rotation.

Figure 5.42 illustrates the radial part of the velocity field in nonradial oscillation viewed from three inclination angles $i = 15°$, $45°$ and $75°$ for some combinations of l and $|m|$, sketched by Telting (1996).

In Figure 5.42, the nodal circles are white; the adjacent patches move in opposite directions. The indices of the spherical harmonics l and m give

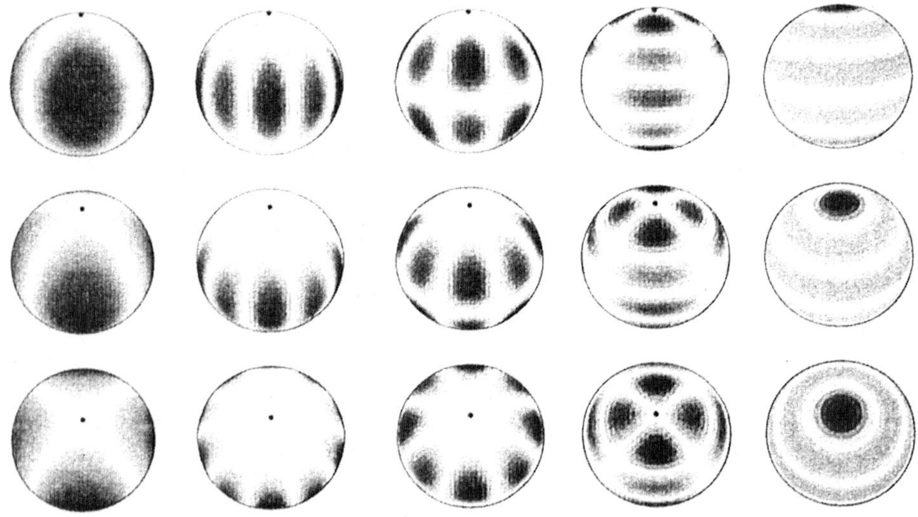

Figure 5.42: The surface pattern of the non-radial pulsation is sketched. The radial part of the velocity field is given for three values of the inclination $i = 15°$ (bottom), $45°$, and $75°$ (top) , with the position of the pole indicated by a dark dot. The nodal circles are white; the adjacent patches move in opposite directions. The indices of the spherical harmonics l and m give the total number of nodal circles and the number of nodal circles along meridians, respectively. Sectoral modes have $l = |m|$ (all nodal circles along meridians); zonal modes have $|m| = 0$ (none of the nodal circles along meridians). The sign of m depicts whether the traveling waves move progradely or retrogradely with respect to the rotation of the star. (From Telting 1996)

the total number of nodal circles and the number of nodal circles along meridians, respectively. Sectoral modes have $l = |\,m\,|$ (all nodal circles along meridians); zonal modes have $|\,m\,| = 0$ (none of the nodal circles along meridians). Sectoral and tesseral oscillations are traveling waves, while zonal oscillation modes are standing waves. The sign of m depicts whether the traveling waves move prograde or retrograde with respect to the rotation of the star.

Various types of oscillations are observed in B-type stars, whose modes are different by the types of stars (Smith 1988). First, β Cep variables are the slowly rotating early B type stars, and ordinarily exhibit the radial oscillation ($l = 0$), but some stars show NRP. Second, B stars seems to show different mode of NRP by different rotational velocities, i.e., moderately rotating stars ($75 < V \sin i < 175$ km s^{-1}) show prograde oscillation with high modal degree ($l = 4$–8), whereas, rapidly rotating stars ($V \sin i > 175$ km s^{-1}) are similar with the moderate rotating stars except the modes are retrograde with respect to the star's surface corotating frame. Third, Be stars also show the NRP with high rate. Like the rapid rotating stars, the NRP of Be stars have almost exclusively retrograde modes.

The NRP of Be stars are characterized as follows: the period of oscillation is around 0.1–3 days, the velocity amplitude ranges from 3 km s^{-1} which is a detection limit, to the order of 20 km s^{-1}, the amplitude of radial motion is within around ± 0.02 of the stellar radius, and the variation of surface temperature due to oscillation is in a range of $\pm$ 2000–3000 K. The traveling of subfeatures as illustrated in Figure 5.41 appears in case of large ℓ modes ($\ell = 4$–8), while the oscillations with small ℓ modes can be detected by the periodic change of line width or asymmetry of profiles. According to Smith (1988), the existence of $\ell = 2$ mode in the NRP of Be stars is important as a cause of the Be star phenomena, since pulsation theory generally predicts that low-ℓ modes extend deeper into the star and contain more pulsational energy, which are enough to drive Be outbursts.

On the formation of Be phenomena, Osaki (1999) suggested a possible role of nonradial oscillation. Large-amplitude nonradial waves give rise through wave-breaking phenomenon to drive the equatorial mass-loss, which forms a viscous decretion-disk around Be stars.

Corotating circumstellar material

Time scales of short-term variations are of the order of one day that is comparable to the periods of rotation so that the explanation by some spots (hot or cold) rotating with stellar surface has been proposed by Harmanec and others. Harmanec (1989) argued the possibility of rotating spots for the Be stars ζ Oph and ε Per by inspecting their rotation periods and variations of line profiles.

Recently, it has been found that the short-term variations in some Be stars are hardly reconciled with the models of NRP. For example, Balona and

Lawson (2001) analyzed the line-profile variation of ε Cap (B3 Ve-shell) and argued as follows.

(1) Mean profiles of the Balmer lines (Hα, Hβ, Hγ) in their observing runs exhibit strong shell absorption in the line centers. Also, the profiles of HeI lines are almost affected by the circumsteller material to some degree.
(2) By period search analysis for line profiles they found two periods of $P = 0.99$ day ($f = 1.01$ cycle day^{-1}) and $P = 0.495$ day ($f = 2.02$ cycle day^{-1}), among which they preferred the former period as more realistic.
(3) They applied the pulsational parameter fitting under the simple NRP models, but failed to find any consistent set of parameters (ℓ, m) in Equation (5.3.11) for two cases of $f = 1.01$ and 2.02 day^{-1} in the two HeI lines at λ 4144 Å and λ 4388 Å.
(4) Some HeI lines, including above, show large difference in the line profile, suggesting that the line profile is formed at various levels above the photosphere. However, the profile variations show very similar pattern for all of HeI lines. This indicates that the periodicity arises not in the photosphere, but some distance above it in the circumstellar material.

In this way, Balona and Lawson suggested the presence of corotating circumstellar material. Balona et al. (2002) also analyzed the results of multisite spectroscopic and photometric observations on the Be star ω Ori (B3 IIIe) and found the periodic variations associated with corotating circumstellar material.

5.3.7 Peculiar Be stars

Allen and Swings (1976) have selected 65 "peculiar" Be stars which are different from classical Be stars in their IR excess (H–K color) and the appearance of forbidden lines. These stars are classified into three groups as summarized in Table 5.20.

Stars in groups 1 and 2 are more or less similar to classical Be stars except showing low-excitation forbidden lines. Among the stars in group 2, HD 45677 is worth to be mentioned by its rapid variability in time scales of hours to days, along with a long-term variability caused by a possible explosive event occurred in around 1950 (de Winter and van den Ancker 1997). Stars of group 3 are characterized by the existence of highly ionized ions such as [OIII], [ArIII], and [NIII], spectroscopically resemble to planetary nebulae.

Be stars that show forbidden lines are usually designated as B[e], then all of the peculiar Be stars belong to B[e] stars. Hence the peculiar Be stars are often the synonym of B[e] stars.

A new classification scheme for B[e] stars has been proposed by Lamers et al. (1998) and Zickgraf (1998) from the view point of evolutionary status of Be stars: (a) B[e] supergiants (sgB[e]), (b) pre-main sequence B[e] stars, (c)

Table 5.20: Groups of peculiar Be stars and sample stars

Group	IR excess $H-K$	Emission lines Permitted	Forbidden	Number of stars	Sample Stars
1	0.7–2.0	Weak emission in Balmer, FeII	[OI]	17	HD 31648 HD50138 HD163296
2	1.0–2.3	Strong emission	Numerous, [FeII],[OI], etc.	26	HD45677 MWC645 MWC349
3	0.4–2.2	High excitation	[OIII], [NeIII] Similar to PN with high excitation (>25 eV)	22	HD51585 HD316248 HD167362

Note: The H–K color takes the values of $-0.1 \sim +0.5$ in ordinary Be stars (Allen and Swings 1976).

compact planetary nebula B[e] stars, (d) symbiotic B[e]-type stars, and (e) unclassified B[e] stars. The unclassified B[e] stars are the stars that show no signature of properties defining the class (a) to (d) so that the stars of this class may be of low luminosity and nearer to the classical Be stars.

Cidale et al. (2001) carried out low-resolution spectroscopic observations in the wavelength range $\lambda\lambda$ 3500–4600 Å for 23 B[e] stars, and showed that some stars exhibit strong Balmer emission lines and forbidden lines, whereas other stars show strong shell-absorption lines up to around H15 in the Balmer series. The optical thickness of the envelope seems to show a large variety from star to star as in case of ordinary Be stars.

5.4 Supergiant emission-line stars

5.4.1 Luminous blue variable

What are LBVs?

The term LBV (luminous blue variable) was first proposed by Conti (1984). Humphreys (1989) defined the LBVs as the bright early-type variable stars with absolute bolometric magnitude brighter than −9 magnitude. LBVs are stars of late O to A in spectral type, and occupy the upper most region of the HR diagram. Humphreys and Davidson (1979) constructed the HR diagram for the brightest part of the stars in the Milky Way and LMC, and found that the approximate upper boundary of the supergiant luminosity is almost similar in both galaxies. This upper limit is now called the Humphreys–Davidson instability limit. Distribution of LBVs on the HR diagram is shown

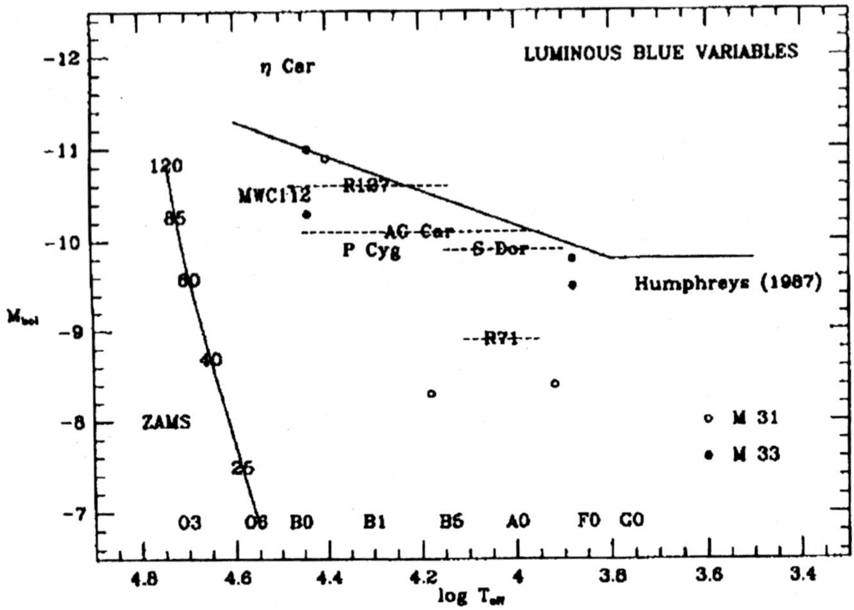

Figure 5.43: The distribution of LBVs on the HR diagram. The LBVs in the Galaxy are indicated by name, those in M31 and M33 are marked by open and filled circles, respectively. The Humphreys–Davidson instability limit is also shown. On the ZAMS curve, stellar mass is designated at the corresponding positions. (From Bohannan 1989)

in Figure 5.43, where the names of typical LBVs are given along with the ZAMS and the Humphreys–Davidson instability limit (Bohannan 1989). Since typical LBVs such as η Car and P Cyg are located near this instability limit, the term of LBV is sometimes used as a synonym of evolved massive stars.

The characteristics of the LBVs can be summarized as follows:

(i) Luminosity—the LBVs are stars of high intrinsic luminosity. They have absolute bolometric magnitude brighter than around −9.5, corresponding to luminosity of the order of $10^6\ L_\odot$.

(ii) Photometric variability—a wide range of variations in luminosity and time scales is observed. Large luminosity variations generally occur in long time scales.

(iii) Spectra—they mostly show prominent emission lines of HI, HeI, FeII, [FeII], often with P Cygni profiles. The spectra are variable, usually corresponding to the photometric variations.

(iv) Stellar temperatures—the spectral and photometric variations are somehow related to the apparent stellar temperature. At visual minimum, or in "quiescent state," the LBVs are generally hot, with temperatures from 12,000 to 30,000 K, whereas at visual maximum, or in "eruptive state," the LBVs tend to have lower temperatures of 7000–8000 K regardless of luminosity.

(v) Mass-loss rates—the P Cygni profiles of emission lines infer large mass-loss rates of the LBVs. In active state, typical mass-loss rates are of 10^{-5}–$10^{-4}\,M_\odot$ per year, which are 10 to 100 times of normal supergiants of comparable luminosity. In quiescent state, mass-loss rates are more or less similar to those of normal supergiants.

The relationship between stellar activity and effective temperature variation was explained by Leitherer et al. (1989), based on the radiatively driven stellar-wind models. They calculated the models for typical LBVs in minimum ($T_{\rm eff} = 12,000$ K) and maximun ($T_{\rm eff} = 8000$ K) states. They paid particular attention to the fact that recombination of iron group elements such as Fe, Cr, Mn, Co, Ni, from doubly to singly ionized stages, occurs when the effective temperature falls below around 10,000 K. Since the singly ionized ions provide a large number of strong lines which lead to efficient momentum transfer to the stellar wind, we can explain why the mass-loss rates increase when LBVs approach their maximum active phase. In contrast, when the effective temperature is higher than 10,000 K, the wind activity declines by the lack of effective ions for wind acceleration, and this corresponds to the minimum activity stage of the LBVs. In this way Leitherer et al. explained the transition between maximum and minimum states of the LBVs by the change of their effective temperature. However, their models are ad hoc nature without explaining the cause of temperature change of the stars in their phase transition. They suggested that the variations of stellar temperatures and photospheric radii must be induced from deeper, subphotospheric regions.

LBV and activity types

In Table 5.21 we give a list of LBV in the Milky Way and LMC selected from Lamers (1989). Column titles are self-explanatory except the type that is given below, and the activity state. The latter is classified as Q (quiescent), Max, and Min (activity maximum and minimum) at the epoch of observation.

LBVs are photometric variables on a very wide range of time scales. The larger the variations, the longer is the time scale so that the types of variations can be classified into the following three types according to their scale of variations (Lamers 1987).

Type 1: The large variations with $\Delta V \geqq 3$ magnitudes, associated with eruptive ejection of a large amount of mass. The time scale between large outbursts is very long over centuries, e.g., η Car, P Cyg.

Type 2: The moderate variations of $\Delta V \sim 1$ magnitude, the time scale is of the order of decades, e.g., AG Car, S Dor.

Type 3: The small-scale irregular variations with $\Delta V \sim 0.1$–0.2 magnitude, occurring on time scales of weeks to months. This type of variations is common among LBVs.

Table 5.21: Selected LBVs in the Milky Way and LMC (adapted from Lamers 1986, 1989)

Star	Type	Activity state	Spectral type	V	$M_{\rm bol}$	$T_{\rm eff}$	R^*	V_∞ (km s)$^{-1}$	log $\dot{M}$ ($M_\odot$ per year)
Milky Way									
ηCar	1	Q	—	6.0	−11.8	30,000	76	700	<−3.0
P Cyg	1	Q	B1Ia+	4.8	−9.9	19,300	76	210	−5.0
AG Car	2	Max	B5Ia	6.1	−11.3	13,700	290	166	−4.6
HR Car			B2eq	—	—	—	—	—	—
LMC									
R 66		Q	B7/Aeq	—	−8.9	12,000	125	200	—
R 71	3	Min	B2.5Iep	10.9	−8.8	16,000	88	170	−6.2
		Max	A1I	9.9	−9.8	10000	330	110	−4.7
R 81		Q	B2.5eq	—	−10.0	20,000	75	330	—
S Dor	2	Max	Aeq	9.5	−9.5	10,000	430	110	−4.5
R127	2	Max	B5I	10.4	−10.6	17000	140	200	−4.2

Notes: Star: S Dor = HD 35343 = R88, R 71 = HD 269006. Activity state: Q = quiescence, Max. = maximum, Min. = minimum. Spectral type of AG Car at minimum activity: classified as Ofpe/WN9 (Walborn 1990), or WN11 (Smith et al. 1994).

It is mentioned that there is some difficulty in finding the variations of type 1 LBVs because of their long time scales over centuries. For instance, the large outburst of P Cygni occurred in AD 1600 and the star itself has been stable within $\Delta V \sim 0.2$ in this century (de Groot 1989). Hence, if we have no observations before nineteenth century, it will be difficult to classify P Cygni as type 1. Similarly, η Car has shown large amplitude variations over the centuries, it has experienced a prominent outburst up to $M_{\rm bol} \sim -14.8$ in 1843, then, once declined rapidly, and gradually brightened up to $M_{\rm bol} = -11.3$ at present time. Due to its long time scale of type 1, there may be an oversight to find it. For instance, the lacking of LBV of type 1 in the LMC is supposed to be due to such oversight effect.

Among the related objects to the LBVs given in Table 5.21 the supergiant B[e] and Hubble–Sandage variable stars (see later) are notable. Humphreys and Davidson (1994) have paid attention to the recurrent irregular eruptions leading to violent mass-loss observed in all of these related objects, and called these phenomena the "astrophysical geysers."

η Car

Eta Carinae is supposed to be one of the most massive stars in our Galaxy having a mass of 120 $M_\odot$, or even more of ~170 $M_\odot$, according Maeder's (1980) model. There is also a binary hypothesis of $P = 5.5$ years, with $M_1 = 80$ $M_\odot$ and $M_2 = 30$ $M_\odot$ (Soker 2001).

As stated above, this star had experienced a bright phase from around 1822 to 1856 (Viotti 1995, Frew 2004), known as the Great Eruption, and ejected

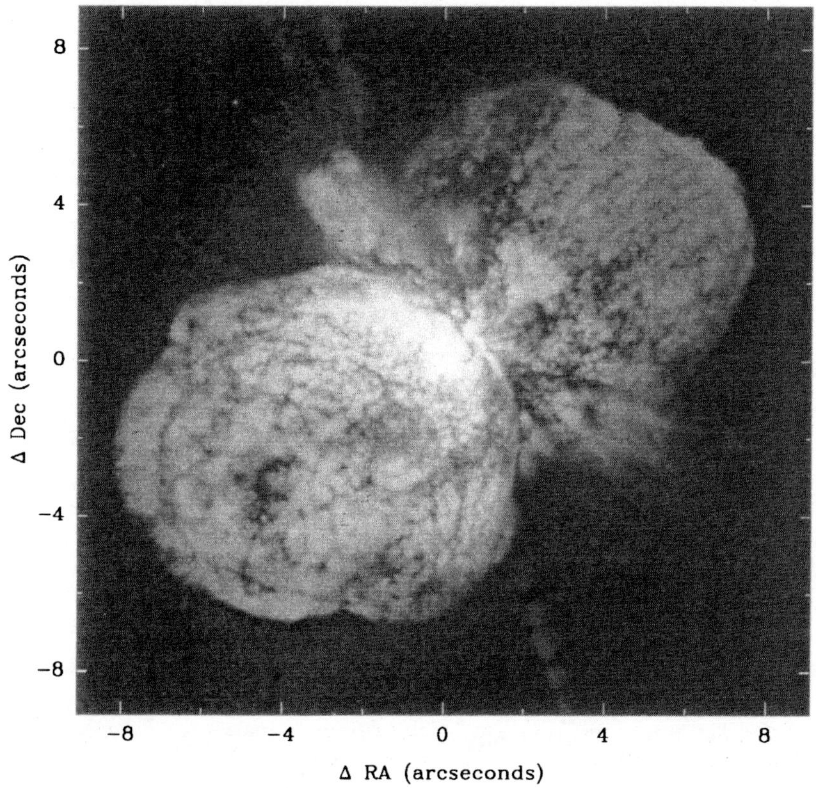

Figure 5.44: Nebulosity around η Car observed with the Planetary Camera (red band) of the HST in 1996. Two distinct lobes are visible on both sides of the unseen central star. In between two lobes, a giant thin equatorial disk is also seen. These lobes are supposed to be gas cloud ejected from the star at the time of great eruption in 1843. The scale $4''$ corresponds 5700 AU at the distance of η Car. (From Morse et al. 1998)

a large amount of gas, which is now observed as a reflection nebula (diameter $16'' = 23{,}000$ AU). Figure 5.44 exhibits the HST image of the nebulosity around η Car, which is composed of two expanding polar lobes and a giant thin equatorial disk (Morse et al. 1998). By making use of the 3.6-m reflector at ESO (La Silla), Viotti et al. (1986) have made high-dispersion spectroscopic observations of this star and its nearby nebulosity in the Hα region. Their spectra are shown in Figure 5.45, where the spectrum named "core" indicates the stellar emission lines characterized by a narrow component and asymmetric broad wing extending to ± 500 km s^{-1}, and the spectrum named "nebula" shows the Hα profile $3''$–$5''$ south of the star. From the emission profile of Hα nebular line, we can estimate the half-width of 200 km s^{-1}, total width of 520 km s^{-1}, and the peak velocity of 102 km s^{-1} receding from the

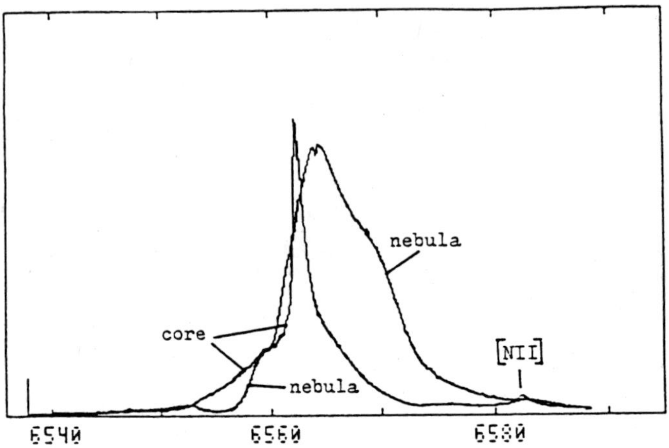

Figure 5.45: Hα emission-line profile in the stellar core and of the nebula 3″–5″ south of η Car. (From Viotti et al. 1986)

star. This infers that the nebula is expanding as a whole with violent inner gas motions. Based on the long-term photometric observations made between 1935 and 1992, van Genderen et al. (1994) showed a secular brightening of around 2.5 magnitudes in the V band during this period. They interpreted this brightening as a decrease of the circumstellar extinction due to the expansion of clumpy dust envelope.

The spectrum of η Car shows rich emission lines from far UV to infrared spectral region, both containing permitted and forbidden lines of HI, HeI, FeII, along with NI, SiII, NaI, MgII, CaII, and AlII, mostly exhibiting P Cygni profiles. A detailed line identification in the wavelength range from λ 3093 to 8956 Å is presented by Wallerstein et al. (2001) in the electronic form. Rich emission lines can be used to the diagnostics of circumstellar envelope of η Car. Hillier et al. (2001) analyzed the emission-line spectrum by non-LTE model fitting and derived some basic properties of the envelopes such as: (1) very high mass-loss rate of $10^{-3}\,M_\odot$ pre year, (2) ionization structure of the stellar wind for H, He, and Fe, with high stratification from inner H^+, He^{++}, F^{++++} region to outer H, He, and Fe^+ region, and (3) chemical abundance, characterized by a high overabundance of N, and a substantial depletion of C and O, and more or less nearly solar abundance for Fe and Mg. The anomaly of chemical abundance may reflect the evolutionary scenario of this star.

5.4.2 P Cygni and P Cyg-type stars

P Cygni

The B type supergiant P Cygni (B1 Ia^+) is a prominent emission-line star belonging to the LBVs, and its spectrum is characterized by line profiles composed of emission component on the red side and absorption component on the violet side, the so-called P Cyg-type profile. These profiles appear in

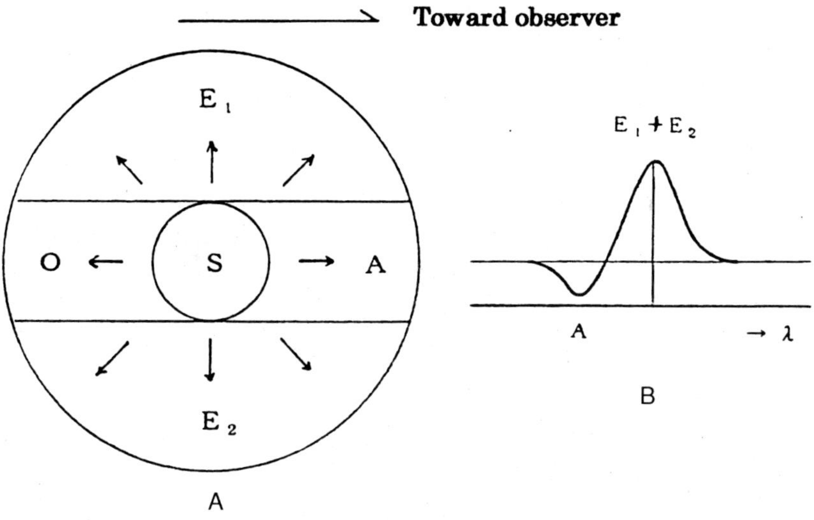

Figure 5.46: Expanding envelope model and the P Cyg profile. **A** Expanding envelope is divided into four parts relative to the star S: A—Violet-shifted absorption-line forming region in front of the star S; E1 and E2—Emission-line forming region in both side of the star S; O—Obscured region behind the star S. **B** P Cygni profile and contributions from each part of the envelope given in **A**.

the permitted emission lines in the Balmer series and metallic ions (FeIII, SII, NII, OII). Meanwhile the forbidden lines such as [FeII], [NII], [TiII] appear as single-peaked weak emission without violet absorption components. These forbidden lines should be formed in the tenuous, low-velocity region lying outside the expanding envelope where the permitted lines are formed.

The atlases of emission-line profiles are provided by Markova and Zamanov (1995) for longer wavelength region (λ 4840– 6760 Å), and by Markova (1994) for shorter wavelength region (λ 3550–4800 Å). In both atlases the spectral resolution is 0.36 Å (inverse dispersion 18 Å mm^{-1}, S/N $\sim$ 50).

P Cygni profile is basically formed in the spherically symmetric, expanding envelope. This was first suggested by Beals (1929). The formation process is schematically shown in Figure 5.46, where panel A illustrates the parts of the expanding envelope, i.e., part A is the absorbing region, parts E_1 and E_2 the emitting regions, and part O the obscured region behind the star S. panel B denotes the P Cyg profile and the contributions from each part of the envelope. If the envelope is in radiative equilibrium, the temperature of envelope gas should be lower than the stellar effective temperature so that a violet-shifted absorption line is formed in part A, whereas the emission component is formed in parts E_1 and E_2. The obscured part O has no contribution for the formation of the profile.

Beals (1951) distinguished the P Cyg profiles into three main types, I, II, and III, as schematically shown in Figure 5.47. Type I is the typical profile

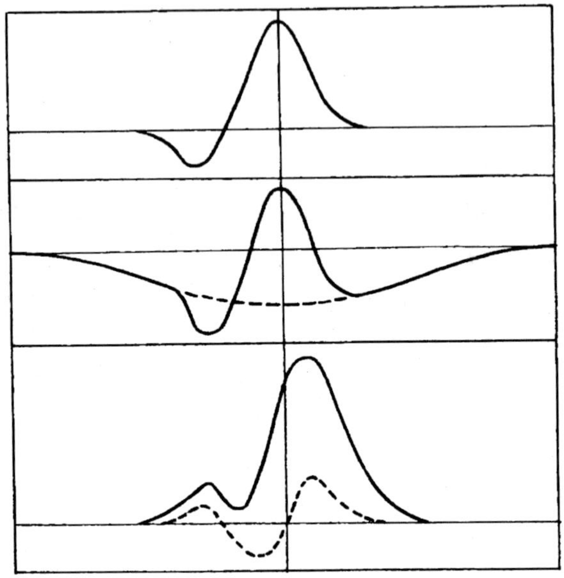

Figure 5.47: Types of P Cygni profile defined by Beals (1951).

formed in a developed expanding envelope as seen in Figure 5.46. Type II is the P Cyg profile superimposed on a broad photospheric absorption line. This profile should be formed in less developed expanding envelope surrounding a star having strong and broadened photospheric absorption lines. In contrast, type III can be seen as a double-peaked emission with V/R less than unity, and hence it does not necessarily represent the existence of an expanding envelope. Actually, Ulrich (1976) showed the possible formation of this type even in contracting envelope. Hence the P Cygni profile in its original sense is to be restricted to types I and II. On the other hand, when an emission line has its absorption component on the red side, it is called the inverse P Cygni profile, and implies the formation of line profile in a contracting envelope.

The star P Cygni was discovered in August 1600 by the Dutch chart maker W. J. Blaeu when it suddenly brightened as a nova to the third magnitude. By 1626 it faded until becoming invisible by naked eyes. After repeating several brightening in late seventeenth century, it gradually increased its brightness from 6 magnitude up to 5 magnitude at present. The visual light history is schematically sketched in Figure 5.48. In twentieth century the brightness is nearly constant with some irregular variations.

According to Markova et al. (2001a,b), photometric and spectroscopic behaviors exhibit different behaviors in the long-term (~7.4 years) and short-term (~3–4 months) variations as follows.

(a) Long-term variation. During 13.8-year observations it was shown that, when the star becomes brighter, its colors become redder and the Hα

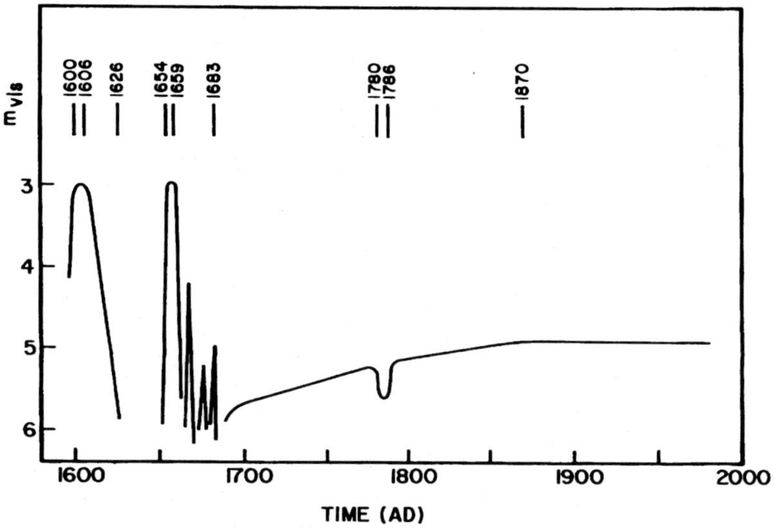

Figure 5.48: Schematic light curve of P Cygni since AD1600. (From Lamers 1986)

intensity (emission equivalent width W(Hα)) is intensified, as shown in Figure 5.49. Large scatter in the values of W(Hα) may reflect the effects of short-term variation.

(b) Short-term variation. Variation is irregular, but there is an apparent anti-correlation between V magnitude and the Hα equivalent width, i.e., there is a tendency when the star brightens, Hα intensity becomes

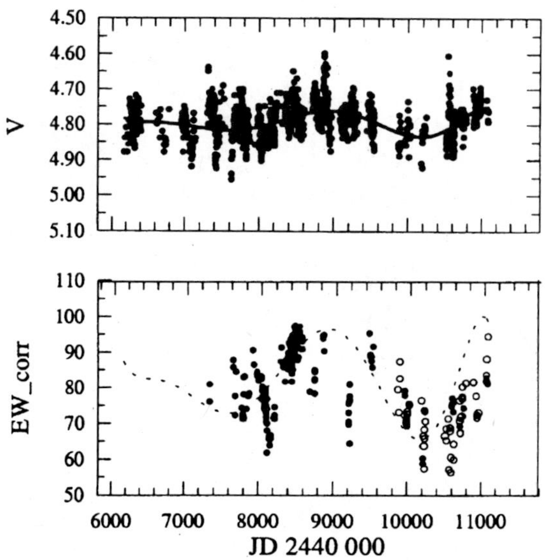

Figure 5.49: Long-term variations of P Cyg in V-magnitude (upper panel) and EW(Hα) (lower panel) observed from July 1988 to January 1999 (10.6 years). The solid and dashed lines represent the nonlinear fitting curves. (From Markova et al. 2001b)

weak. This is in contrast to the case of long-term variation. The amplitude of variation in W(Hα) is around 5–10 Å in short-term variation as compared to the amplitude more than 30 Å in the long-term variations.

Different behaviors in long- and short-term variations infer the existence of different processes in the envelope of P Cygni star. Markova et al. (2001b) attributed the origin of long-term variation of the Hα equivalent width to the change of mass-loss rate, which is possibly connected to the luminosity change of the star. The short-term variability has been explained in terms of stellar nonradial pulsation by Lamers (1986) and by de Groot et al. (2001) who derived a stable quasi-period of 17.3 days. Markova et al. (2001a) also suggested the existence of nonradial pulsation with longer quasi-period of several months as a possible cause of the short-term variations.

P Cyg-type stars

Among the LBVs, the stars that show P Cyg profiles are called the P Cyg-type stars. Or, they are sometimes called S Doradus-type stars named after the variable S Dor that has the similar spectrum with the star P Cygni. This is by the reason to avoid the confusion between P Cyg profiles and P Cygni stars, since P Cyg profile occurs in many types of emission-line stars. The stars P Cgy and S Dor are almost identical as the members of the LBV (Lamers 1986). However, S Dor has a distinct behavior different from P Cygni, in the occurrence of particular period showing the inverse P Cyg profiles in the lines of singly ionized metals such as FeII, ScII, and CrII (Wolf and Stahl 1990). According to Wolf (1994), S Dor showed a large brightening from 1983 to around 1993 and peaked in 1988 with the brightness more than 2 magnitudes above the quiescent level. In the course of brightening, S Dor revealed clear P Cyg profiles in metallic lines and then turned to the period of inverse P Cyg profiles for around 2 years just after the light maximum. Wolf explained this behavior as the ejection of a large amount of gas making an expanding envelope, followed by a falling back of matter to its inner part.

The location of P Cyg-type stars on the HR diagram is shown in Figure 5.50 (Lamers 1986), in which the stars are distinguished as quiescent, brightening, maximum, and unknown, according to their activity. The positions of normal supergiants (Of, OB Ia) are also plotted for comparison. The solid line indicates the Humphreys–Davidson instability limit. The broken line gives the mean brightness for the quiescent phase of the P Cyg-type stars. It is seen in Figure 5.50 that the P Cyg-type stars are generally below the Humphreys–Davidson instability limit, and only reaching that level at their maximum state.

Remarkable characteristics of the P Cyg-type stars are high mass-loss rate, low terminal velocity, and high gas density of the winds. These features derived by Lamers (1986) are shown in Figure 5.51 as functions of stellar luminosity or stellar effective temperature, and compared with those in normal supergiants.

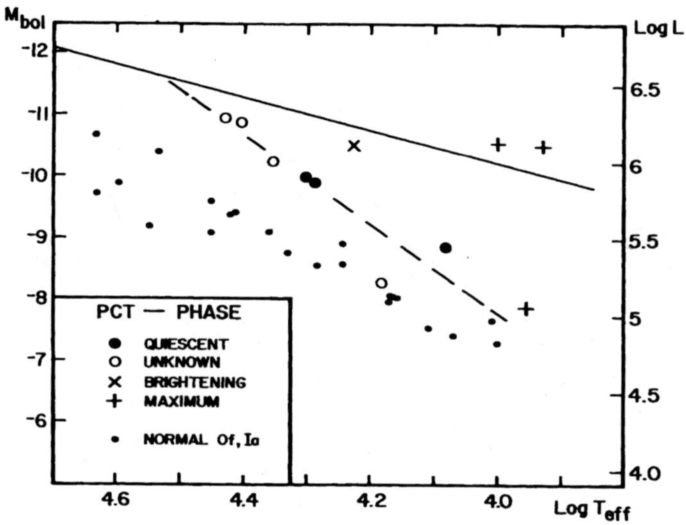

Figure 5.50: HR diagram of the P Cyg-type stars (PCT). Symbols indicate the phase of their activity as shown in the inserted box. The positions of Of stars and normal supergiants are shown by small filled circles for comparison. The solid line indicates the Humphreys–Davidson's upper limit, and the broken line gives the mean brightness for the quiescent phase of the PCT. (From Lamers 1986)

The mass-loss rates are a factor of 3 to 10 higher than those of normal stars, even in the quiescent phase. Terminal velocities of the winds are of the order of 100–300 km s^{-1}, i.e., 3 to 10 times lower, and then the gas density of the winds at the radius 2 R_* is 30 times higher than that in normal supergiants.

The velocity distribution of the winds in the envelope of P Cyg-type stars is also remarkable. The velocity law in the stellar winds is usually written in a parameterized form

$$V = V_0 + V_\infty \left(1 - \frac{R_*}{r}\right)^\beta , \qquad (5.4.1)$$

where V_0 and V_∞ denote the initial and terminal velocity, respectively, R_* the stellar radius and β a parameter expressing the form of velocity distribution in the envelope. When β is small (smaller than around 1), the wind velocity increases rapidly near the stellar surface and then gradually approaches the terminal velocity, whereas, when β is large, wind velocity slowly increases toward terminal velocity. In normal supergiants, β takes the value of around unity, and hence $V = 0.5\, V_\infty$ at $r = 2\, R_*$, whereas, in P Cyg stars as well as in S Dor stars, V is only 0.16 V_∞ at $r = 2\, R_*$ ($\beta \sim 2.6$) (Waters and Wesselius 1985). This infers that the kinetic state of the envelopes of P Cyg stars is quite different from that of normal supergiants.

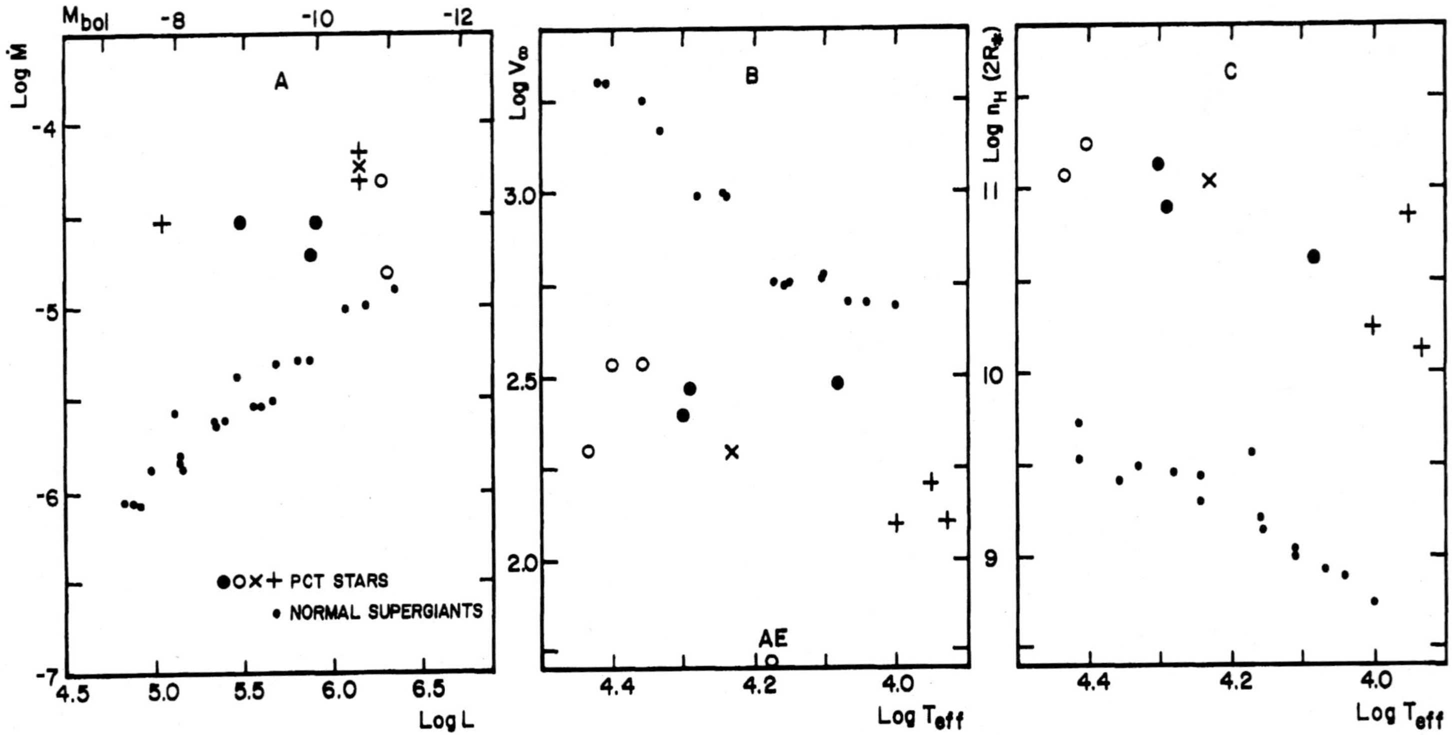

Figure 5.51: The mass-loss characteristics of the P Cyg-type stars compared to those of normal supergiants. **A**: Mass-loss rate $\dot{M}$($M_\odot$ per year) versus stellar luminosity L. **B**: Terminal velocity V_∞ (km s^{-1}) versus effective temperature $T_{\rm eff}$. **C**: Gas density of the wind n_H (cm^{-3}) at 2 R_* versus $T_{\rm eff}$. The symbols of stars are the same as in Figure 5.56. (From Lamers 1986)

Table 5.22: Selected sgB[e] stars in Magellanic Clouds and stellar parameters

Star	Spectral type	*V* mag	*E(B-V)* mag	T_{eff} (10^3 K)	M_{bol} mag	$R/R_\odot$	$M/M_\odot$	log $(L/L_\odot)$
LMC:								
R82	B3	12.0	0.20	18.5	−8.8	50	30	5.4
HenS12	B0.5	12.6	0.20–0.25	23	−8.6	30	25–30	5.3
HenS22	B0-0.5	11.8	0.25–0.30	23–26	−10	49	50–55	5.8–6.0
HenS134	B0	12.1	0.20–0.25	26	−10	45	60	5.9
R126	B0.5		0.25	22.5	−10.5	72	70–80	6.1
R66	B8		0.12	12	−8.9	125	30	5.5
SMC								
R4	B0.5	13.2	0.20	23–26	−8.8–9.3	32	30–40	5.0
R50	B3	11.5	0.15–0.20	17	−9.5	81	40–50	5.7

Note: The value of log $(L/L_\odot)$ is taken from Zickgraf (1998) and others from Zickgraf et al. (1985, 1986).

5.4.3 Supergiant B[e] stars

Supergiant B[e] stars (sgB[e]) are among the peculiar Be stars (Section 5.3.7) and included in the group of LBV by its luminosity and spectral similarity.

The sgB[e] stars are mostly observed in Magellanic Clouds. The great advantage of the MCs is that their distance and hence the luminosities of the B[e] stars are known. The difficulty in finding sgB[e] in the Galaxy mostly lies in the lack of reliable estimation of distance. Zickgraf (1998) listed five galactic sgB[e] including MWC 300 (B1 Ia) and CPD-52°9243 (B3 Ia).

A list of selected sgB[e] stars observed in LMC and SMC is given in Table 5.22, along with some stellar parameters taken from Zickgraf et al. (1985, 1986, 1998). These stars are the supergiants having the mass of 30–80 $M_\odot$ and distributed in the lower part of the Humphreys–Davidson instability limit in the HR diagram, intermingling with normal early-type supergiants (Humphreys and Davidson 1979).

According to Zickgraf (1992), the spectra of sgB[e] stars are characterized by the following properties:

(a) Strong Balmer emission, frequently showing P Cygni profiles. The equivalent width of the Hα emission is as high as 10^2–10^3 Å.
(b) Appearance of narrow emission lines both of permitted and forbidden lines such as FeII, [FeII] and [OI].
(c) Strong infrared excess due to thermal radiation of hot dust envelope ($T_{\rm dust} \sim 1000$ K, $R_{\rm dust} \sim 300\ R_*$).

As an example, the visual spectrum of R50, the sgB[e] star in SMC, is shown in Figure 5.52. The spectrum was taken by Zickgraf et al. (1986) using the Echell spectrograph of 3.6 m ESO telescope (linear dispersion 38 Å mm^{-1}).

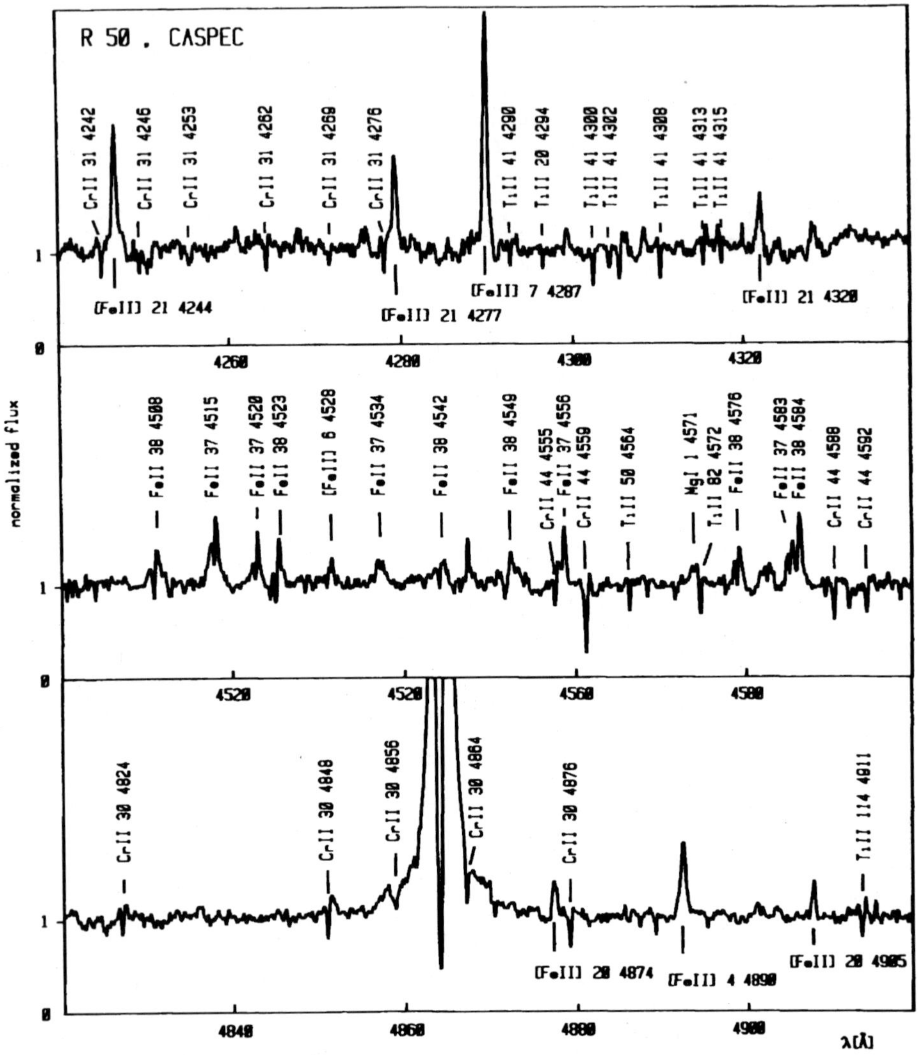

Figure 5.52: Three sections of the blue spectrum of the sgB[e] star R50, observed by ESO 3.6-m telescope. Clear shell absorption lines are seen in the Balmer, TiII, and CrII lines, showing 5 km s^{-1} blue shift relative to the line of [FeII]. (From Zickgraf et al. 1986)

Strong and double-peaked Hβ emission is remarkable. There are also numerous emission lines of ionized metals such as FeII, TiII, CrII, and forbidden lines of [FeII]. While permitted lines exhibit double peaks, forbidden lines are narrow and single peaked. Very sharp "shell type" absorption components are seen in permitted lines of TiII, CrII, and Hβ line, which are blue shifted by -5 km s^{-1} relative to the [FeII] lines.

Table 5.23: Behaviors of the Balmer lines in sgB[e] stars (adapted from the data of Stahl et al. 1985 and Zickgraf et al. 1986)

Star	Spectral type	$H\beta$	n_m(em)	Higher members	Inclination
R50	B3	D(V < R) + w	8	strong shell (H8–H27)	edge-on
R82	B3	D(V < R) + w	6	strong shell (H8–H27)	edge-on
HenS12	B0.5	P (up to H6)	7	H8 (str. abs.) H9(wk abs.)	interm. i
HenS22	B0.5	D(V < R) + w	10	H 10–H11(wk.em)	large i
HenS134	B0	S + br. w	24	H10–H24 (singl.pk.em.)	pole-on
R126	B0.5	S + w	26	H10–H26 (singl.pk.em.)	
R66	B8	P (up to H6)	6	H8–H13 (wk. abs.)	large i

Note: Column Hβ: S and D, single and double peak emission, respectively; P, P Cyg profile V < R (red peak is stronger than violet peak); w, broad wing. Column $n_{\rm m}$(em)—The maximum series number for visible emission component. Column Inclination (angle i)—pole-on (small), intermediate, edge-on (i $\sim 90°$).

The Balmer lines in sgB[e] stars reveal large varieties in their appearance. Some behaviors of the Balmer lines are summarized in Table 5.23, based mainly on the atlases of Stahl et al. (1985) (wavelength range from Hβ to the Balmer series limit) and of Zickgraf et al. (1986) for stars in Table 5.23. One may see that some sgB[e] stars exhibit strong shell absorption lines up to around H26, and general Balmer-line behavior is resembling to the classical Be-shell stars, whereas, some stars exhibit strong emission lines up to around H25 suggesting a very flat Balmer decrement. These Balmer line properties may be explained as an effect of inclination angle from pole-on, where no shell absorption line appears, to equator-on, where strong shell-absorption lines are expected. This view is not contradicting with the arguments of inclination effects suggested by Zickgraf et al. (1986) from spectroscopic behaviors and given in the last column of Table 5.23.

Among the stars in Table 5.22, HenS22 is worth mentioning. When observed in 1974–1975 in the spectral range Hδ–H21 by Muratorio (1978), this star showed a single peak with broad wing in Hδ whereas P Cygni profiles were seen in H8–H20, and the emission components could be traced up to around H24. In 1981–1982, HenS22, contrarily revealed weaker Balmer emission up to around H10 with double peaks (Stahl et al. 1985) as seen in Table 5.23. This change implies that the Balmer line-emitting region experienced some shrinking during this time interval.

Zickgraf et al. (1996) carried out ultraviolet spectroscopic observations with the IUE satellite for three stars (R50, R82, and HenS 22) which are supposed as equator-on or having large inclination angle, and found the P Cygni profiles in FeII and CrII in their spectra. The observations showed that the expanding velocities are unexpectedly very slow with the terminal velocities V_∞ of 75, 100, and 120 km s^{-1} for stars R50, R82, and S22, respectively.

This is about a factor of ten slower than V_∞ of normal B-type supergiants. According to Zickgraf et al. these results are consistent with a two-component wind model. That is, when stars are seen edge-on, the wind velocities from the disklike envelopes are generally very low, whereas, when viewed from high-latitude regions, the velocities of radiation-driven wind are much higher. This is just as in case of classical Be stars.

In addition, the stars in Table 5.22 show remarkable IR excess in the H, K, and L bands, which are supposed to originate from cool and dense equatorial winds with a typical dust temperature of 900–1200 K, and located out of the ionized region in the equatorial winds.

In this way, Zickgraf et al. (1985) suggested a two-component wind model, which consists of a hot and fast radiation-pressure-driven OB star wind in the polar region and a dense, cool, and slowly expanding wind in the equatorial zone forming a disklike envelope. Though this two-component model seems to be similar to the rapidly rotating Be stars, the rotational velocities are not yet measured for sgB[e] stars.

For the galactic B[e] stars, Zorec (1998) lists 13 B[e] stars, among which 3 stars (HD 316285, MWC 314, and MWC 349) are actually sgB[e] stars ($M_{\rm bol} < -8$), and 5 stars are peculiar Be stars. Zorec has considered the distance, kinematics, and distribution of B[e] stars in our Galaxy, and has shown that these stars are strongly concentrated toward the galactic plane, and distribution in the galactic longitude resembles that of WR stars. Zickgraf (2003) analyzed emission-line profiles of the $H\alpha$ and some forbidden lines in 18 galactic B[e] stars including 5 sgB[e] stars. Based on axially symmetric and radially expanding wind models with an equatorial opaque dust disk, Zickgraf calculated emission-line profiles for optically thin lines, and showed that the wind structure has no essential difference between supergiant and nonsupergiant B[e] stars.

5.4.4 Hubble-Sandage stars

Hubble and Sandage (1953) first paid attention to five extragalactic luminous variable stars, one in M31 (Var 19 = AF And) and four in M33 (Vars A, B, C, and 2). These five stars have become known as the Hubble-Ssndage variables (hereafter abbreviated as HS variables). Their characteristic features are as follows:

(1) They are irregular variables with the highest luminosity class in the galaxies.

(2) They show strong, hot continuum in the ultraviolet region.

(3) They show intermediate F-type spectra with strong emission lines in the Balmer and in some ionized metals.

With these behaviors HS variables are distinguished from nonvariable brightest stars or Cepheid variables in the respective galaxy. The light variations of HS variables generally undergo irregularly with the amplitudes of

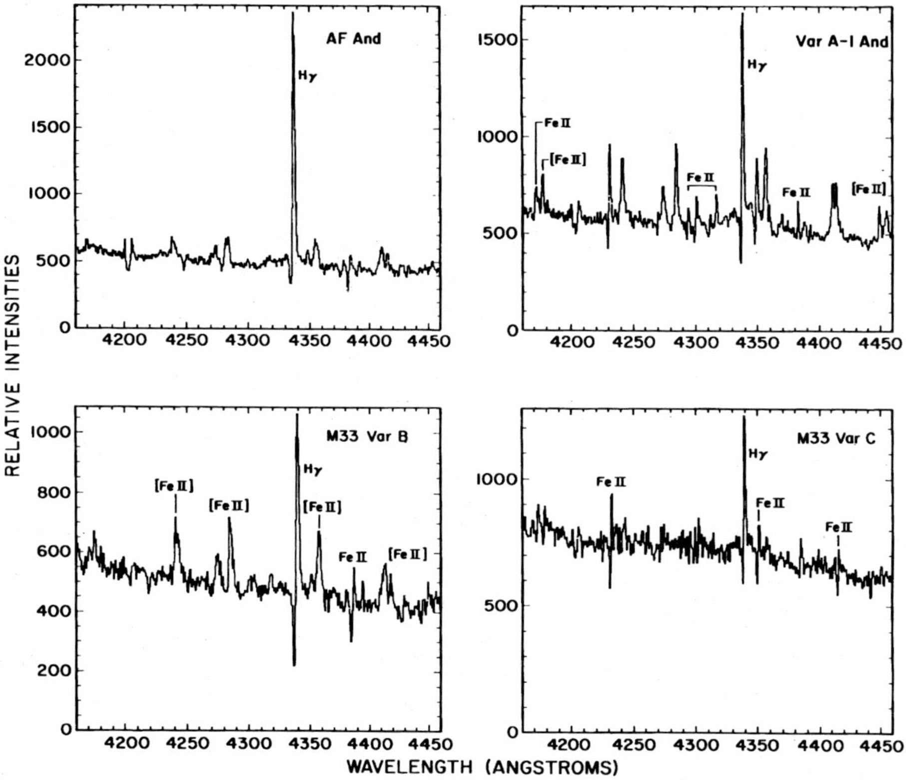

Figure 5.53: Optical spectra over λ 4160–4450 of the HS variables in M31 and M33. Spectral resolution is around 1 Å. P Cygni profiles are visible on the Hγ and some FeII lines. (From Kenyon and Gallagher 1985)

2–3 magnitudes and the time scales of several years to several decades. The absolute magnitudes are as high as $M_{\rm pg} \sim -7$ to -11.5 (Wolf 1986), or $M_{\rm bol} \sim -9.4$ to -11 (Szeifert et al. 1996).

Detailed spectroscopic observations were carried out by Humphreys (1975, 1978), and numerous emission lines were found: permitted lines in HI, HeI, FeII, TiII for all stars, and forbidden lines in [FeII], [NII], [OII], [SII] for some stars. Figure 5.53 shows the spectra near Hγ line taken by Kenyon and Gallagher III (1985) with the Multi-Mirror Telescope (MMT) at Mt. Hopkins Observatory. The P Cygni profile is remarkable in $H\gamma$ and in some FeII lines, while [FeII] lines show single-peaked emission. Overall behaviors are very similar to those of η Car and S Dor.

Based on the HST and ground-based observations, Szeifert et al. (1996) analyzed the photometric and spectroscopic data of the HS variables, and derived their physical parameters and mass-loss rates. The results of model fitting for three stars are summarized in Table 5.24. High luminosity, low

Table 5.24: Physical parameters of the HS variables (adapted from Szeifert et al. 1996)

	Model parameters		Derived parameters			
Star	$T_{\rm eff}$ (K)	log g	$R_*/R_\odot$	V_∞ (km s^{-1})	log $L/L_\odot$	dM/dt($M_\odot$ per year)
AF And	30,000	3.2	63	150	6.5	3 10^{-5}
Var B	9,000	—	440	90	6.05	3 10^{-4} (B) 4 10^{-5} (C)
Var C	13,000	1.6	185	80	5.9	2 10^{-5}

Note: The mass loss rate of Var B: (B) fitting with Balmer profile, (C) fitting with continuum.

terminal velocity, and high mass-loss rate are remarkable, though the mass-loss rate largely depends on the model fitting.

Concerning the relationship between HS variables and LBVs, there are two arguments. One is that HS variables are the same group with LBVs by the reasons (Wolf 1986, Szeifert et al. 1996)

(a) spectral similarity with η Car and S Dor.
(b) similar location of HS variables and LBVs in the HR diagram.

The alternative one is that HS variables are somehow distinguished from LBVs (Kenyon and Gallagher III 1985). The reasons are

(a) Location of HS variables in M31 and M33, outside of major star forming regions.
(b) Lack of evidence for enhanced nitrogen abundance in the stellar winds of HS variables. This enhancement is expected when massive single stars are on the way from luminous OB stars to WR stars as in case of LBVs.

For these particular behaviors of HS variables, Kenyon and Gallagher III suggested the possibility of massive binary nature for the majority of HS variables in M31 and M33. These two arguments are to be reconciled in future.

5.5 Evolutionary status of early-type emission-line stars

5.5.1 Evolution of massive stars and emission-line stars

It is widely known that massive stars with masses higher than 20 $M_\odot$ evolve, after leaving the main sequence, into luminous variable stars such as Of, WR, and LBV, and finally explode as supernovae. In case of single stars, the main parameters to determine the evolutionary path and types of emission-lines are the mass, chemical abundance, and mass-loss rate at the initial stage, i.e., at the ZAMS. In some stars the effects of stellar rotation, magnetic fields play important roles in their evolution. Emission-line stars considered in this chapter are closely related to the evolution of massive stars. We briefly consider the evolutionary state of these stars.

A series of theoretical models carried out by Maeder (1980) and his colleagues has been known as the standard models (Meynet et al. 1994). The parameters they used are the stellar mass (12–120 $M_\odot$), metal abundance Z (0.001, 0.004, 0.008, 0.020, 0.040), and mass-loss rate $\dot{M}(M_\odot$ per year), among which mass-loss rates are taken from the model calculations by Niewenhuijzen and de Jager (1990) in their standard values and twice its value. The evolutionary tracks in the standard models calculated by Schaller et al. (1992) are shown in Figure 5.54 for the case of $Z = 0.020$ (solar metalicity). Upper panel denotes the case of standard mass-loss rate and lower panel shows its twice mass-loss rates. In both panels the stellar initial mass is denoted along the ZAMS, and the evolutionary path for each mass is given with the indication of the position of the different WR subtypes (WNL, WNE, and WC). Some positions along the tracks are also indicated to show the central helium content (in mass fraction). The decrease of this content shows the process of consumption of the helium in the stellar core.

After leaving ZAMS, stars evolve toward cooler stars loosing hydrogen from the surface by the continuous mass loss. In addition, stars begin to show the spectra of helium and heavy-element lines that are brought up from the interiors by convection. This is the stage of hydrogen-deficient WN stars. With the ignition of helium burning, stars turn to move toward hotter regions through various stages of Of, WR, and LBV. In Figure 5.54, remarkable effects of mass loss leading to great luminosity drops during the evolution of the most massive stars ($\sim 120 M_\odot$) are traceable in the lower panel. The importance of metal abundance has been considered by Maeder (1990, 1991) in their grid calculations of stellar evolution and pointed out that the higher the metal abundance, the longer the lifetime of the WR stage. In summary, this standard model predicts that massive stars typically evolving through the following types of stars before reaching the supernova explosion:

$$\mathrm{O} \rightarrow \mathrm{Of} \rightarrow \mathrm{BSG(orH\text{-}rich\,WN)} \rightarrow \mathrm{LBV} \rightarrow \mathrm{WN} \rightarrow \mathrm{WC} \rightarrow \mathrm{SN}$$

where BSG denotes the blue supergiants, and SN the supernova.

Apart from the standard models, there are other evolutionary theories, which additionally consider the dynamical processes in stellar structure. Among these theories, Langer et al. (1994) proposed a new scenario for the evolution of very massive stars. Based on the mass-loss process induced by the violent pulsational instabilities they calculated evolutionary tracks for sample stars of initial mass 60 $M_\odot$ as follows:

$$\mathrm{O} \rightarrow \mathrm{Of} \rightarrow \mathrm{H\text{-}richWN} \rightarrow \mathrm{LBV} \rightarrow \mathrm{H\text{-}poorWN} \rightarrow \mathrm{H\text{-}freeWN} \rightarrow \mathrm{WC} \rightarrow \mathrm{SN}$$

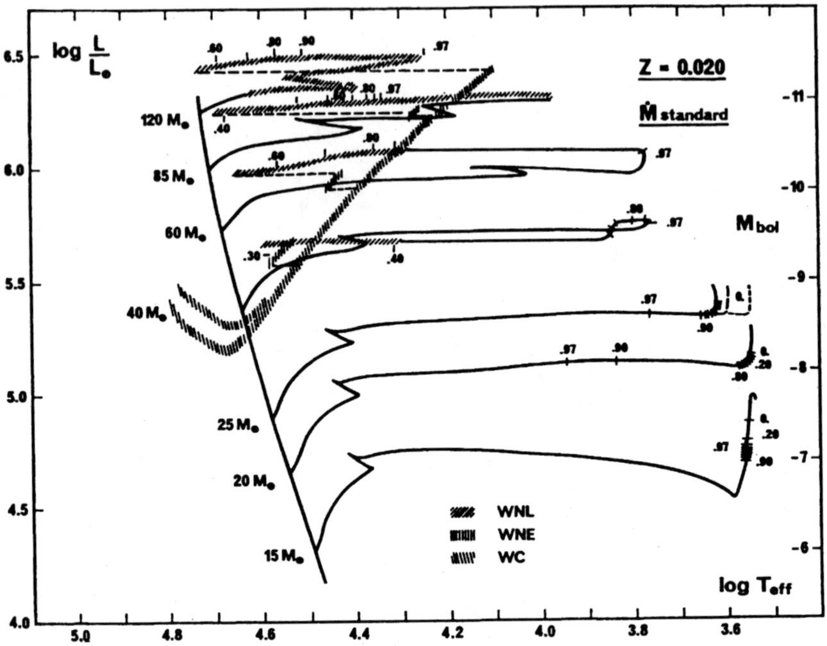

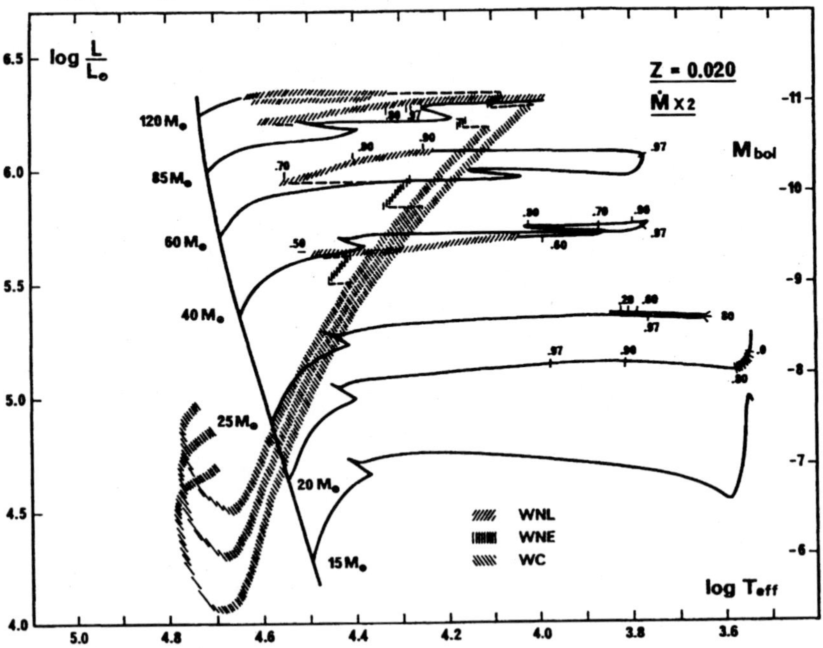

Figure 5.54: The evolutionary tracks of massive stars with $Z = 0.020$ on the HR diagram. The locations of WNL, WNE, and WC stars are indicated, as well as the values of central helium content in mass fraction during the He-burning phase. Upper panel: Case of standard mass-loss rates. Lower panel: Case of twice of the standard mass-loss rates. (From Schaller et al. 1992)

Evolutionary tracks for 30–90 $M_\odot$ are also computed by Stothers and Chin (1996) under particular attention on the evolutionary status of the LBV. They considered the dynamical instability in the outer parts of stellar envelopes as the massive outbursts and applied to the evolution of massive stars. According to them, LBV stage as variable stars appear on the way from cooler to hotter regions in the upper part of the HR diagram. The lifetimes of the LBV phase are very short, as short as 10^3–10^4 years!

The effects of magnetic fields and stellar rotation were considered by Maheswaran and Cassinelli (1994). They calculated theoretical models for the evolutionary changes of the rotation parameter $\alpha = V_{\rm rot}/V_{\rm c}$, where $V_{\rm rot}$ and $V_{\rm c}$ denote the rotational velocity and critical velocity, respectively, and examined the appearance of rapidly rotating supergiants such as B[e] stars. According to them, a star with initial mass of 60 $M_\odot$ evolves to WR phase via B[e] stage and the rotational velocity at the B[e] stage depends on the distribution of angular momentum in the outer part of the star. Rapid rotation at this stage is only possible if the star has substantial polar outflow and differential rotation with angular velocity increasing inward. They showed that the cylindrical shell model gives faster rotation rate at the B[e] stage and beyond, but a spherical shell model yields slower rotation at the same stage. This model supports the two-component wind structure (a low-speed equatorial flow and a high-speed polar flow) of the B[e] stars observed by Zickgraf et al. (1985) as seen in Section 5.4.3.

On the evolution of massive rotating stars, a remarkable progress has been made in late 1990s to early 2000s. Meynet and Maeder (2003) paid particular attention to WR star phase at solar metallicity. Based on revised model calculations, they showed the remarkable effects of rotation in the evolutionary tracks. For comparison we shall consider nonrotating stars ($V_{\rm ini} = 0\,{\rm km\,s^{-1}}$) and rotating stars ($V_{\rm ini} = 300\,{\rm km\,s^{-1}}$) with initial mass of 60 $M_\odot$, where $V_{\rm ini}$ denotes the initial rotation velocity at ZAMS.

(1) Evolutionary sequence. Nonrotation O stars evolve to WR stars through LBV stars, that is, through lower surface-temperature phase. In contrast, rotating O stars directly evolve to WN stars and then to WC stars, because of large mass-loss rate due to rotational instability.

(2) Lifetimes in WR stars. The WR stage of evolved O stars proceeds as

$$\text{WNL(late type WN)} \rightarrow \text{WNE(early type WN)} \rightarrow \text{WN/WC transition} \rightarrow \text{WC}$$

Total duration of this WR stage is 0.4 Myr in nonrotating stars, whereas it takes 0.75 Myr in rotating stars.

(3) Surface abundance in WR stars. Initial cosmic abundance of O stars changes to CNO equilibrium abundance at WN stars, and to C abundant phase in WC stars. This change occurs much gradually in rotation stars as compared to nonrotation stars. This age dependence of surface abundance reflects to the population of WR stars in each phase of WNL, WNE, WN/WC, and WC, which can be compared with observed population of WR stars.

Similar effects of rotation are also seen in their models of other initial stellar mass and initial rotational velocities of O stars. The basic process is the internal mixing of elements by rotation that makes flatter the internal chemical gradients in rapidly rotating models.

5.5.2 Evolution of Be stars

As stated in Section 5.3, classical Be stars are supposed to have two different origins: single stars and binary stars. We first consider the single Be stars in this section and the binary origin will be discussed in Section 5.5.3.

The origin of single Be stars has long been discussed in connection with the distribution of Be stars on the HR diagram. Three different views have so far been proposed.

(1) The first view is that the Be phenomenon occurs during the overall contraction phase that follows the exhaustion of hydrogen in the core (Schild and Romanishin 1976). In this case, one would expect that Be stars appear near the end of the main sequence phase, i.e., Be stars should be brighter than normal B stars as a whole. This view was supported by Singh and Chaubey (1987) based on their photometric data for the distribution of Be stars on the HR diagram.

(2) The second view is that Be stars are the particular phenomena temporarily appearing among rapidly rotating B stars at any point in the main-sequence lifetime. Zorec and Briot (1997) selected B and Be stars of luminosity classes V–III from the BSC, and studied the frequency of Be stars, $P(\text{Be}) = N(\text{Be})/(N(\text{B}) + N(\text{Be}))$ for each spectral subtype and luminosity class, based on the two dimensional BCD classification. They thus found that P(Be) takes two maximum values at B1 and B7 along the spectral subtype, regardless the luminosity class of stars. They suggested that Be stars hardly represent a given stage in the evolutionary track of every B star. In order to see the distributions of B and Be stars in the HR diagram, observations of open clusters are important. Slettebak (1985) constructed the HR diagram for 12 clusters, among which two clusters NGC 3766 and NGC 4755 in southern sky are shown in Figure 5.55. These clusters contain sufficient number of Be stars among their observed clusters. One may see in this figure that the Be stars are well mixed with B stars above the ZAMS. Feinstein (1987, 1990) also considered the characteristics of 124 Be stars in 52 open clusters on the color–magnitude diagrams in support of the conclusion of Slettebak.

If B and Be stars are well mixed on the HR diagram and if we admit the coeval nature of B and Be stars, there are two possibilities on the origin of Be stars. One is that Be stars are the phenomena of transient or temporal nature to be experienced for all B stars in their evolutionary path. In this case some triggering mechanisms such as nonradial pulsation or some angular-momentum transfer inside the star may cause the Be phenomena.

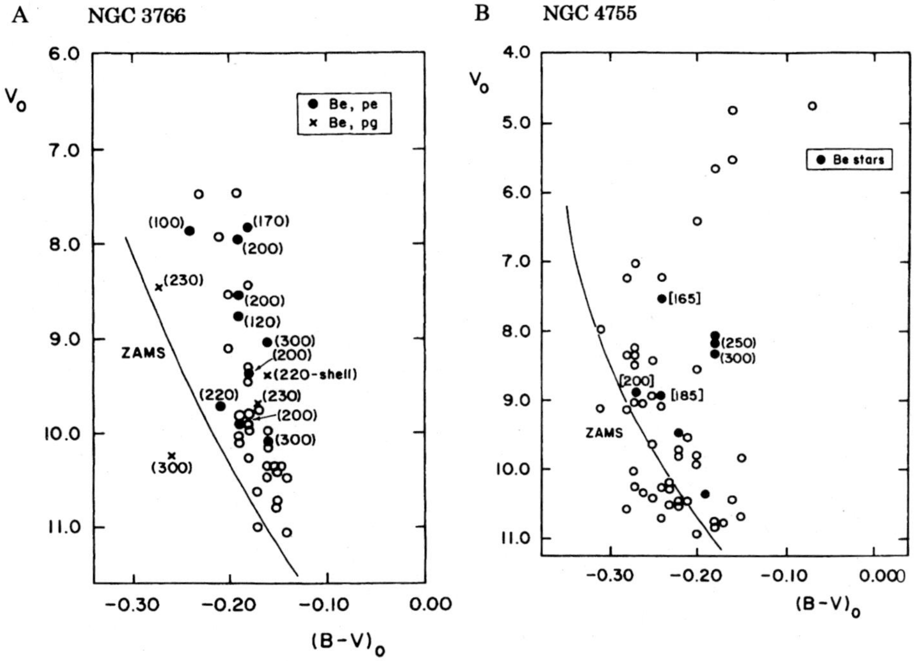

Figure 5.55: Color-magnitude diagram for open clusters. **A**. NGC 3766 (distance = 1.70 kpc, age = 22 Myr). **B**. NGC 4755 (distance = 2.34 kpc, age = 7 Myr). The abscissa denotes the intrinsic color index, the ordinate the V magnitude. The solid line indicates the ZAMS, filled and open circles give the Be and B stars, respectively. The figures in parenthesis in some stars denote the values of $V \sin i$ of that star. (From Slettebak 1985)

An alternative possibility is that the stars that have special features at their ZAMS stage among B stars may cause the Be phenomena. Rapid rotation may be one of such special features, but not the sufficient condition. Some other triggering mechanisms are required.

(3) The third view is to claim the evolutionary effect that appears in the second half of the main-sequence lifetime of a B star. Fabregat and Torrejón (2000) showed the incidence of Be stars in open clusters as a function of the cluster age, i.e., the clusters younger than 10 Myr are almost completely lacking classical Be stars, and the maximum frequency P(Be) reaches in the age interval 13–25 Myr. They interpreted the Be phenomena as to be related to main structural changes happening at the later half evolutionary phase of B stars through the surface enrichment of helium abundance. According to Fabregat and Torrejón, the semiconvection or turbulent diffusion, which is responsible of the surface helium enrichment, is coupled with the high rotational velocity to generate magnetic fields via the dynamo effect and thereby originate the Be phenomena.

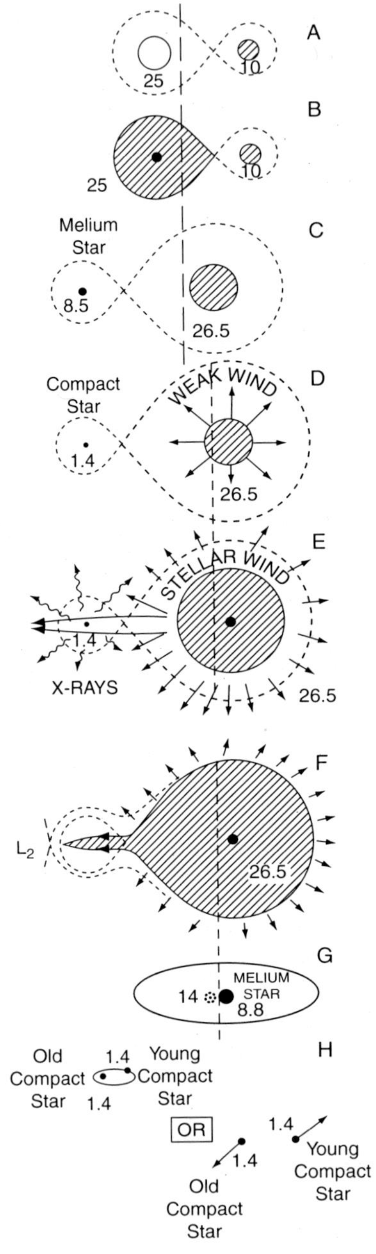

Figure 5.56: Evolution of a close binary system with initial mass of 25 and 10 $M_{\odot}$ and orbital period 5 days. Some typical steps are illustrated from (A) to (H), for each step the duration, binary state, and masses of component stars are given below. The system becomes a wind-powered HMXB (high mass X-ray binary) for some 10^4 years in step (E). Whether the system remains as a binary or disrupted in the final stage (H) depends on the fraction of mass-loss from the system. (From van den Heuvel 1994)

Although there is an argument that all Be stars are formed by binary interaction, this is not fully accepted for all Be star phenomena. The origin of Be phenomena in single stars is still one of the problems remained unsolved.

5.5.3 Evolution of binary systems

The evolution of binary systems is principally determined by the masses of the components, mass ratio, and the orbital period (i.e., separation) in the main-sequence stage. Stellar rotation velocity and mass-loss rate from the binary system may also be important factors. We now consider the binary systems of massive stars (WR stars) and of intermediate mass stars (Be stars).

(1) WR binaries. In massive binary systems, the evolutionary processes are quite complicated, not only by the binary interaction, but also by the stellar mass loss, mass exchange, onset of instability, and supernova explosion in one or both component stars. As an example we consider the evolutionary scenario proposed by van den Heuvel (1994).

Figure 5.56 shows the evolution of a close binary system with initial masses of $M_1^0 = 25\ M_\odot$, $M_2^0 = 10\ M_\odot$ and orbital period $P = 5.0$ days. The different evolutionary steps are given in the order of (A), (B), through (H) and main features of each step are summarized in Table 5.25.

In steps (C) to (D), the original primary and secondary stars are supposed to evolve to a helium star with $8.5M_\odot$ and a WR star with $26.5M_\odot$, respectively, by mass transformation from the primary to secondary star. The helium star will collapse to a compact star through a supernova explosion. In step (E), the WR star expands up to a supergiant with strong wind in an expanding envelope, which produces the strong gas flow onto the compact star giving rise X-ray emission. This step is the so-called HMXB (high mass X-ray binary). In steps (F) to (G), the WR component fills the Roche lobe to start the RLOF

Table 5.25: Main features of each step in Figure 5.56 (adapted from van den Heuvel 1994)

Step	Duration t (yrs) from ZAMS	Orbital period $P(d)$	Mass of components $M_1/M_\odot$	$M_2\ M_\odot$	Binary state
(A)	0	5.0	25	10	ZAMS
(B)	4.71×10^6	5.0	25	10	Start of RLOF
(C)	4.72×10^6	6.84	8.5	26.5	He star + WR?
(D)	5.2×10^6	11.9	1.4	26.5	He star to compact star through SN
(E)	9.00×10^6	11.9	1.4	26.5	M_2 star to supergiant with strong wind
(F)	9.01×10^6	11.9	1.4	26.6	RLOF from M_2 star
(G)	9.02×10^6	0.2	1.4	8.8	M_2 star to helium star
(H)	9.5×10^6	—	1.4	1.4	binary or disruption.

and evolves toward a helium star. Finally, the helium star collapses to a compact star through a supernova explosion. In this step, whether the compact star remains as a binary system or disrupts the system depends on the ejected mass fraction.

(2) Be binaries. It is widely accepted that the binary interaction is one of the basic processes forming Be stars or Be star phenomena. The model of binary interaction has been developed by Harmanec and his group (see Harmanec 1987). According to them, the Be envelope is an accretion disk/envelope structure around the mass gaining component of interacting systems. This is an analogous to the Allgol type binary interaction in that the cool component fills the Roche lobe and the gas flow through the Lagrangean point produces the accretion disk around the mass gaining component. According to Pols et al. (1991), this model has a difficulty in evolutionary time scale. That is, the duration of RLOF phase is too short as compared to the evolutionary time of stars to explain the high fraction of Be stars among all B type stars higher than 10%.

As an alternative model, Pols et al. (1991) proposed the post-mass-transfer model. That is, the transfer of mass and angular momentum to the B star through Roche lobe overflow RLOF gives rise to the spinning up of the B star to high rotation rates. A rapidly rotating B star may become a Be star as in case of single Be stars. After the end of the RLOF, the mass-losing component turns into a helium star or a compact star such as white dwarf or neutron star. The strong stellar winds from Be stars produces the emission of soft X-rays. This step leads to a Be/X-ray binary system. The type of binary system in final phase largely depends on the initial binary mass and mass ratio. Following Pols et al. (1991), we depict the binary evolution.

In Table 5.26 evolutionary scenario of a close binary system and its possible observational counterparts are summarized in the subsequent phases (a) through (e). The initial step is the main sequence binary with masses of M_1^0

Table 5.26: Evolutionary scenario for case B close binary system. Based on Pols et al. (1991)

Step	Binary type	State and counterpart	Example
(a)	Unevolved detached MS + MS binary	B-type spectroscopic binary	
(b)	Semidetached binary with mass transfer	(RLOF phase) Be + cool companion	β Lyr, KX And
(c)	Detached binary spun-up B star + He star	Spectroscopic Be star	φ Per, HR2142(?)
(d)	Case of $M_1^0 > 10$ M. detached binary	Be star + NS classical Be /X-ray binary	X Per
(e)	Case of $M_1^0 < 10$ M. detached binary	Be star + WD low-luminosity Be/ X-ray binary	48 Per? κ Dra(?)

Note: RLOF, Roche lobe overflow; NS, neutron star; WD, white dwarf.

and M_2^0 and the initial mass ratio $q^0 = M_2^0/\ M_1^0(q^\circ$ is smaller than 1, but should be larger than 0.3–0.5). Starting from the zero-age main-sequence binary, the system finally evolves into some binary types of Be stars combined with compact star companion. The system reaches a pair of Be + NS or Be + WD, depending on the initial mass of the primary. Pols et al. (1991) suggested that a large percentage of the known Be stars were formed through close binary evolution as depicted in Table 5.26.

Initial massive star with M_1^0 first experiences the supernova explosion at the step (b) in Table 5.26. If the explosion undergoes nonsymmetric form, it produces a kick motion in the binary system and some part of mass and angular momentum will be left from the system. The evolution of binary systems in which the effects of kick motion are taken into account has been considered by Portegies Zwart (1995) and Van Bever and Vanbeveren (1997). Late-type Be stars are less affected by the presence of a kick, because of smaller initial mass ratio q_0, and tend to remain as binary systems of Be + WD with higher rate. In contrast, early-type Be stars, i.e., high-mass Be stars, suffer the effect of the kick more dominantly resulting the decrease of the number of Be stars with NS companions and the increase of the fraction of disrupted systems. OB stars thus ejected from binary systems are the so-called runaway stars. Portegies Zwart (2000) performed the binary population synthesis for the characteristics of runaway O and B stars, and found that at most 30% of the runaway O stars, but possible all of runaway B stars, obtained high velocities due to supernovae in evolving binaries.

Further reading

Balona, L. A., Henrichs, H. F., and Le Contel, J. M. (eds.) (1994). IAU Symp. No. 162, *Pulsation, Rotation and Mass Loss in Early-Type Stars.* Kluwer, Dordrecht.

Davidson, K., Moffa, A. F. J., and Lamers, H. J. G. L. M. (eds.) (1989). *Physics of Luminous Blue Variables.* Kluwer, Dordrecht.

de Groot, M. and Sterken, C. (eds.) (2001). ASP Conf. Vol. 233. *P Cygni 2000: 400 Years of Progress.* San Francisco, CA.

Hubert, A.M. and Jaschek, C. (eds.) (1998). *B[e] Stars, Proceedings of the Paris Workshop.* 9–12, June, 1997, Kluwer, Dordrecht.

Smith, M. and Henrichs, H. F. (eds.) (2000). IAU Coll. 175, ASP Conf. Ser. Vol. 214. *The Be Phenomenon in Early-Type Stars.* San Francisco, CA.

van der Hucht, K. A. and Williams, P. M. (eds.) (1995). IAU Symp. 163, *Wolf-Rayet Stars, Binaries, Colliding Winds, Pulsation.* Kluwer, Dordrecht.

van der Hucht, K. A., Herrero, A., and Cesar, E. (eds) (2003). IAU Symp. 212, *A Massive Star Odyssey: From Main Sequence to Supernova.* Ast. Soc. Pacific. San Francisco, CA.

References

Abbott, D. C., Torres, A. V., Bieging, J. H., and Churchwell, E. (1986). Radio emission from Galactic WR stars and the structure of WR winds. *Ap. J.*, **303**, 239–261.

Allen, D. A. (1973). Near infra-red magnitudes of 248 early-type emission-line stars and related objects, *M. N. R. A. S.*, **161**, 145–166.

Allen, D. A. and Swings, J. P. (1976). The spectra of peculiar Be stars with infrared excesses, *A.A.*, **47**, 293–302.

Aller, L. H. (1956). Chapter III, Table 3, *Gaseous Nebulae.* Chapman & Hall, London, p. 66.

Andrillat, Y., Jaschek, M., and Jaschek, C. (1988). A survey of Be stars in the 7500–8800 Å region. *A. A. Suppl.*, **72**, 129–149.

Andrillat, Y., Jaschek, M., and Jaschek, C. (1990a). A survey of Be stars in the infrared. II.–Determination of envelope dimensions, *A. A. Suppl.*, **84**, 11–28.

Andrillat, Y., Jaschek, M., and Jaschek, C. (1990b). The Paschen P7 line in Be, Ae and shell type stars, *A. A. Suppl.*, **85**, 855–863.

Andrillat, Y., Jaschek, M., and Jaschek, C. (1994). A study of Be stars in the wavelength region around Paschen 7, *A. A. Suppl.*, **103**, 135–155.

Andrillat, Y., Vreux, J. M., and Dennefeld, M. (1982). Le spectre des étoiles Oe dans le rouge et le proche infrarougem. *Be Stars.* IAU Symp. No. 98, 229–233, D. Reidel Publ., Dordrecht.

Apparao, K. M. V. (1994) X-ray emission from Be star/X-ray binaries. *Ap. Sci. Rev.* **69**, 255–329.

Ashok, N. M., Bhatt, H. C., Kulkarni, P. V., and Joshi, S. C. (1984). Infrared photometric studies of Be stars. *M. N. R. A. S.*, **211**, 471–484.

Baade, D. (1987). Be stars as nonradial pulsators. *Physics of Be Stars.* IAU Coll. No. 92. Cambridge University Press, Cambridge and N.Y., 361–383.

Ballereau, D. and Chauville, J. (1989). Long-term and mid-term spectroscopic variations of the Be-shell star HD 184279 (V1294 Aql) II. Toward a model. *A. A*, **214**, 285–294.

Balona, L. A. and Lawson, W. A. (2001). Short-period line profile variations in the Be star ε Cap. *M. N. R. A. S.*, **321**, 131–142.

Balona, L. A., Handler, G., Aerts, C., Bozic, H., and 5 co-authors (2002). Omega Orionis and the Be stars. *Observational Aspects of Pulsating B- and A Stars*, Sterken, Ch. and Kurtz, D. W.(eds.), ASP Conference Proceedings, Vol. **256**, San Francisco, CA, 9.

Barker, P. K. (1982). The shell episode of 59 Cygni (1974–1975). *Ap. J. Suppl.*, **49**, 89–110.

Batten, A. H., Fletcher, J. M., and Mann, P. J. (1978). Seventh Catalogue of the orbital elements of spectroscopic binary systems. *Publ. Dom. Ap. Obs., Victoria*, **15**, 121–195.

Baum, E., Hamann, W. R. Koesterke, L., and Wessolowski, U. (1992). A semi-empirical model for the X-ray emission from single WR stars. *A. A.*, **266**, 402–408.

Beals, C. S. (1929). On the nature of Wolf-Rayet emission. *M. N. R. A. S.*, **90**, 202–212.

Beals, C. S. (1951). The spectrum of the P Cygni stars. *Publ. Dom. Ap. Obs. Victoria*, **9**, 1–137.

Behr, A. (1959). Intrinsic polarization of Be stars, *Nach. Akad. Wiss. Göttingen* **2**, Math-Phys.KL, No. 7, 185 (= Veröff. Göttingen, No. 126).

Bidelman, W. P. (1976). Concluding remarks. *Be and Shell Stars.* IAU Symp.No. 70. Slettebak, A. (ed.), D. Reidel Publ. Co. Dordrecht, 453–455.

Bjorkman, K. S. (2000). Polarimetry as a diagnostic of circumstellar envelopes. *The Be Phenomenon in Early-Type Stars.* IAU Coll. 175, ASP Conf. Vol. **214**, San Francisco, CA, 384–395.

Bohannan, B. (1989). The distribution of types of LBV's. *Physics of Luminous Blue Variables,* K. Davidson, A. F. J. Moffat, and H. J. G. L. M. Lamers (eds.), Kluwer, Dordrecht, 35–43.

Bond, H. E. (2000). Binarity of central stars of planetary nebulae. *Asymmetrical Planetary Nebulae II: From Origins to Microstructures.* ASP Conf. Ser. Vol. **199**, San Francisco, CA, 115–123.

Briot, D. (1971). Etude d'étoiles Be–décréments Balmer. *A. A.*, **11**, 57–64.

Cassinelli, J. P. and Cohen, D. H. (1994). ROSAT observations of B and Be stars. *Pulsation, Rotation and Mass Loss in Early Type Stars*, L. A. Balona, H. F. Henrichs, and J. M. Le Contel (eds.), IAU Symp., No. 162, Kluwer, Dordrecht, 189–199.

Cassinelli, J. P., Cohen, D. H., Macfarlane, J. J., Sanders, W. T., and Welsh, B. Y. (1994). X-ray emission from near-main-sequence B stars. *Ap. J.*, 421, 705–717.

Castor, J. I. and van Blerkom, D. (1970). Excitation of HeII in Wolf-Rayet stars. *Ap. J.*, **161**, 485–502. see Chapter 4.

Castor, J. I. and Lamers, H. J. G. L. M. (1979). An atlas of theoretical P Cyg profiles. *Ap. J. Suppl.*, **39**, 481–511. see Chapter 4.

Cerruti-Sola, M. and Perinotto, M. (1989). Fast winds in central stars of some planetary nebulae. *Ap. J.*, **345**, 339–345.

Chauville, J., Zorec, J. Ballereau, D., Morrell, N., Cidale, L., and Garcia, A. (2001). High and intermediate- resolution spectroscopy of Be stars. *A. A.*, **378**, 861–882.

Cherepashchuk, A. M. (1991). Wolf-Rayet binaries: Observational aspects. *WR Stars and Interrelations with other Massive Stars in Galaxies.* IAU Symp.No. **143**, K. A. van der Hucht and B. Hidayat (eds.), Kluwer, Dordrecht, 187–199.

Cherepashchuk, A. M. and Moffat, A. F. J. (1994). Cyg X–3 as a benchmark for fundamental properties of WR stars. *Ap. J.*, **424**, L53–L55.

Ciardullo, R., Bond, H. E., Sipior, M. S., Fulton, L. K., Zhang, C.-Y., and Schaefer, K. G. (1999). A Hubble Space Telescope survey for resolved companion of planetary nebula nuclei. *A. J.*, **118**, 488–508.

Cidale, L., Zorec, J., and Tringaniello, L. (2001). BCD spectrophotometry of stars with the B[e] phenomenon. I. Fundamental parameters. *A. A.*, **368**, 160–174.

Cohen, M. (1991). Enrichment by WR stars and other massive stars in gas, dust and energy. *Wolf-Rayet Stars and Interrelations with Other Massive Stars in Galaxies.* IAU Symp.No.143, K. A. van der Hucht and B. Hidayat (eds.), Kluwer, Dordrecht, 323–334.

Cohen, H. and Taylor, A. R. (1994). Difference between Be stars and shell stars from H alpha emission. *Pulsation, Rotation, and Mass-loss in Early-type Stars.* L. A. Balona, H. F. Henrichs, and J. M. Le Contel (eds.), IAU Symp. No. 162, Kluwer, D. Reidel Pub. Co. Dordrecht, 354–355.

Collins, G. W. II (1974). Calculation of theoretical rotational line profiles for the determination of $V \sin i$ in the spectral range O9–F8. *Ap. J.*, **191**, 157–164.

Collins, G. W. II, Traux, R. J., and Cramner, S.R. (1991). Model atmospheres for rotating B stars. *Ap. J. Suppl.*, **77**, 541–606.

Conti, P.S. (1976). On the relationship between Of and WR stars. *Memoires of Soc. Roy. Sci. Liege*, **9**, 193–212.

Conti, P. S. (1984). Basic observational contraints on the evolution of massive stars. *Observational Tests of the Stellar Evolution Theory*, IAU Symp. No. 105. A. Maeder and A. Renzini (eds.), Kluwer, Dordrecht, p. 233.

Conti, P. S. (1986). Parameters of WR stars. *Luminous Stars and Associations in Galaxies*, C. W. H. de Loore, A. J. Willis, and P. Laskarides (eds.), IAU Symp. No. 116. D. Reidel, Dordrecht, pp. 199–214.

Conti, P. S. and Leep, E. M. (1974). Spectroscopic observations of O-type stars. V. The hydrogen lines and λ 4686 HeII. *Ap. J.*, **193**, 113–124.

Conti, P. S. and Frost, S. A. (1977). Spectroscopic studies of O-type stars. VI. The earliest types O3–O5.5. *Ap. J.*, **212**, 728–742.

Conti, P. S., Massey, P., and Vreux, J.-M. (1990). Spectroscopic studies of WR stars. VI. Optical spectrophotometry of near-infrared emission lines in some Galactic stars. *Ap. J.*, **354**, 359–371.

Conti, P. S. and Vacca, W. D. (1990). The distribution of massive stars in the Galaxy. I. The Wolf-Rayet stars. *A. J.*, **100**, 431–444.

Corcoran, M. F. (1996). X-ray emission from colliding wind binaries. Rev. Mex. *A. A.*, **5**, 54–60.

Corcoran, M.F., Stevens, I. R., Pollock, A. M., Swank, J. H., Shore, S. N., and Rawley, G. L. (1996). X-ray variability in V444 Cygni: Evidence for colliding winds? *Ap. J.*, **464**, 434–441.

Coté, J. and Waters, L. B. F. M. (1987). IRAS observations of Be stars. I. Statistical study of the IR excess of 101 Be stars. *A. A.*, **176**, 93–106.

Coyne, G. V. (1976). Polarization in the emission lines of Be stars. *A. A.*, **49**, 89–96.

Coyne, G. V. and Kruszewski, A. (1969). Wavelength dependence of polarization. XVII. Be-type stars. *A.J.*, **74,** 528–532.

Crowther, P. A., Hillier, D. J., and Smith, L. J. (1995). Fundamental parameters of WR stars. II. Tailored analysis of Galactic WNL stars. *A. A.*, **293**, 403–426.

Dachs, J., Rohe, D., and Loose, A. S. (1990). A study of Balmer decrements in Be star emission-line spectra. *A. A.*, **238**, 227–241.

de Groot, M. (1989). P Cygni: The star that started it all. *Physics of Luminous Blue Variables*, Davidson, K., Moffat, A. F. J., and Lamers, H. J. G. L. M. (eds.), Kluwer, Dordrecht. p. 257–258.

de Groot, M., Sterken, C., and van Genderen, A. M. (2001). Cyclicities in the light variations of S Doradus stars. *A. A.*, **376**, 224–231.

De Marco, O., Bond, H. E., Harmer, D., and Fleming, A. J. (2004). Indication of a large fraction of spectroscopic binaries among nuclei of planetary nebulae. *Ap. J.*, **602**, L93–L96.

de Winter, D. and van den Ancker, M. E. (1997). The peculiar B[e] star HD 45677. II. Photometric behavior and spectroscopic properties. *A. A. Suppl.*, **121**, 275–299.

de Zeeuw, P. T., Hoogerwerf, R., de Bruijine, J. H. J., Brown, A. G. A., and Blaauw, A. (1999). A Hipparcos census of the nearby OB associations. *A. J.*, **117**, 354–399.

Doazan, V. (1987). The spheroidal/ellipsoidal, variable mass-loss.decerelating star model. *Physics of Be Stars*, Slettebak, A. and Snow, T. P. (eds.), 384–403.

Doazan, V., Sedmak, G., Barylak, M., and Rusconi, L. (1991). A Be star atlas of Far UV and optical high-resolution spectra. *ESA Sp - 1147.*

Dougherty, S. M. and Williams, P. M. (2000). Non-thermal emission in Wolf-Rayet stars: are massive companions required? *M. N. R. A. S.*, **319**, 1005–1010.

Dougherty, S. M., Pittard, J. M., Kasian, L., Coker, R. F., Williams, P. M., and Lloyd, H. M. (2003). Radio emission models of colliding- wind binary systems. *A. A.* **409**, 217–233.

Drake, S. A. and Ulrich, R. K. (1980). The emission-line spectrum from a slab of hydrogen at moderate to high densities. *Ap. J. Suppl.*, **42**, 351–383. see Chapter 4.

Ergma, E. and Yunletson, L. R. (1998). Cyg X-3 : Can the compact object be a black hole? *A. A.*, **333**, 151–158.

Fabregat, J. and Torrejón, J. M. (2000). On the evolutionary status of Be stars. *A. A.*, **357**, 451–459.

Feibelman, W. A. (1996). An "OIII sequence" for planetary-nebula nuclei. *PAS Pacific*, **108**, 664–667.

Feibelman, W. A. (1999). Detection of OVII λ1522 in IUE spectra of planetary nebua nuclei and other hot stars. *PAS Pacific*, **111**, 221–226.

Feinstein, A. (1968). A survey of southern Be stars. 2. Photometric data. *Zs. f. Ap.*, **68**, 29.

Feinstein, A. (1987). Be stars as members of open clusters. *Physics of Be stars*, A. Slettebak and T. P. Snow (eds.), IAU Coll. No. 92, Cambridge University Press, Cambridge and N.Y., 500–502.

Feinstein, A. (1990). Be stars in open clusters. *Rev. Mex. Astron. Astrofis.*, 12, 373–376.

Frew, D. J. (2004). The historical record of η Carinae I. The visual light curve, 1595–2000. *J. Astr. Data*, **10**, 1–76.

Frost, S. A. and Conti, P. S. (1976). The relationship of the Oe stars to the Be stars. *Be and Shell Stars*, Slettebak (ed.), D. Reidel Dordrecht. 139–147.

Fukuda, I. (1982). A statistical study of rotational velocities of the stars. *PASP*, **94**, 271–284.

Gehrz, R. D., Hackwell, J. A., and Jones, T. W. (1974). Infrared observations of Be stars from 2.3 to 19.5 microns. *Ap. J.*, **191**, 675–684.

Ghosh, K. K., Iyengar, K. V. K., Ramsey, B. D., and Austin, R. A. (1999). Near-simultaneous spectroscopic and broadband polarimetric observations of Be stars. *A. J.*, **118**, 1061–1072.

Hamann, W. R., Koesterke, L., and Wessolowski, U. (1995). Spectral atlas of the Galactic Wolf-Rayet stars (WN sequence). *A. A. Suppl.* **113**, 459–471.

Hanuschuk, R. W., Hummel, W., Dietle, O., and Satorius, E. (1995). V/R variability and global oscillations in Be star disk. *A. A.*, **300**, 163.

Hanuschuk, R. W., Hummel, W., Sutorius, E., Dietle, O., and Thimm, G. (1996). Atlas of high-resolution emission and shell lines in Be stars. Line profiles and short-term variability. *A. A. Suppl.*, **116**, 309–359.

Harmanec, P. (1987). Emission-line stars as interacting binaries. *Physics of Be Stars,* A. Slettebak and T. P. Snow (eds.) IAU Coll. No. 92, Cambridge University Press, Cambridge and N.Y., 339–360.

Harmanec, P. (1989). Towards understanding rapid line-profile and light variations of eaely-type stars. I. General considerations and a critical re-analysis of the data on ζ Oph and ε Per. *Bull. Ast. Inst. Czech.*, **40**, 201–261.

Harmanec, P., Habuda, P., Stefl, S., Hadrava, P., Korcakova, D., and 6 co-authors (2000). Properties and nature of Be stars. XX.Binary nature and orbital elements of γCas. *A. A.*, **364**, L85–L88.

Hillier, D. J. (1987). Modeling the extended atmospheres of WR stars. *Ap. J. Suppl.*, **63**, 947–964.

Hillier, D. J. and Miller, D. L. (1998). The treatment of non-LTE line blanketting in spherically expanding outflows. *Ap. J.*, **496**, 407–427.

Hillier, D. J., Davidson, K., Ishibashi, K., and Gull, T. (2001). On the nature of the central source in η Carinae. *Ap. J.*, **553**, 837–860.

Hirata, R. (1995). Interpretation of the long-term variation in late-type active Be stars. *PAS Japan*, **47**, 195–218.

Hirata, R., Katahira, J., and Jugaku, J. (1982). Spectroscopic study of Pleione in 1977–1979. *Be Stars*, Jaschek, M. and Groth, H. (eds.), IAU, Symp. No. 98. D. Reidel, Dordrecht, 161–165.

Hirata, R. and Kogure, T. (1984). The Be star phenomena. II. Spectral formation and structure of envelopes. *Bull. Ast. Soc. India*, **12**, 109–151.

Horaguchi, T., Kogure, T., Hirata, R., Kawai, N., Doazan, V. Slettebak, A., and 9 co-authors (1994). The Be star Gamma Cassiopeiae: X-ray, far-UV, and optical observations in early 1989. *P.A.S. Japan*, **46**, 9–26.

Howarth, I. D. and Schmutz, W. (1992). Near infrared spectroscopy Of Galactic WR stars. *A. A.*, **261**, 503–522.

Huang, S. S. (1972). Profiles of emission lines in Be stars. *Ap. J.*, **171**, 549–564.

Huang, S. S. (1973). Profiles of emission lines in Be stars. II. Interpretation of the long-period V/R variation. *Ap. J.*, **183**, 541–554.

Hubble, E. and Sandage, A. (1953). The brightest variable stars in extragalactic nebulae. I. M31 and M33. *Ap. J.*, **118**, 353–361.

Hubeny, J., Stefl, S., and Harmanec, P. (1985). How strong is the evidence of superionization and large mass ourflows in B/Be stars? *Bull. Ast. Czech.*, **36**, 214–230.

Hubert-Delplace, A. M., Hubert, H., Chambon, M. T., and Jaschek, M. (1982). Statistical properties of Be stars. *Be Stars*. M. Jaschek and H. Groth (eds.) IAU Symp. No. 98. Reidel Publ. Comp. Dordrecht, 125–130.

Hubert-Delplace, A. M. and Hubert, H. (1979). *An Atlas of Be Stars*. Paris-Meudon Observatory and Haute-Provence Observatory, published from Paris Observatory, Paris.

Hummel, W. and Vrancken, M. (2000). Line formation in Be star disks: Shell absorption and rotation law. *The Be Phenomenon in Early-Type Stars,* Smith, M. A. and Henrichs, H. F. (eds.) IAU Coll. 175. ASP Conf. Vol. 214, 527–530.

Humphreys, R. M. (1975). The spectra of AE Andromedae and the Hubble-Sandage variables in M31 and M33. *Ap. J.*, **200**, 426–429.

Humphreys, R. M. (1978). Luminous variables in M31 and M33. *Ap. J.*, **219**, 445–451.

Humphreys, R. M. (1989). What are LBV's? The characteristics and role in the upper HR diagram. *Physics of Lumious Blue Variables*, K. Davidson, A. F. J. Moffat, and H. J. G. L. M Lamers (eds.) IAU Coll. No. 113, Kluwer, Dordrecht, 3–14.

Humphreys, R. M. and Davidson, K. (1979). Studies of luminous stars in nearby galaxies. III. Comments on the evolution of the most massive stars in the Milky Way and the Large Magellanic Cloud. *Ap. J.*, **232**, 409–420.

Humphreys, R. M. and Davidson, K. (1994). The luminous blue variables: Astrophysical geysers. *PASP*, **106**, 1026–1051.

Ignace, R., Oskinova, L. M., and Foullon, C. (2000). Exospheic models for the X-ray emission from single Wolf-Rayet star. *M. N. R. A. S.*, **318**, 214–226.

Jaschek, M., Slettebak, A., and Jaschek, C. (1981). Be star terminology. *Be Star News Letter*, **4**, 9-11.

Jaschek, C. and Jaschek, M. (1983). The frequency of Be stars. *A. A.*, **117**, 357–359.

Jaschek, C. and Jaschek, M. (1993). A catalogue of radii of Be star line emitting regions. *A. A. Suppl.*, **97**, 807–814.

Jaschek, C., Andrillat, Y., and Jaschek, M. (1996). B[e] stars. III. MWC 645. *A. A. Suppl.*, **120**, 99–105.

Kambe, E., Ando, H., Hirata, R., Walker, G. A., Kennelly, E. J., and Mathews, J. M. (1993). Line-profile variations of Lamda Eridani in emission and quiescence. *P. A. S. P.*, **105**, 1222–1231.

Kato, S. (1983). Low-frequency, one-armed oscillations of Keplerian gaseous disks. *P. A. S. Japan*, **35**, 249–261.

Kenyon, S. J. and Gallagher III, J. S. (1985). Spectroscopy of the winds from Hubble-Sandage stars in M31 and M33. *Ap. J.*, **290**, 542–550.

Kingsburg, R. L., Barlow, M. J., and Storey, P. J. (1995). Properties of the WO Wolf-Rayet stars. *A. A.*, **295**, 75–100.

Koesterke, L. and Hamann, W. R. (1995). Spectral analysis of 25 Galactic WR stars of the carbon sequence. *A. A.*, **299**, 503–519.

Kogure, T. (1969). Contribution à l'étude des profils d'émission d'étoiles Be dites "pole-on." See Chapter 4. *A. A.*, **1**, 253–269.

Kogure, T. (1981). On the statistical nature of Be and Ae stars. *PAS Japan*, **33**, 399–411.

Kogure, T. (1990). A simple model of phase-changing variation between Be stars and shell stars. *Ap. Sp. Sc.*, **163**, 7–22.

Kogure, T. (2000). Balmer decrements and the structure of Be-star envelopes. *Stellar Physics*, L. S. Cheng, H. F. Chau, K. L. Chan, and K. C. Leung (eds.), Kluwer, Dordrecht, pp. 283–290.

Kogure, T., Hirata, R., and Asada, Y. (1978). On the formation of hydrogen shell spectrum and the envelopes of some shell stars. See Chapter 4. *PAS Japan*, **30**, 385–407.

Kogure, T. and Hirata, R. (1982). The Be star Phenomena. I. General properties. *Bull. Ast. Soc. India*, **10**, 281–309.

Kriz, S. (1976). On the V/R variation of emission lines in Be stars. *Bull. Astr. Inst. Czech.*, **27**, 321–325.

Kriz, S. (1979). The interpretation of emission line profiles from envelopes of Be stars. II. The model of an elliptical gaseous rings. *Bull. Astr. Inst. Czech.*, **30**, 83–95.

Kriz, S. and Harmanec. P. (1975). A hypothesis of the binary origin of Be stars. *Bull. Astr. Inst. Czech.* **26**, 65.

Kunjaya, Ch. and Hirata, R. (1995). Multi-band polarimetry of Be stars. I. X Persei. *PAS Japan*, **47**, 589–604.

Kurucz, R. L. (1979). Model atmospheres for G, F, A, B, and O stars. *Ap. J.*, **40**, 1–340.

Lamers, H. J. G. L. M. (1986). P Cygni type stars: Evolution and physical processes. *Luminous Stars and Associations in galaxies*, C. W. H. de Loore, A. J. Willis, and P. Laskarides (eds). D. Reidel, Dordrecht, IAU Symp. No. 116, 157–182.

Lamers, H. J. G. L. M. (1987). Variations in C. W. H. luminous blue variables. *Instabilities in Luminous Early Type Stars*, H. J. G. L. M. Lamers and C. W. H. de Loore (eds.) D. Reidel, Dordrecht, 99–125.

Lamers, H. J. G. L. M. (1989). IAU Coll. No. 113, 135–147, Mass loss from luminous blue variables. *Physics of Luminous Blue Variables*, K. Davidson, A. F. J. Moffa, and H. J. G. L. M. Lamers (eds.).

Lamers, H. J. G. L. M. and Leitherer, C. (1993). What are the mass- loss rates of O stars? *Ap. J.*, **412**, 771–791.

Lamers, H. J. G. L. M., Zickgraph, F. J., de Winter, D., Houziaux, L., and Zorec, J. (1998). A revised classification scheme for B[e] stars. *B[e] Stars*, A. M. Hubert and C. Jaschek (eds.), Kluwer, Dordrecht. 277–279.

Langer, N., Hamann, W. R., Lennon, M., Najarro, F., Pauldrach, A. W. A., and Puls, J. (1994). Towards an understanding of very massive stars. A new evolutionary scenario relating O stars, LBVs and WR stars. *A. A.*, **290,** 819–833.

Leitherer, C., Schmutz, W., Abbott, D. C., and Hamann, W. W. (1989). Atmospheric models for luminous blue variables. *Ap. J.*, **346**, 919–931.

Lépine, S. and Moffat, A. F. J. (1999). Wind inhomogeneities in Wolf-Rayet stars. II. Investigation of emission-line profile variations. *Ap. J.*, **514**, 909–931.

Liu, Q. Z., van Paradijs, J., and van den Heuvel, E. P. J. (2000). A catalogue of high-mass X-ray binaries. *A. A. Suppl.*, **147**, 25–49.

Lundström, I. and Stenholm, B. (1984). Spectroscopy of faint galactic WR stars. *A. A. Suppl.*, **56**, 43–78.

Maeda, Y., Koyama, K., Yokogawa, J., and Skinner, S. L. (1999). X-ray evidence for wind-wind collision in the Wolf-Rayet binary V444 Cygni. *Ap. J.* **510**, 967–973.

Maeder, A. (1980). The most massive stars in the galaxy and the LMC: Quasi-homogeneous evolution, time-averaged mass loss rates and mass limits. *A. A.*, **92**, 101–110.

Maeder, A. (1985). Vibrational instability of WR stars. *A. A.*, **147,** 300–308.

Maeder, A. (1990). Tables for massive star evolution at various metallicities. *A. A. Suppl.*, **84**, 139–177.

Maeder, A. (1991). IAU Symp.No.143, 445–451, Metallicity effects in massive star evolution and number frequencies of WR stars in galaxies. *Wolf-Rayet Stars and Interrelations with Other Massive Stars in Galaxies,* K. A. van der Hucht and B. Hidayat (eds.).

Maheswaran, M., and Cassinelli, J. (1994). Rotation and Magnetic fields during the evolution of massive stars through B[e] and Wolf-Rayet phases. *Ap. J.*, **421**, 718–732.

Marchenko, S. V., Moffat, A. F., and Koenigberger, G. (1994). The WR binary V444 Cygni under the spectroscopic microscope. I. Improved characteristics of the components and their interaction seen in HeI. *Ap. J.*, **422**, 810–822.

Marchenko, S. V., Moffat, A. F., Eemens, P. R. J., Cardona, O., Echevarria, J., and Hervieux, Y. (1997). The WR binary V444 Cyg under the spectroscopic microscope. II. Physical parameters of the WR wind and the zone of wind collision. *Ap. J.* **485**, 826–838.

Markova, N. (1994). How did the spectrum of P Cygni look in 1985? Spectral atlas with complete line identifications in the wavelength from 3550 to 4800. *A. A. Suppl.*, **108**, 561–569.

Markova, N. and Zamanov, R. (1995). P Cygni-spectral atlas with complete line identification in the wavelength range from 4840 to 6760 Å. *A. A. Suppl.*, **114**, 499–508.

Markova, N., Scuderi, S., de Groot, M., Markov, H., and Panagia, N. (2001a). Simultaneous Hα and photometric observations of P Cygni. *A. A.*, **366**, 935–944.

Markova, N., Morrison, N., Kolka, I., and Markov, H. (2001b). P Cygni in a short S Doradus phase. Spectroscopic and photometric evidences. *A. A.*, **376**, 898–906.

Marlborough, J. M. and Peters, G. J. (1986). UV observations of Be stars. II. Superionization in Be stars of different spectral types. *Ap. J. Suppl.*, **62**, 875–897.

McDavid, D. (1999). Multicolor polarimetry of selected Be stars: 1995–1998. *PAS Pacific*, **111**, 494–509.

Méndez, R. H., Kudritzki, R. P., Herrero, A., Husfeld, D., and Groth, H. G. (1988). High resolution spectroscopy of central stars of planetary nebulae. I. Basic atmospheric parameters and their interpretation. *A. A.*, **190**, 113–135.

Mendoza, E. E. (1958). A spectroscopic and photommetric study of the Be stars. *Ap. J.*, **128**, 207–218.

Mennickent, R. E., Vogt, N., Barrera, L. H., Covarrubias, R., and Ramirez, C. (1994). On the rotation properties of Be stars and their envelopes. *A. A. Suppl.*, **106**, 427–439.

Merrill, P. W. (1933). The distribution in spectral type and galactic position of stars with atmospheres of glowing hydrogen. *PASP*, **45**, 198–200.

Merrill, P. W. and Burwell, C. G. (1933). Catalogue and bibliography of stars of classes B and A whose spectra have bright hydrogen lines. *Ap. J.*, **78**, 87–140.

Merrill, P. W. and Burwell, C. G. (1943). Supplement to the Mount Wilson catalogue and bibliography of stars of classes B and A whose spectra have bright hydrogen lines. *Ap. J.*, **98**, 153–184.

Merrill, P. W. and Burwell, C. G. (1949). Second supplement to the MWC and bibliography of stars of classes B and A whose spectra have bright hydrogen lines. *Ap. J.*, **110**, 387–419.

Merrill, P. W. and Burwell, C. G. (1950). Additional stars whose spectra have a bright Hα line. *Ap. J.*, **112**, 72–89.

Meynet, G., Maeder, A., Schaller, G., Scharrer, D., and Charbonnel, C. (1994). Grids of massive stars with high mass loss rates. V. From 12 to 120 $M_\odot$ at $Z = 0.001, 0.004, 0.008, 0.020$ and 0.040. *A.A. Suppl.*, **103**, 97–105.

Meynet, G. and Maeder, A. (2003). Stellar evolution with rotation X: Wolf-Rayet star populations at solar metallicity. *A. A.* **404**, 975–990.

Miroshnichenko, A. S., Bjorkman, K. S., and Krugov, V. D. (2002). Binary nature and long-term variations of γ Cassiopeoae. *PASP*, **114**, 1226–1233.

Miyamoto, S. (1949). On the radiation field of Be stars. I. See Chapter 4. *Jap. J. Astr*, **1**, 17–25.

Miyamoto, S. (1952a). On the radiation field of Be-stars. II. See Chapter 4. *PAS Japan*, **4,** 1–10.

Miyamoto, S. (1952b). On the radiation field of Be-stars. III. Theoretical Balmer decrement See Chapter 4. *PAS Japan*, **4,** 28–36.

Moffat, A. F. J. (1983). WR stars, progenitors of supernovae. *Proceedings of the Workshop held at the Observatoire de Paris-Meudon, April 20–22.* **111**, 13.

Moffat, A. F. J. and Shara, M. M. (1986). Photometric variability of a complete sample of northern WR stars. *A. J.*,**92**, 952–975.

Moffat, A. F. J. and Robert, C. (1991). Instrinsic wind variability of Wolf-Rayet stars. *Non-isotropic and Variable Ortflow from Stars*, Drissen, L., Leitherer, C., and Natta, A. (eds.). ASP Conf. Ser. Vol. 22, San Francisco, CA, 203–213.

Moffat, A. F. J. (1994). Variability of WR stars: Constraints on the properties of hot-star winds. *Impact of Long-Term Monitoring on Variable Star Research*, Sterken, Ch. and de Groot, M. (eds.). Kluwer, Dordrecht. p. 117–140.

Morris, P. W., Brownsberger, K. R., Conti, P. S., Massey, P., and Vacca, W. D. (1993). Spectrophotometry of Wolf-Rayet stars. I. Continuous energy distribution. *Ap. J.*, **412**, 324–340.

Morse, J. A., Davidson, K., Bally, J., Ebbets, D., Balick, B., and Frank, A. (1998). Hubble Space Telescope Wide Field Planetary Camera 2 Observations of Eta Carinae. *A. J.*, **116**, 2443–2461.

Muratorio, G. (1978). Spectrophotometric study of the LMC emission line star Hen S22. *A. A. Suppl.*, **33**, 125–139.

Negueruela, I., Reig, P., Coe, M. J., and Fabreat, J. (1998). Large-scale perturbations in the circumstellar envelopes of Be/X-ray binaries. *A. A.*, **336**, 251–262.

Negueruela, I., Steele, I. A., and Bernabeu, G. (2004). On the class of Oe stars. *Astr. Nachrichten*, **325**, 749–760.

Niedzielski, A. (1994). Stratification in WR envelopes as seen in HeII lines. *A. A.*, **282**, 529–534.

Niedzielski, A. and Rochowicz, K. (1994). Low resolution IUE spectra of Wolf-Rayet stars. *A. A. Suppl.*, **108**, 669–694.

Nieuwenhuijzen, H. and de Jager, C. (1990). Parametarization of stellar rates of mass loss as functions of the fundamental stellar parameters *M*, *L*, and *R*. *A. A.*, **231**, 134–136.

Okazaki, A. (1991). Long-term V/R variations of Be stars due to global one-armed oscillations of equatorial disks. *PAS Japan*, **43**, 75–94.

Okazaki, A. (1996). Emission-line profiles from Be star envelopes with $m = 1$ perturbation patterns. *PAS Jap*, **48**, 305–316.

Osaki, Y. (1974). Nonradial pulsation theory of massive stars. *PAS Pacific*, **98,** 30–32.

Osaki, Y. (1999). Non-radially pulsating hot stars: Non-radial pulsation and Be phenomena. *Variable Stars and Non-spherical Stellar Winds in Luminous Hot Stars*, Wolf, B., Stahl, O., and Fullerton, A.W. (eds.). IAU Coll. No. 169, Springer-Verlag, Berlin and Heiderberg, 329–336.

Owocki, S. P. (1990). Winds from hot stars. *Rev. Modern Astron.*, **3**, 98–123.

Owocki, S. P., Castor, J. I., and Rybicki, G. B. (1990). Time-dependent models of radiatively driven stellar winds. I. nonlinear evolution of instabilities for a pure absorption model *Ap. J.*, **335**, 914–930.

Panagia, N. ane Felli, M. (1975). The spectrum of the free-free radiation from extended envelopes. *A. A.*, 39, 1–5.

Peña, M., Stasinska, G., and Medina, S. (2001). Galactic planetary nebulae with WR nuclei II. A consistent observational data set. *A. A.*, **367**, 983–994.

Penny, L. (1996). Projected rotational velocities of O type star. *Ap. J.*, **463**, 737–746.

Percy, J. R. (1987). Observations of rapid variabilities in Be stars. *Physics of Be Stars*, A. Slettebak and P. T. Snow (eds.). IAU Coll. No. 92, Cambridge University Press, Cambridge and N.Y., 49–63.

Percy, J. R. and Bakos, A. G. (2001). Photometric monitoring of bright Be stars. IV. 1996–1999. *PAS Pacific*, **113**, 748–753.

Perek, L. and Kohoutek, L. (1967). *Catalogue of Galactic Planetary Nebulae*, Publishing House of Czech, Academy of Science, Prague.

Plavec, M. and Pollidan, R. S. (1976). The Algols, red spectra, Be stars, and even neutrinos. *Structure and Evolution of Close Binary Systems*, P. Eggleton, S. Mitton, and J. Whelan (eds.). IAU Symp. No.73, D. Reidel Pub. Dordrecht, 289.

Poe, C. H., Friend, D. B., and Cassinelli, J. P. (1989). A rotating, magnetic, radiation-driven wind model for Wolf-Rayet stars. *Ap. J.,* **337**, 888–902.

Poeckert, R. and Marlborough, J. M. (1977). Linear polarization of Hα in the Be star Gamma Cassiopeiae, *Ap. J.*, **218**, 220–226.

Poeckert, R. and Marlborough, J. M. (1978). A model for γ Cassiopeiae. *Ap. J.*, **220**, 940–961.

Pollock, A. M. T. (1987). The Einstein view of the WR stars. *Ap. J.*, **320**, 283–295.

Pols, O. R., Coté, J., Waters, L. B. F. M., and Heise, J. (1991). The formation of Be stars through close binary evolution. *A. A.*, **241**, 419–438.

Portegies Zwart, S. F. (1995). The formation of Be stars in close binary systems. The importance of kicks and angular-momentum loss. *A. A.*, **296**, 691–699.

Portegies Zwart, S. F. (2000). The characteristics of high-velocity O and B stars which are ejected from supernovae in binary systems. *Ap. J.*, **544**, 437–442.

Porter, J. M. and Rivinius, T. (2003). Classical Be stars. *PAS Pacific*, **115**, 1153–1170.

Pottasch, S. R. (1961). Balmer decrements: II. The Be stars. *Annales d'Astrophys.*, **24**, 159–167.

Pottasch, S. R. (1984). *Planetary Nebulae,* D. Reidel, Dordrecht.

Quirrenbach, A., Bjorkman, K. S., Bjorkman, J. E., Hummel, C. A., and 5 co-authors (1997). Constraints on the geometry of circumstellar envelopes: optical interferometric and spectro- polarimetric observations of seven Be stars. *Ap. J.*, **479**, 477–496.

Rauw, O., Vreux, J. M., Gosset, E., Hutsemekers, E., Magain, P., and Rochowicz, K. (1996). WR22 : the most massive WR star. *A. A.*, **306**, 771.

Rinehart, S. A., Houck, J. R., and Smith, J. D. (1999). Mid-infrared spectra of Be stars. *A. J.*, **118**, 2974–2987.

Rivinius, Th., Baade, D., Stefl, S., and Maintz, M. (2001). Evolution in circumstellar envelopes of Be stars: From disks to rings? *A. A.*, **378**, 257–269.

Robinson, R. D., Smith, M. A., and Henry, G. W. (2002). X-ray and optical variations in the classical Be star gamma Cas: The discovery of a possible magnetic dynamo. *Ap. J.*, **575**, 435–448.

Rountree, J. and Sonneborn, G. (1991). Criteria for the spectral classification of B stars in the ultraviolet. *Ap. J.*, **369**, 515–528.

Schaller, G., Shaerer, D., Meynet, G., and Maeder, A. (1992). New grids of stellar models from 0.8 to 120 $M_\odot$ at $Z = 0.020$ amd $Z = 0.001$. *A. A. Suppl.*, **96**, 269–331.

Schild, R. and Romanishin, W. (1976). A study of Be stars in clusters. *Ap. J.*, 204, 493–501.

Schulte-Ladbeck, R.A., Eenens, P. R., and Davis, K. (1995). The structure of WR winds. I. Observation of ionization stratification in WR 6 and WR 111. *Ap. J.*, 454, 917–926.

Seggewiss, W. and Altmann, M. (2002). Long-term *UBV* obsrvations of the episodic dust maker WR 140. *Interacting Winds form Massive Stars,* Moffat A. F. L. and St-Louis N. (eds.). ASP Conf. Ser., **260**, San Francisco, CA, 417–421.

Singh, M. and Chaubey, U. S. (1987). Spectrophotometric observations and evolutionary status of ten Be stars. *Ap. Sp.Sci.*, **129**, 251–260.

Slettebak, A. (1982). Spectral types and rotational velocities of the brighter Be stars and A–F type shell stars. *Ap. J. Suppl.*, **50**, 55–83.

Slettebak, A. (1985). Be stars in open clusters. *Ap. J. Suppl.*, **59**, 769–784.

Slettebak, A. (1986). H-alpha and near-infrared spectra of late-type Be and A F-type shell stars. *PASP*, **98**, 867–871.

Slettebak, A. (1994). Ultraviolet spectral classification and stellar winds in a sample of Be and standard stars. *Ap. J. Suppl.*, **94**, 163–182.

Slettebak, A., Collins II, G.W., and Traux, R. (1992). Physical properties of Be star envelopes form Balmer and FeII emission lines. *Ap. J. Suppl.*, **81**, 335–376.

Smith, L. F. and Kuhi, L. V. (1981). An atlas of Wolf-Rayet line profiles. *JILA Report* No. 117, University of Colorado, 1–49.

Smith, L. F. and Maeder, A. (1989). The bolometric correction and the dM/dt :L relation for Wolf-Rayet stars. *A. A.*, **211**, 71–80.

Smith, L. J., Crowther, P. A., and Prinja, R. K. (1994). A study of the luminous blue variable candidate He 3–519 and its surrounding nebula. *A. A.*, **281**, 833–854.

Smith, M. A. (1988). Nonradial pulsations and early Be stars: Angular momentum considerations. *Pulsation and Mass Loss in Stars*, Stalo R. and Wilson L. A. (eds.), Kluwer, The Netherlands. pp. 251–274.

Smith, M. A. (1995). Dynamic processes in Be star atmospheres. 3. Rapid multiwavelength variability in Gamma Cassiopeiae. *Ap. J.*, **442**, 812–821.

Smith, M. A., Murakami, T. Ezuka, H., Anandarao, B. G., Chakrabolty, A., Corcoran, M. F., and Hirata, R. (1997). Dynamic processes in Be star atmospheres. VI. Simultaneous X-ray, UV, and optical variation in λ Eri. *Ap. J.*, **481**, 479–488.

Smith, M. A. (1998). The X-rays from Cassiopeia's lap. *Mercury*, **27**, May–June, 26–30.

Snow, T. P. (1981). Stellar winds and mass-loss rates from Be stars. *Ap. J.*, **251**, 139–151.

Soker, N. (2001). The departure of η Carinae from axisymmetry and the binary hypotheses. *M. N. R. A. S.*, **325**, 584.

Stahl, O., Wolf, B., Leitherer, C., and de Groot, M. (1985). Atlas of high-dispersion spectra of peculiar emission-line stars in the Magellanic Clouds. *A. A.Suppl.*, **61**, 237–258.

Stella, L., White, N. E., and Rosner, R. (1986). Intermittent stellar wind accretion and the long-term activity of population I binary systems containing an X ray pulsar. *Ap. J.*, **308**, 669–679.

Stothers, R. B. and Chin, C. W. (1996). Evolution of massive stars into luminous blue variables and WR stars for a range of metallicities : Theory versus observations. *Ap. J.*, **468**, 842–850.

Struve, O. (1931). On the origin of bright lines in spectra of stars of class B. *Ap. J.*, **73**, 94–103.

Szeifert, Th., Humphreys, R. M., Davidson, R., Jones, T. J., Stahl, O., Wolf, B., and Zickgraf, F. J. (1996). HST and groundbased observations of the 'Hubble-Sandage' variables in M31 and M33. *A. A.*, **314**, 131–145.

Tarasov, A. E. (2000). Be stars in Roche-lobe interacting binaries. *The Be Phenomena in Early-Type Stars*, IAU Coll. **175**, ASP Conf. Ser. Vol. 214. San Francisco, CA 644–655.

Telting, J. (1996). Be-star disks and non-radial pulsations in rotating stars. *PhD Thesis*, University of Amsterdam.

Thom, C., Granes, P., and Vakili, F. (1986). Optical interferometric measurements of Gamma Cassiopeia's envelope in the H-alpha line. *A. A.*, **165**, L13–15.

Townsend, R. H. D., Owocki, S. P., and Howarth, I. D. (2004). Be-star rotation: how close to critical? *M. N. R. A. S.*, **350**, 189–195.

Tycner, Ch., Lester, J. B., Hajian, A. R., Armstrong, J. T., Benson, J. A., and 4 co-authors (2005). Properties of the Hα -emittiong regions of Be stars. *Ap. J.*, **624**, 359–371.

Tylenda, R., Acker, A., and Stenholm, B. (1993). Wolf-Rayet nuclei of planetary nebulae. Observations and classification. *A. A. Suppl.*, **102**, 595–602.

Ulrich, R. K. (1976). An infall model for the T Tauri phenomena. *Ap. J.*, **210**, 377–391.

Underhill, A. B. (1983). Possible evidence for the driving of the winds of hot stars by Alfvén waves. *Ap. J.*, **268**, L127–L130.

Underhill, A. B. (1991). IAU Symp.No.143, 255, Key WR binary stars. *WR Stars and Interrelations with Other Massive Stars in Galaxies.* van der Hucht and Hidayat (eds.).

Underhill, A. B. (1995). A study of seven Of stars.I.Radial velocities. *Ap. J. Suppl.*, **100**, 433–460.

Underhill, A. B., Gilroy, K. K., and Hill, G. M. (1989). The Of emission lines near 4650 Å. *A. J.*, **98**, 1063–1068.

Underhill, A. B., Gilroy, K. K., Hill, G. M., and Dinshwa, N. (1990). A possible meaning of the spectrrum of HD 191765, WN6. *Ap. J.*, **351**, 666–674.

Underhill, A. B. and Gilroy, K. K. (1990). Notes on the radial velocities of 14 O stars and the emission lines in their spectra. *Ap. J.*, **364**, 626–636.

Uesugi, A. and Fukuda, I. (1982) Revised catalogue of stellar rotational velocities, Department of Astronomy, Kyoto University, Kyoto.

Vacca, W.D. and Torres-Dodgen, A.V. (1990). Color excess, intrinsic colors, and absolute magnitudes of Galactic and LMC Wolf-Rayet stars. *Ap. J. Suppl.*, **73**, 685–708.

van Bever, J. and Vanbeveren, D. (1997). The number of B-type binary mass gainers in general, binary Be stars in particular, predicted by close binary evolution. *A. A.*, **322**, 116–126.

van den Heuvel, E. P. J. and Rappaport, S. (1987). X-ray observations of B-emission stars. *Physics of Be Stars*, A. Slettebak and P. T. Snow (eds.). IAU Coll. No. 92, Cambridge University Press, Cambridge and N.Y., 291–307.

van den Heuvel, E. P. J. (1994). Massive close binaries :Observatinal characteristics. *Sp. Sci. Rev.*, **66**, 309–322.

van der Hucht, K. A. (1992). Wolf–Rayet stars. *A. A. Rev.*, **4**, 123–159.

van der Hucht, K. A. (1996). Observational properties of population I Wolf-Rayet stars. *Ap. Sp. Sci*, **238**, 1–15.

van der Hucht, K. A. (2001). The VIIth catalogue of galactic Wolf-Rayet stars. *New Astron. Rev.*, **45**, 135–232.

van der Hucht, K. A., Conti, P. S., Lundström, I., and Stenholm, B. (1981). The sixth Catalogue of galactic WR stars, their past and present. *Space Sci. Rev.*, **28**, 227–306.

van der Hucht, K. A., Morris, P. W., Williams, P. M., Setia Gunawan, D. Y. A., Beintema, D. A., and 10 co-authors (1996). ISO-SWS spectrophotometry of glactic Wolf-Rayet stars : preliminary results. *A. A.*, **315**, L193–L196.

van der Hucht, K. A., Schrijver, H., Stenholm, B., Lundstrom, O., Moffat, A. F. J., and 7 co-authors (1997). The Hipparchus distance determination of the WR system γ^2 Vel. *New Astron.*, **2**, 245–250.

van der Hucht, K. A., Williams, P. M., and Morris, P. W. (2001). Active dust formation by population I Wolf-Rayet stars. *The Promise of the Herschel Space Observatory*, G. L. Pilbratt, J. Cernicharo, A. M. Heras, T. Prusti, and R. Harris (eds.), 273–276.

van Genderen, A. M., Verheifen, M. A. W., van der Hucht, K. A., de Loor, C. W. H., and 10 co-authors (1991). (Intrinsic) variations of WR stars. *Wolf-Rayet Stars and Interrelations with Other Massive Stars in Galaxies*, K. A. van der Hucht and B. Hidayat (eds.). IAU Symp. No. 143, Kluwer, Dordrecht, 129–145.

van Genderen, A. M., de Groot, M., and Thé, P. S. (1994). New Astrophysical Insights based on historical and modern data of Eta Carinae 1935–1992. *The Impact of Long-Term Monitoring on Variable Star Research*, C. Sterken and M. de Groot (eds.), Kluwer, Dordrecht, 19–30.

van Kerkwijk, M. H., Charles, P. A., Geballe, T. R., King, D. L., Miley, G. K., and 4 co-authors (1992). Infrared helium emission lines from Cygnus X-3 suggesting a WR star companion. *Nature*, **355**, 703–705.

van Kerkwijk, M. H., Waters, L. B. F. M., and Marlborough, J. M. (1995). Hα emission and infrard excess in Be stars: Probing the circumstellar disc. *A. A.*, **300**, 259–268.

van Kerkwijk, M. H., Geballe, T. R., King, D. L., van der Klis, M., and van Paradijs, J. (1996). The Wolf-Rayet counterpart of Cygnus X-3. *A. A.*, **314**, 521–540.

Viotti, R. (1995). A hstorical overview of the η Carinae problem. *Rev. Mex. A. A.*, (Ser. de Conferencias), **2**, 1–10.

Viotti, R., Rossi, L., Altamore, A., Rossi, C., and Cassatella, A. (1986). New results on Eta Carinae. Evidence for an asymmetric, inhomogeneous winds. *Luminous Stars and Associations in Galaxies*, IAU Symp. 116. C. W. H. de Loore, A. J. Willis, and P. Laskarides (eds.), D. Reidel Dordrecht, 249–252.

Vreux, J. M. (1985). Emission-line variability in WR stars: short periods and nonradial pulsation? *PASP*, **97**, 274–279.

Vreux, J. M., Dennefeld, H., Andrillat, Y., and Tochowicz, K. (1989). Near infrared spectra of galactic and Magellanic WR stars. *A. A. Suppl.*, **81**, 353–365.

Vreux, J. M., Gosset, E., Bohannan, B., and Conti, P. (1992). WR 134 line-profile variability revisited. *A. A.*, **256**, 148–158.

Walborn, N. R. (1971). Some spectroscopic characteristics of the OB stars: An investigation of the space distribution of certain OB stars and the reference frame of the classification. *Ap. J. Suppl.*, **23**, 257–282.

Walborn, N. R. (1990). Digital optical morphology of OB spectra. *Properties of Hot Luminous Stars*, C. D. Garmany (ed.). ASP Conf. Ser. Vol. 7, San Francisco, CA, 23–29.

Walker, M. F. (1953). The light-variations of HD 217050. *Ap. J.*, **118**, 481–488.

Wallerstein, G., Gilroy, K. K., Zethson, T., Johansson, S., and Hamann, F. (2001). Line identifications in the spectrum of η Carinae as observed in 1990–1991 with CCD detectors. *PAS Pacific*, 113, 1210–1214.

Waters, L. B. F. M. (1986). The density structure of discs around Be stars derived from IRAS observations. *A. A.*, **162**, 121–139.

Waters, L. B. F. M. and Wesselius, P. P. (1985). The infrared and radio spectrum of early type stars with mass loss. III.–The effect of coronae, hot blobs and clumping on the IR and radio spectrum. *A. A.*, **138**, 25–35.

Waters, L. B. F. M., Coté, J., and Lamers, H. J. G. I. L. M. (1987). IRAS observations of Be stars. II. Far-IR characteristics and mass loss rates. *A. A.*, **185**, 206–224.

White, R. C. (1985). Synchrotron emission from chaotic stellar winds. *Ap. J.*, **289**, 698–708.

White, N. E., Swank, J. H., Holt, S. S., and Parmar, A. N. (1982). A comparison of the X-ray properties of X Persei and γ Cassiopeiae. *Ap. J.*, **263**, 277–288.

Williams, P. M., van der Hucht, K A., Sandell, G., and The, P. S. (1990). Millimeter and infrared observations of the wind from the WR star γVelorum. *M. N. R. A. S.*, **244**, 101–104.

Williams, P. M. and van der Hucht, K. A. (1994). Long-term infrared photometry of WC stars making circumstellar dust. *Impact of Long-Term Monitoring on Variable Star Research*, C. Sterken and M. de Groot (eds.), Kluwer Dordrecht 85–100.

Williams, P. M., Kidger, M. R., van der Hucht, K. A., Morris, P. W., and 8 co-authors (2001). Episodic dust formation by HD 192641 (WR 137). *M. N. R. A. S.*, **324**, 156–166.

Wolf, B. (1986). S Doradus type; Hubble-Sandage variables. *Luminous Stars and Associations in Galaxies*, C. W. H. de Loore, A. J. Willis, and P. Laslkarides (eds.). IAU Symp. 116, D. Reidel Dordrecht, 151–156.

Wolf, B. (1994). Investigation of LBV of the Magellanic Clouds during the past decade with LTPV, CASPEL and IUE. *The Impact of Long-Term Monitoring on Variable Star Research*, C. Sterken and M. de Groot (eds.), Kluwer Acad. Pub. Drodrecht 191–301.

Wolf, B. and Stahl, O. (1990). Inverse P Cygni-type profiles in the spectrum of the luminous blue variable S Doradus. *A. A.*, **235**, 340–344.

Wood, K., Bjorkman, K. S., and Bjorkman, J. E. (1997). Deriving the geometry of Be star circumstellar envelopes from continuum spectropolarimetry. I. The case of ζ Tauri. *Ap. J.*, **477**, 926–939.

Yudin, R. V. (2001). Statistical analysis of intrinsic polarization, IR excess and projected rotational velocity distributions of clasical Be stars. *A. A.*, **368**, 912–931.

Zhekov, S. A. and Skinner, S. L. (2000). X-ray emission from colliding wind shocks in the Wolf-Rayet binary WR 140. *Ap.J.*, **538**, 808–817.

Zickgraf, F. J. (1992). Observations of B[e] stars. *Nonisotropic and Variable Outflows from Stars*, L. Drisen, C. Leithere, and A. Nota (eds.), ASP Conf. Ser. Vol. 22, San Francisco, CA, 75–80.

Zickgraf, F. J. (1998). Current definition of B[e] stars. *B[e] Stars, Proc. Paris Workshop*, Hubert, A. M. and Jaschek, C. (eds.), Kluwer, Dordrecht, 1–19.

Zickgraf, F. J. (2003). Kinematical structure of the circumstellar environments of galactic B[e] stars, *A. A.*, **408**, 257–285.

Zickgraf, F. J., Wolf, B., Stahl, O., Leitherer, C., and Klare, G. (1985). The hybrid spectrum of the LMC hypergiant R126. *A. A.*, **143**, 421–430.

Zickgraf, F. J., Wolff, B., Leitherer, C., Appenzeller, I., and Stahl, O. (1986). B[e] supergiants of the Magellanic clouds. *A. A.*, **163**, 119–134.

Zickgraf, F. J., Humphreys, R. M., Lamers, H. J. G. L. M., Smolinski, J., Wolf, B., and Stahl, O., (1996). Spectroscopic study of the outflowing disk winds of B[e] supergiants in the Magellanic Clouds. *A. A.*, **315**, 510–520.

Zorec, J. (1998). Distance, kinematics and distribution of B]e] stars in our Galaxy. *B[e] Stars*, Hubert A. M. and Jaschek C. (eds.), Kluwer, Dordrecht, 27–45.

Zorec, J. and Briot, D. (1997). Critical study of the frequency of Be stars taking into account their outstanding characteristics. *A. A.*, **318**, 443–460.

Chapter 6

Late-Type Stars and Close Binaries

6.1 Late-type stars and chromospheric activity

Late-type stars are generally characterized by the chromospheric structure showing active phenomena observable by the formation of emission lines and emission of X-ray and UV radiations. In the optical region, most ubiquitous emission lines are CaII H, K lines, while Hα line appears in fully developed chromospheres. In this section we consider the basic chromospheric activities by focusing to the CaII H, K emission lines.

6.1.1 Emission-line intensities

As an indicator of the chromospheric activity, Vaughan and Preston (1980) defined the activity index S by the following intensity ratio of CaII H and K lines relative to the adjacent continuum,

$$S = \alpha_1 \frac{N_{\rm H} + N_{\rm K}}{N_{\rm V} + N_{\rm R}}, \tag{6.1.1}$$

where $N_{\rm H}$ and $N_{\rm K}$ denote the emission intensities of CaII H, K lines measured by count numbers, $N_{\rm V}$ and $N_{\rm R}$ the intensities of continuum emission in the violet and red side, respectively, measured by count numbers per unit wavelength, and α_1 is the normalization factor to adjust the observation conditions. In Figure 6.1 we show the correlation between the index S and the color index B–V for the stars in the vicinity of the Sun.

Figure 6.1 exhibits the general correlation that the redder the star, the higher the chromospheric activity. In closer inspection we can divide the stars into the following four groups, A, B, C, and D, which are schematically divided in Figure 6.1.

Group A. The stars which occupy the upper right of the diagram and are characteristic of strong Hα emission (mostly dMe stars).

Group B. The K to M type stars with $(B$–$V) > 1.1$, which have strong emission in CaII H, K lines, but weak or absent for the Hα emission (mostly giant stars).

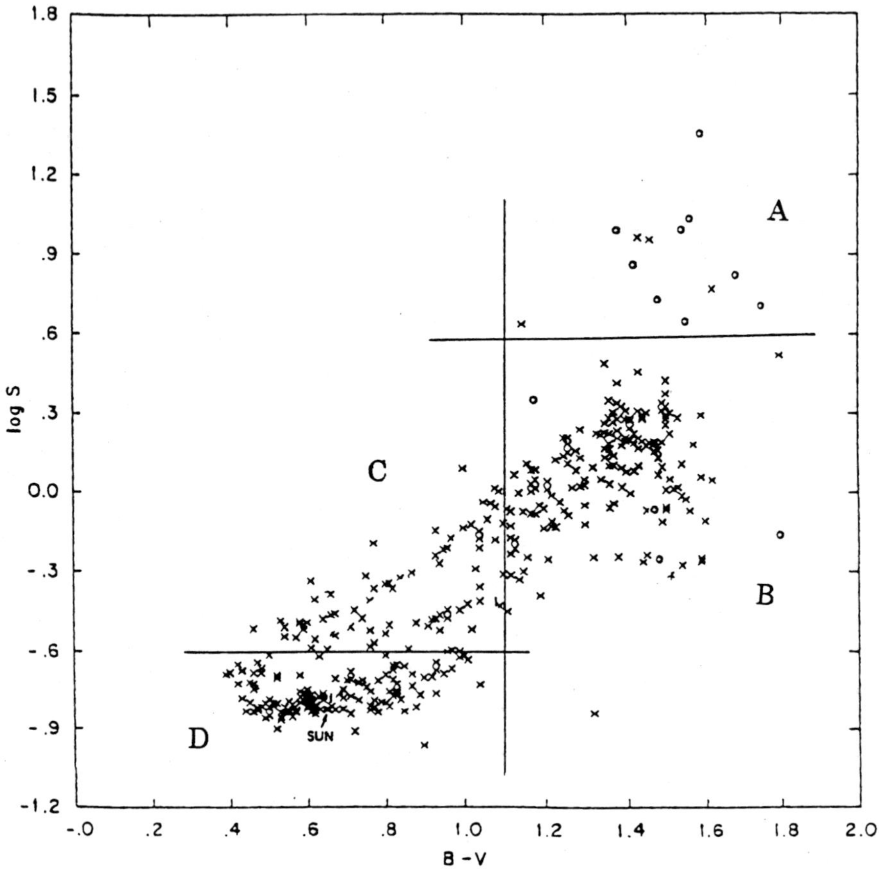

Figure 6.1: Activity index S versus color (B–V) diagram for solar nearby stars. Open circles are dMe stars and cross marks are other late-type stars. The position of the Sun is indicated. On the division of areas A, B, C, D, see text. (Based on the data of Vaughan and Preston 1980).

Group C. The F–G type stars with $(B–V) < 1.1$ and log $S > -0.6$, which have strong emission in CaII H, K lines. They show no appreciable Hα emission. Most of these stars show distinctive activity in their chromospheres.

Group D. The F–G type stars with log $S < -0.6$. Mostly they are chromospherically inactive main sequence stars including the Sun. The log S value of the Sun is taken from the quiescent period of the solar cycle. The stars of this group are not included in the category of emission-line stars.

Since M type stars are generally faint particularly in the ultraviolet region, observations of CaII H, K lines are often difficult. Therefore the Hα emission has been adopted as the activity index instead of CaII lines (Herbig 1985). Stars with Hα emission are generally called as dMe stars so that the surveys

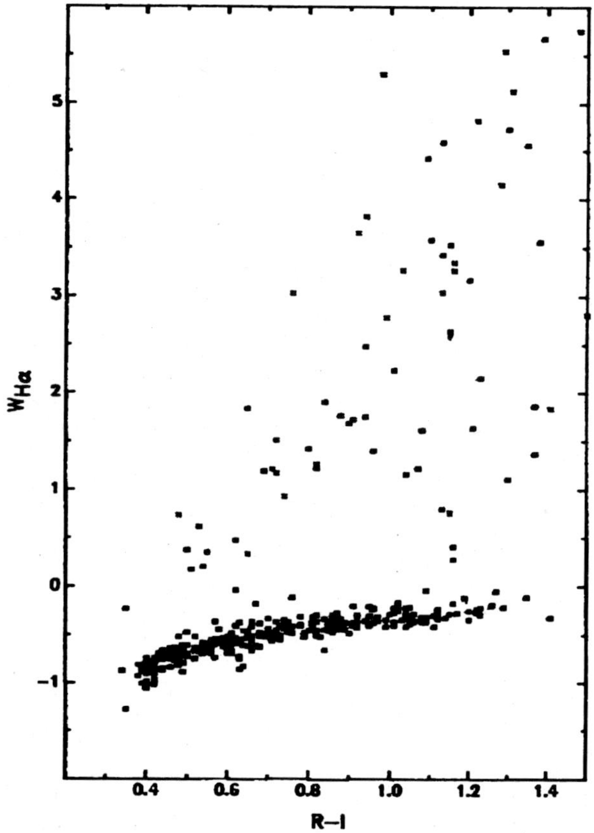

Figure 6.2: Equivalent width EW (Hα) versus color index (R–I) for dwarf K and M stars (from Herbst and Miller 1989). The color index corresponds to the spectral type: R–I = 0.60, 0.80, 1.10, 1.25; Spectral type = M0 M2 M4 M5.

made in this way are weighted toward the stars of high activity. Based on the multicolor photometry and spectroscopic observations, Stauffer and Hartmann (1986) and Herbst and Miller (1989) compared the dM and dMe stars in the Hα intensity. Their result is shown in Figure 6.2 giving the correlation between color index (R–I) (spectral type) and the Hα equivalent width $EW(H\alpha)$ (negative for the absorption). It is seen that most of the stars are dM type with the equivalent width slowly decreasing to redder side. The fraction of dMe among dM stars and the strength of Hα emission both exhibit the increasing trends with the advance of the spectral type.

The reason why chromospheric activity is higher for redder stars may be explained by the depths of convection layers which are more developed and penetrate more deeply into the star's interior in cooler stars as compared to the hotter stars. The developed convection layer causes the increase of mechanical energy flux and arouses the chromospheric activity in redder stars.

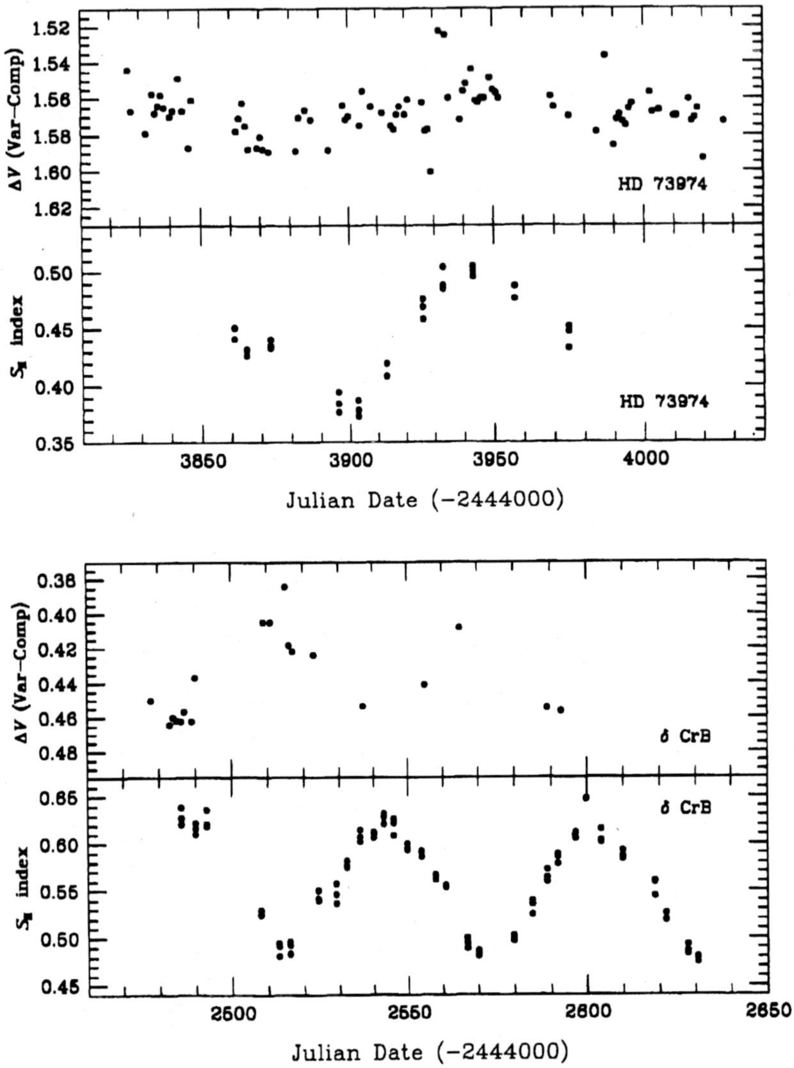

Figure 6.3: Photometric variation ΔV and S index in two bright G–K giants, HD 73974 (upper panel) and δ CrB (lower panel), observed by Choi et al. (1995).

The variation of Ca II emission lines in S index defined by Equation (6.1.1) has been monitored by Choi et al. (1995) for 8 years in 12 bright G–K III stars, along with the photometric observations in V band. The monitoring was aimed at finding the presence of rotationally modulated magnetic activities such as dark photospheric spots or bright chromospheric plages. The results of their observations are illustrated in Figure 6.3 in two stars of HD 73974 (K0 III, $P_{\rm rot} = 112$ days) and δ CrB (G5 III-IV, $P_{\rm rot} = 59$ days), where $P_{\rm rot}$ indicates the rotation period. Among two stars, δ CrB exhibits an inverse correlation between photometric brightness and Ca II flux during the rotation period. On the other hand, HD 73974 showed periodic Ca II signals, but no detectable photometric variability during the rotation period. The region of CaII emission, i.e., magnetically active region, may or may not correlate with the rotational modulation. Some stars show no clear rotational modulation in either Ca II or the photometry. The variability seems different from star to star.

6.1.2 Emission-line width

The relationship between emission-line width and stellar luminosity was first pointed out by Wilson and Bappu (1957), showing a clear correlation for G, K, M stars, for a wide range of stellar luminosity. This relation is called Wilson–Bappu relation and shown in Figure 6.4 in its original form constructed by them for 185 late-type stars.

In Figure 6.4 we can see a clear correlation between log W and M_V for a span of 15 magnitudes, but the relation to the spectral types was not clear. In addition, log W has no clear relation with the emission-line intensities. Wilson (1959) expressed the empirical relation as

$$M_V = 27.59 - 14.94 \log W. \tag{6.1.2}$$

Similar relation has been obtained for MgII h(λ 2802.3 Å), k(λ 2795.4 Å) in the ultraviolet region. According to Vladilo et al. (1987), the empirical relation for MgII k line of G, K-type stars in the range of $M_V = 7$ to -6 was

$$M_V = 41.36\,(\pm 0.78) - 19.58\,(\pm 0.40)\ \log W, \tag{6.1.3}$$

where the figures in the brackets denote the mean error.

Wilson and Bappu (1957) temporarily attributed the line broadening of H, K emission to the turbulent broadening in the emitting chromospheric layer. As seen, however, the Wilson–Bappu relation does not depend on the emission-line intensities and profiles, but depends only upon the stellar luminosity, i.e., the combination of the surface gravity and effective temperature. This leads to the idea that the Wilson–Bappu relation is caused by the chromospheric density and temperature structure. From this point of view, several model calculations have been carried out (Engvold and Rygh 1978, Kneer 1983). Recently, Cheng et al. (1997) carried out a numerical simulation of the Wilson–Bappu relation, based on the sound-wave heated model atmospheres of late-type stars. These model-atmosphere calculations showed that the emission-line width depends on the column density of Ca^+ ions rather than turbulent motions, and hence the Wilson–Bappu relation is explained by the increase of column density with the increase of stellar luminosity. The growth of line width can also be seen in the case of basal atmospheres for the $H\alpha$ and CaII H and K lines, as shown in Figure 4.19.

Özeren et al. (1999) have extended the Wilson–Bappu relation for chromospherically active binaries using the Mg II h and k lines of 41 RS CVn stars (see Section 6.5) observed with the IUE satellite. Though the observed points

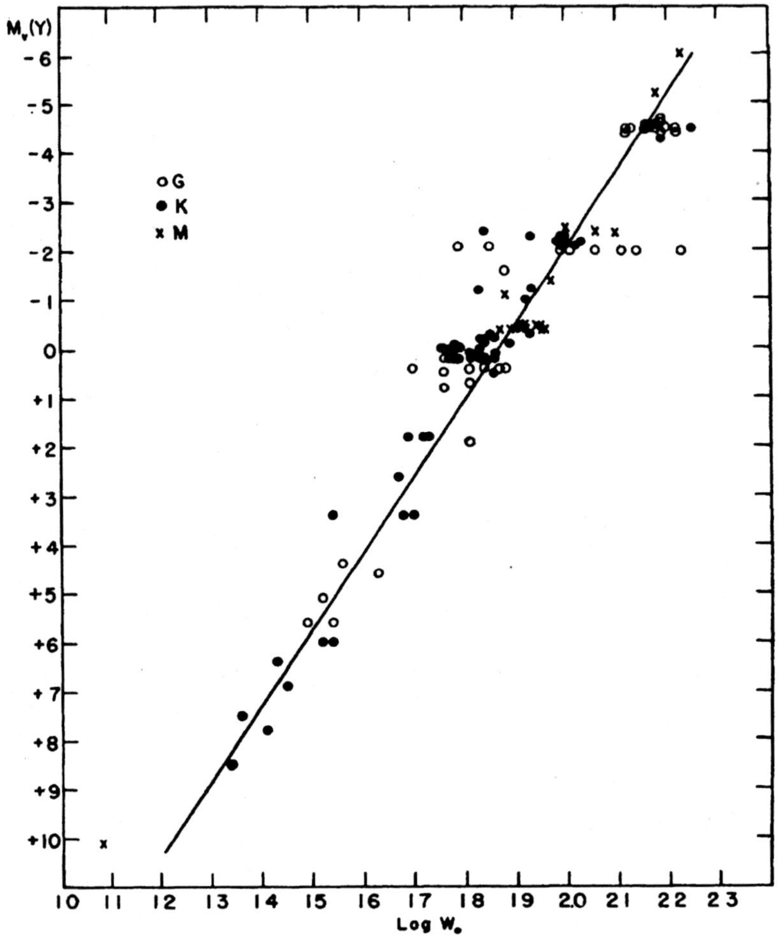

Figure 6.4: Wilson–Bappu relation for late-type stars. The abscissa denotes the logarithm of emission-line width of CaII H, K lines, log W (km s^{-1}), and the ordinate gives the absolute magnitude M_V in the Yerkes Observatory system. The spectral types are shown by different symbols as given inside. (From Wilson and Bappu 1957)

on the ($\log W$, M_V) plane are more scattered as compared to the normal stars seen in Figure 6.4, they derived the following empirical relation

$$M_V = 26.41\,(\pm 1.82) - 12.01\,(\pm 0.89)\ \log W. \tag{6.1.4}$$

They also showed that the line widths W is mainly sensitive to surface gravity ($W \sim g^{-1/4}$) but less sensitive to effective temperature. Within a luminosity class, more active objects tend to show larger line widths. Large scatter in line width might reflect a variety of turbulent velocity, in addition to the density variation in the active regions, of stars.

6.1.3 Excitation degree of emission lines

Emission lines in the optical region

The typical chromospheric lines of CaII H, K, Hα, and NaI D lines are widely observed in emission in the optical region of late-type stars. However, the excitation potentials of these lines are largely different as shown in Table 4.7. This implies that these emission lines are formed in different regions with different temperature. Generally speaking, the Hα line with high excitation potential are formed in the regions hotter than about 8000 K, whereas, CaII H and K lines are less than 6000 K and NeI D line is formed in the cooler region less than 4000 K. It is then necessary to distinguish the appearance of emission lines in CaII and Hα lines from the viewpoint of line excitation.

Coronal X-ray emission

Late-type stars having developed chromospheres usually possess coronae. Coronal temperatures are more or less higher than 10^6 K and emit soft X-rays. Then, X-ray radiation is another indicator of the development of coronae. Young et al. (1989) studied the correlation among the stellar luminosity ($L_{\rm bol}$), the luminosity of excess Hα emission ($\rm L_{H\alpha}$), and the X-ray luminosity ($L_{\rm X}$). Among these, they found that if the stellar luminosity increases, the Hα emission and X-ray luminosity both increase in an almost parallel fashion. Figure 6.5 exhibits the correlation between X-rays and the excess Hα luminosity. It is clear that, even though the X-rays and chromopheric Hα lines are formed in different regions, both regions disclose a good correlation with the bolometric luminosity of the stars. This infers that the ultimate thermonuclear energy gushing out from deep layer once reserved in these regions with strong interaction and gives a measure of stellar activity in some different forms.

Transition layer

In stars having chromosphere–corona structure, there are always transition layers between them where the temperature suddenly increases. In the Sun, the transition layer is optically thin as compared to the chromophere and there is essentially no impact on coronal heating. In contrast, dMe stars, particularly actively flaring stars, generally have developed transition layers where some observable emission lines are formed. Pettersen (1988) noted that the highly excited emission lines of CIV, SiIV, NV, etc., can originate from such transition layer, since they are formed in the region just between chromphsere and corona where the electron temperature is about 25,000–250,000 K.

6.2 Emission-line red dwarfs and flare stars

6.2.1 Emission-line red dwarf stars (dMe)

As stated in Section 6.1.1, the red dwarf stars (dM) often exhibit CaII H, K as emission lines, whereas the red dwarf emission-line star (dMe) is

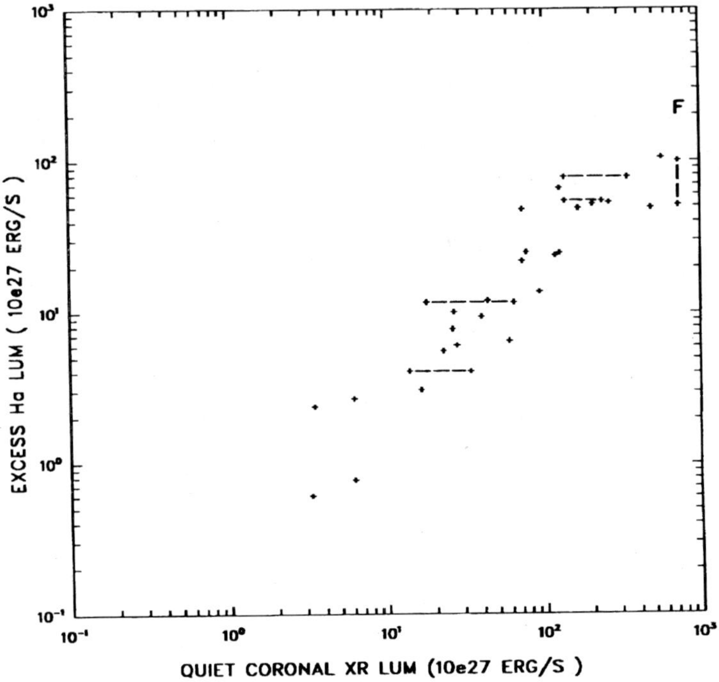

Figure 6.5: The correlation between Hα excess emission and X-ray intensity of chromospheric active M dwarf stars. The abscissa denotes the quiet coronal X-ray luminosity in unit of 10^{27} erg s^{-1} and the ordinate is the $H\alpha$ intensity in unit of 10^{27} erg s^{-1}. The value of Hα in a flare state AU Mic is shown by a mark F. Horizontal lines show ranges of values of X-ray emission when more than two observations are reported. (From Young et al. 1989)

defined by the presence of Hα emission as a sign of higher chromopheric activities.

Stauffer and Hartmann (1986) carried out spectroscopic observations for 260 nearby dwarf M stars at the Mt. Hopkins Observatory and obtained the statistical trend in the equivalent widths of the Hα emission $EW(\mathrm{H}\alpha)$ as shown in Table 6.1. In this table plus and minus signs denote the emission and absorption, respectively, in the column of $EW(\mathrm{H}\alpha)(\text{Å})$. If we define the stars with $EW(\mathrm{H}\alpha) > 0$ as the dMe stars, then the rate of dMe reaches 38% in total. In addition, the stars with $EW(\mathrm{H}\alpha) = -0.1$–$-0.5$ could be regarded to weak emission-line stars having the emission components under the level of nearby continuum. If these stars can be included, the rate to emission-line stars will exceed 80%. That is, the formation of emission lines of the Hα is a general trend even among dM stars. In Figure 6.6, we show some examples of the spectra near the Hα of dMe stars. Since the intensity of the Hα emission is often variable, clear distinction between dM and dMe is difficult as a whole.

Table 6.1: The distribution of the Hα emission equivalent widths for nearby dM stars. The plus and minus signs in EW(Hα) column indicate the lines of emission and absorption, respectively (based on Stauffer and Hartmann 1986)

EW($H\alpha$) (Å)	Number of stars	Rate of existence (%)
1–8	26	10.0
0–1	73	28.0
−0.1 to −0.5	123	45.3
−0.5 to −1.0	38	14.6
Total	260	100

As seen in Figure 6.2, the Hα emission becomes stronger and appears to have a higher rate toward later spectral types. In addition, Stauffer and Hartmann (1986) pointed out the following properties.

(1) Rotational velocities in dMe stars. Around half of 29 dMe stars with measured rotational velocities show the values of $V \sin i > 10$ km s^{-1}, and a large part of the remaining are also supposed to have more than 5 km s^{-1} in V sin i. In contrast, in dM stars only a few stars yield the measurable values of V sin i among 170 stars. This confirms Bopp and Fekel's (1977) suggestion that one of the requirements to be dMe stars is their rotational velocities larger than 5 km s^{-1}.

(2) dMe stars are spectroscopic binaries at a remarkably high rate. This infers that the binary systems have a tendency to form developed chromospheres.

(3) The X-ray intensity of dMe exhibits a good correlation with the Hα emission luminosity, as shown in Figure 6.5. Doyle (1989) derived the empirical relation as

$$\log\ L_{\rm X} = 1.11\ \log L_{\rm H\alpha} - 2.56. \tag{6.2.1}$$

This implies an approximate relation of $L_{\rm x} \propto L_{\rm H\alpha}$. Cram (1982) suggested that back-heating from coronal X-rays may be an important component of chromospheric heating. Doyle (1989) estimated that about 50% may come from X-ray back-heating for the active dMe stars, with less contribution for the less active dM stars. However, back-heating may not be sufficient to explain the Relation (6.2.1).

The formation of the Hα line has been considered in the case of a basal atmosphere in Chapter 4 (Section 4.4.3). The mechanism of line formation is the same in principle for the active dMe stars. The basic parameters for defining the intensities and profiles of emission lines are the mass column density of the chromosphere and temperature structure inside the chromosphere. Houdebine et al. (1995) calculated a grid of chromosphere models for dM and dMe stars in various levels of activity. They thus reproduced all types of

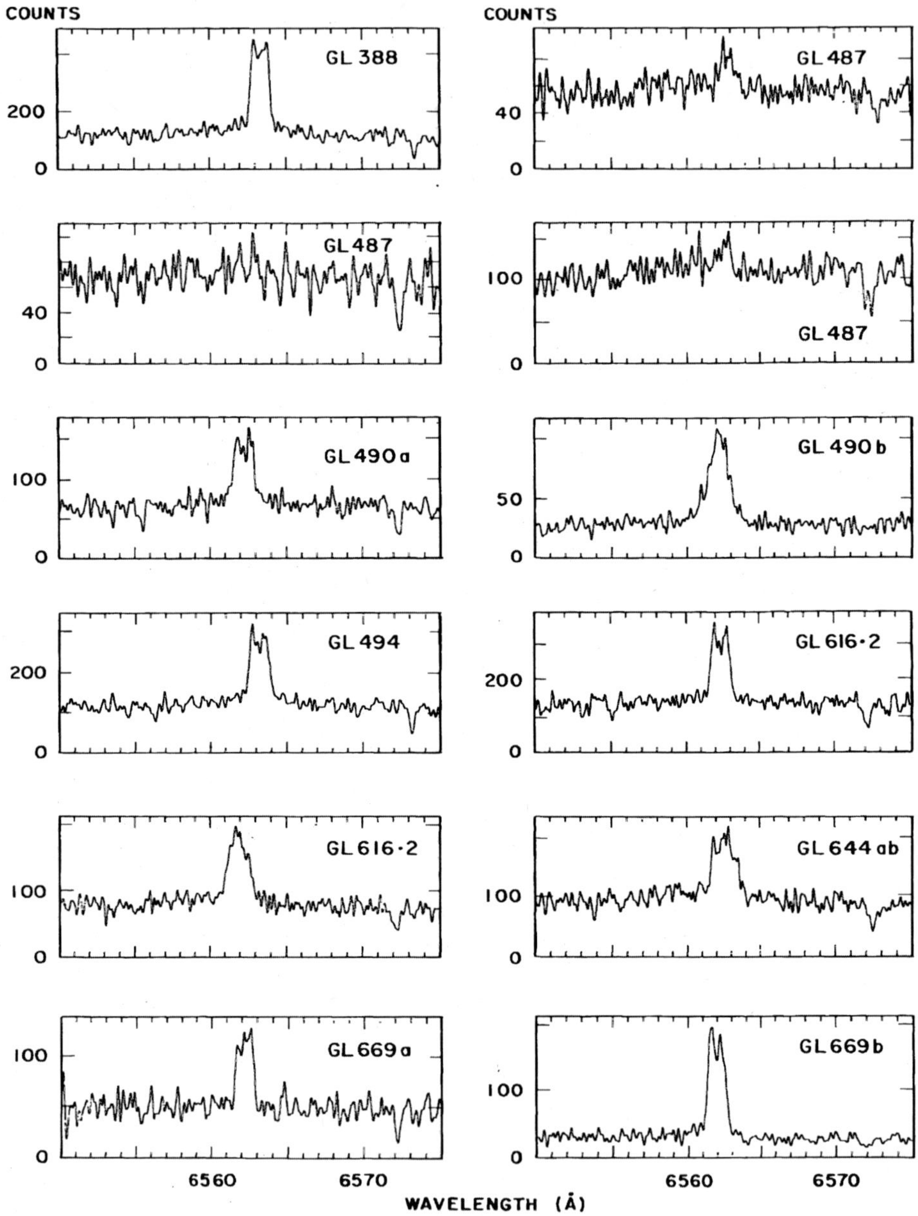

Figure 6.6: Samples of $H\alpha$ emission lines in dMe stars. The stars are mainly designated by Gliese's catalogue number. The same stars in different panels show the spectra taken at different époques. (From Stauffer and Hartmann 1986)

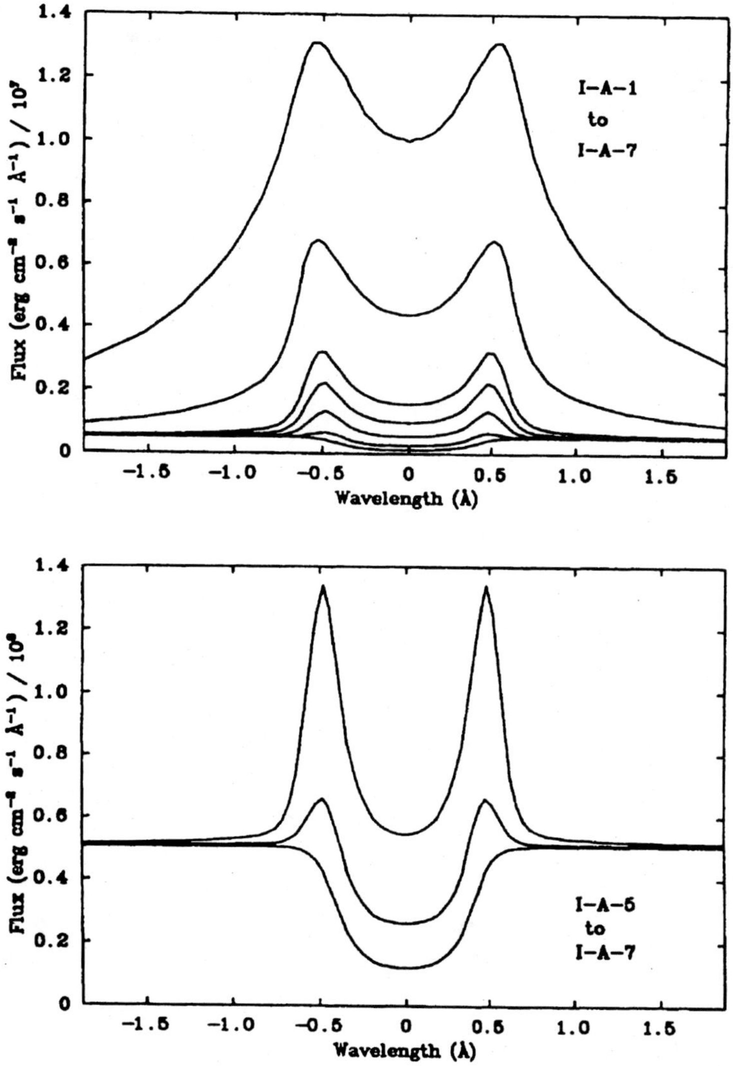

Figure 6.7: $H\alpha$ line profiles in the model calculation for dM and dMe stars. Upper panel indicates the change of profiles caused by the change of column density (from log $(M) \sim -5$ to log $(M) \sim -3$). Lower panel shows the enlarged profile to change from absorption to emission. (From Houdebine et al. 1995)

observed Hα profiles, from weak and narrow absorption in low activity up to strong emission with weak self-reversal. A part of their model profiles is shown in Figure 6.7, where the minimum temperature (temperature at the photospheric surface) is taken as 3000 K and the column density (M) is adopted as a free parameter. In the upper panel of Figure 6.7, the Hα profiles in seven cases of column densities log $(M) = -5.00, -4.50, -4.30, -4.15, -4.00, -3.50,$

and $-3.00(\mathrm{g\ cm}^{-2})$ are shown from the bottom (absorption) up to the top (strong and broad emission). Lower panel illustrates the enlarged profiles in order to show the change of profiles from absorption to emission in the three lowest column densities. Short and Doyle (1998) revised this model to include both Hα and Na I D lines and applied to five dM stars of low-to high-activity level.

6.2.2 Flare stars

Types of flare stars

Flare stars are mostly contained in the group of dMe stars. Table 6.2 gives a list of selected flare stars, where the stellar data are taken from Gliese's (1969) catalogue and the Hα-emission equivalent widths in quiescent state are from Young et al. (1989) (emission is given by plus and absorption by minus sign). Some of stars illustrate absorption lines in a quiescent stage, but all of the stars exhibit strong emission in their outburst stages.

The outburst of flare stars generally occur as a sudden brightening of light, retained at maximum brightness for several seconds or for several hours, and then gradually declining. As in the case of the Sun, outbursts are observable in visible, radio, X-ray and other wavelengths, though the samples of multiwave

Table 6.2: A list of selected flare stars

Star name	Catalogue Gl	V	Mv	Spectral type	EW(Hα) (Å)
DM+43°44	15A	8.09	10.33	dM2.5	−0.3
DM+66°34	22A	10.51	10.42	dM2.5	−0.1
FF And	29.1	10.38	8.7	dM1e	+2.3
UV Cet	65B	12.95	15.8	dM6e	—
Ross 15	82	12.1	11.7	dM4e	+4.7
G036-031	109	10.60	11.13	dM3.5e	−0.3
DM-21°1377	229	8.15	9.35	dM2.5	−1.1
YY Gem	278C	9.07	8.26	dM0.5e	+4.0
YZ CMi	285	11.20	12.29	dM4.5e	+7.48
AD Leo	388	9.42	10.98	dM3.5e	+3.43
SZ UMa	424	9.32	9.70	dM1.5	−0.40
DT Vir	494	9.79	9.1	dM1.5e	+2.12
DM+1602708	569	10.20	10.1	dM2e	+3.58
CR Dra	616.2	9.97	8.38	dM1.5e	+1.81
Wolf 630A	644A	9.76	10.79	dM4.5e	+1.73
V1285 Aql	735	10.07	9.9	dM3e	+1.69
EV Lac	873	10.05	11.56	dM4.5e	+3.36

Notes: Gl, Gliese's (1969) catalogue number; EW(Hα); equivalent widths of $H\alpha$ emission in quiescent stage (Young et al. 1989), plus and minus signs denote the emission and absorption, respectively.

length observations, in photometry and spectroscopy, are still limited. The flare stars are generally divided into two groups according to their brightness duration (Haro and Parsamian 1969).

(a) Fast flare (UV Ceti type). The brightness increases within 30 minutes (mostly several seconds to several minutes). When flaring, star gets much brighter in ultraviolet to yellow bands, along with the rapid increase of the Hα emission intensity. After the maximum light, emission lines become weaker and sometimes disappear in the quiescent state. Gershberg et al. (1999) prepared a catalogue of the UV Cet-type flare stars, which includes 463 stars in the solar vicinity. It contains astrometric, spectral, and photometric data as well as information on the infrared, radio, and X-ray properties and general stellar parameters.

(b) Slow flare. The brightening takes more than 30–40 minutes. In quiescent stage, no Hα emission is seen. The Hα line appears getting stronger during brightening and becomes conspicuous at the maximum light. During the brightening, the red continuous radiation also becomes stronger. The decline is gradual and the Hα emission is seen for several days after the maximum. According to Ambartsumian (1954), slow flares occur when excess energy is released in deeper layer of the star. The number of observed slow flares is low, indicating that the probability of flare action in deep layers is much smaller than that of rapid flares (Parsamyan 1971).

Observations of flare stars

UV Ceti (dM 6e) The flare burst occurred in December 23, 1985, was observed with UBV band photometry, spectrophotometry, as well as X-ray and radio observations under the international collaboration (de Jager et al. 1989). The light variations and spectral change are shown in Figure 6.8.

This flare of UV Cet in December, 1985, reached its maximum brightness in only 2–3 minutes (Upper panel of Figure 6.8), and after a rapid declining, it gradually returned to the original brightness. The amplitudes of brightening were different by color. It was 2.5 magnitudes in V color while it reached 5 magnitudes in B color. After onset of the burst, the star becomes rapidly bluer by the time of the light maximum, then, it turns to redder to the original color. The remarkable flare lasted about 20 minutes. The spectra of the star at light maximum and subsequent three times during the rapid declination within five minutes are exhibited in Figure 6.8 (Lower panel). The identification of main emission lines are given in the upper edge of this panel, strong emissions in the Balmer and HeI line are notable.

UV Cet is a binary system composed of two dM5.5e stars (UV Cet A and B) with similar masses and radii ($M \sim 0.1 M_\odot, R \sim 0.15 R_\odot$). With the spatially resolved X-ray observations by the Chandra satellite, Audard et al.

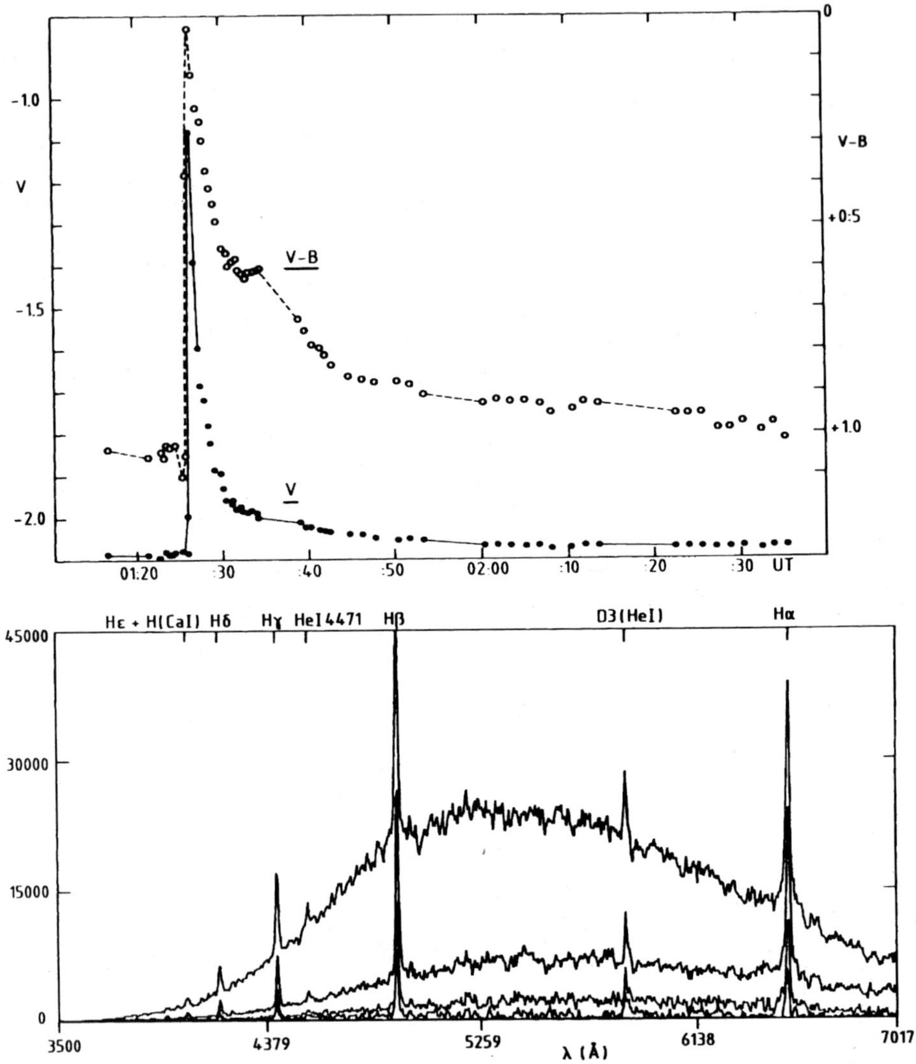

Figure 6.8: Flare burst of UV Ceti on December 23, 1985. Upper panel shows the brightness (modified V band) and (V–B) light curves. Lower panel shows spectra of the flare. The diagram shows successive spectra of the flare component (preflare spectrum subtracted) in declining phase from top to bottom (flare maximum at the top). The times of observations are approximately at 01:26 (at max.), 01:28, 01:29, and 01:30 UT, from top to bottom. (From de Jager et al. 1989)

(2003) showed that both components show similar X-ray luminosity during the low-level state, but the B component exhibits a higher degree of variability, resulting in a high average X-ray luminosity. B component might cause the flare activity in this system.

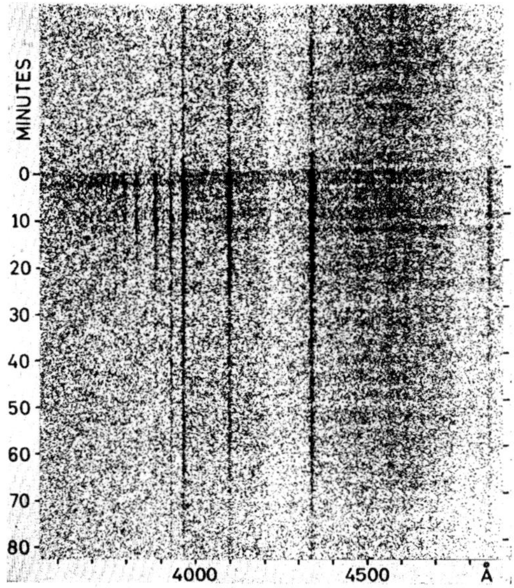

Figure 6.9: Spectral variation of EV Lac at the flare December 11, 1965, observed by 91-cm telescope. The spectral holder of the spectrograph is moved at 3.18 mm s^{-1} perpendicular to the dispersion. The scale in the left-hand side denotes the time after the flare burst in minutes. (From Kunkel 1970)

EV Lac (dM4.5e) The flare event in EV Lac was detected on December, 1965 by photoelectric photometer (U, B, V) and low-dispersion spectrograph (260 Å mm^{-1}) (Kunkel 1970). Spectral variation in this event is shown in Figure 6.9, the ordinate is the time (in minuets) after the onset of the flare, and the abscissa is the low-dispersion spectrum (λ 3600–4900 Å). The brightened continuum and strong emissions in Hγ, Hδ, and CaII H($H\varepsilon$) and K lines could be traced for more than 1 hour. The variation of the observed Balmer and CaII emission lines were measured and summarized in Table 6.3.

The behavior of the Balmer decrements in Table 6.3 is sharply different from the nebular case (Case B), with the maximum intensities appearing around $H\gamma-\delta$ lines and showing an inverted decrement. After about 20 minutes from the onset of burst, the members higher than $H\zeta$ declined rapidly.

The flare event of EV Lac in September 2001 was observed in multiband from radio to X-ray (Osten et al. 2005). In this event, the optical flare showed coincidence with radio flare, but with the lack of an observable X-ray enhancement, somewhat puzzling in its nature. Flat Balmer decrement was also apparent in this event.

AD Leo (dM3.5) By using UBV color photometry, 12 flare events were observed in February 1972, by Ichimura et al. (1973) at the Okayama

Table 6.3: The variation of the Balmer and CaII K emissions in the flare of EV Lac on December 11, 1965 (Hβ is taken as unity for all lines) (adapted from Kunkel 1970)

Time after the burst\ (minutes)	Hβ	Hγ	Hδ	Hζ	Hη	H10	CaII K
0	1.0	1.24	1.48	1.22	1.17	0.94	0.47
5	1.0	1.04	1.16	0.92	0.63	0.64	0.59
8	—	1.10	1.28	1.10	0.90	0.67	0.68
13	1.0	1.13	1.06	0.76	0.54	0.52	0.68
61	1.0	1.15	0.90	—	—	—	0.76

Astrophysical Observatory. All of them were fast flares with duration less than several minutes and the brightening in the U band was remarkable. For example, the brightening of the flare in February 19 in U band reacted $\Delta U = 2.56$, as compared to $\Delta B = 0.88$ and $\Delta V = 0.29$.

As another example, we show the flare event of AD Leo occurred on March 28, 1990. The multiwavelengths' spectral monitoring was undertaken by Rodono (1990). The results were summarized as follows.

The ultraviolet, infrared, and radio continuums in flux or in magnitude are shown from top to bottom in Panel (a) of Figure 6.10. The flare burst was most conspicuous in U band. The K band exhibited rather absorption and the radio continuum showed different behaviors. The complicated structure of the flare in the time variations of optical emission lines appeared as shown in Panel (b), which disclosed the following properties in relation with the ionization potential and in the time sequence.

He Iλ 4026 Å (He: $IP = 24.6$ eV)—after flare burst, strong emission was seen shortly for around 5 minuets.

Balmer lines (H: $IP = 13.6$ eV)—after the burst, sharp rise and gradual decline of strong emission were seen for about 30 minutes much longer than that of He I. The second peaks were seen on the light curves within around 10 minutes after the burst.

CaII K (Ca: $IP = 6.1$ eV)—emission started to appear at the burst, and gradually strengthened for 40 minutes. Peak emission occurred after the peaks of the Balmer lines.

The different behaviors of emission lines can be basically explained by the difference of the excitation levels of the atoms, i.e., the time variation of electron temperature in the flare. In the early phase of the flare, electron temperature is high and excites the He and H, and then with the decrease of the temperature, Ca ions are excited to form the emission in K line. As seen in Panel (b), the Balmer decrement seems flat as in case of EV Lac (Figure 6.9). Such flat decrements suggest in some part the similarity with the case of CV (Section 6.6.6). In addition, Houdebine et al. (1990) observed a high-velocity

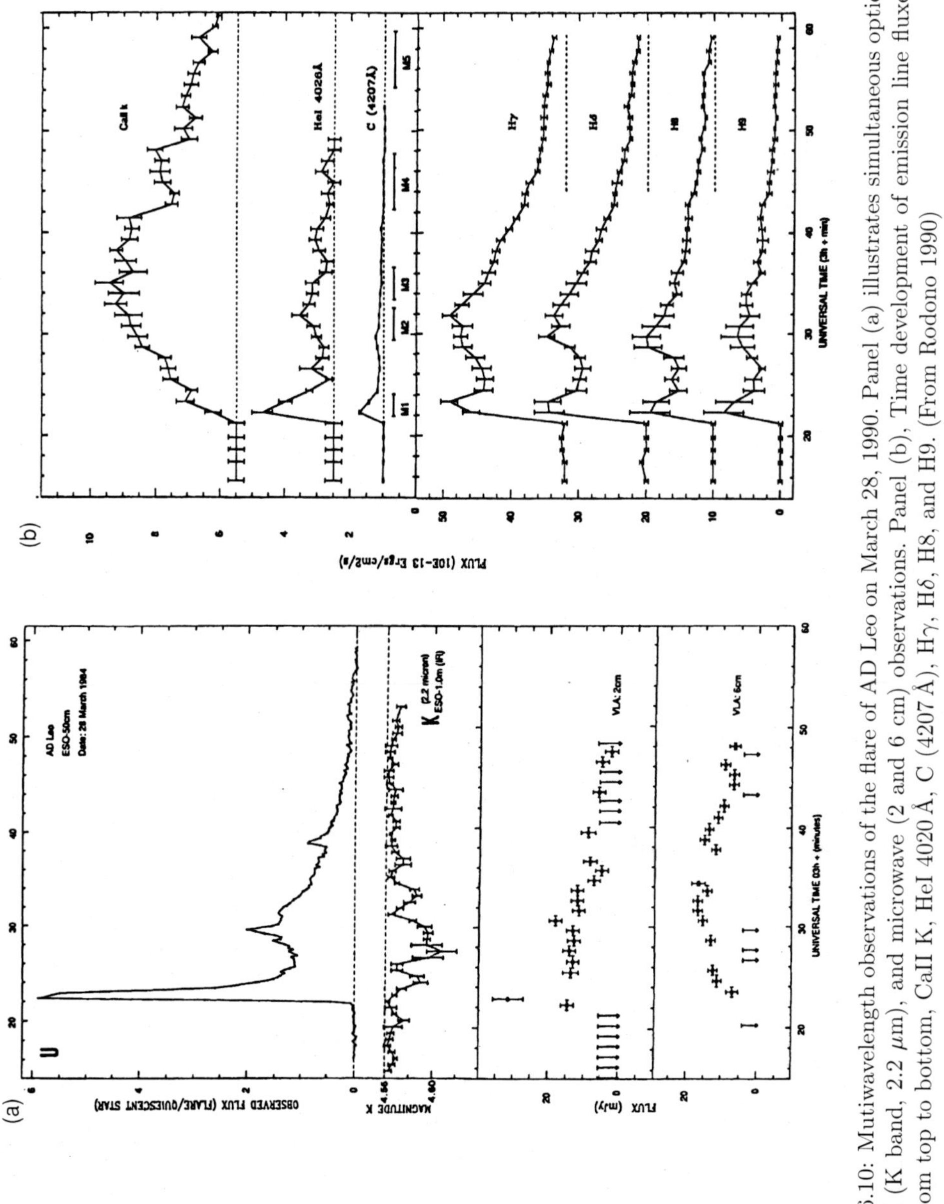

Figure 6.10: Mutiwavelength observations of the flare of AD Leo on March 28, 1990. Panel (a) illustrates simultaneous optical (U band), infrared (K band, 2.2 μm), and microwave (2 and 6 cm) observations. Panel (b), Time development of emission line fluxes during the flare. From top to bottom, CaII K, HeI 4020 Å, C (4207 Å), Hγ, Hδ, H8, and H9. (From Rodono 1990)

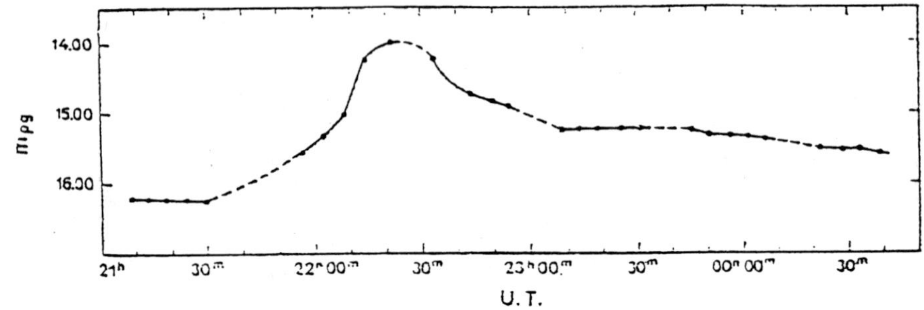

Figure 6.11: Light curve of the slow flare observed on September 2, 1970, in star No. 103 in the Pleiades cluster. (From Parsamian 1971)

gas stream in the ultraviolet absorption line of the Hγ line, as high as 5800 km s^{-1} in early phase of the outburst.

As compared to the fast flares (UV Cet type), slow flares are scarce in number and the samples of spectroscopic observations are also few. According to Parsamian (1971), there are 312 flare stars in the Orion Association but the slow flare events occurred only in 7 stars. In the Pleiades cluster, only one slow flare has been observed among 221 flare stars. Figure 6.11 illustrates the light curve of the slow flare in star No. 103 in Pleiades cluster observed on September 2, 1970. In this star, four flare events have been observed in 1968−1970, among which the last one was the slow flare. This flare took 40 minutes for brightening, and 5 hours for the total duration.

It may be interesting to compare the slow flare events in Orion and Pleiades:

(1) In every event, the maximum appeared in around 40 minutes or more after the beginning of brightening. The decay of light continued for several hours depending on the amplitude of the flare.
(2) Slow flare stars in Orion are of earlier spectral type (K7–M0) than in the Pleiades (M3s–M4e).
(3) The average amplitude of slow flares is larger in Orion than in Pleiades.

These differences somehow reflect the effect of the evolution of stars and clusters (see Figure 6.13).

The Balmer decrements of dMe and flare stars

The Balmer decrements of flare stars are getting flatter in burst time, and even inverted. Flat feature may be seen in Figures 6.9 (Hβ–H10) and 6.10 (Hγ–H9). In contrast, the Balmer decrements in the quiescent stage are steeper than those in burst stage. An example of the decrement in quiescent stage is as shown in Table 6.4 (Geschberg 1974).

In order to get the flatter or inverted decrement as compared to the decrements in quiescent phase, following models have been proposed:

Table 6.4: The Balmer decrements for some flare stars in quiescent stage (adapted from Gershberg 1974)

Star name	Hα	Hβ	Hγ	Hδ
EV Lac	—	1.0	0.72	0.52
AD Leo	1.4	1.0	0.70	0.39
YZ CMi	2.9	1.0	0.76	0.54
EQ PegA	1.5	1.0	0.66	0.40

(i) LTE model. Under the LTE assumption for the distribution of level populations of hydrogen atoms, Kunkel (1970) derived the theoretical decrements, with a fixed parameters of $N_e = 3.2 \times 10^{13}$ cm^{-3}, $V_{turb} = 20\,\mathrm{km\,s^{-1}}$, for two cases of T_e = 20,000 K and 25,000 K. Free parameter was the optical depth for the Hα line, τ(Hα). By comparing the theoretical decrements with his low-dispersion spectra of EV Lac (Figure 6.9 and Table 6.3), he obtained the best-fitting values in two cases: near the light maximum when optical depth reached $\tau(\mathrm{H}\alpha) \approx 10^3$ and 1 hour later the light maximum when optical depth decreased to the value of about $\tau(\mathrm{H}\alpha) \approx 10^2$. In this way Kunkel derived the flat decrements by assuming a dense gas ($Ne \sim 3 \times 10^{13}$ cm^{-3}) with high excitation temperature ($Te \sim$ 20,000−25,000 K) in the flare region. Rapid decrease of the optical depth τ (Hα) around a factor 10 during 1 hour may show the existence of large expanding motion of flare gas which cause the deviation of radiation field from LTE state.

(ii) Moving envelope model. Gershberg and Shnol (1974) applied the method of the escape probability by motion (see Section 4.2.1) for the envelopes of late-type stars. They adopted the collisional ionization and excitation processes under the chromospheric high-electron temperature, instead of assuming photoionizing radiation that is not sufficient in late-type stars. Assuming 30 energy levels for hydrogen they solved the equations of statistical equilibrium, the adopted parameter are the physical state of gas (T_e, N_e) and the escape probability β_{21^0} for the Lyman α line. The results of calculations are shown in Table 6.19 in Section 6.6.6.

According to this model, inverted decrements $F(\mathrm{H}\alpha) < F(\mathrm{H}\beta)$ occur only when electron density is high ($>10^{13}$ cm^{-3}) and the escape probability is small ($\beta_{21^0} < 10^{-3}$). If we apply this model to the flare event, we need to assume a gas blob having high gas density and large optical thickness at the beginning of the outbursts. The blob expanded rapidly with time.

(iii) Multiple escape probability by scattering. Since no sufficient large scale gas motion has been observed in the atmospheres of flare stars, Bruevich et al. (1990) assumed a static atmosphere and applied the escape-probability-by-scattering method (Section 4.1.3), but assuming the multiscattering process. They derived the escape probability by scattering from a plane parallel slab as follows: Consider that the source function is everywhere constant in this slab, and let the optical thickness for the Lα radiation be τ_o. Then the probability, λ_{12}, that a Lα photon can survive as Lα photon by one scattering, is given as

$$\lambda_{12} = \frac{A_{21}}{A_{21} + N_e \alpha_{21}}, \tag{6.2.2}$$

where α_{21} denotes the downward collisional coefficient. The probability of this photon to escape from the atmosphere after n scatterings, β_{12}, is expressed as

$$\beta_{12} = 1 - (1 - \lambda_{12})\, n, \tag{6.2.3}$$

where n is a function of the slab's optical thickness τ_0. The escape probabilities calculated for the respective transitions are inserted into the equations of statistical equilibrium, where it is assumed that each hydrogen atom has 30 energy levels and there is no incident radiation from outside. Electron collisions are the only exciting process inside this slab. Parameters are the electron temperature and electron density. The results are partly shown in Figure 6.12.

It is seen from Figure 6.12 that the Balmer decrements become flatter when electron density is high and the layer is optically thick. Katsova (1990) applied this method to the Balmer decrements of red dwarfs in their quiet state and during outburst for 16 red dwarfs. In the quiet state, the decrements were generally steep and did not notably differ with each other, while during the flare the decrement becomes remarkably flatter, particularly when the electron temperature takes $T = 10{,}000 - 15{,}000$ K with high electron density of $n_e = 10^{14}$ cm^{-3}. Figure 6.12 illustrates the flat decrements in the case of large optical depths.

(iv) Escape probability by scattering. Garcia-Alvarez et al. (2002) directly applied the method of escape probability by scattering to the flare event in AT Mic observed in August 1985. The decrements $F(Hn)/F(H\gamma)$ (n = 5–10) had a steep slope at the very beginning and became flatter during the burst stage. Thereafter the slope steepened reaching similar values to that in the quiescent phase. According to their calculation, this behavior can be qualitatively explained mainly by the variation of electron density Ne as seen in Figure 4.6. That is, the electron density suddenly increased up to a range of $Ne \sim 10^{12} - 10^{14}$ cm^{-3} and stayed in dense state during the burst phase, and then gradually returned to the steep slope in quiescent phase.

The results of above theoretical model calculations exhibit the general tendency for steep decrements to appear in low-density gas less than around

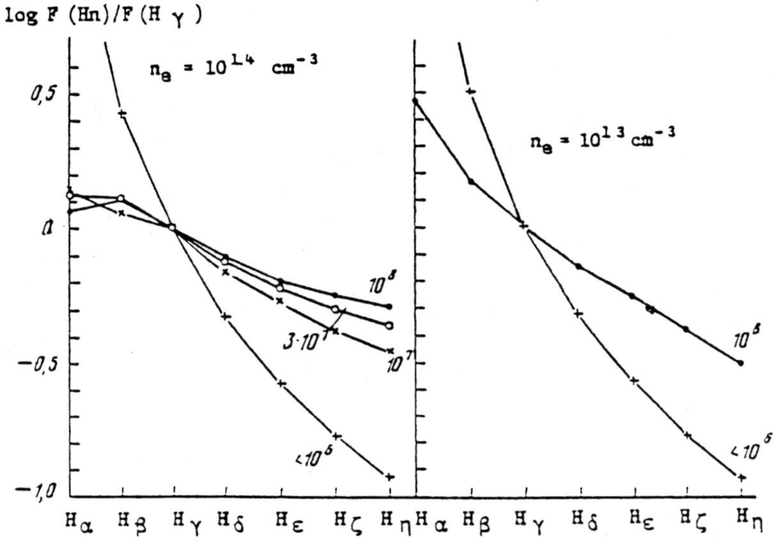

Figure 6.12: The Balmer decrements calculated by the escape probability by multiple scattering. The ordinate is the logarithmic intensity relative to Hγ. The electron temperature is taken as 10,000 K; the left and right panels indicate the cases of electron density 10^{14} and 10^{13} cm^{-3}, respectively. The optical depth of the layer at the center of the Lα line is indicated near the respective curves. The curves with $\tau(\mathrm{L}\alpha) < 10^6$ describe the layer optically thin for Balmer lines, and represent the case of steep decrements in the quiescent stage. (From Bruevich et al. 1990)

$N\mathrm{e} = 10^{12}$ cm^{-3}, whereas flat decrements need higher electron density. Particularly inverted decrements appear when $N\mathrm{e} > 10^{12}$ cm^{-3} in the moving envelope, or when $N\mathrm{e} > 10^{13}$ cm^{-3} in the static envelopes. In any case the gas of flare bursts must be in high electron density and high optical thickness at their impulsive phase. Spectroscopically, we can imagine that the flare burst suddenly occurs as a result of an abrupt concentration of gas with high temperature and large optical thickness. After the light maximum, the gas begins to disperse toward a low electron density and low optical thickness with some fluctuation of physical parameters. It may be an interesting problem how to reconcile the spectroscopic and gas dynamical processes before and during the flare outbursts.

Evolutionary state of flare stars and emission-line red dwarfs

Since 1920s, a number of flare stars have been discovered in the solar vicinity and in some clusters and associations. In the Orion association, Haro and Morgan (1953) surveyed flare stars, and the similarity was disclosed between flare stars and T Tauri stars in spatial distribution as pointed out by Ambartsumyan (1954). Since then flare stars have been recognized as members of young stars.

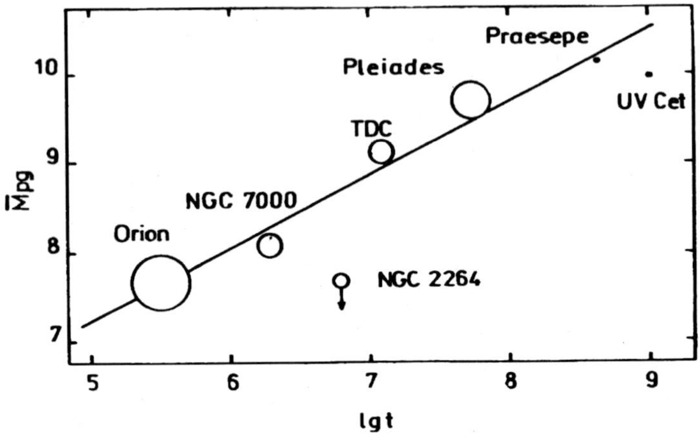

Figure 6.13: Correlation between the mean absolute photographic magnitude (M_{pg}) of flare stars and the age of the parent clusters or associations in years. The circle areas are proportional to the number of stars involved. TDC means Taurus Dark cloud. The ages of the UV Cet stars (Solar vicinity) are taken as 10^9 years. (From Mirzoyan 1990)

Comparing flare stars and T Tauri stars, Haro (1957) suggested that the flare stars are rather older than T Tauri stars by the age of the parent associations or clusters. In addition, Mirzoyan (1990, 1993) and Mirzoyan et al. (1995) studied the relationship between flare stars and dMe stars in and near the main sequence.

These arguments may be summarized as follows:

(1) No statistical difference was found between flare stars in the solar vicinity and those belonging to clusters and associations as to their light curves, stellar colors, and positions on the HR diagram. Fast and slow flares appear at almost the same rate in both groups of flare stars. Stellar spectra in both groups reveal no appreciable difference.

(2) There is an indication that the luminosity of flare stars varies with the age of parent cluster or association, as shown in Figure 6.13. The abscissa is the estimated age of the clusters of association. The ages of flare stars in the solar vicinity (UV Cet type stars) are adopted as 10^9 years. The ordinate is the mean absolute photographic magnitude. The scale of stellar brightness is downward. The circled areas are proportional to the number of stars involved (Mirzoyan 1990). The Orion association, the largest, contains 473 flare stars, while Presepe, the smallest cluster, contains only 14 stars, and the UV Cet type nearby stars have 16 stars. Stars in the association NGC 2264 yield the mean luminosity markedly deviated from the normal correlation. This has been attributed to errors in the determination of either age or distance of this system. In this diagram, one may see an apparent age effect, i.e., the younger the cluster or association, the brighter the flare stars in those systems.

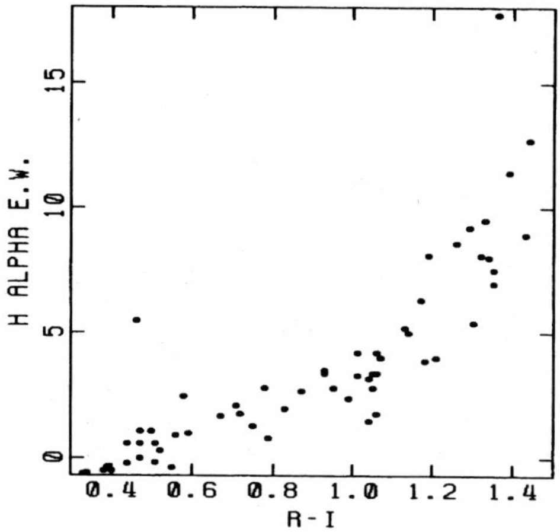

FIG. 23a

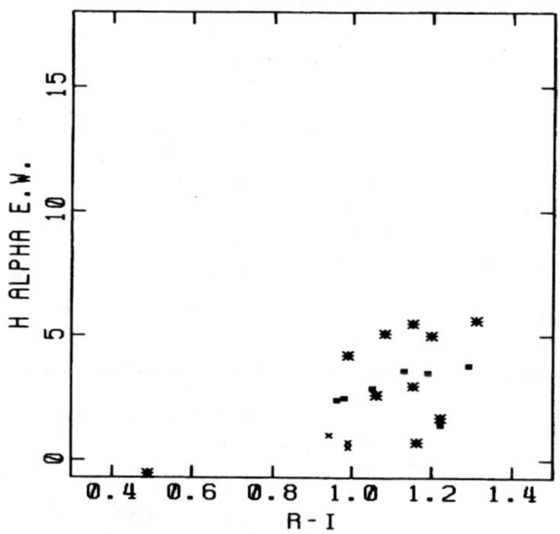

Figure 6.14: Correlation between Hα equivalent width and R–I color index for M dwarfs in the Pleiades (upper panel) and the Hyades clusters (lower panel). For the Hyades panel, small symbols indicate low-quality observations. Crosses indicate upper limits. (From Stauffer and Hartmann 1986)

Age effect can also be seen in dMe stars. Compare the dMe stars in the Pleiades and Hyades clusters. Figure 6.14 exhibits the correlation between color index R–I and the Hα emission equivalent width for dM and dMe stars in both clusters (Stauffer and Hartmann 1986). In the Pleiades cluster (upper panel), dM stars ($EW(\mathrm{H}\alpha) \leqq 0$) are almost concentrated in the blue stars, while the remaining stars become dMe stars with their Hα emission getting stronger along the spectral sequence. In the Hyades cluster (lower panel), stars are mostly late type (R–$I > 1$) with weak Hα emission. It is notable that there are no stars with strong Hα emission ($EW(\mathrm{H}\alpha) > 6$. The difference in distribution of dMe stars in both clusters is clear. The ages of clusters are estimated as 7.8×10^7 years for the Pleiadies, and 6.6×10^8 years for the

Hyades. From this we can infer that the existence of emission-line stars and emission intensities are more or less related to the ages of clusters so that younger clusters have high distribution of dMe stars with strong Hα emission.

6.3 Red giants and long-period variables

6.3.1 Red giants

As stated in Section 6.1.2, the Wilson–Bappu relation can be traced for a wide brightness range from main-sequence to supergiants. In Figure 6.15, the spectrograms of the main parts of the stars are shown as they appeared in the original Wilson–Bappu relation. These spectra were taken by 91-cm reflector

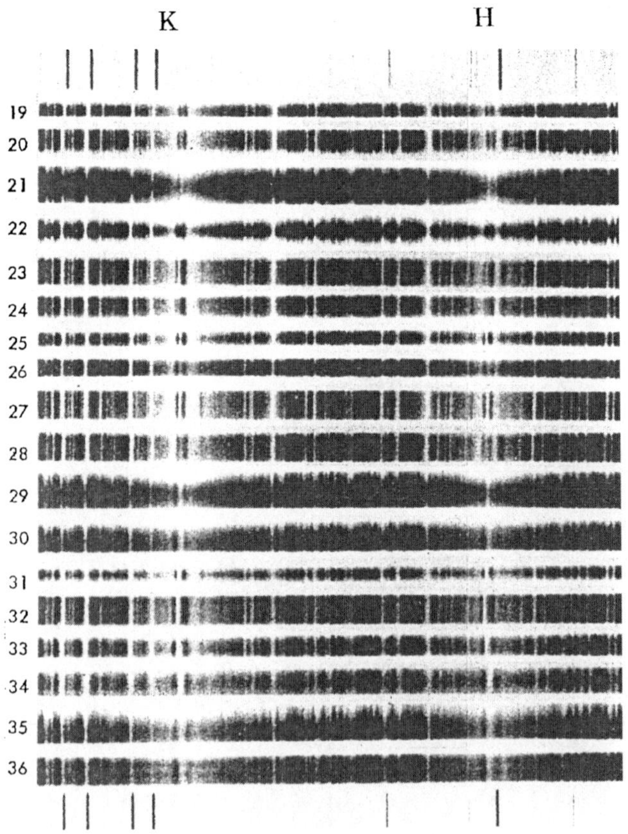

Figure 6.15: Stellar spectra arranged from giants to supergiants in G–M types. Stars denote the brighter parts in Wilson–Bappu relation in Figure 6.3. The star number in the left-hand side is the same as the star number in Table 6.5. K and H denote the CaII K and H lines, respectively. (From Wilson and Bappu 1957)

Table 6.5: List of red giants. The star number is the same as in Figure 6.15. The emission-line intensities of CaII H and K lines are also shown (adapted from Wilson and Bappu 1957)

Star number	Name of star	HD	mv	Mv	Spectral type	Emission intensity of CaII H, K
19	75 Leo	98116	5.4	+0.4	M0 III	3
20	β UMi	131873	2.2	+0.1	K4 III	3
21	η Psc	9270	3.7	−0.2	G8 III	1
22	α Lyn	80493	3.3	−0.4	K0	3
23	β And	6860	2.4	−0.9	M0 III	4
24	π Leo	86663	4.9	−0.9	M2 III	3
25	μ Gem	44478	3.2	−1.3	M3 III	4
26	γ And	12533	2.3	−1.6	K3	2
27	β Peg	219706	2.6	−1.6	M2 II-III	4
28	γ Aql	186791	2.8	−2.1	K3 II	3
29	ε Leo	84441	3.1	−2.2	G0 II	2
30	ω Gem	52497	5.2	−3.0	G5 II	2
31	π Aur	40239	4.6	−3.0	M3 II	4
32	ε Peg	206778	2.5	−4.6	K2 Ib	4
33	ζ Cep	210745	3.6	−4.9	K1 Ib	3
34	α Sco	148478	1.5	−4.9	M1	3
35	α Aqr	209750	3.2	−5.5	G2 Ib	2
36	β Dra	159181	3.0	−5.7	G2 II	3

Notes: Star number is the same as in Figure 6.15. Emission-line intensity of CaII H, K is given by eye estimation in four steps (weak 1 to strong 4)

at Mt. Wilson Observatory. The stars are arranged by increasing brightness from top to bottom. The positions of CaII H and K lines are indicated in the upper site. Table 6.5 yields the magnitude and spectral type of the red giant stars given in Figure 6.16. The intensities of emission lines of CaII H and K eye-estimated into four steps are also listed in the last column of the table.

The emissions of CaII H and K lines display various line profiles, single peak, double peaks, and/or asymmetric profiles, as seen in Figure 6.15. Although the CaII emissions are formed in developed chromospheres, emission profiles differ from line to line, suggesting a relationship different from the Wilson–Bappu relation. Double-peaked emission could be formed even in a single "basal" atmosphere as shown in Figure 4.19. But strong double-peaked or strong single-peaked emission may show important effects of stellar rotation and inclination angle. The effect of binary interaction may not be ignored in case.

These effects have been statistically studied by Strassmeier et al. (1994) for G and K giant stars. They measured the radiation flux (erg cm^{-2} s^{-1}) of the emission component of CaII K line using narrow band photometry, and defined the emission flux $F'(\mathrm{K}) = F\ (\mathrm{K}) - F^*(\mathrm{K})$, where $F(\mathrm{K})$ is the observed K line

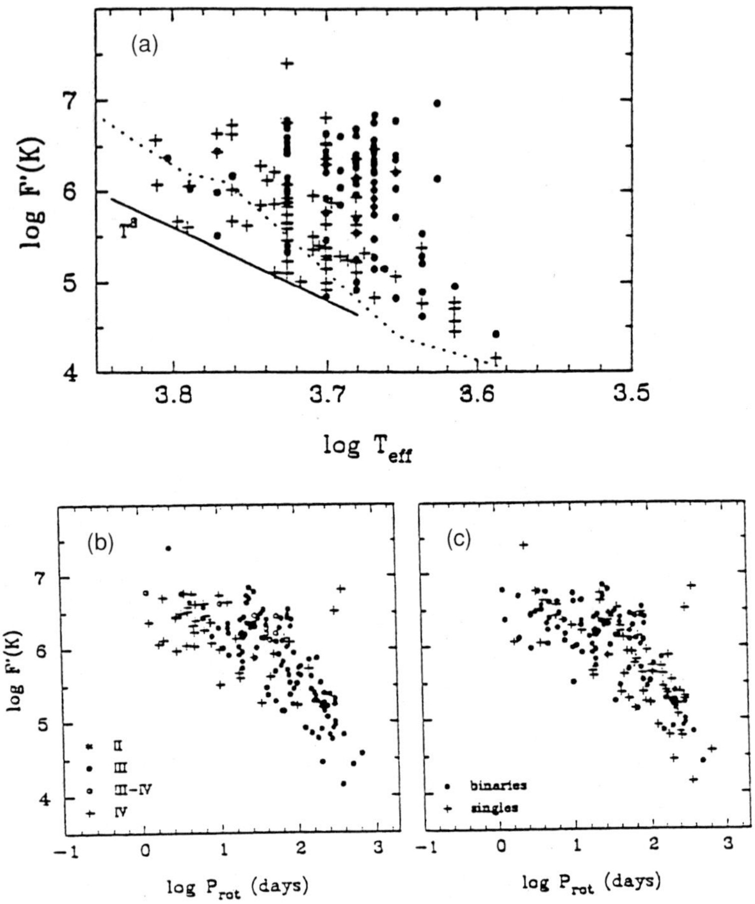

Figure 6.16: A, The emission flux of K line as a function of stellar effective temperature. The dotted lines is the theoretical prediction by the basal atmosphere theory, the full line denotes the empirical lower limit. Panels B and C yield the relation between emission flux of K line and the rotational period, where Panel B distinguish the stellar luminosity, and C the distinction whether singles or binaries. (From Strasmeier et al. 1994)

flux in a narrow band, $F^*(\mathrm{K})$ the theoretical line flux derived by the radiative equilibrium model of the same spectral type. Hence the difference $F'(\mathrm{K})$ yields the emission flux of the K line. The comparison between $F'(\mathrm{K})$ and the stellar properties for 59 stars in GOIV–K5III are delineated in Figure 6.16A, B, and C. In Panel A, the abscissa is the stellar effective temperature, the ordinate is the emission flux $F'(\mathrm{K})$, the dotted line is the basal flux corresponding to K line (Section 4.4.2.), and the full line segment shows the empirical lower limit of observations. One may see that the single stars are distributed nearly along the basal atmospheres, while binary stars generally have much higher emission fluxes, suggesting the effects of binary interaction. In Panels B and C, the abscissas are the rotational period (days) in common, and the ordinates are both the emission flux $F'(\mathrm{K})$ as in Panel A. Stellar symbols are different in both figures. In both panels B and C, it is clear that the stars of shorter rotational periods (rapid rotators) yield larger emission fluxes suggesting the large effect of stellar rotation. In Panel B stars are distinguished by the stellar luminosity. This shows a global tendency that the stars of higher luminosity show a higher emission flux. Finally in Panel C single stars and binary stars are distinguished, but no particular difference is seen between them. Throughout these panels, it is clear that the rotational velocity is an important parameter for the formation of emission lines.

The correlation between rotational velocity and stellar luminosity is also seen in emission lines in the far UV region. For 50 G–K giants, Böhm-Vitense (1992) measured the rotational velocity $V \sin i$ and surface flux of the emission line C II λ 1335 Å, and C IV λ 1550 Å, formed in the transition layers from chromospheres to coronae. According to her measurements, the mean rotational velocity $V \sin i$ rapidly drops from 31.1 km s^{-1} (15 stars) for G0–G9 stars to 6.1 km s^{-1} (7 stars) for (G9–K2) stars. The surface flux of emission lines also decreases in a parallel fashion. Gray and Endal (1982) suggested that the rapid decrease is a result of evolutionary effect. They argue that the expanding stars evolving to red giants gradually deepen their convection zones and at some stage there occurs a magnetic breaking or an uncoupling of the connective layer and the magnetic nucleus. A new uniform specific angular momentum in the convective layer is developed causing the slowdown of stellar rotation. As the result chromospheric activity also declines accompanying the decrease of emission-line intensities.

6.3.2 Long-period variables

Mira-type variables

Mira-type variables, known as long-period (150–400 days) pulsating late-type stars, reveal conspicuous emission lines in phases around maximum light. Spectral type is usually defined by the absorption spectrum at maximum light, though the spectral type changes markedly with the pulsation phase, and both absorption and emission lines vary with phase. Among Mira-type variables, the majority are M type (826 stars), followed by C type (53 stars) and S type (47 stars) (Hoffmeister et al. 1985). Keenan (1966) catalogued Mira spectra for 253 stars, and this was extended by Keenan et al. (1974) for 7445 mainly northern Me and Se stars. Crowe and Garrison (1988) extended these catalogues to the southern sky (72 Mira variables). The catalogue compiled by Kharchenko et al. (2003) gives spectroscopic data for 1103 Mira variables.

Lists of selected M-type Mira variables are given in Table 6.6, and those of S- and C-type Mira variables are in Table 6.7.

Table 6.6: A list of selected M-type (Me) Mira variables

			Spectral type		
Star name	Magnitude M_{pg}	Period (days)	GCVS	Fox	CG
S Car	6.9–11.0	149.40	M0e	M0e	K7-M6.5
T Col	6.6–12.6	225.27	M4e	M5e	M5.5-M9
RR Sco	5.0–12.2	279.45	M6e	M6e	M6-M8
R Aql	5.1–12.0	284	M7e	M6.5e	—
R Car	5.6–11.1	309.15	M5e	M5e	M6.5-M9
o Cet	2.0–10.1	331.48	M6e	M5.5e	M7-M9
RR Sgr	5.5–14.0	334.56	M5e	M6e	M6-M9
S Scl	6.3–13.4	365.53	M6e	M6e	M6.5-M9
W Vel	9.5–15	394.45	M7e	M7e	M7-M8

Note: GCVS (Kukarkin and Parenago 1970), Fox (Fox et al. 1984), and CG (Crowe and Garrison, 1988).

Table 6.7: A List of main S-type (Se) and C-type (Ce) Mira variables

			Spectral type	
Name of stars	Magnitude M_{pg}	Period (days)	GCVS	Knapp 1985
R And	5.0–15.3	408.87	Se	S4.6e-8.8e
R Cyg	5.6–14.6	425.44	Se	S3,9
W And	6.5–14.3	397.05	M8e	S6.1-8.2e
V Hya	6.0–12.5	532	N	C6.3e
T Dra	7.2–13.9	432.02	Ne	C6.2e–8.3e
V Cyg	6.8–13.8	419.64	Ne	C7.4e

Note: Magnitude, period, spectral type are taken from GCVS (1970).

Formation of emission-lines in Mira-type variables

In Mira variables, Balmer lines (Hα–H10) and neutral metallic lines (SiIλ 4102.95 Å, MgIλ λ 3829.32, 3832.30, 3838.29, 4571.10 Å, FeIλ λ 4202.03, 4307.91 Å, etc.) appear in emission near the maximum light. Intensities and profiles vary not only with phase but also with different pulsational cycles even in the same phase. Examples on Balmer lines are shown in Figure 6.17(a)

Figure 6.17: The phase change of the Balmer emission of the Mira variable S Scl (M6e). Panel A: line profiles of Hγ, Hδ, Hζ, and Hη, as a function of phase. The flux F_r is plotted vertically, with the zero level for each spectrum being indicated by a tick mark on the ordinate. The phase of the observation is indicated near each tick mark. Phase 0.0 corresponds to the first light maximum observed. The phase 4.22 for the upper most profiles denotes the phase +0.22 at different cycle. The abscissa of panel A denotes the line-of-sight velocity relative to the stellar photosphere, and plus sign denotes the outflow, negative signs the infalling motions. Panel B: absolute Balmer line flux plotted against the phase in S Scl. (From Fox et al. 1984)

(a)

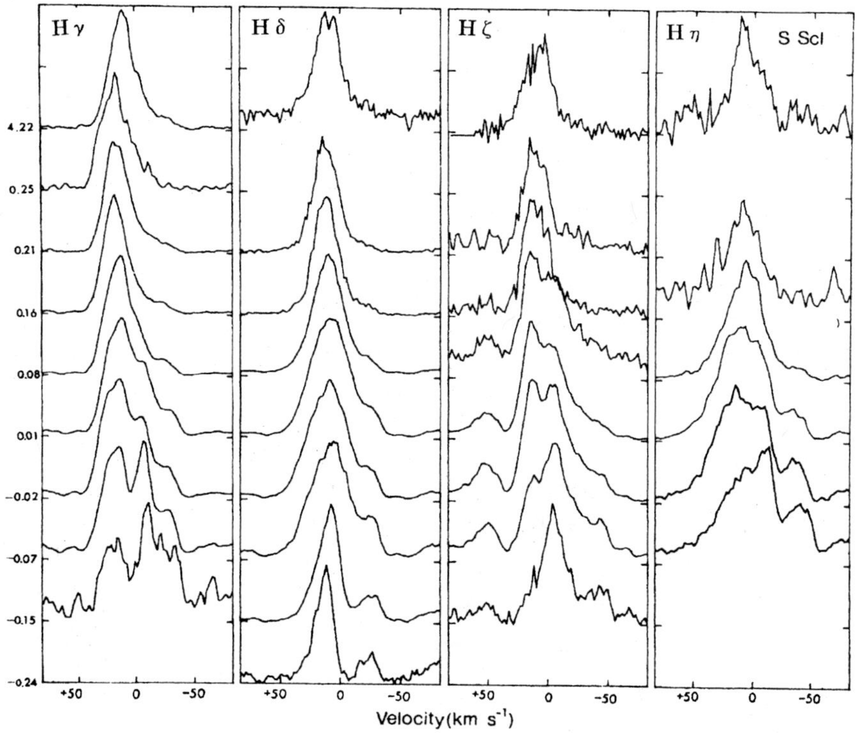
H γ
H δ
H ζ
H η
S Scl
4.22
0.25
0.21
0.16
0.08
0.01
−0.02
−0.07
−0.15
−0.24
F_λ
+50
0
−50
Velocity(km s^{-1})

(b)

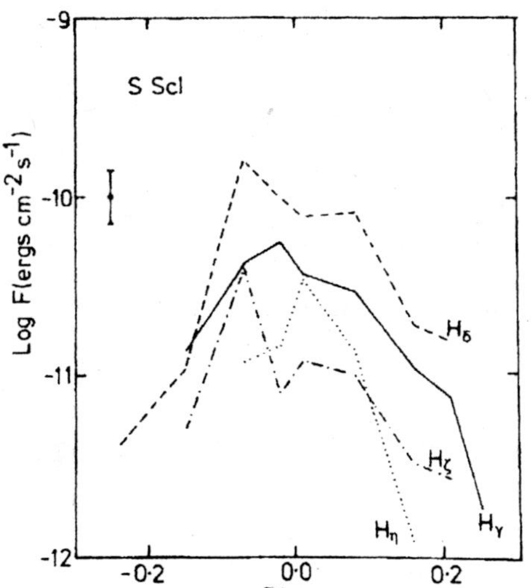
S Scl
-9
-10
-11
-12
Log F(ergs cm^{-2}s^{-1})
H_δ
H_ζ
H_η
H_γ
-0·2
0·0
0·2
Phase

Table 6.8: Averaged Balmer decrement for Mira variables as compared to the nebular decrement of Case B (Te = 10,000 K) (adapted from Fox et al. 1984)

	Hγ	Hδ	Hζ	Hη
Mira variables (mean)	0.7	1.0	0.7	0.4
Nebula (Case B, Te = 10,000 K)	1.81	1.00	0.39	0.28

and (b) as observed by Fox et al. (1984) at Mt. Stromlo Observatory. Each panel of Figure 6.17(a) exhibits the phase variations of emission-line profiles and intensities in the near-maximum phase (phase −0.25 to +0.25, phase 0 corresponds the maximum light) for the Mira variable S Scl (M6e). The abscissa of Panel (a) denotes the line-of-sight velocity relative to the stellar photosphere. Plus sign denotes the outflow, while minus sign denotes the infalling motion. The ordinates of Panel (a) indicate the emission-line flux, with the scale corresponding to the level of continuum. In addition, the phase of the observation date is given on the zero level of the each line profile. Panel (b) exhibits the phase variations of Balmer emission lines. It can be seen that the emission lines exhibit maximum intensity near the light maximum. The variations of line profiles in Panel (a) show that the profiles in the early phase (~ -0.02) are extended upon both sides to around 40 km s^{-1}, making a double-peaked like profiles. After the light maximum, the line centers gradually shift toward plus velocities, inferring the predominance of outward gas motion.

The mean Balmer decrement (Hγ: Hδ: Hζ: Hη) measured by Fox et al. (1984) is shown in Table 6.8 for some Mira variables, along with the standard nebular decrement for comparison in its very flat nature.

It is generally accepted that the emission lines in Mira variables are formed by shock waves. Atmosphere of Mira variables consists of a normal pulsating atmosphere and a shock powered expanding outer atmosphere, and the Hα emission is supposed to be formed in the relaxation layer behind the shock waves.

Fox and Wood (1985) considered the shock structure in the Mira type stellar atmospheres. They supposed a one-dimensional stationary shock propagating in the Mira-type atmosphere. Shock waves are composed of the precursor, shock front (discontinuous surface), and the relaxation layer. The basic equations are the conservations of mass, momentum, and energy, and these equations are combined with the equation of radiative transfer taking into account the effects of ionization and dissociation. The system was numerically solved to derive the structure of shock waves and emission-line intensities. Calculations have been carried out by taking three parameters: T_o, ρ_o, the temperature and density of the undisturbed gas, and the velocity of the shock front, v_0. The models are designated as B, D, I by the following combinations of parameters:

Model	T_0(K)	ρ_0(g cm^{-3})	v_0(km s^{-1})
B	2000	10^{-10}	20, 30, 40
D	3000	10^{-10}	20, 30, 40
I	3000	10^{-12}	20, 30, 40

If we take the model D as typical, then model B denotes the low-temperature model, and model I is the low-density model. The structure of shock waves for models D40 and B40, where the number 40 indicates the velocity v_0 (km s^{-1}), are shown in Figure 6.18A and B. The abscissa is the logarithmic distance (cm) behind the shock front. The four panels for each model show the physical parameters top to bottom in this order: first panel (T, ρ, v), second panel (α = the fraction of hydrogen atoms in ionized states, while L^+ and L^- are the up- and downstream Lyman continuum photon fluxes), third panel (2 and 6 are the population densities N_2 and N_6 of hydrogen in the state $n = 2$ and $n = 6$ levels) and fourth panel (α^+, δ^+ = upstream photon fluxes of Hα, Hδ, and α^-, δ^- = down streams of the respective photons). In each panel the scale of the ordinate is taken from 0 to 1 and the maximum ordinate values are given at the foot of Figure 6.18.

The results of model calculations for the relative emission lines of Hδ/Hγ, Hζ/Hδ, and Hη/Hδ are shown in Figure 6.19 for models B (low temperature), D (typical), and I (low density). Relative intensities are plotted for three different velocities of the shock front. In model D, the relative intensities nearly approach nebular approximation (Case B) for the higher shock velocities. When shock velocity is small, the relative intensities show large scatter, while for high-velocity shock, the scatter gets smaller. Among these models, it is only model I that shows the sufficiently flat decrement I (Hδ/H$\gamma > 1$, Hζ/Hδ, and Hη/H$\delta > 1$) in case of low-shock velocity ($\sim$20 km s^{-1}). In model D it is difficult to get the flat decrement for any shock velocities. In Fox and Wood's model, flat or inverted decrements are realized in very limited cases of parameter selection.

Gillet (1988) has carried out high-resolution spectrophotometric observations of the bright and hot Mira star S Car and obtained the Balmer emission (Hα–H12) at the light maximum. The Hα and Hβ lines show double-emission structure whereas Hγ–H12 present strong asymmetric profiles. These emissions are interpreted in the framework of a single spherical shock, outward propagating above the photosphere. He calculated the line profiles and relative intensities under two models that he designated geometric model and self-reversal model. They are characterized as

(1) Geometric model. In this model the spherical shock propagates far from the stellar photosphere. The emission lines are formed in the relaxation zone behind the shock front that is divided into two hemispheres, one is advancing toward an observer and the other is receding. By this geometric effect, complicated line profiles are formed.

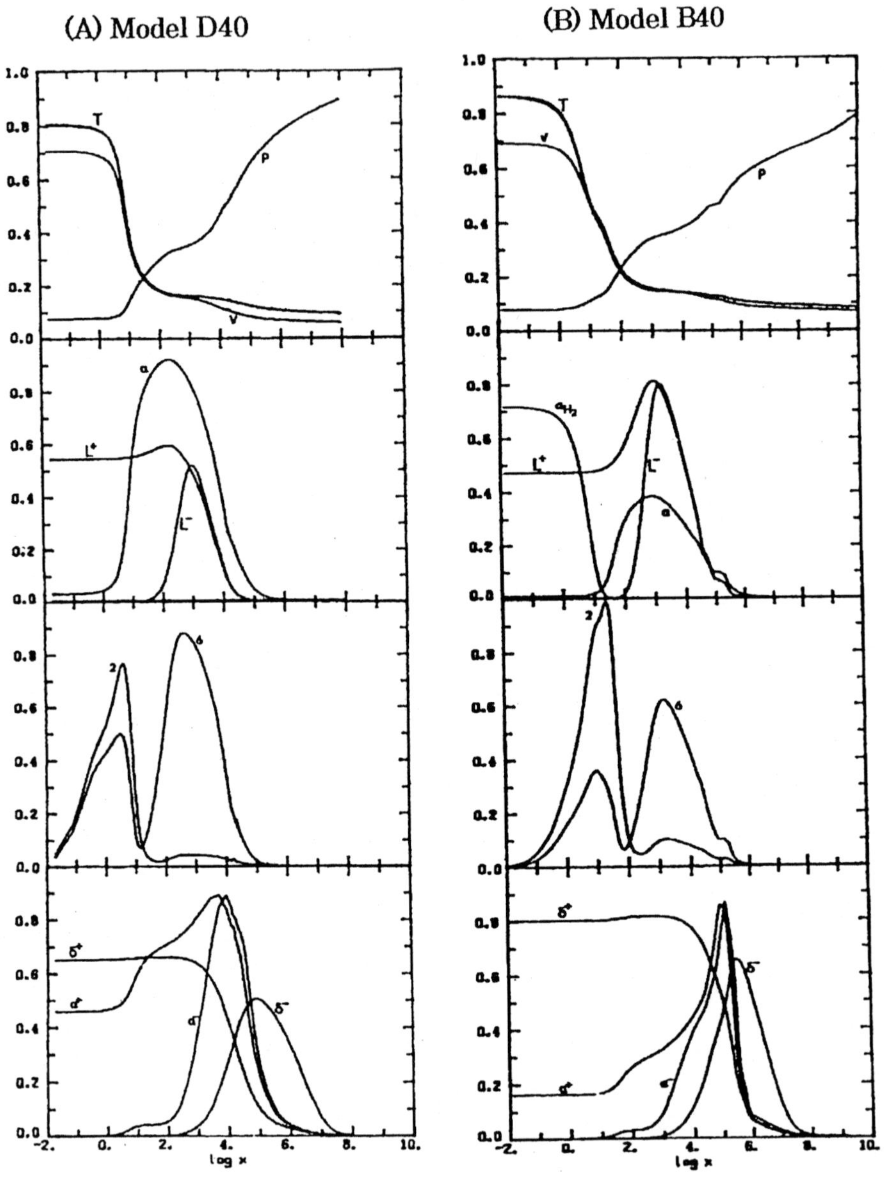

Figure 6.18: Theoretical models of shock waves in the Mira variables (see text for each of the curves) (from Fox and Wood, 1985). Model D40 (the maximum ordinate values): Panel 1 (T = 60,000 K, v = 15 km s^{-1}, $\rho = 5 \times 10^{-9}$ g cm^{-2}); panel 2 (α = 0.5, L = 5×10^{18} erg cm^{-2} s^{-1}); panel 3 ($N_2 = 3 \times 10^{11}$ cm^{-3}, $N_6 = 3 \times 10^{9}$ cm^{-3}); panel 4 (F(Hα) = 10^{20} cm^{-2} s^{-1}, F(Hδ) = 3×10^{19} cm^{-2} s^{-1}). Model B40: Panel 1 (T = 60,000 K, v = 15 km s^{-1}, $\rho = 5 \times 10^{-9}$ g cm^{-2}); panel 2 (α = 0.5, L = 5×10^{17} erg cm^{-2} s^{-1}); panel 3 ($N_2 = 5 \times 10^{10}$ cm^{-3}, $N_6 = 10^{9}$ cm^{-3}); panel 4 (F(Hα) = 10^{20} cm^{-2} s^{-1}, F(Hδ) = 2×10^{19} cm^{-2} s^{-1}).

Figure 6.19: The flux ratios of Hδ/Hγ, Hζ/Hδ, and Hη/Hδ for the models B, D, and I. The ratios for Case B nebular approximation are also shown, together with mean observed ratios near maximum light for Mira and for S Car in particular. (From Fox and Wood 1985)

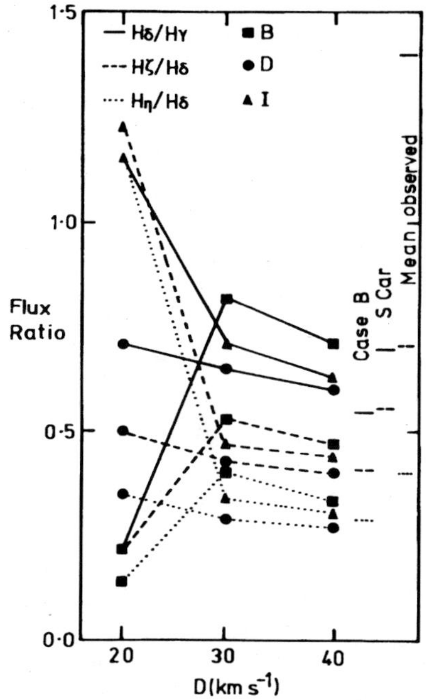

(2) Self-reversal model. This model assumes that the spherical shock wave is propagating just above the photosphere so that there is a large amount of gas located above the shock in the precursor region. By this optically thick gas, hydrogen suffers self-reversal to make the complicated line profiles.

The relative intensities of the Balmer emission lines relative to $H\beta$ based on the above two models are given in Table. 6.9. In both cases the highly inverted Balmer decrements are similarly obtained in spite of large difference in the models. For comparison, the observed decrement for S Car at light maximum is given. It is to be mentioned that the hydrogen gas in both

Table 6.9: Model calculations of the Balmer decrements of Mira variable and observed decrement (adapted from Gillet 1988)

Balmer Decrement	Hα/Hβ	Hγ/Hβ	Hδ/Hβ
Geometrical model	0.44	1.33	—
Self-reversal model	0.60	1.27	—
S Car (at light maxim)	0.45	1.17	1.52

models must be sufficiently heated by shock waves in order to get the inverted decrements.

Dynamical structure of the atmospheres of Mira variables has been considered by numerical modeling by Bowen (1988). The behavior of periodic shock waves has been derived such as seen in Figure 3.8. Based on Bowen's model, Luttermoser and Bowen (1992) reproduced the inverse Balmer decrement in the order of $F(\mathrm{H}\alpha) < F(\mathrm{H}\beta) < F(\mathrm{H}\gamma)$ in the Mira variable S Car. On this star Luttermoser and Castelaz (2003) obtained a far ultraviolet spectrum using the FUSE satellite (Far Ultraviolet Spectroscopic Explorer, wavelength range 905–1187 Å) and found that the Balmer lines are formed in the inner, hot shocks, whereas the Lyman lines are formed in the outer shocked region, showing the decrement of $F(\mathrm{L}\beta) > F(\mathrm{L}\gamma) > F(\mathrm{L}\delta)$. The different behavior in the decrements of Lyman and Balmer lines remains as a problem.

Binary systems in the Mira-type variables

In the Mira-type variables, the rate of occurrence of binary systems is quite low as compared to other types of stars. The well-known Mira Ceti (o Cet) is almost the only example. The Mira variables are supposed to be the precursors to planetary nebulae, many of which have binary systems so that the presence of binary companions is expected in many Mira variables. To search for such companions, Shawl and Bord (1991) have carried out spectroscopic observations with the 1-m and 1.5-m telescopes at CTIO (Cerro Tololo Inter-american Observatory) for more than dozens of Mira variables near minimum light. But no obvious companions were detected. At present, the discovery of Mira companions is highly anticipated.

The binary nature of Mira Ceti was first discovered by Joy (1954). The angular distance is as small as 0.6 arcsec, and the companion itself is very faint, with observations being limited in its minimum light. In spite of many attempts, the nature of the companion is still an enigma. Since the spectrum exhibits many emission lines in Balmer lines, with P Cygni profiles, and in FeII and CaII, etc. (Yamashita and Maehara 1978), the companion (Mira B) was first supposed to be a Be star with an outflowing envelope. However, Yamashita and Maehara (1979) have suggested that a Be star as a companion would be too faint and it might be a white dwarf showing the Be star phenomena through accretion flow from the primary. Subsequently, Chembers (1995) suggested white dwarf, while Jura and Helfand (1984) and Karovska et al. (1996) suggested the B type star. In addition, Karovska (1992) suggested the existence of the third body from orbital analysis. Thus the nature of Mira B still remains unsolved.

The ultraviolet spectra of Mira B have been obtained by spectrographs on board IUE, HST, and FUSE satellites (Wood et al. 2002, Wood and Karovska 2003). Remarkable behaviors found by UV observations are (1) existence of accretion flow onto Mira B, implying that the optical and UV emission might be entirely from the accretion rather than the star itself; (2) the P Cyg profiles of the Mg II lines suggest that the accretion flow drives a warm and fast outflow; and (3) the UV spectrum showed dramatic changes in the continuum and line emissions during the spans of observations, suggesting large variation of accretion rate onto Mira B.

6.4 Eclipsing binary systems

In close binary systems there are many different groups showing a strong emission-line phenomena. In this book we divide the close binary systems into four groups: eclipsing binaries, RS CVn stars, CV (including novae) and symbiotic stars. We shall consider them mainly from the point of view of emission-line formation.

The basic properties of close binary systems are determined by the size and gas density of the Roch lobes, for which there are three types of configurations: detached, semidetached, and contact systems, as shown in Figure 6.20 (Kopal 1959).

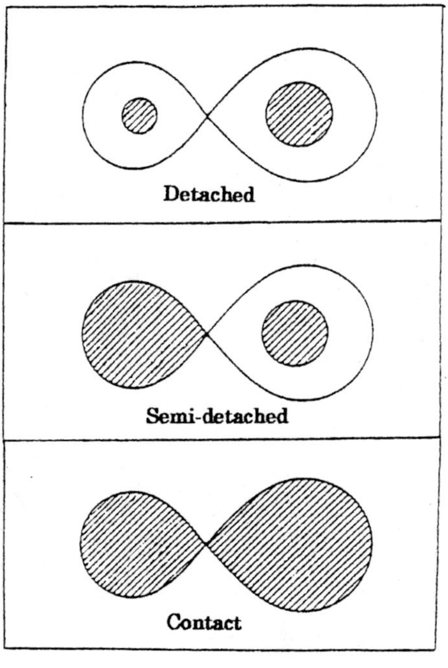

Figure 6.20: Roche lobes and types of close binaries.

6.4.1 Algol-type eclipsing binary systems

Often, there is confusion in terminology of concerning Algol-type binary systems and Algol-light curve binary systems. Historically, Algol-type light curves represent light curves with well-defined starting and ending times of eclipses, and nearly constant brightness outside the eclipses. This notion does not concern with the configuration of the binary system. Statistically, most of the eclipsing systems with Algol light curves turns out to be detached systems while only a small fraction of them are having semidetached configuration.

According to classification given in Figure 6.20, the Algol-type system belongs to the semidetached type. Most of their primaries are B or early A type main-sequence stars and their companions are G or K giants or subgiants fulfilling the Roche lobe.

Furthermore, semidetached binaries are classified into three cases, A, B, and C, according to their evolutionary stage. This is distinguished by the point in time when the initial massive primary fulfils its Roch lobe on the evolutionary path.

Case A. The primary fulfils its Roche lobe during the main sequence (hydrogen-core burning stage).

Case B. The primary fulfils its Roche lobe on the way to the giant stage (hydrogen shell burning stage), after leaving the main-sequence stage.

Case C. The primary fulfils the Roche lobe after reaching the helium-core and hydrogen-shell burning stage, when the primary already has arrived at the red-giant stage.

The names cases A and B were given by Kippenhahn and Weigert (1967) and case C was named by Lauterborn (1969). Statistically, most Algol-type systems found belong to case B.

In the General Catalogue of Variable Stars (GCVS) (Kukarkin and Parenago 1970), eclipsing binary systems were classified by only shape of their light curves. The classification does not convey the notion of a specific configuration/structure of a binary system. The GCVS divide the eclipsing systems into the three groups: EA, EB, and EW. The light curves of EA systems have long duration of constant or nearly constant out-of-eclipse brightness. The shape of the light curves is very similar to those of Algol so that this group should be designated as Algol-type light-curve systems instead of commonly called "Algol systems"! Most of the EA are found to be detached systems. Only a few systems are found to be semidetached Algol systems. Nevertheless, the name of Algol system is widely used to express the semidetached eclipsing systems. Budding (1984) prepared a catalogue of 414 Algol systems (as a group of EA) and their candidates.

The light curves of EW systems, W UMa systems, show continuous out-of-eclipse light variations that are close to sinusoidal shape. They also have equal or near equal eclipse depths. Their light curves are similar to that W UMa. Most EW systems are contact systems and a small fraction is found to be semidetached. Historically, contact systems were found only in late-type systems with F or G type components. In recent decades early-type contact systems are also found to exist (Leung 1992).

6.4.2 Formation of emission-lines in Algol-type systems

The occurrence of emission lines in Algol systems more or less depends upon the periods and phases. Plavec and Polidan (1976) surveyed the occurrence rate of the Hα emission for 46 systems. In total, they detected emission in more than 50% of the stars, with the emission ordinarily appearing near the light maximum or light minimum in their orbital motions. The Hα emission tends to appear in longer orbital period systems. Peters (1980) has found that the rate of emission appearance is higher for the Algol systems having orbital periods longer than 6 days. According to Plavec and Polidan's (1976) data, the detection rate is as given in Table 6.10.

For short-period Algol systems, Kaitchuck and Honeycutt (1982) surveyed the presence of emission lines for 18 systems with orbital periods shorter than 4 days around the epoch of eclipses. Among the 42 eclipses observed, emission lines in $H\beta$–Hγ were detected only 15 times for three stars, and thence the detection rate was 16.7%. Comparing the short-period systems ($P < 4.5$ days) with the long-period systems ($P > 5$ days), Kaitchuck et al. (1985) suggested that in long-period systems the gas stream from the secondary moves around the trailing hemisphere of the primary and forms a permanent disk that forms emission lines. In contrast, the gas stream of the short-period systems directly

Table 6.10: Detection rate of the Hα emission in Algol systems based on the data of Plavec and Polidan (1976)

Orbital period (days)	<5	5–10	10–20	20	
Number of Stars					Total
E_m	8	5	11	2	26
A_b	13	3	0	1	17
Total	21	8	11	3	43
$E_m/(E_m + A_b)$	0.38	0.62	1.00	0.66	0.60

Note: E_m, H$_\alpha$ emission detected at maximum, minimum, or both phase; A_b, H_α emission is not detected.

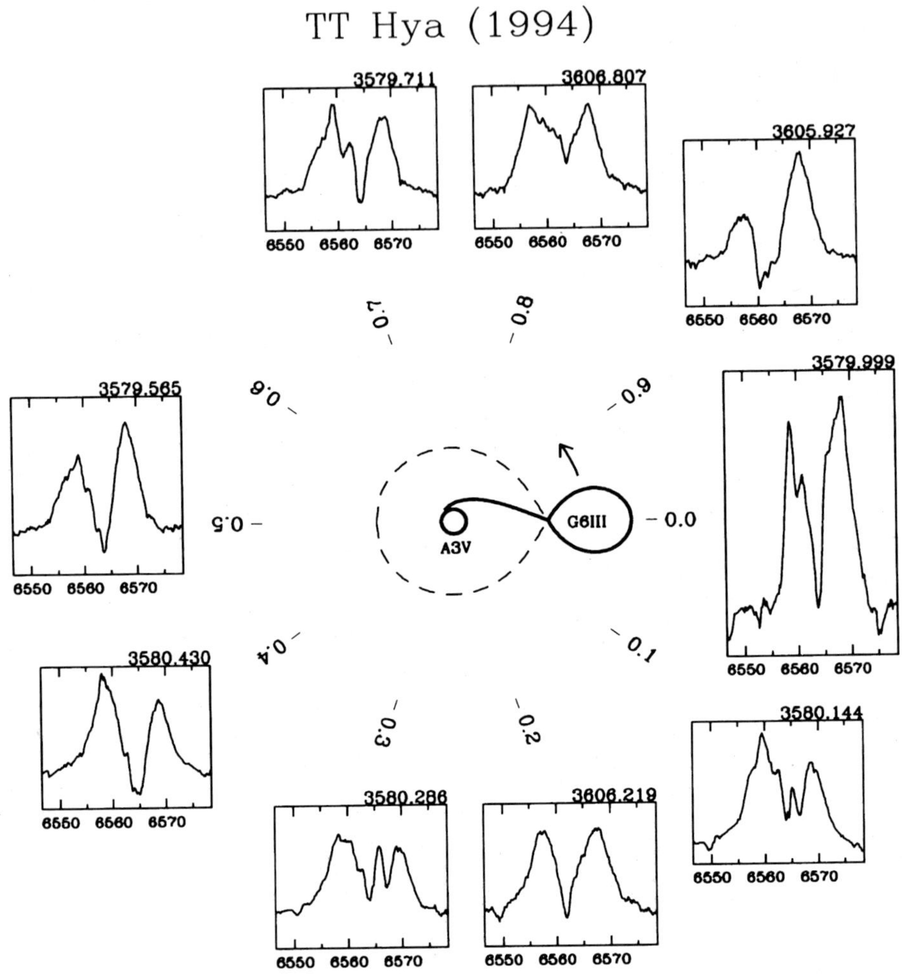

Figure 6.21: The phase variation of the Hα line profile for the long-period Algol system TT Hya ($P = 6.953429$ days, $V = 7.5$–9.5 magnitude). (From Richards and Albright 1999)

strike the primary stars, and the disk, if formed, is transient and unstable, making the formation of emission lines difficult.

Richards and Albright (1999) carried out time-resolved, high-dispersion spectroscopic observations for 18 Algols in the orbital period $P = 1.18$–11.11 days, and studied the profile variations of the Hα line. In order to determine the general morphologies and physical structure of the accretion regions in Algol systems, they derived the difference profiles by subtracting a composite theoretical photospheric spectrum from the observed spectrum. The phase variation of Hα difference profile is shown in Figure 6.21 for the long-period

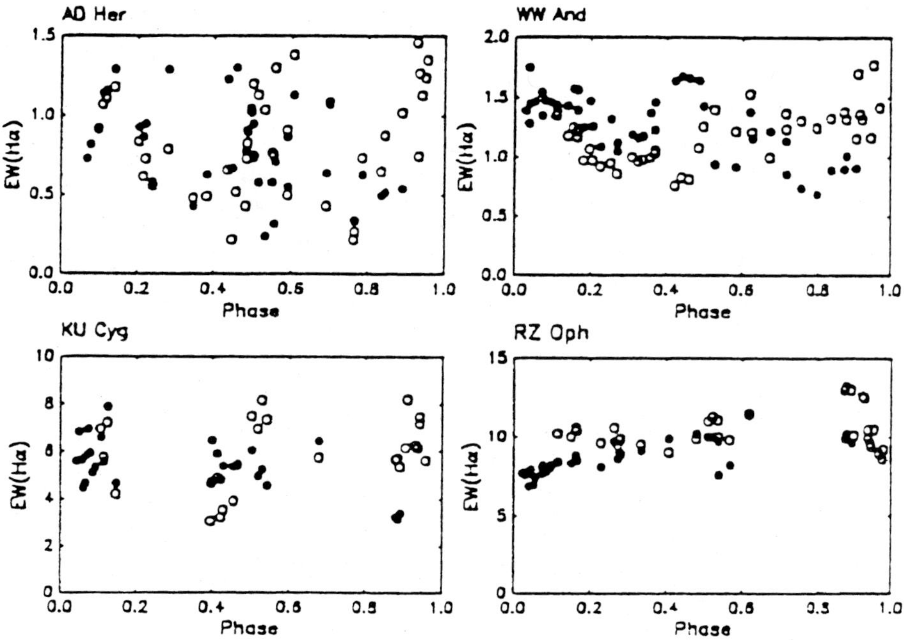

Figure 6.22: Hα EW variation as a function of orbital phase in four Algol systems. The abscissa is the orbital phase, and the ordinate is the EW of Hα for red peak (○) and violet peak (•) in double-peaked profiles (from Olson and Etzel 1995). The data for each star are given below by Budding (1984):

Name	Orbital period (d)	Spectral type	Mass of the primary $M_\odot$	Mass ratio
AD Her	9.7666	A4V + K2III	3.93	0.40
WW And	23.2853	A5 + (F6IV)	2.92	0.30
KU Cyg	38.4393	F0peIII + K5eIII	2.81	1.13
RZ Oph	261.943	F3eIV + K5II	18.46	0.60

Algol TT Hya (P = 6.95 days). The long-period systems ($P > 6$ days) exhibit strong emission lines, mostly double peaked, at all epochs indicating the existence of accretion disks slightly variable but permanent, similar to those found in CV (Section 6.6). Double peak may show the effect of rotation of the disk. Contrarily, the short-period systems predominantly show single-peaked emission with blue or red shifted centroid of emission along the orbital phase. Richard and Albright explained this feature by the existence of accretion flows between two stars.

The long-period Algols ($P \gtrsim 10$ days) generally exhibit double-peaked *Hα* emission with variable profiles along the orbital phase. In Figure 6.22 the profile changes observed by Olson and Etzel (1995) are shown for four stars arranged in the order of increasing orbital period, and the emission intensities are separately shown for red (open circle) and violet (filled circle) peaks. The equivalent widths (EW) of both peaks are plotted as a function of the orbital phase. Emission lines of both red and violet components are seen in all stars. It can be seen that the shorter the periods, the larger the variations. The phase dependence is also shown in Figure 6.22. AD Her, the shortest period star, has almost no phase dependence. The strongest correlation is seen in WW And. In this star, the violet peak becomes strongest in phase 0.45 (the primary passes in front of the secondary at phase 0.5), and the red peak becomes the maximum strength at phase 0.95, just before the secondary comes to the front. This change can be explained by an asymmetric gas flow from secondary to the primary.

We have already stated on their similarity between Algol systems and Be stars in Section 5.3.3. Looking at Be stars, Be binaries with orbital periods shorter than 30 days show close similarity to Algol systems. Those with orbital period longer than 30 days statistically occupy the highest rotational velocity $V \sin i$ among B stars, and exhibit different behavior than nonemission B binaries (see Figure 5.18). On the other side, orbital periods of the Algol systems reach as long as 300 days and most of long-period Algols exhibit emission lines almost permanently. Thus clear distinction between Algol systems and Be stars is generally difficult.

6.4.3 Binary system with an atmospheric eclipse

Among the eclipsing binaries, there are some remarkable systems that show long lasting eclipses of more than 100 days. They are composed of a cool giant or supergiant star and a hot companion, which periodically passes behind the larger star, and shines through the much extended, rarefied atmosphere for a long time. Such eclipses give us opportunities to observe the time-sequential change of absorption lines produced by the cool extended atmosphere and then enable us to study their stratified structure. These are called "atmospheric eclipses" and the main systems are collected in Table 6.11 (Hack 1992).

The binary systems with atmospheric eclipses can be classified into three categories: The first is the ζ Aur group, containing ζ Aur, 31 Cyg, and 32 Cyg in Table 6.11. In these systems, the eclipses occur when hot B type main-sequence stars revolve behind the G or K type supergiant components. The second is the VV Cep group which are composed of M-type supergiants and A or B type stars. The third group, ε Aur is an extremely peculiar system where an early-type supergiant suffers a long total eclipse of around 330 days by its unseen companion. Since the ingress and egress take 140 days each, the total length of eclipse reaches 610 days. The spectrum of the companion cannot be obtained so the nature of the companion is still unknown. Hack (1992) suggested that the companion might be an early-type star surrounded by an optically thick disk or ring.

Table 6.11: Binary systems with atmospheric eclipses (adapted from Hack 1992)

Name	Magnitude (m_{pg})	Orbital period (days)	Spectral type	Mass ($M_\odot$)	Radius ($R_\odot$)
22 Vul	5.7v	249.111	(1) G7 II	5.2	40
			(2) B9 V	3.0	2.9
ζ Aur	5.0–5.6	972.162	(1) K4 Ib	8	154
			(2) B8 V*	5.8	4.3
32(o^2)Cyg	5.6–5.8	1147.8	(1) K5 Iab	19:	215
			(2) B4 IV-V	10:	3
31(o^1)Cyg	3.8–3.9	3784.3	(1) K1 Ib	9.2	169
			(2) B4 V	4.2	3.2
VV Cep	6.6–7.4	7450	(1) M2 Ib	—	1600
			(2) A0II*	—	13
ε Aur	3.7–4.5	9890	(1) A9 Ia	—	214
			(2) B:(binary?)		
			+ disk or ring	—	1221

Note: Spectral type is given as (1) cool supergiant, and (2) hot companion. The spectral type marked * has another designation given as follows: ζ Aur, the spectral type of companion is sometimes classified as B4-5; VV Cep, classification A0II for companion is due to the ultraviolet luminosity (Hack et al. 1989). Emission lines suggest early Be stars.

We now consider VV Cep that exhibits remarkable emission lines. The optical spectrum of VV Cep can be distinguished by the following components (Kawabata et al. 1981):

(a) the absorption spectrum of the primary (M type supergiant)
(b) Balmer emission from the ionized region around the hot companion (B star)
(c) forbidden lines formed in a common envelope
(d) stellar wind spectrum observed from the lines of the UV emission and highly ionized ions

In Figure 6.23 the phase change of the Hα profiles in VV Cep observed at the Okayama Astrophysical Observatory is shown. The figure is composed of (a) 8 panels (mainly before the light minimum) and (b) 8 panels (after the light minimum). The dates of observations are given in each panel and the numeral under the date yields the époque of observations (in days, and $\pm$ denotes before ($-$) and after ($+$) the light minimum). It is noted that this star needs 104 days for the ingress and 205 days for the egress. As seen in this figure, the Hα profile shows a strong double peak with violet stronger ($V/R > 1$) off the eclipse. Near the light minimum, both component (V and R) markedly weaken, with the V component first rapidly declining in the ingress then recovering earlier to a high level in the egress. From these observed profile

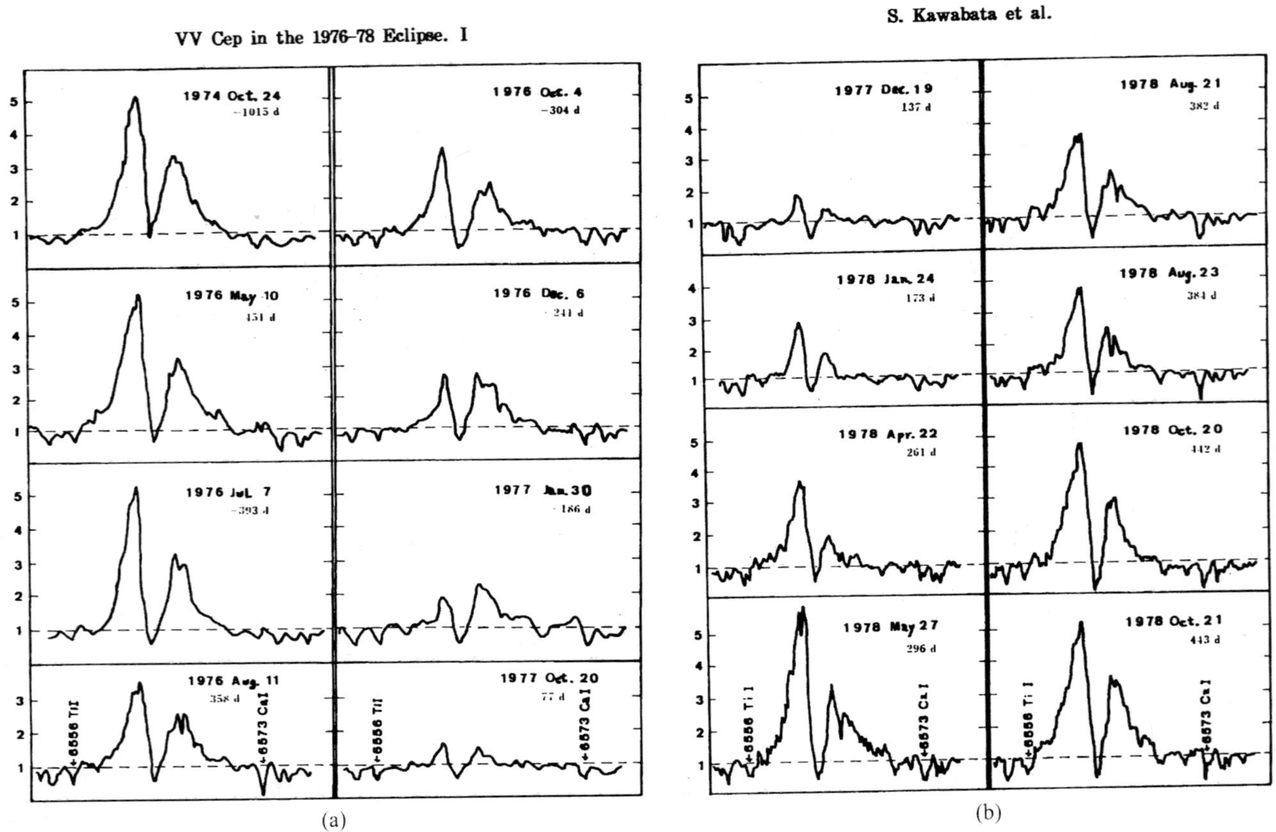

Figure 6.23: Phase change of the $H\alpha$ profile of VV Cep for the phases outside eclipse (phase −1015 days approximately), in ingress (phase −451 to −186 days), on totality (phase 77–137 days) and in egress (phase 173–443 days). The numeral under the date denotes the days from the light minimum (−before, +after). (From Kawabata et al. 1981)

variations Saijo (1981) proposed a rotating disk model surrounding the B star with radius around $500R_\odot$ where the Hα emission is produced.

VV Cep also exhibits remarkable forbidden lines in the optical spectra. Kawabata and Saito (1997) identified many forbidden lines in the wavelength range 3500−5600 Å in spectra obtained from 1976 through 1984 during the eclipse and outside. Main lines are [FeII] (13 lines), [NiII] (3 lines), [CuII] (1 line), etc., and the equivalent widths are in the order of 0.1–1 Å. The line-of-sight velocities of these forbidden lines are almost independent of the orbital motions of the system, showing an expanding motion with an average velocity around 20 km s^{-1}. By making use of the values of mean density and radial velocity, the mass-loss rate is estimated as $10^{-6}M_\odot$ per year.

The relative sizes of the M supergiant and the envelope of B stars are schematically shown in Figure 6.24, along with the outline of the orbit. The

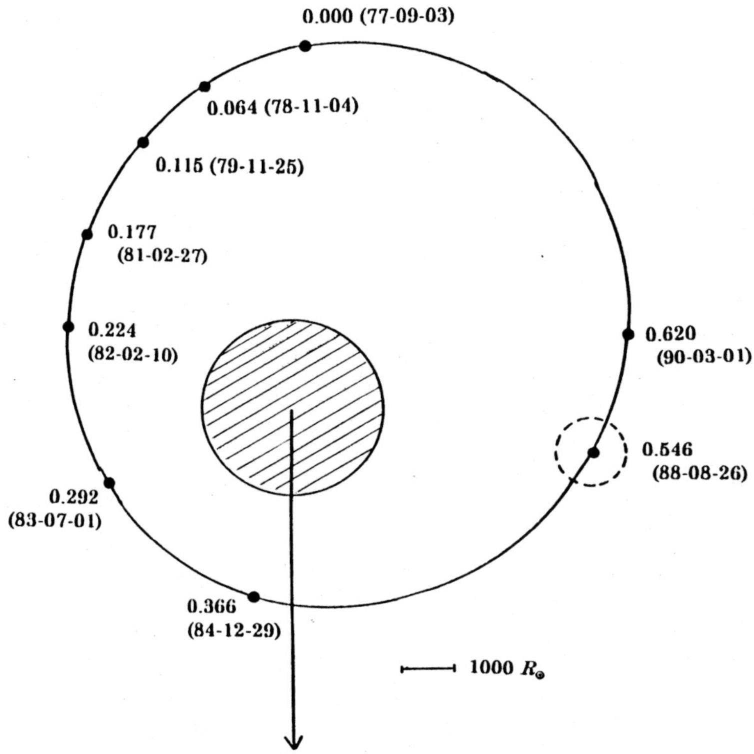

Figure 6.24: Schematic diagram of VV Cep with relative sizes of the M supergiant and the envelope of the B star, along with the outline of the orbit. The size of the B star itself is much smaller than the mark on the orbit. Approximate scale is given in the lower corner. The arrow from the M star denotes the direction of the observer. The marks on the orbit indicate the phase (phase 0.0 is the light minimum) and the date in brackets. On the mark (88-08-26) the relative size of the envelope of B star is shown (Based on Bauer et al. 1991).

size of the B star itself is much smaller than the mark. The marks on the orbit denote the phase and date. Phase 0.0 is corresponding to the light minimum.

The first IUE observations of VV Cep was made at the time of egress during the 1976–1978 eclipse (Hagen et al. 1980). Since then a number of IUE observations have been accumulated, though mostly outside the eclipse (Hack et al. 1989, Stencel et al. 1993, Bauer and Bennett 2000). Hack et al. (1989) using the high dispersion spectra, made the identification of spectral lines, and Bauer and Bennett (2000) classified the lines into the following five categories:

(1) Emission lines. Most conspicuous are MgII h and k which exhibit strong double-peaked profiles with strong red components (R $>$ V). This is just the opposite of $H\alpha$ line (see Figure 6.23).
(2) Absorption lines from neutral atoms (e.g., CI, NI, Cl I). Many of them disappear as the hot component emerges from the eclipse.
(3) Broad absorption from highly ionized ions (e.g., SiIV, CIV, FeIII). They arise from the hot component or nearby hot circumstellar material.
(4) "Shell" absorption lines with variable profiles. Most of the lines arise from singly ionized iron-group elements. Because of the rapid variation, shell-absorption system must occur in material nearby (and probably accreting into) the hot component.
(5) Narrow circumstellar absorption lines. These lines are likely formed in the extended circumstellar envelope of the M supergiant.

Based on the analysis of the IUE spectra, Stencel et al. (1993) suggested that the primary component might be fully filling the Roche lobe near periastron, and trigger mass transfer and accretion into the B star companion. However, if the companion is an early-type B star with a wind, a wind–wind collision may occur and cause some extra output in emission lines. At present, reconciling the optical and IUE data to build a more reliable model is still in its early stages.

6.5 RS Canes Venatici (RS CVn) type stars

RS CVn stars are detached binary systems with at least one late-type component showing CaII H and K emission in the spectra. The strength of these emission lines is an indicator of chromospheric activities. The $H\alpha$ emission is often strong, and chromospheric activities are also linked with coronal activities emitting strong X-ray and radio waves. The spectral types of the components cover the wide range of F–M, and the so-called RS CVn stars are classified more specifically into two types:

(1) RS CVn star—the binary systems composed of a G-type main sequence and an evolved F-type subgiant.
(2) BY Dra type—the systems including K and M type dwarfs.

Table 6.12: A list of emission-line stars of RS CVn types, selected from CABS (adapted from Strassmeier et al. 1993)

CABS	Name	Maximum brightness V_{max}	Spectral type	Orbital P. (days)	Group	X-ray intensity ($\times 10^{31}$ erg s^{-1})
6	FF And	10.38	dM1e + dM1e	2.170 day	1	0.034
10	AY Cet	5.47	WD + G5 III	56.815	3	1.001
19	CC Eri	8.76	K7Ve + dM4	1.5614	1	<0.0001
25	LX Per	8.14	G0 IV + K0 IV	8.0382	2	0.589
29	V711 Tau	5.7	G5 IV + K1 IV	2.8377	2	2.377
42	V833 Tau	8.16	* + dK5e	1.7877	—	0.042
71	YY Gem	9.07	dM1e + dM1e	0.8143	1	0.026
73	σ Gem	4.14	* + K1 III	19.140	3	2.67
94	DF UMa	10.14	dM0e + [dM5]	1.0338	1	0.017
109	RS CVn	7.93	F5 IV + G9 IV	4.797	2	1.186
137	ε UMi	4.23	A8-F0V + G5 III	39.4809	3	0.458
156	BY Dra	8.07	K4 V + K7.5 V	5.975	1	0.028
178	V1396 Cyg	0.13	M2 V + M4 Ve	3.2761	1	0.017
189	RT Lac	8.84	G5: + G9 IV	5.0740	—	0.416
195	FK Aqr	9.05	dM2e + dM3e	4.0832	1	0.013
202	SZ Psc	7.2	F8 IV + K1 IV	3.9658	2	2.01

Note: Grouping by Montes et al. (1996): 1, dwarf + dwarf; 2; pair of subgiants; 3, others (single-lined binary).

Strassmeier et al. (1993) compiled a catalogue of the 206 binary systems as the Chromospherically Active Binary System (CABS) catalogue. This catalogue contains both types, and summarizes information on the photometric, spectroscopic, orbital, and other physical parameters. Some RS CVn and BY Dra stars, selected from the CABS catalogue, are listed in Table 6.12. Grouping by Montes et al. (1996) shown below is also given in the table.

Montes et al. (1996) divided their 73 sample stars in three groups according to the assigned luminosity class of the active components.

Group 1: Both components are dwarf stars (21 systems). When both components have the same or very similar spectral type, the observed emissions are also very similar (YY Gem, BF Lyn, AS Dra, etc.). However, when the components of the system have different spectral types, the hot component tend to be more active (DH Leo, BY Dra, KT Peg, etc.), or even only hot components are active (MS Ser, V815 Her, V775 Her, etc.).

Group 2: Two-subgiant systems (23 systems). In most cases the cool component is responsible for the activity. In 7 systems, CaII H and K emissions are observed from both components, although the cool component tends to be more active in the system (RS CVn, V711 Tau, etc.).

Group 3: Mainly single-lined binaries with line emissions in CaII H and K (29 systems).

There are also other systems where the hot component is an A-type star (RZ Eri, 93 Leo, ε UMi), or a white dwarf star (AY Cet, DR Dra).

The degree of chromospheric activities is expected to be dependent on the binary interaction, stellar effective temperature, and stellar rotational velocity. By taking the CaII H and K emission intensity as an index of activity, Montes et al. (1996) examined the correlation on the above parameters. Figure 6.25 show the results.

Figure 6.25 (upper panel) shows the correlation between EW of excess CaII K emission and the effective temperature ($T_{\rm eff}$) for each type of luminosity (insert). The binary systems in which only one star is chromospherically active are designated as single stars and marked by larger open circles. There is a slight trend of increasing EW(CAII K) with the decrease of effective temperature. The activities seem to be higher in average for luminosity V, IV stars than III, II stars. The active binary components are generally stronger emitters than·single active stars.

Figure 6.25 (lower panel) denotes the correlation between the CaII K surface flux, $F_{\rm s}$(CaII K), and the effective temperature. A slight trend to decreasing $F_{\rm s}$(CaII K) with lower effective temperature can be seen. In this figure dashed lines indicate constant $F_{\rm s}$(CaII K)/$F_{\rm bol}$ ratios. The observed points are found to lie mostly in the range 10^{-5} to 10^{-6}. The large spread can be due to the fact that CaII K indicators can change with orbital phase or with intrinsic large-scale variations. The effect of rotational velocities is also clearly seen.

The flare activity of the RS CVn stars releases a large amount of energies in X-ray and radio wavelengths. Table 6.12 exhibits the X-ray intensity as high as 10^{29}–10^{31} erg^{-1}.

Consider the case of CABS 29 in Table 6.12 (V711 Tau = HR1099 = HD22468) as an example of emission-line analysis. This star was observed in 1993 by Zhai and Zhang (1996) at Xinglong Station of the Beijing Observatory using the Cassegrain spectrograph attached to the 2.16-m reflector. This system is the binary of G5 IV+ K1 IV (Group 2), showing remarkable emission lines among RS CVn stars and exhibits strong $H\alpha$ emission in all phase of orbital motion (Period 2.8377 days). Figure 6.26 illustrates the spectral phase variation in the $H\alpha$ of this system.

In this figure phase $\varphi = 0$ corresponds to the primary minimum. The solid and thin markers on each spectrum denote the orbital velocity of the primary (K1 IV) and companion (G5 IV), respectively. One may see that the strong $H\alpha$ emission always exists and belongs to the primary, while the companion reveals weak absorption lines in $\varphi = 0.6$–0.9, and no trace of absorption in $\varphi = 0.1$–0.3. Since this system is a double-lined binary, the individual masses of 1.1 $M_{\odot}$ (G5IV) and 1.4 $M_{\odot}$ (K1IV) can be obtained from their velocity curves. There is no eclipse in this system, so the light variation is supposed to originate from the primary's surface structure as bright or dark spots. By

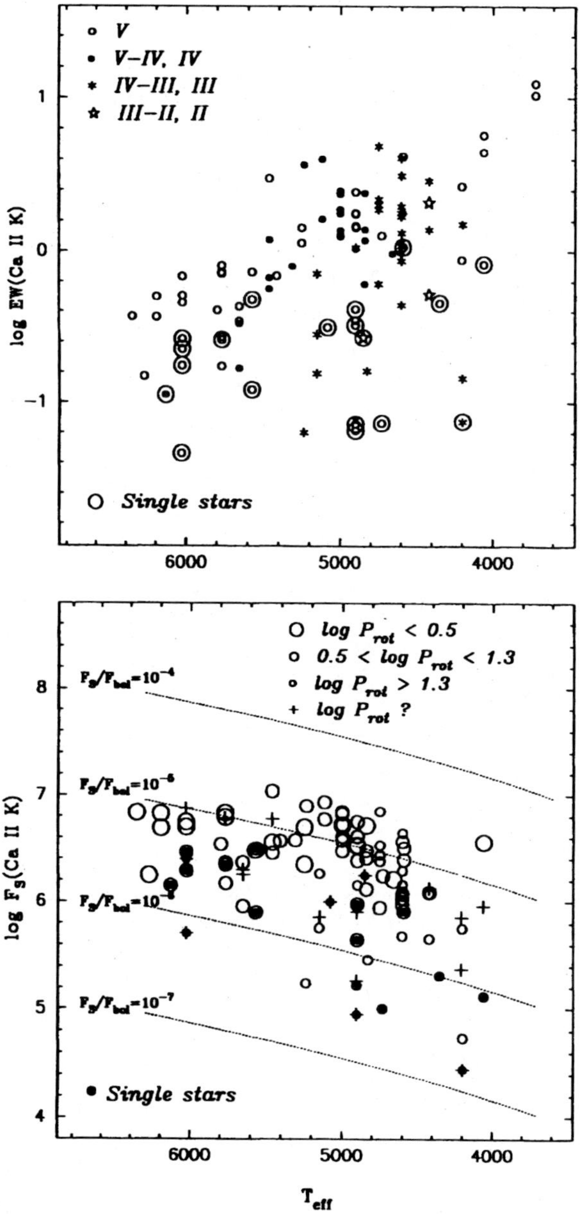

Figure 6.25: Chromospheric activities of RS CVn stars in CaII K line. Excess CaII K emission equivalent width, EW (CAII K) (upper panel), and CaII surface flux, F_S(CaII K) (lower panel). Both panels are plotted against effective temperature. In the upper panel different symbols are used to represent the stars with different luminosity class. In the lower panel the size of the symbols is inversely proportional to the rotation period (P_{rot}). The dashed lines indicate constant F_S (CaII K)/F_{bol} ratios. (From Montes et al. 1996)

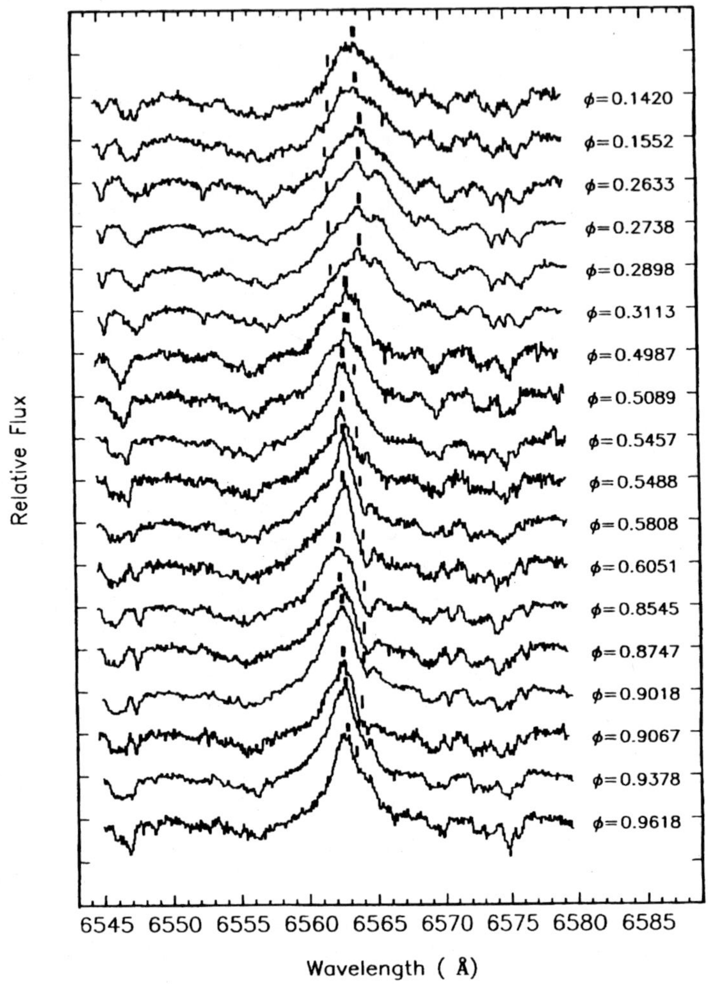

Figure 6.26: Phase variation of the Hα profile of the RS CVn type binary V711 Tau (HR1099). Phase $\varphi = 0$ corresponds to the primary minimum (K1 star is just in front of the secondary G5 star). The solid and thin marks on the spectra indicate the orbital velocities of the primary and secondary, respectively. (From Zhai and Zhang 1996)

analyzing the phase variations of the line profiles shown in Figure 6.26, Zhai and Zhang (1996) proposed the binary model as given in Figure 6.27. Their model can be summarized as follows.

(a) Existence of gas stream. In a detached binary such as this star, gas flow from K1 to G5 star through Lagrangean point L_1 is not ordinarily realized. However, they argued that, since the K1 star has already

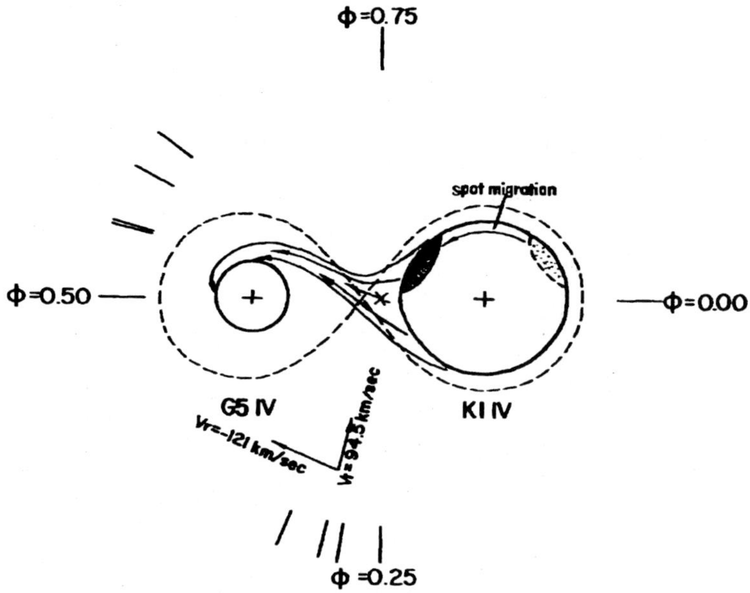

Figure 6.27: A mass-transfer model for the RS CVn type binary V711 Tau (HR 1099). The observed heliocentric radial velocities V1 = 94.5 km s^{-1} at $\varphi = 0.29$ and $V1 = -121$ km s^{-1} at $\varphi = 0.58$ are indicated. A remarkable spot migration from 1992 September to 1993 January is also illustrated in the figure. (From Zhai and Zhang 1996)

expanded nearly 80% of its Roche lobe, the gas stream could possibly exist.

(b) Spot migration. Large displacement of a bright spot on the K1 surface from September 1992 to January 1993 has been observed as shown in Figure 6.27. This spot is thought to be a quiet-prominence like ionized cloud formed above the active region (Zhai et al. 1994).

(c) Long-term variation. Though not shown in Figure 6.27, long-term variations have been observed in the brightness and in Hα intensity in a time scale of several years. These variations suggest the unstable nature of the gas stream, and Hα emission and/or spot formation.

For RT Lac (CABS 189 in Table 6.12), Çakirli et al. (2003) carried out long-term photometric observations covering the period 1978–2000, and analyzed the light curve under a model that assumes the existence of stellar spots on both components. They thus found that the more massive G5 primary appears to be more active with a short-term activity cycle of a period 8.4 years, and a possible long-term cycle of 33.5 years. The spots are concentrated at high latitudes (above 45°). The G9 IV secondary did not show any evidence for an activity cycle; its spot coverage appears rather constant at about 10% of its surface.

During the multiwavelength observations of the RS CVn star, II Pegasi, Byrne et al. (1998) found an unusual steep Balmer decrement, i.e., strong Hβ absorption appeared against strong Hα emission. Short et al. (1998) calculated a series of semiempirical atmospheric models, by making use of the iterative non-LTE calculations in a hydrostatic atmospheric structure. As the result they showed that the steep decrement can be reproduced with suitable model parameters, mostly due to the effect of lower surface gravity on the formation of the spectrum.

6.6 Cataclysmic variables and novae

6.6.1 Cataclysmic variable stars

Among close binaries, stars that show eruptive brightening are called CV stars. Though CVs reveal diverse form of activities, all of them are composed of a cool star and a white dwarf, as schematically illustrated in Figure 6.28. The basic process is that gas stream from the cool star onto the white dwarf gives rise to eruptive phenomena at the surface of the white dwarf or on the accretion disk around the white dwarf (Hoffmeister et al. 1985). Based on the form of activities, CVs are ordinarily classified into the following four categories:

Classical novae	CN
Recurrent novae	RN
Dwarf novae	DN
Nova-like variables	NL

Beside these categories there is also an important group called magnetic CVs. This type occurs when white dwarfs have a strong magnetic polar field, and is called the polars, since the gas accretes onto the white dwarfs along the magnetic lines of force of the polar fields. When the magnetic field is somewhat weaker, the stars are called intermediate polars (see Warner 1995).

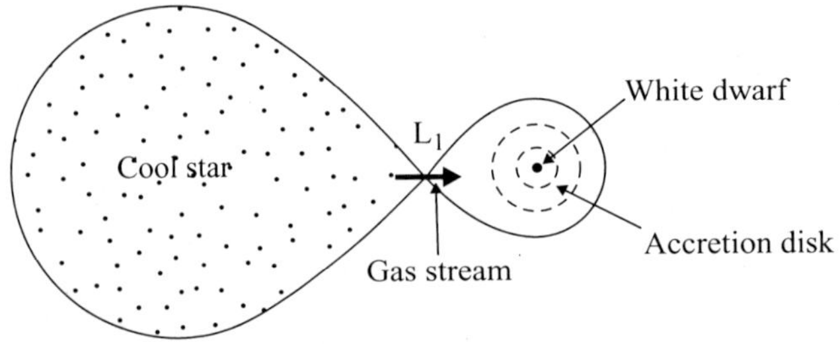

Figure 6.28: Schematic picture of CVs in the orbital plane. Cool star expands to fill the Roche lobe, and a gas stream flows onto the accretion disk of the white dwarf. (From Hoffmeister et al. 1985)

In this book we consider the nonmagnetic CVs mainly from the viewpoint of the formation of emission lines. Most of CVs indicate emission lines in quiescent as well as in active (explosive) phases. In general, emission lines are formed mainly in accretion disks or in envelopes near hot components in quiet phase. In active phase, the formation of emission lines turns out to be in the gas shells ejected by explosion.

6.6.2 Classical novae

Classical novae (CNe) are the stars whose nova explosion was historically recorded only once, and with a brightening ΔV ranging from 6 to 19 magnitudes. Their light curves are classified by the speed of light declining after the light maximum as follows:

Na—fast nova. The rise is very steep and the light maximum lasts one or, at the most, just a few days, and the decline takes around 100 days or generally much sooner, to reach 3 magnitudes below the maximum.

Nb—slow nova. The 3-magnitude decline takes more than 100 days. In many cases, a deep and wide temporal minimum occurs several months after the maximum (Figure 6.29). After recovering to a magnitude

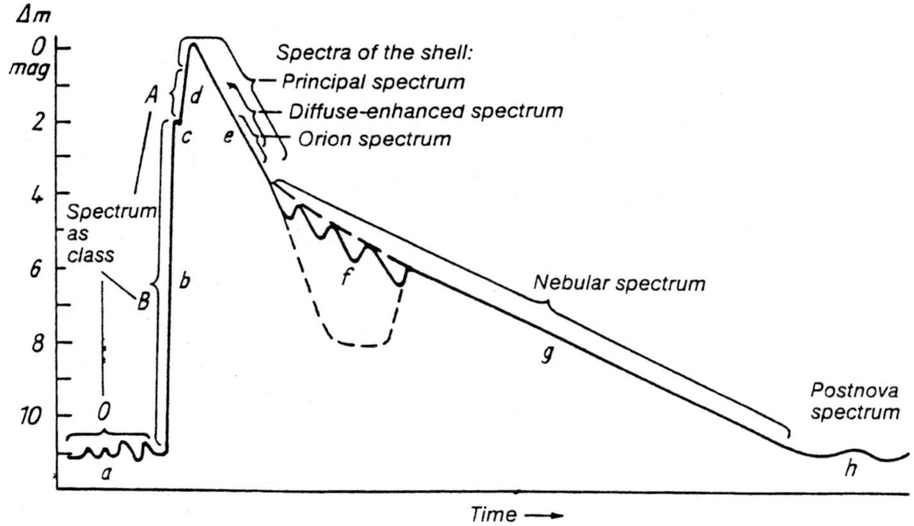

Figure 6.29: Schematic diagram of the luminosity, phase, and spectra for understanding the nomenclature of nova phenomena. The spectral characteristics before and after the maximum phase are also shown. The main signs are mainly based on Hoffmeister et al. (1985) as follows: a, prenova; b, first rise; c, pre-maximum pause; O, B, A prenova to burst spectrum; d, final rise; e, initial decline; f, transition stage; g, final decline; h, postnova. For spectral phase, see text.

Table 6.13: A selected list of classical novae observed by 1975

Nova	Name of star	Brightness Max	Brightness Min	Amplitude Δ m	Type of nova	Orbital period	Spectral type
Per 1901	GK Per	0.2	14.1 V	13.9	Na	1.99680	K0/4
Gem 1912	DN Gem	3.6	15.8 pg	12.2	Nb	—	—
Aql 1918	V603 Aql	−1.1	11.97 B	13.1	Na	0.138154	—
Cyg 1920	V476 Cyg	2.0	17.1 pv	15.1	Na	—	
Pic 1924	RR Pic	1.2	12.8	11.6	Nb	0.145026	—
Her 1934	DQ Her	1.3	15.6 pg	14.3	Nb	0.193621	M3v
Lac 1936	CP Lac	2.1	15.6 pg	13.5	Na	—	—
Mon 1939	BT Mon	4.5	16.8 pg	12.3	Na	0.333814	K5-7
Pup 1942	CP Pup	0.5	<18 pg	>17.5	Na	0.06115	—
Her 1963	V533 Her	3.0	14.9 pg	11.9	Na	0.28	—
Del 1967	HR Del	3.70	12.38 B	8.68	Nb	0.214167	—
Ser 1970	FH Ser	4.5	16.1 pg	11.6	Na	—	—
Cyg 1975	V1500Cyg	1.8	<21 B	>19.2	Na	0.139613	—

Type of nova: Na, fast nova; Nb, slow nova. Brightness and type of nova are taken from Pettit (1982), and orbital period and spectral type are taken from Ritter (1987).

approximately equal to that expected in an undisturbed decline, this star again continues its declining.

Nc—very slow nova. The prototype is the star RT Ser, which started in 1915 to rise slowly to 10.5 magnitude, remained at this level for almost 10 years, and then faded away very slowly reaching 14 magnitude in 1942. The number of stars in this type is small and there is some possibility that they are related to P Cyg variations (Section 5.4.2) or symbiotic novae (Section 6.7.4).

In Table 6.13 we show a selected list of the CNe that have been observed by 1975. Since 1980s, the rate of discovery of novae has been greatly increased with the expansion of telescope sites including sophisticated amateur astronomers.

Light curve and nova spectrum

Depending on the phase, novae exhibit remarkable spectral change. For nomenclature, we schematically show in Figure 6.29 the names of phases and spectra for a typical light curve.

The nova spectra are usually divided into three epochs of prenova, nova outburst, and postnova. In principle, postnova is the prenova to the next nova outburst so that both epochs should show the same spectra. According to Robinson (1975), among the novae whose light curves were studied before and after the burst for a long time span, more than half showed the same magnitudes. The remaining light curves showed significant changes before eruption, usually a small increase in luminosity for months to years. An example of prenova variation is V1500 Cyg (Nova 1975). This star was not visible on the blue (mag $\geqq 21$), or red (mag > 20) in 1950s, but a month before maximum it had reached 16 mag and stayed this magnitude until the onset of the fast rise (Seitter 1990). Spectral observations in prenova stage are also few and recorded only for stars such as V603 Aql (1918) (19 years), V533 Her (1963) (2 years), and HR Del (1967) (7 years), where the years prior to maximum light are indicated in the respective brackets. These spectra are all low dispersions showing bluish continuum without indicating clear absorption or emission lines (Seitter 1990).

In the outburst epoch, we have full spectroscopic data. We can easily trace the spectral variations during the burst processes. In many cases, since absorption and emission line spectra exist in parallel, we consider them separately. The reviews of Payne-Gaposchkin (1957), McLaughlin (1965), and Seitter (1990) are useful to the study of developing nova spectrum.

Absorption-line systems

The absorption-line system in the outburst epoch occurs in a nearly similar sequence for every nova. McLaughlin (1965) called them premaximum, principal, diffuse enhanced, and Orion spectra (see Figure 6.29). These names are still in use.

(a) Premaximum spectrum. In Figure 6.29, the marks O, B, A indicate the approximate spectral types of the premaximum spectrum. The spectrum is similar to a blue star characterized by strong blue light and blue-shifted absorption lines. This blue shift can be explained by an expansion at high velocity, situated in front of the hot stellar surface. The expansion velocity is of some 100–1000 km s^{-1}. Near the maximum luminosity the lines of neutral atoms such as OI, CI become conspicuous. The premaximum spectrum lasts until shortly after light maximum and then rapidly fades. The absorption features become stronger and narrower. In some cases the expansion accelerates, and in other cases it decelerates.

(b) Principal spectrum. This spectrum appears as if it replaces the premaximum spectrum at about the time of light maximum. The spectrum resembles that of supergiants of class A–F. Spectral lines show prominent P Cygni profiles with 200 to 2000 km s^{-1} of Doppler shifts. Main lines are OI, CI, CaII, and MgII. The emission features are wide because they arise from the still transparent expanding shell. Spectral lines are getting weaker as the light declines, while the absorption lines arising from quasi-stationary levels such as FeII, TiII, MgI are relatively strengthened, suggesting the dilution effects for the incident light of the expanding shell. The duration of principal spectrum depends on the novae.

(c) Diffuse-enhanced spectrum. The next stage appears even before the principal spectrum declines. This third system has a stronger blue shift than the previous system. Hydrogen and CaII are particularly conspicuous as compared to FeII, OI, and NaI. This system appears in more developed form in slow novae accompanying the lines of TiII and CrII. In this system, the lines are very wide and very diffuse, presumably as a result of strong turbulence in the expanding gas cloud.

(d) Orion spectrum. The Orion spectrum is dominated by absorption lines that are characteristics of "Orion stars," i.e., class B stars found in the Orion Association, and reveal HeI, OII, NII and CII lines relatively strengthened. Hydrogen lines may be either weakly present or absent. The blue shifts are higher than those of the diffuse-enhanced spectrum. The Orion spectrum also contains emission lines that are wide and diffuse. They are most prominent at the transition minimum in the light curves.

(e) Postnova spectrum. As the gas shell ejected from nova is diffused out, the above absorption system gradually declines and the nova spectrum begins to resemble the spectrum of planetary nebula. We consider the emission lines in the next paragraph.

In a number of novae the absorption systems are not clearly distinguished, as sometimes the distinction is difficult and sometimes different absorption systems simultaneously appear. In Figure 6.30, the light curves of the nova DN Gem 1912 (Nb type) are illustrated for around 25 days from just before light maximum along with the radial velocities of observed absorption systems. One may see in this figure that the gas shells are ejected from the star successively with gradually higher expanding velocities.

Emission lines and variability

Emission lines are observable in almost all periods from just after the outburst up to the postnova stage, and their species and behaviors exhibit some systematic changes in most cases. Generally speaking, since emission lines are formed in the ejected gas by nova bursts (denser in early phase and gradually rarefied), emission lines in early phase are mainly permitted lines while they turn to form forbidden lines in rarefied gas. Thus the formation of emission lines in nova spectra can be classified into the following four stages depending on the physical state of nova ejecta (Williams 1990).

P: Permitted-line phase,
A: Auroral-line phase,
C: Coronal-line phase,
N: Nebular-line phase.

In P phase, gas density is higher than the critical electron density for any forbidden line so that the permitted lines of hydrogen Balmer lines and of FeII, MgII, NII, OII lines are observable. These lines are recombination lines

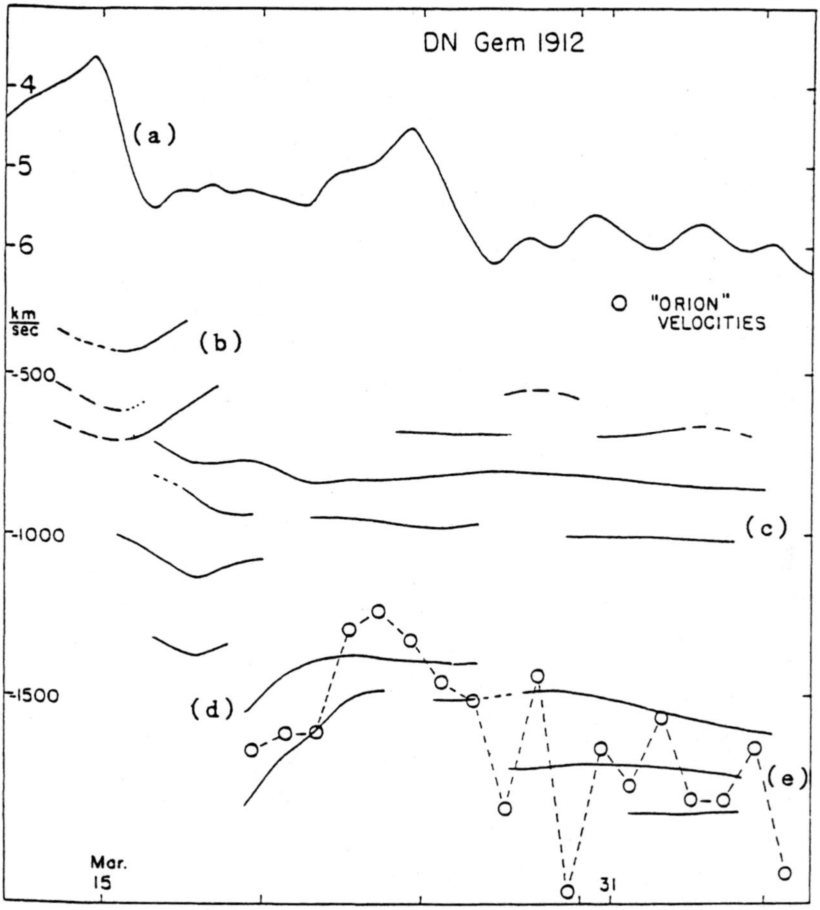

Figure 6.30: The light curve of the nova DN Gem 1912 (Nb type) for around 25 days from just before the maximum, and the variation of the radial velocities of the observed absorption systems. (a) The light curve; (b) radial velocities of pre-maximum spectrum ($V_r \sim -500\,\mathrm{km\ s^{-1}}$); (c) radial velocities of the principal spectrum ($V_r \sim -800\,\mathrm{km\ s^{-1}}$); (d) radial velocities of the diffuse-enhanced spectrum ($V_r \sim -1400\,\mathrm{km\ s^{-1}}$); (e) radial velocities of the Orion spectrum ($V_r \sim -1600\,\mathrm{km\ s^{-1}}$), which are indicated by open circles and broken lines. (From McLaughglin 1965)

due to ionizing ultraviolet radiation. We can also observe emission lines of neutral atoms, OI and NI, which are formed in a neutral region surrounding the ionized region. In this phase, the shell is optically thick, and the ionized region is confined in the inner part of the shell.

In A phase, the forbidden lines in auroral transitions appear. The auroral lines such as [OIII]λ 4363 Å, [OII]λ 7325 Å are the forbidden lines created by transitions between excited levels with higher spontaneous transition probability A_{21} and smaller collision strength α_{21}. As a result the critical electron

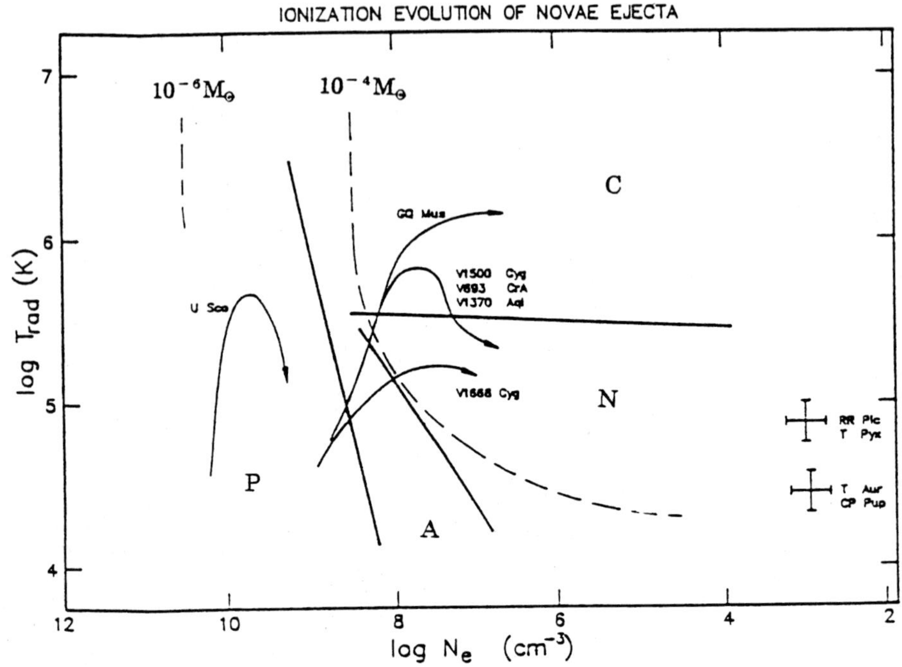

Figure 6.31: Schematic diagram of nova emission-line evolution on the radiation temperature–electron density diagram of nova ejecta. The different phases of evolution which characterize the emission spectra are demarcated as follows: P (permitted-line phase), A (auroral-line phase), C (coronal-line phase), and N (nebular-line phase). The dashed lines denote the boundary at which the ejecta change from being optically thick to being optically thin to ionizing radiation. The boundary depends on the amount of mass of gas ejecta. Two cases of 10^{-4} and 10^{-6} $M_{\odot}$ for the ejected mass are shown. The ionization evolution for some nova ejecta is also shown. (From Williams 1990)

density $N_e^c = A_{21}/\alpha_{21}$ takes the values of the order of 10^7 cm^{-3} which are higher than nebular lines (see Table 4.5). In A phase these auroral lines appear mixed with some permitted lines.

With the expansion of the gas shell, emission lines enter the C or N phase, both of which are the phases of low gas density in fully expanded shells. In this stage, the surface of the central white dwarf is heavily heated by accretion flow, emitting strong UV radiation that causes the formation of highly excited spectral lines. When radiative temperature is higher than around 10^5 K, coronal emission lines such as [FeX]λ 6375 Å, [FeVII]λ 6078 Å are formed and the nova enters the C phase. When radiative temperature is lower than 10^5 K, nebular lines such as [OIII]λ 5007, 4958 Å, [OII]λ 3727 Å become dominant and the nova enters the N phase where the spectra are as a whole resemble planetary nebulae.

A general scheme of nova emission-line evolution is illustrated in Figure 6.31 (Williams 1990), as a function of the radiative temperature and

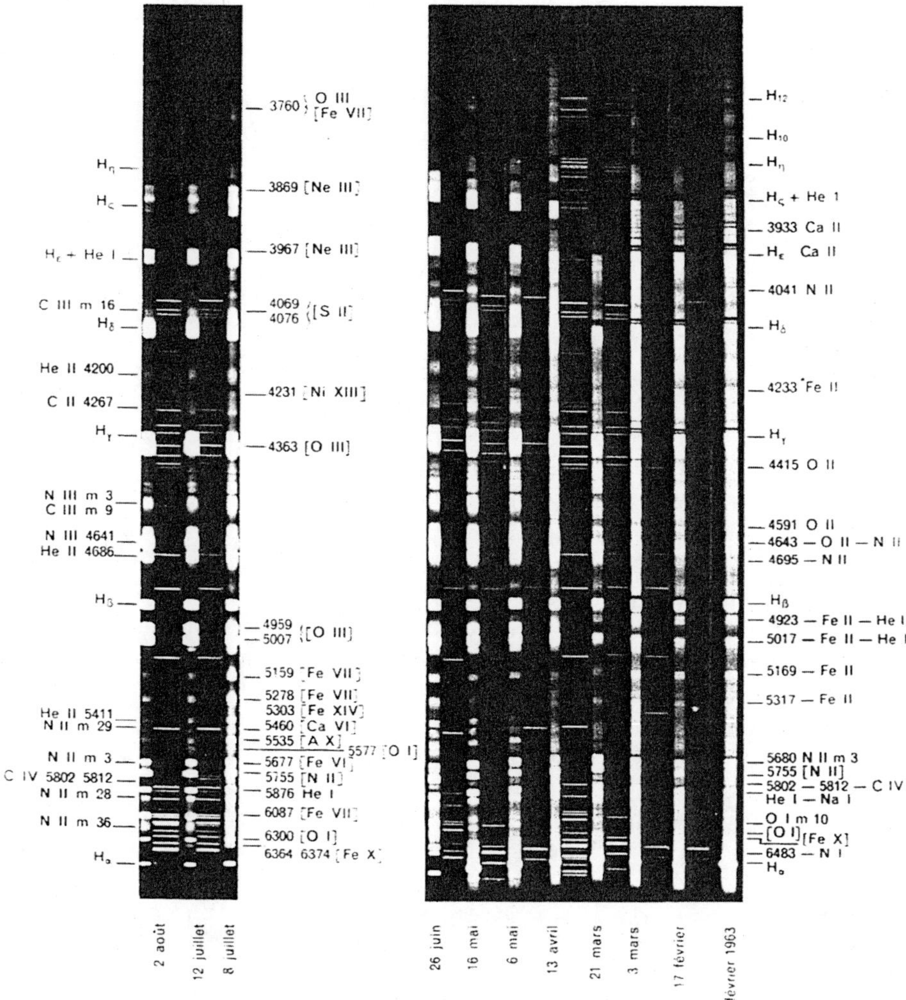

Figure 6.32: Spectral variation of the Nova Her 1963, obtained with a prism spectrograph attached to the 120-cm telescope at Haute-Provence Observatory. Dispersion is 76 Å mm^{-1}. The dates of observation are shown in the low side (from bottom to up in the order of observations). (From Bloch and Chalonge 1965)

the electron density. The paths of ionization state for some novae are also traced on the diagram.

As an example of spectral variation we consider the nova V533 Her (1963). This nova reached the maximum light of $m_{\mathrm{pg}} \sim 3$ mag in around January 30, 1963, and thereafter rapidly declined. Spectroscopic observations began on February 8 at Asiago Observatory and on February 11 at Haute-Provence Observatory after about 1.5 magnitude decline. In Figure 6.32 a series of spectrograms obtained by Bloch and Chalonge (1965) at Haute-Provence Observatory using the 120-cm reflector are shown. The spectrograms ($\sim \lambda 3700$ Å to $H\alpha$) exhibit the variation of spectra for around 6 months.

As seen in the figure, the $H\alpha$ line has a strong emission from the very first. The emission lines were in P phase in February and March of 1963 and the following emission lines were confirmed.

NI—λ 4138, 6483, 6723 Å	OI—λ 6156-57-58 Å
FeII—λ 4233,5018,5169,5317 Å	MgII—λ 4181 Å
OII—λ 4415,4591-96 Å	HeI—λ 5876 Å

Auroral emissions [OI]λ 6300 Å began to appear in February and [OIII]λ 4363 Å, [OII]λ 7325 Å began in March, indicating the phase transition from P to A. In April the spectrum moved to phase N andC in a mixed form. Nebular lines [OIII]λ 4959, 5007 Å and [NeIII]λ 3868, 3967 Å along with transauroral lines [SII]λ 4069, 4076 Å, [NII]λ 5755 Å were visible up to September, whereas coronal lines [FeVII]λ 5159, 5278, 6068 Å, [FeX]λ 6364, 6374 Å, [FeXIV]λ 5303 Å, [NiXIII]λ 4231 Å appeared in April and began to be conspicuous in June and July as the C phase overlapped in the weak N phase.

Chincarini and Rosino (1965) observed this nova from April to September. They measured the intensity variations of main emission lines as shown in Table 6.14 One may see that the excitation degrees of emission lines gradually gets higher overlapping C and N phases of coronal and nebular lines.

Emission lines of novae are distinguished in four phases as stated above. In addition, Williams (1992) divided the early outburst spectra, corresponding nearly to the P phase of novae, into two classes, based on the relative strength of two groups of emission lines, FeII type and He/N type. In FeII type, lines of lower ionization (NaI, OI, MgI, CaII) appear showing P Cygni absorption components. The emission spectrum of this type moves from P to A phase showing auroral lines of relatively low ionization such as [NII]λ

Table 6.14: Intensity variation of main emission lines of Nova Her 1963 (based on Chincarini and Rosino 1965)

Line λ(Å) Type	[FeVII] 6086 C	[NII] 5755 TA	[OIII] 5007 N	[OIII] 4959 N	HeII 4686 P	[OIII] 4363 A	[NeIII] 3869 N
Date							
4/19	—	30	50	15	15	20	3
5/21	5	40	60	20	20	40	6
6/17	8	50	100	35	25	60	12
7/23	12	80	200	60	25	100	15
9/14	25	100	450	150	30	120	20

Note: The scale of emission intensity is taken so as $H\beta = 50$, $H\alpha = 300$. The type of emission lines is designated as C, coronal; P, permitted; A, auroral; N, nebular; and TA, transauroral types, respectively.

5755 Å, [OII]λ 7325 Å, [OIII]λ 4363 Å, [OI]λ 6300 Å. Novae in He/N type reveal lines of higher ionization (HeIIλ 4686 Å, HeIλ 5876 Å, NIIλ 5679/5001 Å, NIIIλ 4640 Å, etc.) with larger expansion velocities. This type moves to C phase in forbidden lines, showing coronal lines of [FeX]λ 6375 Å, [FeVII]λ 6087 Å, [NeIII]λ 3869 Å, etc. The emission lines of this type generally exhibit broad and flat-topped profiles without P Cyg type absorption. Samples of both types in the P phase are shown in Figure 6.33. In both cases the most prominent emissions are the Balmer lines. Note that a few novae change their type during this early P phase, evolving from a FeII type to a He/N type before forbidden lines appear. They are the so-called hybrid objects. Nova Her 1963 shown in Figure 6.32 is an example of hybrid objects.

These two types of spectra have been interpreted by Williams (1992) in terms of a two-component gas consisting of a discrete shell and a continuous wind. That is to say, the narrower FeII type spectrum is formed in the wind, while the broader He/N type spectrum is formed in the shell ejected from the white dwarf at maximum light.

Imaging and emission lines

The expanding clouds ejected by the eruptions of CNe become optically visible after sufficient time of expansion. Deep imaging observations were carried out for 13 CNe by Slavin et al. (1995) using the William Herschel Telescope in the Canary Islands (ORM). They resolved the optical images and found that the faster novae tend to show spherically symmetric shells with clumps, while slower novae produce more ellipsoidal remnants with one or several rings of enhanced emission. In case of a spheroidal shell, Gill and O'Brien (1999) calculated emission-line profiles under the assumption that the nova shells are transparent for the lines, and line intensities are proportional to the column density of gas. They applied this method to an expanding ellipsoidal-shell encircled by three rings as shown in Figure 6.34. The model shell is assumed to have the inclination angle 60°. The line profile calculated for this model is illustrated also in Figure 6.34, where total line profile is decomposed into the three ring components. The observed emission-line profiles of V705 Cas for five lines are shown in Figure 6.35 after suitable scaling of the peak intensities. When compared with model profile in Figure 6.34, observed profiles are suggestive of complicated structure of nova shell such as an ellipsoid with several rings.

Mechanisms of nova outbursts

New stars, suddenly brightened and fade away after a several months, enraptured the minds of peoples as heavenly mysterious phenomena from the ancient. However, the physical understanding of their explosion mechanism came out only after the 1950s. The first clue was the discovery of nova DQ Herculis (1934) as an eclipsing binary by Walker (1954), and then the supposition of Walker (1957) and Kraft (1959) that all novae are binary systems composed of white dwarfs and red faint stars. The second clue was the physics

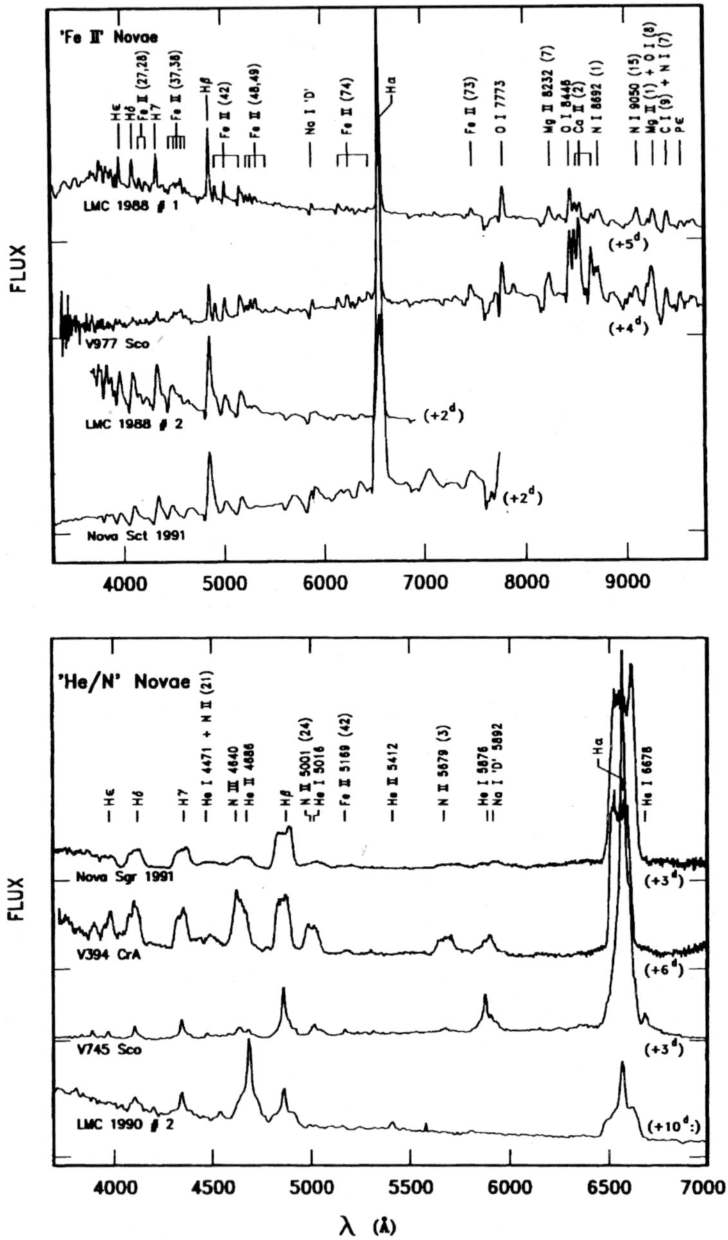

Figure 6.33: Two types of permitted-line phase spectra. Upper panel: FeII type spectra; Lower panel: He/N type spectra. In each spectrum, the name of nova is indicated in left side, and the days from the light maximum are given in right-hand side in the bracket. (From Williams 1992)

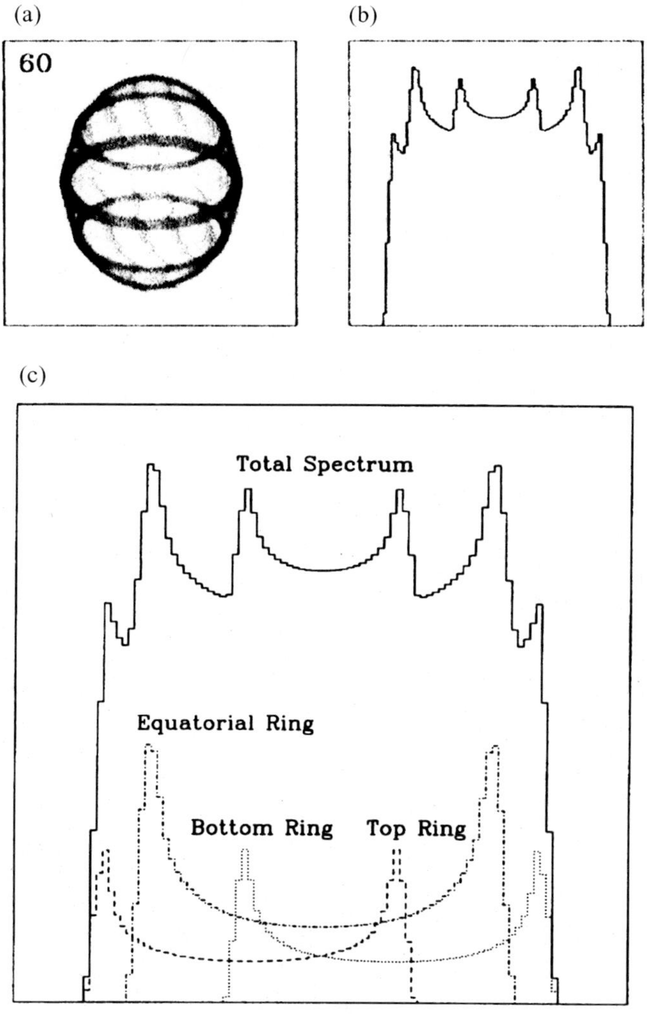

Figure 6.34: Panel (a), A model image of ellipsoidal shell (inclination 60°) with an equatorial and two polar rings of enhanced brightness in classical nova. Panel (b) and (c), Calculated profile and its decomposition to the three ring components. (From Gill and O'Brien 1999)

of thermonuclear runaway process on the surface of white dwarfs discovered by Starrfield et al. (1972, 1974). The mechanism of nova outbursts was thus recognized and many models have been put forwarded.

The basic process can be stated as follows. Consider a close binary system, as a result of binary evolution, composed of a cool star fulfilling the Roche lobe and a white dwarf (Figure 6.28). The white dwarf is surrounded by an accretion disk formed by the flow fed from the cool star through the Lagrangean

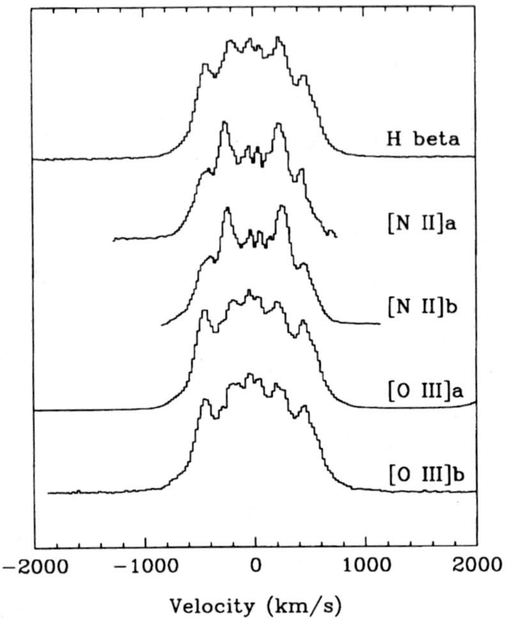

Figure 6.35: Emission-line profiles of V705 Cas taken on 1996 August for H$\beta\lambda$ 4861.3, [NII]aλ 6549, [NII]bλ 6584, [OIII]aλ 4959, and [OIII]bλ 5007 Å. All profiles are scaled to the same peak intensity and corrected for the systematic velocity of the star. (From Gill and O'Brien 1999)

point L_1.Gases inside the disk further accrete onto the surface of the white dwarf losing the angular momentum. The white dwarf is a high-density star of degenerate gas, where hydrogen is already burnt out. On the other hand the fallen gases are rich in hydrogen so that the surface of the white dwarf is covered by a thin layer of nondegenerate hydrogen-rich gas. If this nondegenerate gas is sufficiently compressed and heated by a strong pull of gravity of the white dwarf at the surface of the degenerate gas layer, the boundary layer will suddenly explode by the so-called thermonuclear runaway. Nova outburst is triggered in this way by the release of nuclear energy. The parameters governing this process are the mass of the white dwarf, the accretion rate $M_\odot$ per year on the white dwarf and the metal abundance of infalling gas. Conditions that cause the CNe to explode are (i) the large mass of the white dwarf (larger than one solar mass), (ii) the accretion rate onto the white dwarf is moderate, and (iii) the high hydrogen abundance of infalling gas. If the accretion rate is high, small explosions are repeated in a short time scale and become RN which we shall consider below. Theoretical problems of a nova outburst are concisely reviewed by Shara (1989).

6.6.3 Recurrent novae

RN and the burst mechanism

Stars historically repeating outbursts are called recurrent novae (RNe). The time interval is about 20 to 80 years and the magnitude range in an outburst

Table 6.15: Selected recurrent novae

Star	Maximum brightness mv	Magnitude difference Δ mv	Outburst year (year)	Mean interval (year)	Nova type	Orbital period	Spectral type
T CrB	1.8–2.0	8.6	1866, 1946	80	fast	227.53 d	M4 III
RS Oph	4.5–5.4	6.9–7.8	1898, 1933, 1958, 1967, 1985	23	fast	230 d	K5.7 II-III
T Pyx	6.6–7.5	7.0–7.9	1890, 1902, 1920, 1944, 1967	19	slow	2.3783 h	—
V1017 Sgr	7.2	7.1	1901, 1919 1973	36	slow	—	—
U Sco	8.8–9.0	10.0–10.2	1863, 1906, 1936, 1979	39	fast	1.2344 d	G 3-6

References: Nova data (Hoffmeister et al. 1985); orbital period (Livio and Truran 1992); spectral type (Warner 1995). According to Webbink et al. (1987) the spectral type of RS Oph is M0-2 III.

is around 7–11 magnitudes. Their speed class spans the range from very fast (e.g., U Sco, T CrB) to slow (T Pyx), not directly depending on the binary orbital period. The time it takes to decline to 3 magnitude from the maximum (T_3), known as decay time, was 6 days in T CrB, and 9 days in RS Oph. The slow nova T Pyx showed a light variation similar to slow nova. The RN T CrB has shown the brightest maximum luminosity among RNe so far observed; it reached $m_v = 2.0$ during the burst in 1866, and $m_v = 1.8$ in 1946. Selected RNe are shown in Table 6.15.

It is apparent that the RNe, which are simply distinguished from CNe and dwarf novae (DNe) by the time interval of outbursts, constitute a very heterogeneous class of objects in their observational behaviors. Webbink et al. (1987) showed that, in spite of their diversity, RNe could be classified into two types according to their outburst mechanisms: (a) those powered by thermonuclear runaway on white dwarf companions (T Pyx, RS Oph and U Sco) and (b) those powered by accretion from a red giant to a companion (T CrB). Notice that DNe (Section 6.6.4) are accretion-powered events, while CNe are powered by thermonuclear runaway on the surfaces of white dwarfs. In contrast, RNe are supposed to have both types of outburst mechanism. These types are outlined as follows.

(a) RNe caused by thermonuclear runaways. The requirements for this type are (i) very massive white dwarfs (close to the Chandrasekhar limit) and (ii) high accretion rates, $\dot{M} \geq 10^{-8} M_\odot$ per year, in order to produce short recurrence times. If the accretion rate is high and the mass of the white dwarf is high, the duration to raise the mass of the gainer up to the limiting mass is short so that relatively small-scale explosions are repeated. T Pyx, RS Oph and U Sco belong to this type, although there is a considerable difference in outburst characteristics between them. Observationally, the outbursts of U Sco are more violent such as in the large magnitude difference in the brightening ($\Delta m \sim 10$ magnitude in U Sco, while $\Delta m \sim 8$ magnitude in T Pyx), and in higher ejection velocities (7500–10,000 km s^{-1} in U Sco, while 850–2000 km s^{-1} in T Pyx). In spite of such difference, these stars are distinguished from accretion-powered type of RNe or CNe by their light curves, color variations in and out of outbursts, and the CNO abundances.

(b) RNe caused by accretion events. T CrB belongs to this type. RS Oph is sometimes classified in this type. Both systems are very similar in orbital period (~230 days), in the type of secondary star (an M giant), and in spectral evolution through the outburst. They show coronal emission lines sometime following maximum light. These systems are often confusing with symbiotic stars, and indeed T CrB and RS Oph are frequently classified as symbiotic stars (see Section 6.7). Their spectra in quiescent stage exhibit the composite of M giant and many highly excited emission lines characteristic of symbiotic stars (Iijima et al. 1994, Dobrzycka et al. 1996). However, they are distinguished by (i) shock-type dynamic ejection of mass at high velocities, and (ii) light curves exhibiting an extremely rapid rise followed by a smooth decline. In this type the bursts are supposed to occur in the accretion flow from red giants to companions due to some instability, though it is still not clear whether the instability occurs inside the accretion flows, or in the accretion disks around the companions.

Spectroscopic observations of RS Ophiucus

Spectral variation of RNe at the outburst resembles those of CNe. RS Oph has repeated several outbursts since 1898 (Table 6.15). The first spectroscopic observations of this nova were carried our for the 1958 outburst at Haute-Provance Observatory (Dufay and Bloch 1965, Dufay et al. 1965). Figure 6.36 illustrates the spectral variations of this outburst observed in the period from July 13, one day after the light maximum, through the October 18 (98 days after the maximum) (Dufay et al. 1965).

The nova in July 1958 revealed the spectra to have a strong continuum in blue to red regions. The Hα line appeared as a strong P Cyg type emission, the Doppler velocity at the absorption edge reached -3620 km s^{-1}. The emission lines HeIλ 5876, 7065, 6678, 4472 Å, FeIIλ 4923, 5017 Å in optical region and OIλ 7774, 8244 Å, NIλ 7468, 8216, 8680 Å in near infrared region were also remarkable. All of them were permitted lines.

The absorption lines in the early phase consist of three components with different line-of-sight velocities. The narrow -115 km s^{-1} lines corresponding to the principal spectrum were visible during July 14–17, and then faded out.

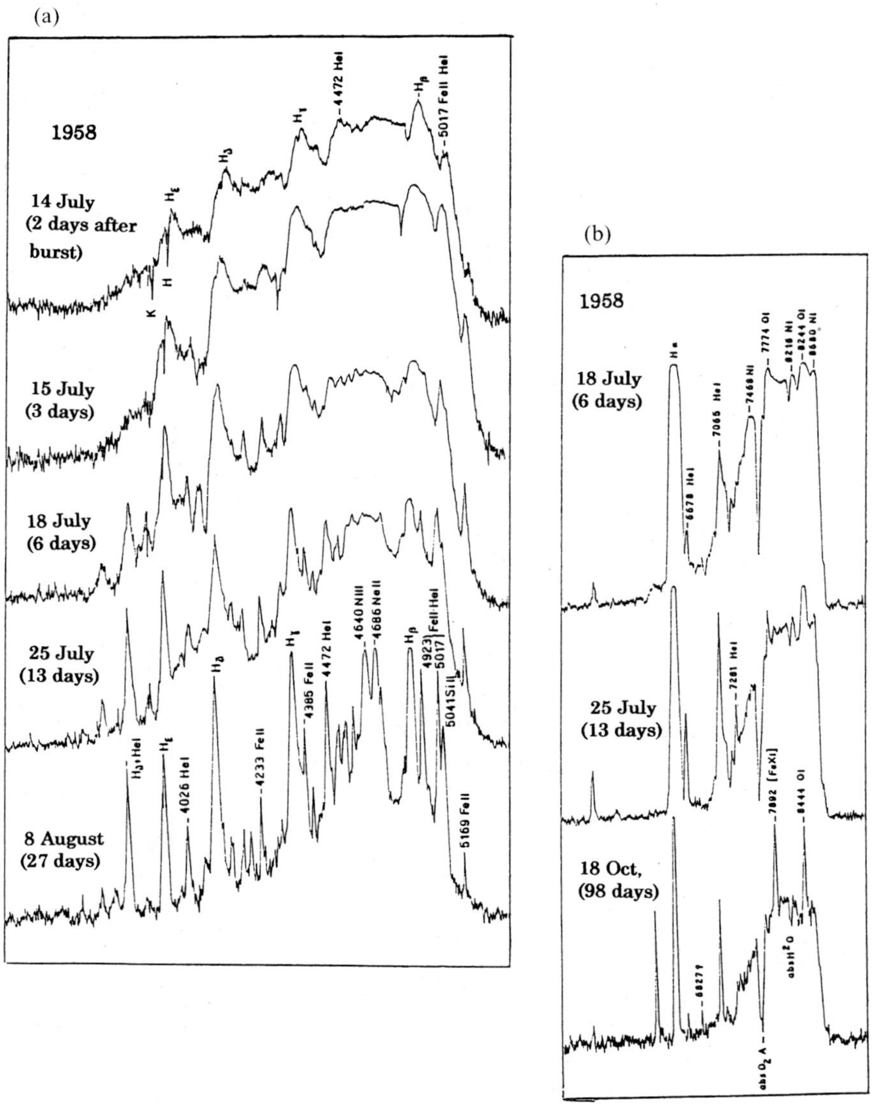

Figure 6.36: Spectral variation of the RN RS Oph in 1958 outburst, observed with 120-cm telescope at Haute-Provence Observatory. Panel (a) denotes the optical-near ultraviolet spectral region and panel (b) the red-near infrared spectral region. (From Dufay et al. 1965)

The -2000 km s^{-1} lines corresponding to the diffuse-enhanced spectrum were first observed in July 17 as -2142 km s^{-1} absorption and disappeared in July 23 after the deceleration down to -1799 km s^{-1} absorption. The third velocity system having variable velocity components of -3637 to -4080 km s^{-1}, appeared in almost the same epoch as the -115 km s^{-1} system. CaII H and

K lines appeared as sharp absorption with nearly constant radial velocities of -46.5 km s^{-1} (H) and -45.1 km s^{-1} (K) and then disappeared in late July.

The emission-line spectrum varied through permitted-line phase, auroral-line phase, coronal-line phase, and finally entered the nebular-line phase in the following year. In the permitted-line phase, besides the strong Balmer line emission, main emission lines were, FeIIλ 4233.2, 4385.4, 5316.8, 5534.9 Å, and HeIλ 4026.2 Å, 4471.5, 5875.6 Å. In addition, emission lines of NeII, SiII, OI, NI, and of NIII λ 4640 Å were observable. These line behaviors indicate that this star can be classified as the hybrid of FeII type and He/N type of Williams (1992), though some weight is to be given to a higher ionization state of He/N type.

In August, forbidden lines and coronal lines appear almost in parallel. The auroral lines of [O III]λ 4363 Å, [N II]λ 5755 Å and transauroral lines of [S II]λ 4068, 4076 Å first appeared and then moved to the nebular phase with the lines of [O III]λ 4959, 5007 Å, [Ne III]λ 3968, 3869 Å along with a number of [Fe II], [Fe III] emission lines. On the other hand, coronal emission lines became stronger in September. Main lines were [Fe X]λ 6375 Å, [Fe XIV]λ 5303 Å, [A X]λ 5535 Å, [A XI]λ 6919 Å, [Ni XII]λ 4231 Å (blended with FeII 4233 Å), and [Ni XV]λ 6702 Å.

The next outburst in 1985 of RS Oph was observed in a wide wavelength range from optical to X ray, ultraviolet, infrared, and radio waves (see Bode 1987). The spectroscopic observations for this outburst were carried out by Rosino and Iijima (1987) at Asiago Observatory and by Anupama and Prabhu (1989) at Vainu Bappu Observatory. Combined IUE and optical spectroscopic observations were carried out by Shore et al. (1996).

The spectral variations in the 1985 burst proceeded similarly to the 1958 burst (Anupama and Prabhu 1989). At around 30 days after the light maximum, forbidden and coronal emission lines began to be conspicuous. In particular, the spectrum after 60 days showed the strong coronal emission. A spectrum taken at this epoch at V. Bappu Observatory (dispersion 132 Å mm^{-1}) is shown in Figure 6.37. Strong coronal lines of [Fe XIV], [A X], along with the permitted lines of Balmer, HeII, FeII, and forbidden line [OIII] are remarkable.

For the 1985 outburst of RS Oph, Anupama and Prabhu (1989) estimated the physical parameters of the expanding shell through a series of spectroscopic observations. Some parameters obtained in a period from 32 through 108 days after the burst are shown in Table 6.16. The expanding velocity was estimated by the width of Hα emission, which rapidly decreased with time. Similar decrease can also be seen in 1958 burst as shown in Figure 6.38. This is interpreted as a braking effect by low-velocity stellar winds ejected in prenova stage. The theoretical values of expanding velocity in Table 6.16 are those calculated by Bode and Kahn (1985) in their spherically symmetric model for nova explosion. In the columns (4) and (5), the electron density and the mass of gas shell are those derived by the volume integral of the *H*α emission under the assumptions that the shell is optically thin for the *H*α line and has the

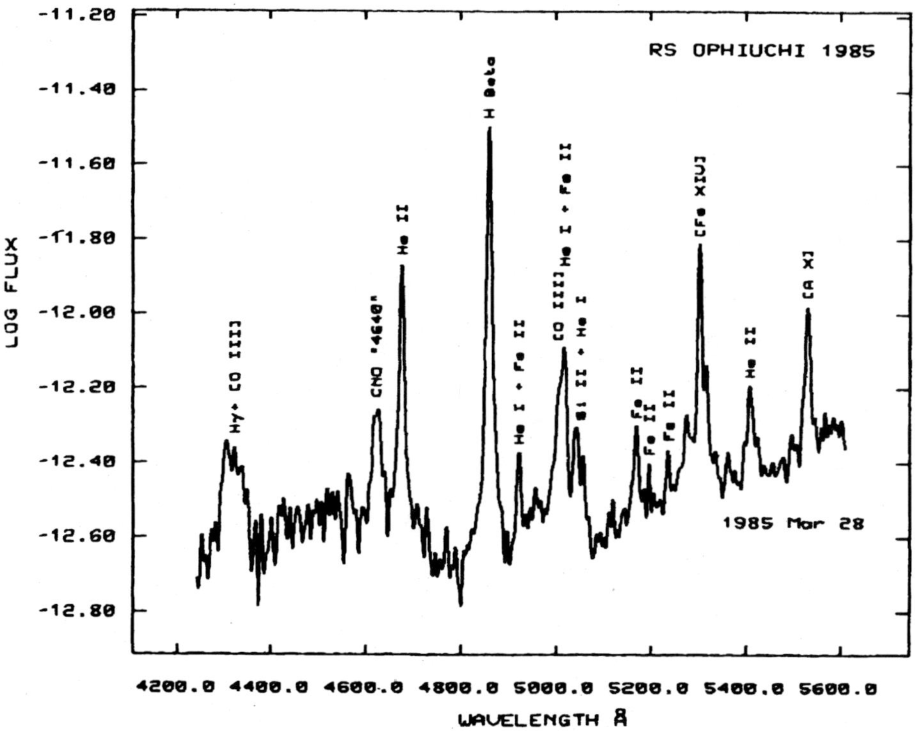

Figure 6.37: Spectrum of RS Oph on 28 March, 1985 (60 days after the maximum), in the range 4250–5600 Å. Note the strengths of coronal and HeII lines. Flux in units of erg cm^{-2} s^{-1} $Å^{-1}$. (From Anupama and Prabhu 1989)

volume and filling factor estimated from the radio VLBI observations. On the other hand, the coronal region is supposed to be formed by shock heating in the shell where the electron temperature is derived by the relative intensity of [FeXII]/[FeX] as given in column (5).

According to Bode and Kahn's model, coronal emissions and X-ray emission are produced by the collisions between slowly expanding shell ejected in the preburst stage and the high-velocity gas shell ejected in the burst stage. Contini et al. (1995) applied a shock-wave model to the nova ejected in the 1985 burst of RS Oph. Assuming a plane-parallel, stationary shock wave propagating in a nova shell, they calculated the structure of the shocked region and derived the relative emission-line intensities, which well reproduced the line intensities observed at day 201 after the outburst. They also derived the high helium abundance and very low oxygen abundance in agreement with the thermonuclear runaway scenario for RNe. Based on the high-resolution optical and ultraviolet line profiles, Shore et al. (1996) showed the existence of two components in the formation of emission lines. One is a broad-line component produced by emission from the high-velocity ejecta, and the other is a narrow-line component produced in the red giant wind that is ionized by

Table 6.16: The variations of parameters of gas shell after explosion of nova RS Oph (adated from Anupama and Prabhu 1989)

	Expansion velocity (km sV)		Electron density	Mass of gas shell	Electron temperature of coronal region from [FeXI]/[FeX]
Days[a]	Measured[b]	Theory[c]	$(10^8 \mathrm{cm}^{-3})$ [d]	$10^{-6} M_\odot$ [d]	10^6 K[e]
32.0	836 ± 192	792	30.6	3.7	1.50
59.9	470 ± 107	643	8.8	3.5	1.34
90.7	424 ± 127	560	3.4	3.0	0.98
108.4	431 ± 9	527	1.8	2.3	1.11

[a] Days after the burst.
[b] measured from the width of $H\alpha$ emission line.
[c] model calculation by Bode and Kahn (1985).
[d] estimated by the intensities of Hα emission.
[e] estimated by emission line ratio.

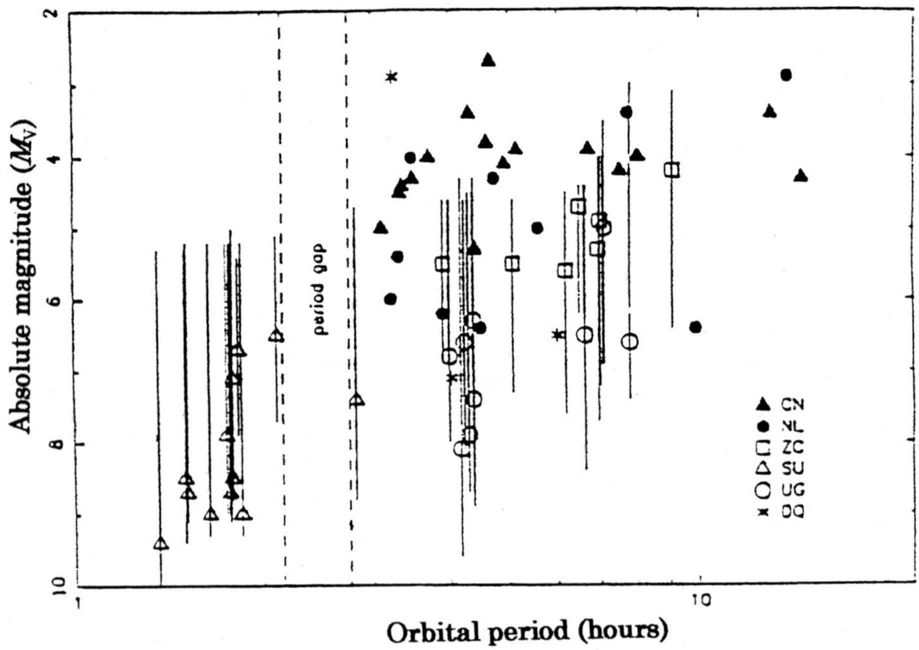

Figure 6.38: Absolute magnitudes and orbital periods of DNe (including some type of CVs). The abscissa is the logarithmic scale of orbital period in hours; the ordinate is the absolute magnitude of stars with variation range from minimum (lowest end) to maximum (highest end). For comparison, some types of CVs are also included. Stellar marks (see insert): ZC, Z Cam; SU, SU UMa; UG, U Gem; CN, classical novae; NL, nova-like variables; and DQ, Nova DQ Her. (From Horne 1991)

the UV pulse from the explosion or by radiation from the shock-heated region of the wind.

6.6.4 Dwarf novae

Types of dwarf novae

Dwarf novae also repeat outbursts like RNe, but with smaller scales as follows:

scale of outbursts—typically 2–5 magnitudes
intervals of outbursts—10 through several 100 days
duration of outburst—around 2 to 20 days

In the quiescent stage, the stars exhibit strong ultraviolet continuum and Balmer emission lines. During the outburst, emission lines are weakened or often disappeared and instead sometimes the He II line appears. A spectral atlas in the wavelength range λ 4000–5000 Å for 48 DNe in outburst state is provided by Morales-Rueda and Marsh (2002).

DNe are classified into the following three subtypes (the number of stars given in bracket for each type is taken from Warner 1995).

U Gem type or SS Cyg type. The interval from burst to burst varies from several days to several years. After an abrupt brightening of 2 to 8 magnitudes, it stays at maximum for 1 to 2 days and declines over several days or weeks (29 stars).

Z Cam type. This type is characterized by the standstill phenomena, i.e., an outburst cycle is interrupted by a phase of more or less constant standstill brightness between the outburst maxima and minima. The duration at standstill varies greatly from star to star, and with different instances of the same star, ranging from a few days up to more than a year (12 stars).

SU UMa type. In this type there are two distinct types of outbursts: normal outburst typically lasting for a few days, and superoutburst occurring after 3 to 10 normal outbursts, and lasting longer with larger amplitudes for about 2 weeks. The orbital periods are very short, mostly within 2 hours (34 stars).

A list of selected DNe is given in Table 6.17.

The DNe included in these types have different orbital periods and absolute magnitudes. We show the orbital period versus absolute magnitude diagram in Figure 6.38 (other types of CVs are also included) (Horne 1991). The stellar types are distinguished by different notations inserted in the diagram. It is evident that there is a gap in orbital period between 2 and 3 hours, and SU UMa type stars are the only stars found within it. Other types of stars lie in the region longer than 3 hours.

DNe are the binaries composed of the primary (late-type main sequence star or subgiant, filling Roche lobe) and white dwarf companion. Surrounding

Table 6.17: Selected dwarf novae with known orbital periods (adapted from Warner 1995)

Star	Brightness m_v (Min)	m_v (Max)	Orbital period $P_{\rm orb}$ (hours)	Spectral type	Subtype
CN Ori	14.2	11.9	3.917	M4V	U Gem
U Gem	14.6	9.4	4.2458	M4.5V	U Gem
SS Cyg	11.7	8.2	6.6031	K5V	U Gem
RU Peg	12.7	9.0	8.990	K2-3V	U Gem
BV Cen	12.6	10.7	14.6428	G5-8 IV-V	U Gem
GK Per	13.2	10.3	47.9233	K0IV	U Gem
KT Per	16.0	11.7	3.903	M3.3	Z Cam
RX And	13.6	10.9	5.037	—	Z Cam
HL CMa	14.5	10.5	5.148	—	Z Cam
V426 Oph	13.4	10.9	6.847	K2-4V	Z Cam
Z Cam	13.6	10.4	6.956	K7V	Z Cam
EM Cyg	14.2	12.0	7.178	K5V	Z Cam
SY Cnc	13.7	11.1	9.12	G8-9V	Z Cam
WZ Sge	14.9	—	1.36051	—	SU UMa
VW Hyi	13.3	9.5	1.78250	—	SU UMa
WX Hyi	14.7	12.5	1.79551	—	SU UMa
SU UMa	14.8	12.2	1.8324	—	SU UMa
YZ Cnc	14.5	11.9	2.0862	—	SU UMa

the white dwarf an accretion disk is formed by the accretion flow passing through the Lagrangean point L_1 from the primary. In contrast to CNe and RNe, the outbursts of DNe are triggered by the instability of the accretion disks. Two types of instabilities for the outburst mechanism have so far been proposed: the mass transfer instability (Bath 1975) and disk instability (Osaki 1974, 1996). Osaki (1996) discussed the burst model in detail, particularly based on disk instability. According to him, the unified explanation for the difference of DN subtypes can be obtained by two parameters of the binary orbital period and the mass exchange rate.

Spectral variations of SS Cyg in U Gem type

We now consider the spectral variation of SS Cyg as a typical sample of U Gem type DNe. Martinez-Pais et al. (1994, 1996) and Horne (1991) carried out a time-resolved optical spectroscopy in quiescence and outburst stage, and portrayed the characteristic behaviors as follows.

The spectrum of SS Cygni in quiescence is a combination of an absorption spectrum of K5 dwarf and strong emission of Balmer and HeI, HeII lines. The emission region probably is the accretion disk and the accreting flow. The intensities and profiles of these emission lines exhibit phase changes according to the orbital period of 6.60 hours as shown in Figure 6.39. The upper panel

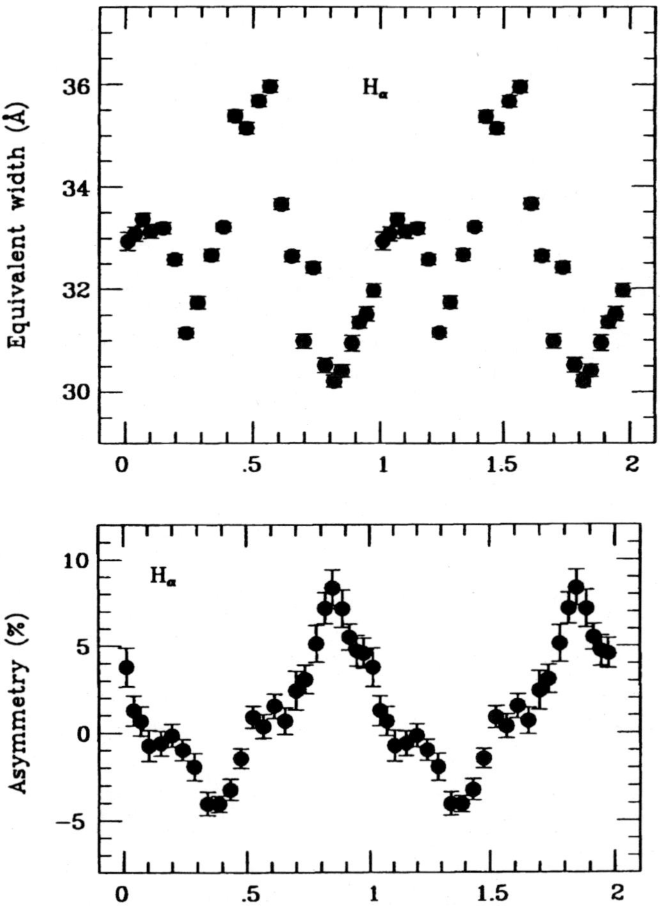

Figure 6.39: Phase variation of the Hα line in the quiescent phase of the dwarf nova SS Cyg. The abscissa is the phase in the orbital period of 6.60 hours. Upper panel: Emission equivalent width in Å. Lower panel: Asymmetry of double peak profile expressed in relative intensities of V (violet) and R (red) components. In the ordinate "plus" denotes V < R and "minus" denotes V > R. (From Martinez-Pais et al. 1994)

depicts the phase variation of the emission equivalent width of the Hα line, the lower panel shows the asymmetry of the double peaked Hα emission (V/R variation). The ordinate is "plus" when $V/R > 1$, and "minus" when $V/R < 1$, the asymmetry is expressed by the percent of V and R peak relative intensities. These phase variations of the Hα profile can be roughly explained as follows.

The equivalent width $W\alpha$ takes its maximum value at the phase $\varphi \sim 0.1$ and $\sim$0.5, where $\varphi = 0$ is defined as the superior conjunction (the cool star is nearer to the earth). At these phases the Hα emission shows nearly symmetric

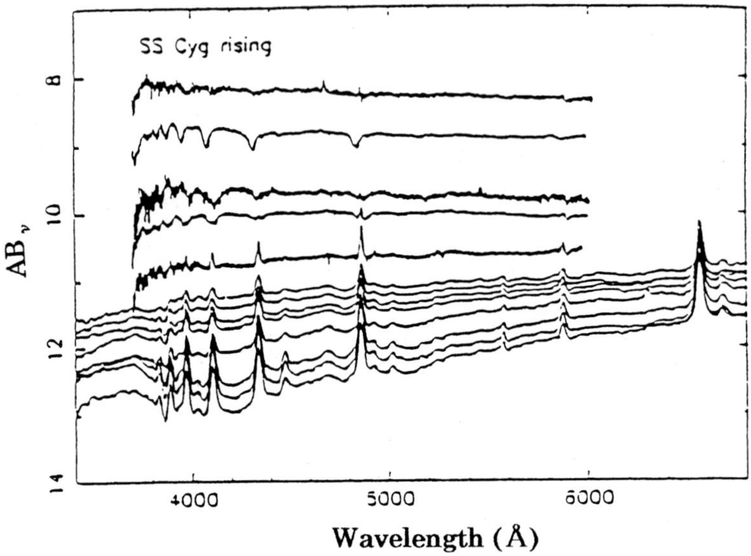

Figure 6.40: Spectral variation of SS Cyg in rising phase from quiescent (bottom) to maximum (top) state observed in 5 days. The ordinate is the intensity expressed by absolute magnitude scale. (From Horne 1991)

profile (V/R ~1). That is to say, the binary system is almost axisymmetric in line-of-sight velocity when viewed from the line of conjunction. The minimum value of $W\alpha$ appears at the phase φ ~0.25 and ~0.8, the direction nearly perpendicular to the line of conjunction, and the profile becomes asymmetric in the sense, V/R > 1 at φ ~0.25 and V/R < 1 at φ ~0.8. The emission region is thus dominant in approaching motion at φ ~0.25 and is dominant in receding motion at ~0.8. These profile changes suggest the asymmetric and non-Keplerian motion of the disk around the white dwarf.

The spectral variation of SS Cygni in the rising phase has been observed by Horne (1991) as shown in Figure 6.40. It exhibits the spectral change in 5 days from the quiescent state (bottom) through the outburst at maximum (top). In the rising phase the continuous spectrum became stronger and bluer, and the emission lines, which were strong in quiescence, had been weakend or disappeared by the strengthening of continuum. Near maximum light, however, another species of highly excited lines such as HeII appear as narrow emission lines, different from those in quiescence. The formation of highly excited emission lines could be explained either by the increase of ionizing radiation due the heating of accretion disk or by ionization and excitation due to electronic collisions in heated disk. In any case, highly excited emission lines are formed in the inner region of the disk. In the burst stage the spectrum of red dwarf becomes invisible due to the strong continuous radiation.

Spectra of other types of DNe

KT Per of Z Cam type was observed by Ratering et al. (1993) for both quiescent and outburst phases. The strong emission in the quiescent phase markedly weakened in the outburst phase resembles SS Cygni, but the emission lines do not show any appreciable dependence on the orbital phase (orbital period = 3 hour 55 minutes). The mean equivalent widths (Å) of KT Per in quiescence are Hα (46.4 ± 10.4), Hβ (44.7 ± 5.1), Hγ (40.6 ± 6.7), Hδ (31.6 ± 5.9), i.e., strong emission and slow Balmer decrements are notable.

YZ Cnc is an example of SU UMa type of DNe. The spectra at quiescence and at outburst are observed by Shafter and Hassman (1988). In quiescent phase, strong emission in the Balmer and HeI, FeII lines are remarkable along with the emission of HeII, as in case of Z Cam type. In burst phase, a strong ultraviolet continuum with strong absorption lines in the higher Balmer members appear, and emission remains only in Hα and the central part of Hβ line. In Figure 6.41 the spectra of YZ Cnc in quiescence and outburst are shown. Panel A displays the spectral change from outburst (top) to quiescence (lower), which occurred in two days. Panel B shows the spectrum at quiescence.

For seven SU UMa type stars, Tappert et al. (2003) analyzed the orbital variations of the Hα line profiles based on the method of Doppler tomography (see Section 4.2.4). They found that VW Hyi exhibited more or less nonorbital variations such as the existence of isolated emission sources like bright spots and of gas streams in the two-dimensional emission distribution in the binary systems. Two stars (AQ Eri, HS Vir), however, showed suborbital variations which are not appropriate to apply the Doppler tomography. In spite of their data sets obtained in medium spectral resolution and low S/N ratios, they could derive the nonorbital distribution of various types of emission sources in some stars. This implies that the Doppler tomography is a useful tool in spectroscopic analysis of binary stars by its careful application.

6.6.5 Nova-like variables

Types of nova-like variables

Among the CVs, there are some groups that do not show eruptive phenomena. We call them the nova-like variables (NLs), which are mainly divided into the following two subtypes designated by the appearance of emission spectra.

RW Tri type. Stars reveal almost pure emission spectra with few absorption lines.

UX UMa type. Stars are characterized by pertinent broad absorption with central narrow emission components.

Warner (1995) lists 53 NLs. Table 6.18 gives a selected sample of NLs in the above two subtypes.

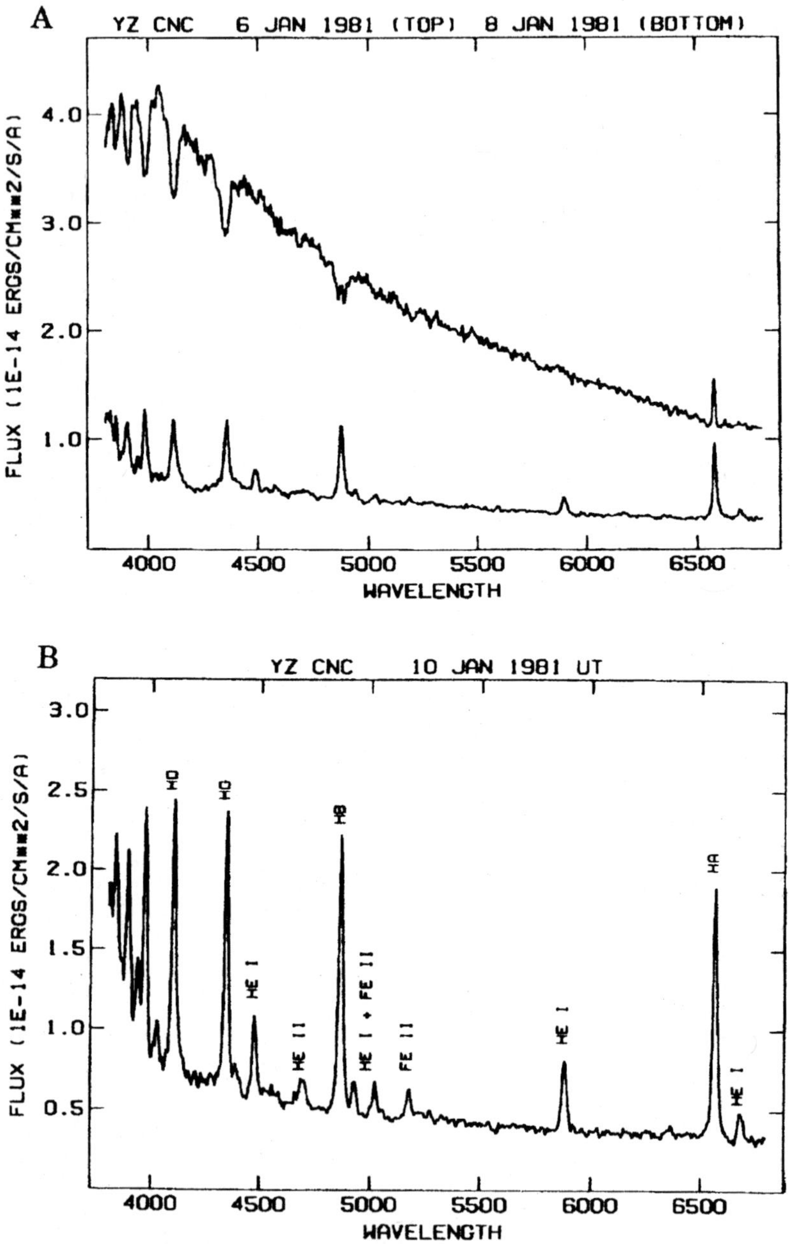

Figure 6.41: Low-resolution spectra of the dwarf nova YZ Cnc (SU UMa type) obtained by 152-cm telescope at Mt. Lemon Observatory. Panel A displays the spectrum at outburst (top) and at quiescence (lower), which is obtained two days after the burst. Panel B shows the spectrum at quiescence. Strong emissions in Balmer, HeI and HeII are notable. (From Shafter and Hessman 1988)

Table 6.18: Selected nova-like variables (adapted from Warner 1995)

Name	Subtype	Orbital period (hours)	V-magnitude	Spectral type
BK Lyn	UX	1.80:	14.1–15.1	—
MV Lyr	UX	3.201	12.1–17.7	M5V
DW UMa	RW	3.2786	14.9–17	—
UU Aqr	RW	3.9259	13.3–14.0	K7-M0
UX UMa	UX	4.7201	12.7	K8-M6V
V3885 Sgr	UX	5.191	9.6–10.5	—
RW Tri	RW	5.5652	12.5–13.4	K7
RW Sex	UX	5.8817	10.5–10.8	—
AC Cnc	RW	7.2115	13.8–14.4	G8-K2V
V363 Aur	RW	7.7098	14.2	K0V
RZ Gru	UX	10.01:	12.3–13.4	—

Subtypes: UX, UX UMa type; RW, RW Tri type. V-magnitude: range of brightness variations.

Spectral features for each subtype

As an example of RW Tri type, the spectra of the eclipsing NL variable DW UMa (PG 11030 + 590) observed by Shafter et al. (1988) during and out of eclipse are exhibited in Figure 6.42. The inclination angle of the orbit is supposed to be about 80°. The ordinate is the absolute scale of energy. The development of the Balmer, HeI, and HeII emission lines is remarkable and no absorption line is seen in either spectrum. During the eclipse, the weakening of high-excitation emission lines such as CIIλ 4267 Å, CIIIλ 4640 Å, and HeIIλ 4686 Å is notable. This suggests that the high-excitation emission lines are formed near the inner edge of the disk, and so are seriously affected by the eclipse. The Balmer decrements are Hα:Hβ:Hγ:Hδ = 1.09:1.00:0.90:1.46 (out of eclipse) and 1.14:1.00:0.98:1.54 (in eclipse). In both phases the decrements are generally flat and the high strength of Hδ is remarkable.

As an example of UX UMa type, the spectrum of RZ Gru observed by Stickland et al. (1984) is shown in Figure 6.43. As noted above, the broad absorption feature persistently exists. Strong and narrow emission lines are seen in the Balmer series from Hα to around H14 (with flat decrement) and in HeI, but HeII lines are not visible.

It may be interesting to compare the spectral features of DW UMa and RZ Gru. DW UMa is an eclipsing binary and then the inclination angle of the disk is estimated to be around 80°. In contrast, RZ Gru is supposed that its inclination angle is less than 20° by the narrow width of the emission lines. The difference of inclination angle may give the difference in the profiles of emission lines, i.e., if we suppose that both systems are composed of a cool star, a white dwarf, and a developed accretion disk, the white dwarf will produce the broad hydrogen absorption lines, while the optically thick disk

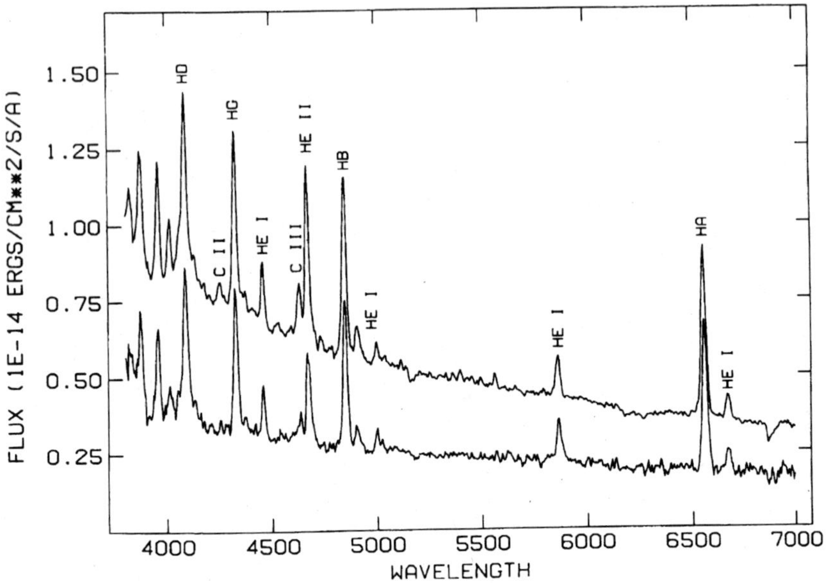

Figure 6.42: Low-resolution spectra of the NL DW UMa observed at the Mt. Lemon Observatory. The absolute fluxes (erg cm^{-2} s^{-1} Å^{-1}) both in (lower) and out (upper) of eclipse are plotted as a function of wavelength. Note the weakening of the flux of HeIIλ 4686 and CIIλ 4267 during the eclipse. (From Shafter et al. 1988)

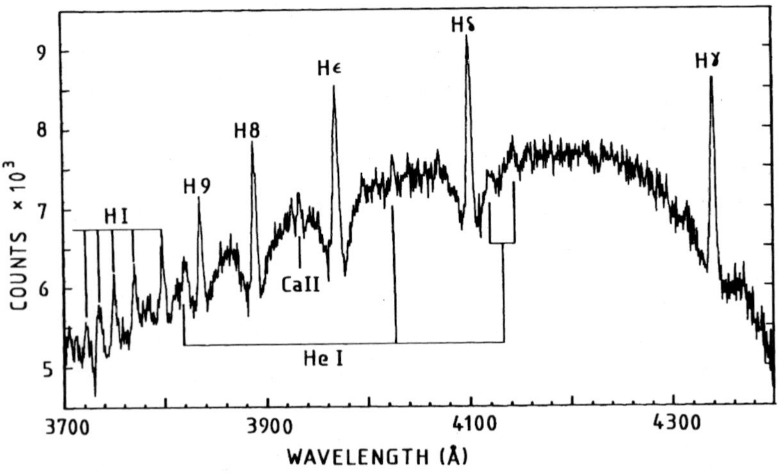

Figure 6.43: Optical spectrum of RZ Gru (UX UMa type NL) observed with 1.9-m telescope at the South-African Observatory. The superposition of strong emission and broad absorption in the Balmer series is conspicuous. (From Stickland et al. 1984)

will emit strong continuous radiation. If we observe this system with small inclination angle, we may observe the atmosphere of the white dwarf, forming broad absorption lines as in case of RZ Gru. According to la Dous (1991), strong Hα emission in RW Tri type NLs is well concentrated in stars with small inclination angles. On the contrary, if we observe this system with large inclination angle, the white dwarf will be embedded behind the disk and we only observe the strong continuum and emission lines, formed in the disk as in case of DW UMa. The subtypes of NLs may reflect the variations in the inclination angle and in the development of the disk.

6.6.6 Balmer decrements of CVs

Observations

The CVs exhibit a large variety of the Balmer decrements from very steep to very flat, or even inverted.

For the CNe and RNe, Downes et al. (2001) collected from the literature, unpublished data, and from their own observations the data on emission-line fluxes of [OIII], Hα, and Hβ in 96 nova shells years and even decades after their outbursts. When the Hα and Hβ emission-line fluxes are available in the same epoch of the observations, we can derive the Balmer decrements $D_{34} = F(\mathrm{H}\alpha)/F(\mathrm{H}\beta)$. Time variations of D_{34} are also observed for some novae. Examples of time variation for fast nova (V443 Sct) and RN (RS Oph) are shown in Figure 6.44.

The general trend of the decrements may be summarized as follows:

(i) Among the CNe, fast novae tend to concentrate in steep decrements (24 among 36 stars showed steep decrement of $D_{34} > 3$), whereas slow novae exhibited flat or inverted decrements in observed two stars.

(ii) When long-term variations are observed, most of flat or inverted decrements ($D_{34} < 3$) appear temporarily at some epochs in their history of variations.

(iii) RNe also show a variety from steep to inverted decrements (three among five stars showed steep decrements). Among these, U Sco exhibited an abnormally inverted decrement of $D_{34} = 0.02$ at 11.7 days after the light maximum ($t_3 = 4$ days). At this moment the forbidden line of [OIII] was absent, suggesting a high electron density. RS Oph showed the decrement from flat to steep as seen in Figure 6.44. In the quiescent phase, this star exhibited steep decrement with a large scatter from $D_{34} = 5$ up to $\sim$30 (Anupama and Mikolajewska 1999).

For DNe and NLs, flat nature of the decrements is seen from Figure 6.40 for SS Cyg (DN in quiescence), Figure 6.41 for YZ Cnc (DN), Figure 6.42 for DW UMa (NL), and Figure 6.43 for RZ Gru (NL).

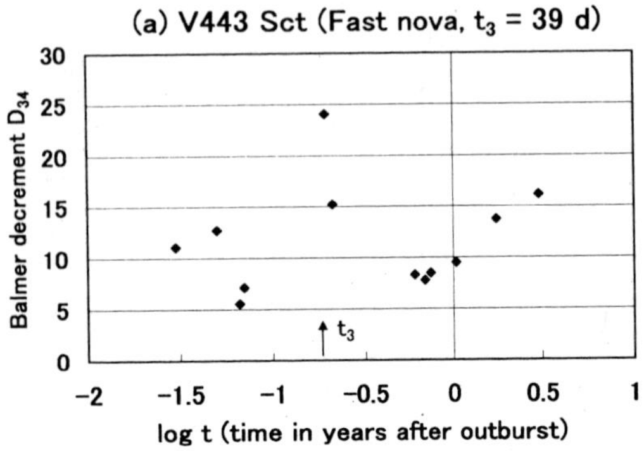

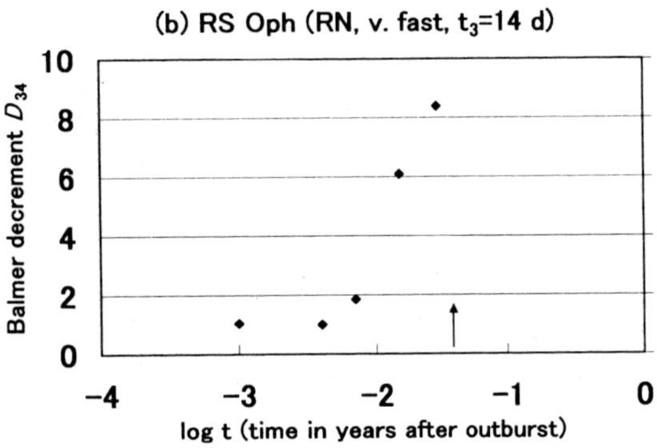

Figure 6.44: Time variations of the Balmer decrements D_{34} in novae. A, V443 Sct (F, $t_3 = 39$ days $= 0.1068$ year); B, RS Oph (RN, $t_3 = 14$ days $= 0.0383$ year). (Plotted based on the data of Downes et al. 2001)

Large scatter of the Balmer decrements in CVs is suggestive of sensitive dependence of these decrements on the physical parameters of the nova shells, particularly of electron density, during expansion phases.

Theoretical decrements

There are several theoretical approaches in deriving the Balmer decrements applicable to the CVs. Main models are summarized as follows.

(a) Moving envelope model. Gershberg and Shnol (1974) considered a moving envelope collisionally excited under a uniform electron temperature T_e. Using the escape probability by motion they solved the equations of statistical equilibrium for hydrogen atoms simplified for 30 energy levels and derived the Balmer decrements. The adopted parameters are the physical state of gas (T_e, N_e) and the escape probability β^0_{12} for the Lyman-α radiation. A part of the model calculations is shown in Table 6.19 for the two cases of $T_e = 10{,}000$ and 20,000 K. A large

Table 6.19: Balmer decrements in the moving-envelope model of Gershberg and Shnol (1974)

β_{21}^0		1	10^{-2}	10^{-4}	10^{-6}	10^{-8}	$\to 0$
$T_e = 10{,}000$ K							
$n_e = 10^8$	D_{34}	6.17	8.20	7.66	9.50	239	3.14
	D_{54}	0.33	0.28	0.30	0.28	0.53	0.31
10^{10}	D_{34}	6.18	8.25	9.41	37.5	3.45	3.12
	D_{54}	0.32	0.27	0.28	0.39	0.30	0.32
10^{12}	D_{34}	6.38	10.2	12.6	2.53	1.97	1.95
	D_{54}	0.32	0.27	0.25	0.55	0.67	0.69
10^{14}	D_{34}	4.99	5.15	3.20	0.72	0.71	0.71
	D_{54}	0.37	0.36	0.39	1.07	1.08	1.08
$T_e = 20{,}000$ K							
$n_e = 10^8$	D_{34}	4.13	5.56	5.21	87.7	2.66	2.45
	D_{54}	0.39	0.33	0.35	0.65	0.34	0.36
10^{10}	D_{34}	4.14	5.61	6.99	2.62	2.45	2.45
	D_{54}	0.39	0.32	0.34	0.35	0.36	0.36
10^{12}	D_{34}	4.41	7.42	2.10	1.83	1.83	1.83
	D_{54}	0.37	0.30	0.53	0.65	0.66	0.66
10^{14}	D_{34}	3.47	2.95	0.53	0.52	0.52	0.52
	D_{54}	0.44	0.44	1.22	1.27	1.27	1.27

variation of the decrement D_{34}, from very steep to very flat or even inverted, is notable depending on the values of parameters. When the envelope is optically thin ($\log \beta_{12^0} \sim 0$), the decrement takes a typical value in collisionally excited thin gas, i.e., $D_{34} \sim 5$–6 ($T_e = 10{,}000$ K) and ~ 4 ($T_e = 20{,}000$ K). Very steep decrements appear in the envelopes with low gas density ($Ne < 10^{12}$) and moderately opaque ($\log \beta_{12^0} = -6 \sim -8$), whereas the inverted decrement only appears in the highest electron density $N_e > 10^{13}$ cm^{-3} with β_{12^0} lower than 10^{-5} ($T_e = 10{,}000$ K) or 10^{-3} ($T_e = 20{,}000$ K).

When compared with the observational decrements of Downes et al. (2001), this model seems promising for explaining the large variation of observed values of D_{34}. For example, steep decrements of V443 Sct in Figure 6.44, though scattered in a range $D_{34} = 5$–25, can be explained by a larger value of β_{12^0}($>10^{-6}$) (optically thin) and a lower value of electron density ($N_e < 10^{10}$) in case of $T_e = 10{,}000$ K. In contrast, the change of D_{34} of RS Oph in Figure 6.44, from very flat ~ 1 to a steep one ~ 9 in its early burst phase ($t < t_3$) can be understood by a rapid expansion of gas shell from a dense state ($N_e > 10^{13}$) with sufficiently small value of β_{12^0} ($<10^{-8}$) (optically thick) toward a rarefied state

Table 6.20: Balmer decrements from accretion disk model (adapted from Williams 1980)

Accretion rate ($M_\odot$ per year	10^{-9}	10^{-10}	10^{-11}	10^{-12}
Disk luminosity (erg s^{-1})	2×10^{34}	2×10^{33}	2×10^{32}	2×10^{31}
$H\beta$ equivalent width (Å)	0.2	2.0	12	65
Line	Relative intensities			
Hα	0.94	1.05	1.09	1.23
$H\beta$	1.00	1.00	1.00	1.00
Hγ	0.89	0.86	0.84	0.75

($N_e < 10^{10}$) with still small value of β_{12^0}. In some novae, strong Balmer emission lines coexist with strong forbidden lines of [OIII]. This infers that the [OIII] lines are formed in the outer low-density ionized region of nova shells, since the Balmer lines are generally formed in the inner part of the shell where the electron density is higher than the critical density of [OIII] lines.

(b) Accretion disk model (Williams 1980). Let an accretion disk surrounding a white dwarf be heated by an accretion flow and emit the ultraviolet continuum in the inner borders of the disk. This UV radiation photoelectrically excites the outer parts of the disk and forms the emission lines. Williams (1980) calculated the fluxes of the Balmer lines under the condition of assumed LTE state in the disk. A part of his model calculation is shown in Table 6.20, calculated in the case of stellar mass $M_* = 1.0\ M_\odot$, stellar radius $R_* = 6 \times 10^8$ cm, and the outer radius of the disk $R_{\rm disk} = 4 \times 10^{10}$ cm. The decrements thus derived are generally very flat and the Hβ line becomes stronger than Hα when accretion rate is high. Williams (1980) argued that these results are in rough agreement with the observed Balmer spectra of DNe at minimum light.

(c) Nebular model with high gas density. As stated in Chapter 4, Drake and Ulrich (1980) extended the nebular approximation to the case of high gas density by using the escape probability by scattering. In this scheme, the Balmer decrement D_{34} takes values in a wide range from very flat to very steep depending on the electron density and the optical depth $\tau(H\alpha)$. Very flat decrement appears when the electron density is higher than around 10^{13} cm^{-3} as seen in Figure 4.6.

Drake and Ulrich's model has been applied to various types of CVs: for WZ Sge (dwarf nova) by Mason et al. (2000), for SU UMa (dwarf nova) by Echevarria et al. (1996), and for RS Oph (recurrent nova) by Dobrzycka et al. (1996). In particular, Echevarria (1988) applied this model to 148 DNe including various types of novae and nova-like stars, and statistically showed that the line ratios Hβ, Hγ, and Hδ can be reproduced by this model applied for a static layer of hydrogen gas at a high density.

(d) Nebular approximation with strong incident radiation. It is pointed out by Elitzur et al. (1983) that, if strong continuous radiation enters a gas layer, stimulated emission operates to modify the line ratios and finally flattens the Balmer decrement. They solved the radiation fields of plane parallel layer by making use of the escape probability by scattering and taking into account the effect of incident radiation to stimulate the induced emission. This effect is represented by the new parameter U, defined as

$$U = \frac{4\pi I_0}{hcN_e}, \tag{6.6.1}$$

where I_0 is the intensity of incident radiation at the ionization limit, N_e is the electron density.

The effect of the strong incident radiation can be seen by the following sample case with the parameters: $T_e = 10^4$ K, $N_e = 10^{11}$ cm^{-3}, $\Delta v = 10$ km s^{-1} (thermal Doppler velocity), τ (Lα) = 150 (optical thickness for the Lyman-α line). Elitzur et al. (1983) calculated the Balmer decrement D_{34} as a function of the parameter U, taken in a range from 10^{-3} up to 10^3. According to this model, the decrement D_{34} is close to the nebular approximation (Case B) when $U \ll 1$, and it tends to flatten as U increases. Finally the decrement turns to be inverted when U exceed several hundreds, which correspond to an opaque layer of τ(Hα) > 100. In this way, Elitzur et al. (1983) explained the observed flat decrements of the fast nova V603 Aql, which is D_{34} = 1.0–1.17 according to Downes et al. (2001).

6.7 Symbiotic stars

6.7.1 Symbiotic stars and classification

The name of "Symbiotic stars" was first given by Merrill (1958) to the "stars of composite spectra" in the emission-line star catalogue (Merrill and Burwell 1933). Actually, "stars of composite spectra" include many different types of stars showing different ionization stages so that "symbiotic stars" are confined to stars having particular spectral features as defined by Kenyon (1992). According to him, a symbiotic star displays

(i) a red continuum and absorption features of a late-type giant star (spectral type K or M), which includes CaI, FeI, H_2O, CO, and TiO lines, among others
(ii) a blue continuum with bright HI and HeI emission lines and either
  (a) additional emission lines from ions such as HeII, [OIII], [NV], and [FeVII] with an equivalent width exceeding 1 Å, or
  (b) an A- or F-type continuum with additional absorption lines from HI, HeI, and singly ionized metals.

Table 6.21: A selected list of symbiotic stars

Star	HD	Magnitude Max/Min[a]	Spectral type[b]	Variable type[c]	Infra red type	Orbital period (days)[d]	Distance (Kpc)[e]
AX Per	—	10.8/13.0	M5.2 II-III	ZAnd	S	682	—
RX Pup	69190	11.1/14.1	≧ M5	M(332.5)	D	—	1.0–1.3
RW Hya	117970	9.7/10.9	M2 III	ZAnd	S	376	1.3
AG Dra	—	9.1/11.2		ZAnd	S	554	—
YY Her	—	11.7/13.2	M2 II-III	ZAnd	S	150–200	3.6–8.3
CH Cyg	182917	7.1/9.1	M6.5 II	SR(100)	S	5750	—
CI Cyg	—	10.8/13.0	M4-5 III	NL	D	855	1.5
V1016Cyg	—	11.3/17.5	M4-7 III	Zand	D	3467	1.3–6.0
AG Peg	207757	6.0/9.4	M2-3 III	ZAnd	S	733–827	1.3
Z And	221650	8.0/12.4	M3-6 III	NL,ZAnd	S	756	1.1
RAqr	22800	5.8/11.5	M7III	M(386.9)	D	44 years	0.2–0.3

[a] Photographic or visual magnitude at max/min (Kukarkin and Parenago 1970, GCVS).
[b] Spectral type due to Meier et al. (1994).
[c] Variability type is taken from GCVS. ZAnd, Z And type (symbiotic binary star), NL, nova-like; M, Mira type (with period in days in bracket); SR, semiregular type (variability scale in days is given in bracket).
[d] Infrared type—S, ordinary M star type; D, dust type (Allen 1979).
[e] Distance is due to Meier et al. (1994).

Symbiotic stars are generally irregular variables and often give rise to outburst phenomena. The appearance of spectral features depends on the star or on the state of activity (quiescence or outburst) for a star. The stars that exhibit outbursts resembling very slow novae (Nc in CNe) are called symbiotic novae, and will be considered in Section 6.7.3. Some stars, such as T CrB and RS Oph show RN-like explosion and often classified as RNe or as symbiotic stars by different observers.

In Merrill and Burwell's catalogue (1933), prototype stars are limited to several stars including AX Per (MWC 411), RW Hya (MWC 412), CI Cyg (MWC 415), and Z And (MWC 416), all of which have numbers in Mt. Wilson Catalogue for B type emission line stars. Allen (1984) listed 144 galactic symbiotic stars and Belczynski et al. (2000) catalogued 188 stars with 30 suspected objects. In the latter catalogue the data on V, K magnitude, as well as radio, IR, IUE, and X-ray data are included. Munari and Zwitter (2002) presented a multiepoch spectrophotometric atlas of 130 symbiotic stars in electronic form. Table 6.21 yields a selected list of symbiotic stars.

Based on the color indexes H, K, and L bands, Allen (1979) classified symbiotic stars into two types S and D according to the infrared features in late-type stars (J–H $> \sim$1.0). Later this classification was extended to earlier type stare (J–H $\sim <$ 1.0) and designated as D' type. The typical ranges of color indexes for S, D, and D′ types are given in Table 6.22 based on Munari et al. (1992).

Table 6.22: Color index ranges of symbiotic stars (based on Munari et al. 1992)

Type of stars	J–H	H–K	K–L	Number of stars
S	0.8–1.5	0.1–0.3	0.1–0.6	13
D	1.2–2.2	1.0–2.0	1.0–2.2	17
D′	0.4–1.0	0.2–0.8	0.5–2.0	4

As seen in Table 6.22, S type stars have color indices similar to the ordinary red giants, D type stars exhibit infrared excesses in H–K and K–L, and D′ type stars are also characterized by high H–K, and K–L colors. By far-infrared observations (12–100 μm) with the IRAS satellite, Anandarao et al. (1988) detected infrared excesses in both S and D (D and D′) types indicating the existence of dust shell. The dust temperature, however, is as high as 800 K in S type stars, while as low as 300 K in D type stars.

6.7.2 Spectral features

Optical spectra

Symbiotic stars generally exhibit strong emission lines in the optical spectra in every type of S, D, and D′. Two samples of optical spectra observed by Oliverson and Anderson (1982) are illustrated in Figure 6.45—AX Per, a typical S type star (panel a), and CI Cyg, a typical D type star (panel b). In both stars upper panels are low-dispersion spectra in the blue region, and the lower panels show high-dispersion spectra near the Hα region.

Both in AX Per and CI Cyg, strong Balmer lines, high excitation HeII, and forbidden lined such as [OIII], [NeIII] are remarkable. However, closer inspection discloses some different features in S (AX Per) and D (CI Cyg) types.

In S type stars, the permitted lines such as Balmer lines, HeI, HeII appear strong, while forbidden lines like [OIII], [NeIII] are weakly present or almost fade away. The Balmer decrement is generally steeper than Hα/H$\beta \sim 2.8$ (Case B).

On the other hand, in D type stars, forbidden lines are well developed in a wide range of excitation from [OI] up to [FeVII] and [ArIV] depending on stars. The Balmer lines are strong and the decrements are also appreciably steeper than the nebular case. The wide range of ionization levels may express either some stratification or strong inhomogeneity.

In the catalogue of symbiotic stars, Allen (1984) presented low-dispersion spectra of 127 stars (101 S stars, 22 D stars, and 4 D′ stars). The wavelength range is λ3400–7500 Å, including the Balmer continuum emission in the blue end, and some of the stronger TiO bands in the red end. The slight difference between S and D type stars is as stated above by Oliverson and Anderson (1982). D′ stars exhibited emission lines similar to D type stars, except UV

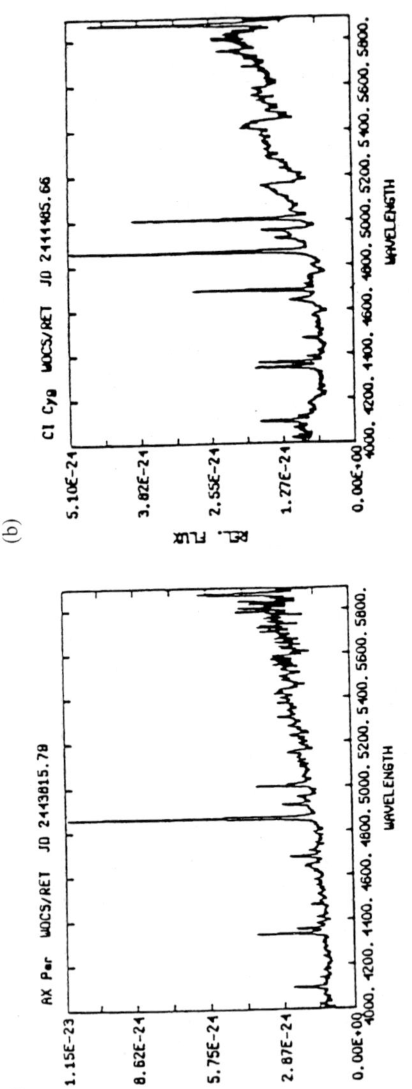

Figure 6.45: Samples of optical spectra of symbiotic stars observed with the Echelle spectrograph attached to the 90-cm telescope at Washington Observatory. Panel (a) AX Per (M5II) (Z And type) (infrared S type), (b) CI Cyg (M4-5 III) (nova-like type) (infrared D type), In both stars upper panels indicate low-dispersion. lower panels are the high-dispersion spectra near Hα. (From Oliverson and Anderson 1982)

Aur, which showed emission line only in $H\alpha$, and strong absorption in other Balmer lines and HeI lines.

The Balmer decrements of symbiotic stars are generally steeper than the nebular decrement (Case B) both in S and D types. The observed decrements D_{34} take the values of 4.07–22.89 for 6 D-type symbiotic stars (Gutiérrez-Moreno and Moreno 1996), and of 5.48–22.89 for 14 S-type stars (Gutiérrez-Moreno et al. 1996), If the interstellar reddening is corrected, these values will reduce to some degree, even so, the decrements may still be higher than the nebular case (D_{34} ~2.8).

If we apply the moving envelope theory of Gershberg and Shnol (1974), observed steep decrements may be explained by an optically thin envelope (β_{12^0} ~0–10^{-4}), or by an envelope with low electron density ($N_e < 10^{12}$ cm^{-3}) in a range of T_e = 10,000–20,000 K (see Table 6.19).

Schwank et al. (1997) carried out non-LTE calculations for an expanding red giant's atmospheres ionized from outside by the radiation of a nearby hot radiation source. They considered an S-type symbiotic binary composed of a hot white dwarf of typically R_* ~0.1 $R_\odot$, T_* ~100,000 K, and a late-type giant, typically MIII. The red-giant atmosphere is divided into HII and HI regions by an ionization front. One of their models is schematically shown in Figure 6.46. They found that the observed Balmer emission lines are predominantly formed close to the recombination zone near the ionization front, and

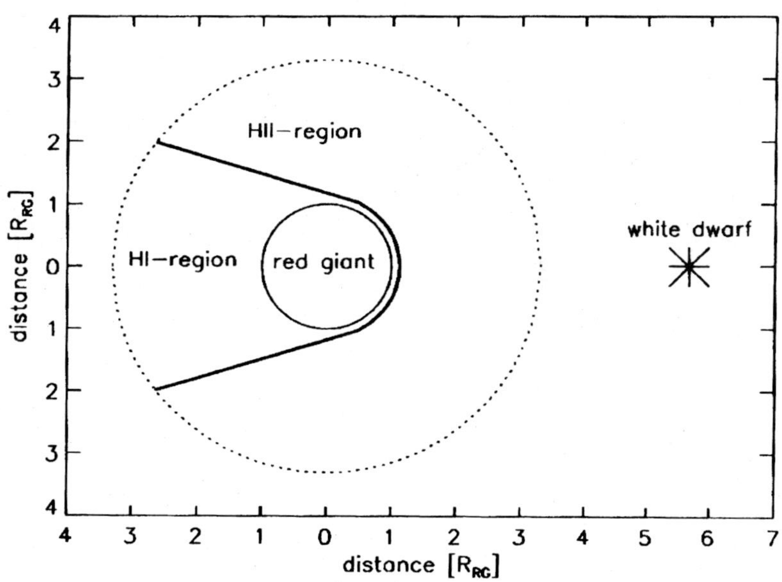

Figure 6.46: Schematic geometry of a Symbiotic star composed of a red giant and a white dwarf. Distance unit is the radius of red giant (R_{RG}). Ionization front separating HI and HII regions in the atmosphere of the red giant is also shown. (From Schwank et al. 1997)

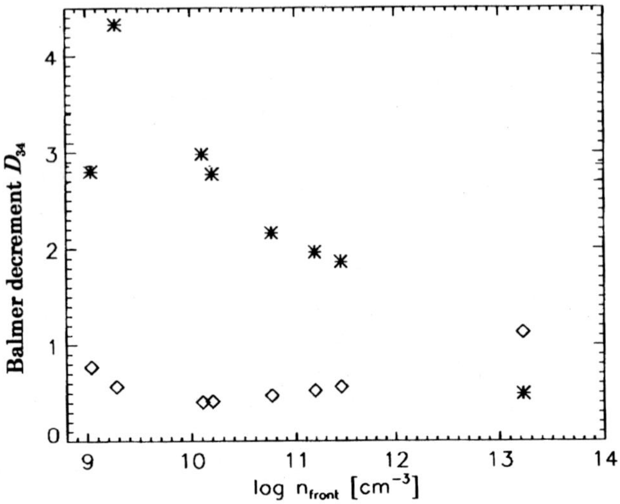

Figure 6.47: Theoretical Balmer decrement for a series of model calculations. The ordinate is the calculated decrement D_{34} ($\star$) and D_{54} ($\diamond$), and abscissa is the gas density of the corresponding recombination zone. Different point gives different model. (Partly reproduced from Schwank et al. 1997)

Balmer decrements of the model calculations depend on the densities in the recombination zone of the radiatively ionized wind. This picture is illustrated in Figure 6.47 where the decrement D_{34} is plotted as a function of electron density calculated for a series of theoretical models. In order to explain the steep Balmer decrement D_{34} of symbiotic stars, low gas density lower than around 10^{10} cm^{-3}, is required in these model calculations.

Under the similar situation (a red giant + a hot white dwarf system), Contini (2003) analyzed the emission-line spectra of AG Peg and derived the Balmer decrement D_{34}. In order to explain the steep decrement ($D_{34} \sim 5$), Contini emphasized the role of shock heating coupled with the photoionization from the hot star. The shocks may be created by collision of winds from both stars.

Spectral features in ultraviolet and radio-wave regions

The atlas of far-ultraviolet spectra obtained with the IUE satellite for 32 symbiotic stars are given by Meier et al. (1994). The spectral range is λ1200–2000 Å, where a number of emission lines with a wide range of ionization stage is identified, including permitted, forbidden, and semiforbidden lines. In Figure 6.48, the IUE spectra of the symbiotic star AG Dra in outburst and quiescence are shown. The left two panels denote the burst spectra, while right two panels the quiescence spectra. The upper and lower panels illustrate the spectra in different dates. In burst phase, strong ultraviolet continuum

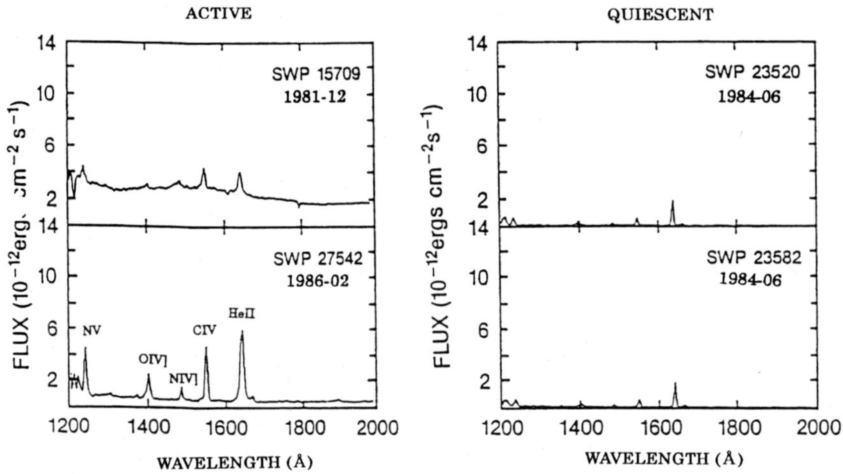

Figure 6.48: The IUE spectra of symbiotic stars AG Dra. The left two panels denote the burst spectra, while right two panels the quiescence spectra. The upper and lower panels give the spectra in different dates. The wavelengths of lines in low-dispersion spectrum in the lower left panel are as follows: NV λ 1239,1243; OIV] λ 1397, 1400, 1401, 1405; NIV]λ 1483, 1486; CIVλ 1548, 1551 ; HeII λ 1640; OIII]λ 1661, 1666. (From Meier et al. 1994)

and strong emission in permitted (e.g., NV, CIV, HeII) and semiforbidden lines (e.g., OIII], OIV], NIV]) are conspicuous. The formation of ultraviolet continuum and strong emission lines suggests high effective temperature 50,000–100,000 K of UV radiation and high electron density ($N_e \geqq 10^7$ cm^{-3}) in the emitting regions.

In radio wave regions, Seaquist et al. (1984) carried out survey observations with VLA (very large array) at 4.885 GHz band (partly containing the 1.465 GHz band) for 59 symbiotic stars, among which radio sources were detected for 17 stars. In every case radio spectrum is the thermal radiation from ionized gas. For most of detected stars, the strength of Hβ emission and the radio flux densities reveals a positive correlation. In addition, the radio intensity increases with the advance of spectral type from M1 to M8. Infrared features denote that 12 stars are S type, 4 are D type, and 1 unknown type among 17 radio-detected stars. It is noticed that some stars showing strong emission lines in the optical and ultraviolet region, such as AX Per, T CrB, AG Dra, and CI Cyg, are not included in the radio-detected stars.

The SEDs of 34 symbiotic stars from optical to radio wavelengths were presented by Ivison et al. (1995). They showed that the spectral characteristics of each IR type are distinct form one another: the SED peaks appear at 1$-$2 μm for S type, at 5$-$15 μm for D type, and 20$-$30 μm for D$'$ type, stars. In general, the D and D$'$ type stars are noticeably brighter at all wavelengths beyond the near IR. From these behaviors it is generally accepted that S type stars are the first-ascent giants, D type stars are the asymptotic giant branch (AGB) stars, and D′ type stars are yellow giants.

6.7.3 CH Cygni, spectrum, and its variation

As an example of well observed symbiotic stars we consider CH Cyg, which was found to be a binary system by the spectroscopic observations of Yamashita and Maehara (1979).

This system is a long-period binary (period ~15.7 years), composed of an M-type supergiant which is a semiregular pulsating star (period 90–100 days) and a white dwarf. This binary has been observed for over 100 years since 1885 (Mikolajewski et al. 1990, Bode et al. 1991). Before 1963 no remarkable activity was observed. Thereafter CH Cyg experienced four outbursts with different scale prior to 1986.

Mikolajewski et al. (1990) proposed for CH Cyg a model of a magnetic white dwarf with accretion from the M giant wind. This system is separated into three different phases according to the hot component's activity:

(1) Quiescent or inactive state. This is the phase of accumulation of material in a ring or disk above the white dwarf's magnetosphere by capturing the gas from the wind of the giant component rather than from accretion flow. The light variation in this phase is mostly caused by the variation of the red giant.
(2) Propeller state or low active state. Accretion of the ring matter onto the surface of magnetosphere is the main process. A part of the accreting gas is recoiled by the propeller action (see below) and forms a jet stream in both side of the star.
(3) High state. The matter of magnetosphere accretes onto the white dwarf's surface through accretion columns. When the amount of accreting matter exceeds some critical value, there occurs an outburst as an NL.

The propeller action is a mechanism proposed by Illarionov and Sunyaev (1975) to explain the accretion disk of neutron stars (Figure 6.49). If there is a rapidly rotating magnetosphere inside the accretion disk, the magnetosphere acts like a rotating propeller to recoil some parts of accretion disk gas outwardly and forms jet-like structure in two directions. The high-velocity component as high as 700 km s^{-1} appear in the Balmer emission lines in the propeller state of CH Cyg. Leedjärv and Mikolajewski (1995) suggested that these velocity components are formed in the gas ejected by the propeller action in this star.

6.7.4 Symbiotic novae

Among symbiotic stars, some stars explode once in a historical time as in CNe. Unlike ordinary CNe, their outbursts proceed very slowly, just like very

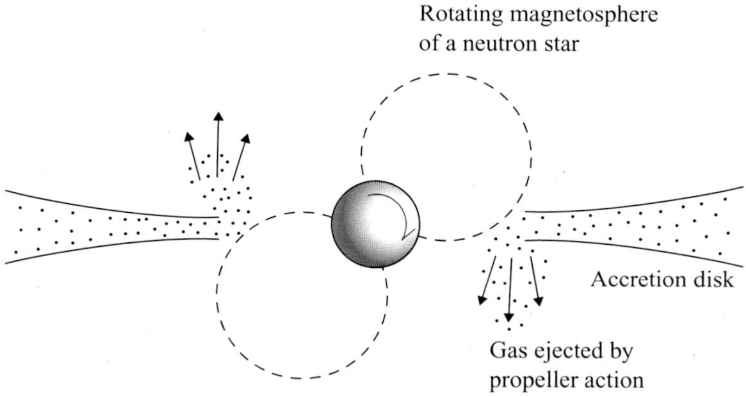

Figure 6.49: Schematic picture of propeller action in an accretion disk. An inner part of accretion disk matter is recoiled into two directions by a rapidly rotating magnetosphere around a neutron star.

slow novae (Nc) in CNe. These outbursts are called symbiotic novae, though sometimes confused with classical Nc or RNe. A list of selected symbiotic novae is given in Table 6.23.

In Figure 6.50 we show the samples of light curves of symbiotic novae (Arkhipova et al. 1990). The abscissa is 1000 days for one scale and the origin is taken at a point on the light curve just before the maximum luminosity for each star. In the figure, HR Del, the CN (slow nova, Nb), is added for

Table 6.23: A list of selected symbiotic novae

	Year of explosion		Visual magnitude					
Star[a]	Start	Maximum	Pre.	Max.	Post	Spectral type at Max[b]	Infrared type[c]	Note[d]
AG Peg	1855	1871	9	6	8.3	—	S	SB
RT Ser	1909:	1923:	>16	9.5	13	A8	S	RN
RR Tel	1944	1967	14v	6	11	F5	D	Mira
V1016 Cyg	1964	1967	15	11	11	em	D	Mira
V1329 Cyg	1964–66	1967	14v	11.5	13–14	em	S	EB
HM Sge	1975	1975	18	11	11	em	D	Mira
PU Vul	1978	1982–3	15v	8.8	9	A7	S	

Note: [a]Stars are selected from Viotti (1990).
[b] Spectral types due to Kenyon (1986) are as follows; AG Peg (A or later); RT Ser (F supergiant); RR Tel (F supergiant); V1016 Cyg (Planetary nebula); HM Sge (planetary nebula).
[c] Infrared type is the same meaning as in Table 6.21.
[d] SB, spectroscopic binary; RN, recurrent nova; Mira, Mira type infrared variability; EB, eclipsing binary.

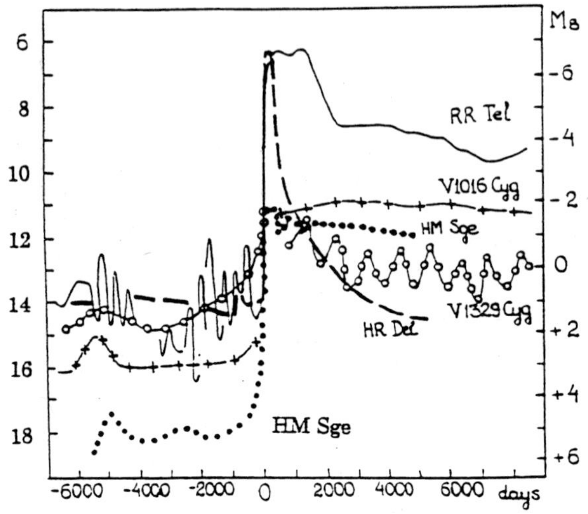

Figure 6.50: Comparison of the blue light curves of the symbiotic novae RR Tel. V 1016 Cyg, V1329 Cyg, and HM Sge shifted in time to coincide at the time of rise to maximum. Small-scale light oscillations are smoothed throughout. The light curve of HR Del (classical slow nova, Nb) is also shown for comparison. The scale of the ordinate in the left-hand side denotes the apparent B magnitude of HM Sge, and that in the right-hand side is the absolute B magnitude. (From Arkhipova et al. 1990)

comparison. Short-term variability is smoothed out throughout the delineated light curves.

Though the observations prior to the outbursts are not well sampled, the stars demonstrate the characteristics of symbiotic stars repeating irregular variations of 2 to 3 magnitudes before the bursts. The duration of brightening seems to differ from star to star ranging from less than 1 year up to several years. After the burst maximum, the 3-magnitude decline, (T_3), takes 7–10 years (RT Ser, RR Tel) to 12–40 years (AG Peg, V1329, Cyg, PU Vul) repeating light fluctuations of 1–2 magnitudes. The spectra at light maximum are estimated as F-type supergiants (RT Ser, RR Tel), or planetary-nebula like (V1016 Cyg, HM Sge), as predicted by thermonuclear runway model of symbiotic novae (Kenyon and Truran 1983).

As an example of spectroscopic observations, we consider the outburst in V1329 Cyg (HBV 475). This star gradually brightened in 1964–1966 reaching the maximum light in 1967, and then slowly declined as seen in Figure 6.50. The first spectroscopic observations were carried out in 1969 by Baratta and Viotti (1990), when numerous emission lines including forbidden lines were already seen. According to Baratta and Viotte, the emission-line profiles in this epoch can be separated into the following three types:

(i) The narrow emission lines. Besides the strong Balmer lines, permitted lines of singly or doubly ionized metals of C, N, O, and Ne are visible with the radial velocities of −10 to −30 km s^{-1}. These lines are supposed to be formed in the dense part of the stellar winds of M supergiant and ionized by ultraviolet radiation from the hot component.

(ii) The broad emission lines. The Balmer emission lines of $H\alpha$ through $H\zeta$ reveal wide wings. The highly excited emissions of HeIIλ 4686 Å, [NeIII], and [OIII] are composed of several separate components with radial velocities ranging from −300 to +300 km s^{-1}. These lines are supposed to be formed by cloudlets ejected from the hot star with different velocities at the time of outburst. These cloudlets are characterized by low electron density of the order of 10^6 cm^{-3} and high electron temperature around 8000 K.

(iii) WR-type emission lines. This system is composed of broad and weak emissions similar to the spectra of WN5 stars with an expansion velocity of 2300 km s^{-1}. This types of lines disappeared in the 1970s. These WR-like emissions are supposedly formed in the surface of the white dwarf companion temporally entered into a WR state just after the outburst.

The spectrum of V 1329 Cyg has shown complicated, time-variable line profiles. Figure 6.51 exhibits the phase variations of the $H\alpha$ and [OIII] emission profiles, observed by Ikada and Tamura (2000 a,b) during 1983 and 1994 at the Okayama Astrophysical Observatory using the Coudé spectrograph attached to the 188-cm reflector. They carried out the Gaussian deconvolution of emission profiles into three components and analyzed the kinematic structure of the ionized region in V1329 Cyg. It is found that this star is a binary system composed of a mass-losing giant ($M_1 \approx 2.1\,\mathrm{M}_\odot$) and a hot white dwarf ($M_2 \approx 0.65\,\mathrm{M}_\odot$) with mass function $f(M) = 1.2\ M_\odot$ At phase $\varphi = 0.93(-0.07)$, the cool star is at the inferior conjunction (i.e., between the observer and the hot star). The inclination angle of the orbital plane is almost 90° (edge-on). Figure 6.52 illustrates a schematic picture of the structure and mass flow in the system. The diameter of the binary orbit may be within 5 AU.

Based on the spectroscopic observations of 1987, Schmid and Schild (1990) analyzed the relative intensities of emission lines and estimated the physical parameters and chemical compositions of gas cloudlets. The results showed that nitrogen is around 10 times overabundant as compared to the Sun, but no such overabundance was detected in He, C, and O. This property seems common in symbiotic novae, since similar abundance was found in HM Sge and V1016 Cyg. On the other hand, CNe reveal the overabundance of He and CNO elements and they are supposed to be produced by nuclear reaction at the surfaces of white dwarfs. Thus Schmid and Schild suggested that the outburst processes of symbiotic novae are different from those in CNe (see next section).

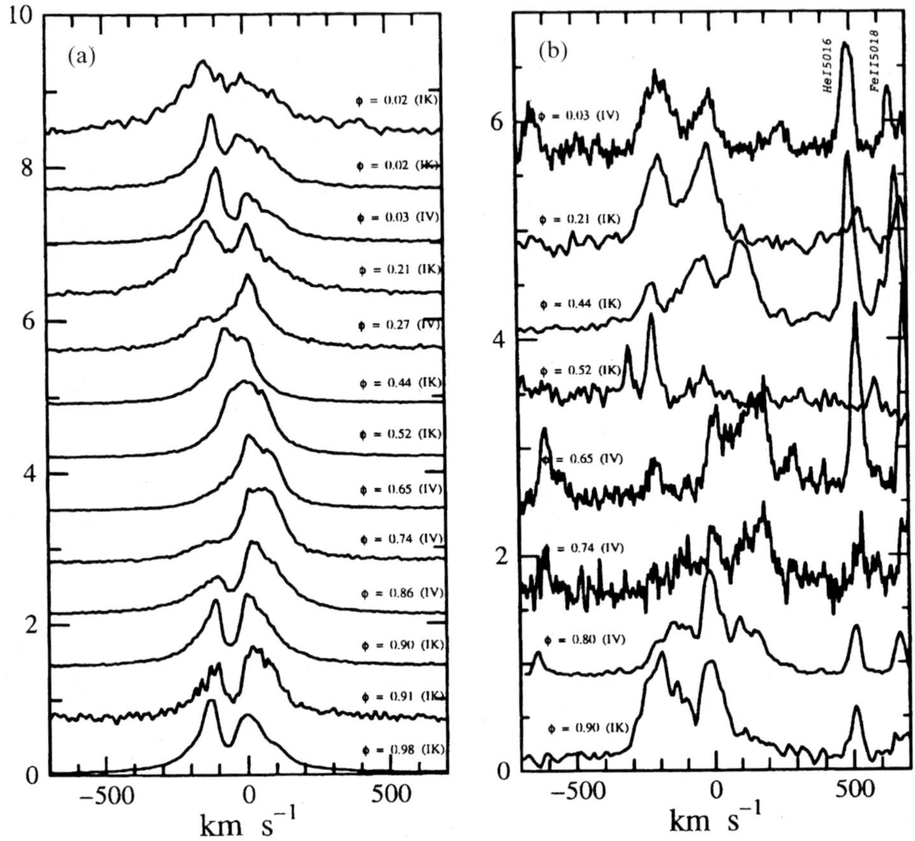

Figure 6.51: The phase variations of the emission-line profile of the symbiotic nova V1329 Cyg (HBV 475): Panel (a) $H\alpha$ and panel (b) [OIII]. Horizontal axis denotes heliocentric radial velocity, and vertical axis relative intensity, each line is normalized at the intensity = 1.0. Phase $\varphi = 0$—hot star in front, $\varphi = 0.5$—giant star in front. (From Ikeda and Tamura 2000b)

6.7.5 Binary nature and evolutionary state of symbiotic stars

Binary nature

When Merrill (1958) introduced the name of symbiotic stars and reviewed the nature of their peculiar spectral behaviors, the controversy on whether the symbiotic phenomena is related to a single star or to a binary system was not conclusive, due to the difficulty of confirming the orbital parameters. According to a single-star hypothesis, a symbiotic star is a red giant surrounded by a hot and dense coronal envelope (e.g., Gauzit 1955). In the last two decades, however, photometric, spectroscopic, as well as satellite observations

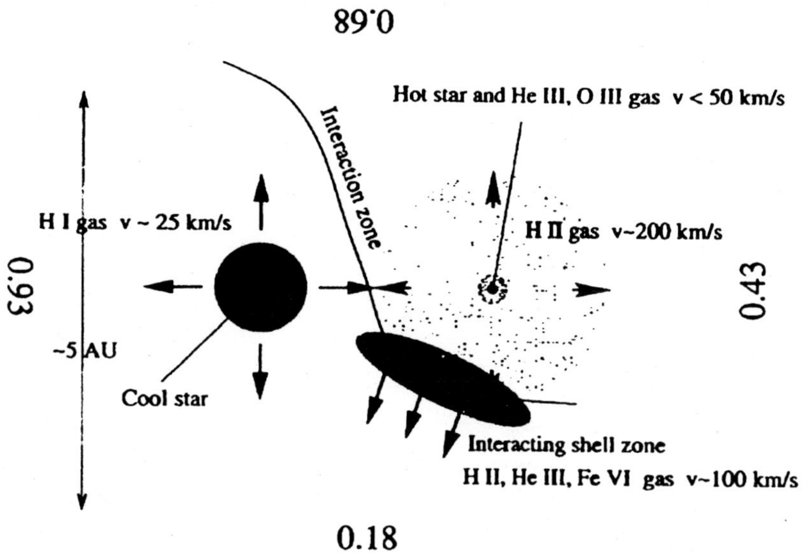

Figure 6.52: A sketch of the structure of gaseous nebulae around V1329 Cyg (HBV 475), viewed face-on. The numbers (= 0.18, 0.43, 0.68, and 0.93) represent the phases. The observer is in the direction of the number at each phase. (From Ikeda and Tamura 2000b)

have accumulated the data supporting the binary hypothesis. Binary system for the symbiotic phenomena is now widely accepted.

According to Kenyon (1992), the symbiotic binaries are broadly classified into three types of semidetached, detached (I), and detached (II), as summarized in Table 6.24.

(a) Semidetached binary. This type of symbiotic binaries consists of a Roche-lobe filling red giant and a main-sequence star companion, and most closely related to other types of interacting binaries. The accretion disk around the hot star is partly ionized and produces most emission lines. Some high-ionization lines could be formed in a bipolar structure at large distance from the disk plane. During an eruption, the accretion disk brightens dramatically by the increase of mass-flow rate from 10^{-5} to $10^{-3} M_\odot$ per year.

(b) Detached binary (I). This is the most abundant type of symbiotic binary and contains a fairly normal red giant star that underfills its Roche lobe and loses material via a stellar wind. The hot component is essentially a white dwarf, having temperature ranging about 30,000–200,000 K and radius of 0.01–1 $R_\odot$. The hot component is surrounded by a photo-ionized nebula that has a radius of 1–200 AU, where emission lines are formed by UV radiation from the hot component.

(c) Detached binary (II). The primary component is a Mira variable instead of a nonpulsating red giant star. Roughly 20% of all known

Table 6.24: Types of symbiotic binary stars (based on Kenyon 1992)

	Stellar system				
Binary type	Primary	Secondary	Orbital period (years)	Infrared type	Sample stars
Semidetached	Red giant	Main-sequence Dwarf star	1–5	S	CI Cyg
Detached (I)	Red giant	White dwarf	1–5	S	RW Hya AG Peg Z And
Detached (II)	Mira variable	White dwarf	Decades	D	V1016 Cyg HM Sge

Note: For the infrared types S and D, see Section 6.7.1.

symbiotic binaries belong to this type. Major difference between type (I) and (II) of detached binaries lies in the large mass-loss rate of the Mira, ranging 10^{-6}–10^{-5} $M_\odot$, 1–2 orders of magnitude larger than those for nonpulsating red giants. High mass-loss process produces an optically thick dust shell around the Mira and forms the infrared excess in D type. In some cases, a shock-excited region is formed by the collision of winds from both components (see Formiggini et al. 1995 in case of HM Sge).

Symbiotic novae can occur in both type (I) and (II), particularly in type (II). Roughly half of the known symbiotic novae have erupted in Mira type symbiotic binaries, and their behavior was essentially identical to other symbiotic novae.

Evolutionary state of symbiotic stars

CNO abundance: Typically symbiotic stars are binary systems composed of a cool star, a hot star, and a nebula. The nebular materials are supposed to originate from the winds of cool stars so that the relative C, N, and O abundance in nebulae derived from emission lines will yield the information on the chemical state of the primary stars. Nussbaumer et al. (1988) deduced the C/N and O/N ratios for 24 symbiotic stars in all semidetached and detached types. For this purpose, they measured the strong emission lines CIII, CIV, NIII, NIV, and OIII, observable with the IUE, under the assumption that the regions emitting these lines are radiatively ionized from a source with $T^* > \sim 10^5$ K, $T_e \sim 12{,}000$ K, and $N_e \sim 10^6$–10^{10} cm^{-3}, and in this case, the ion densities obey the relation:

$$\frac{N(C^{+2}) + N(C^{+3})}{N(C)} \approx \frac{N(N^{+2}) + N(N^{+3})}{N(N)} \approx \frac{N(O^{+2})}{N(O)} \approx 1, \tag{6.7.1}$$

where $N(\mathrm{X}^{+\mathrm{m}})$ is the population density by number of the m-times ionized element X, and $N(\mathrm{X})$ is the total density of the element X. Using this relation for the observed emission line fluxes, they derived $N(\mathrm{C})/N(\mathrm{N})$ and $N(\mathrm{O})/N(\mathrm{N})$

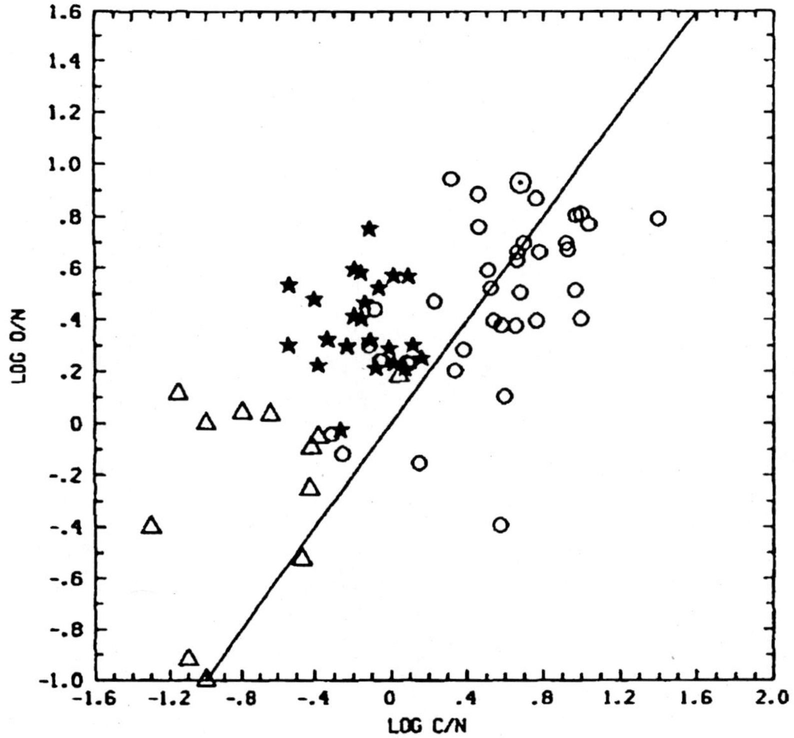

Figure 6.53: Relative abundance O/N against C/N for symbiotic stars ($\star$), planetary nebulae ($\circ$) and novae ($\triangle$). The solar abundance is given by $\odot$. (From Nussbaumer et al. 1988)

ratios, and compared with those available in other types of objects including red giants, planetary nebulae and novae. As an example, Figure 6.53 shows the relative abundance O/N against C/N for symbiotic stars, planetary nebulae and novae.

Evolutionary state. We now consider the evolutionary state of symbiotic stars, based on the relative abundance of O/N and C/N in Figure 6.53 by Nussbaumer et al. (1988).

(a) Symbiotic stars. It is noticed that symbiotic stars are centrally concentrated in Figure 6.53 as compared to novae and planetary nebulae, indicating that the symbiotic phenomenon only occurs in a well-defined evolutionary stage. This region is where the N element is in overabundance around one order of magnitude larger than C and O, and represents the product of CNO cycling giving an enhancement of N at the expense of C and O (e.g., Maeder 1987). Nussbaumer et al. also point out that the location of all symbiotic stars in Figure 6.53 is almost coinciding with the location of M giants (including Miras), supporting the idea that the symbiotic nebular gas is fed by M giants.

(b) Novae. CNe occur in close binary systems, containing a mass-losing star and an accreting white dwarf. The outbursts occur as a result of thermonuclear runaway in CNO-cycle hydrogen burning, and then their ejecta show products of heavy CNO processing. This effect can be seen in the location of novae on the C/N–O/N plane in Figure 6.53, where novae are well separated from symbiotic stars. The enhancement of N by CNO-cycling is more advanced in novae than in symbiotic stars. The nova-like outbursts in symbiotic stars are known to have similar C/N/O abundance in quiescent symbiotic stars. Therefore there is a fundamental difference between symbiotic phenomena and CNe in the requirement of thermonuclear runaway at the surface of accreting star, i.e., the CNO abundance in symbiotic stars and symbiotic novae do not require additional thermonuclear processing.

(c) Planetary nebulae. As seen in Figure 6.53, planetary nebulae occupy an area much wider than symbiotic stars on both side of $C/O = 1$ line. It is therefore unlikely that symbiotic stars as a whole are direct progenitors of planetary nebulae. Probable candidates of the progenitors are now supposed to be AGB stars, which occupy the higher luminosity region than normal red giants on the HR diagram (see D'Antona and Mazzitelli 1992).

It has thus become evident that, from the viewpoint of C/N/O abundance, symbiotic stars including symbiotic novae indicate a common evolutionary stage, and closely related with the stage of red giants and Mira variables in their evolution.

Concerning the links between symbiotic and planetary nebulae, López et al. (2004) noticed that some nebulae around symbiotic stars have been misidentified as planetary nebulae because of the similarity of HST images. As a binary system, a symbiotic star consists of a mass-losing Mira or red giant component, which has not reached the planetary phase yet, and a white dwarf component, which has already experienced it. This implies that the symbiotic binaries are not the direct progenitors of planetary nuclei. Iben and Tutukov (1996) have considered the evolution of symbiotic binaries with accreting degenerate dwarfs in detail.

CNs and symbiotic stars are mostly short-period binaries; a world-wide combination of photometric and spectroscopic observation will greatly contribute to the understanding of their nature and evolutionary status.

Further reading

Cassatella, A. and Viotti, R. (eds.) (1990). *Physics of Classical Novae.* IAU Coll. No. 122, Springer-Verlag, Berlin.

Corradi, R. and Mikolajewska, J. (eds.) (2002). *Symbiotic Stars, Probing Stellar Evolution.* ASP Conf. Vol. 303, San Francisco, CA.

Havnes, O., Petterson, B. R., Schmitt, J. H. M. M. and Solkeim, J. E. (eds.) (1988). *Activity in Cool Star Envelopes*, Kluwer, Dordrecht.

Mirzoyan, L. V., Petterson, B. R., and Tsventkov, M. K. (eds.) (1990). IAU Symposium, No. 137. *Flare Stars in Star Clusters, Associations and the Solar Vicinity*, Kluwer, Dordrecht.

Sahade, J., McCluskey, G. E., and Kondo, Y. (eds.) (1993). *The Realm of Interacting Binary Stars*, Kluwer, Dordrecht.

Warner, B. (1995). *Cataclysmic Variable Stars*, Cambridge University Press, Cambridge, New York.

References

Allen, D. A. (1979). Symbiotic stars at optical, infrared and radio wavelengths. *Changing Trends in Variable Star Research*, Bateson, F. M., Smak, J., and Ruch, I. H. (eds.). IAU Coll. No. 46, University of Waikato, Hamilton, N.Z., 125–147.

Allen, D. A. (1984). A catalogue of symbiotic stars. *Proc. Australian Soc. Astr.*, **5**, 369.

Ambartsumian, V. A. (1954). Phenomena of stellar continous emission and origin of stellar energies. *Comm. Byurakan Obs.*, **13**, 3–26.

Anandarao, B. G., Taylor, A. R., and Pottasch, S. R. (1988). Dust emission from symbiotic stars—An interpretation of IRAS observations. *A. A.*, **203**, 361–366.

Anupama, G. C. and Prabhu, T. P. (1989). The 1985 outburst of RS Ophiuchi: spectroscopic results. *Astrophys. Astr.*, **10**, 237–255.

Anupama, G. C. and Mikolajewska, J. (1999). Recurrent novae at quiescence: System with giant secondaries. *A. A.*, **344**, 177–187.

Arkhipova, V. P., Belyakina, T. S., Dokuchaeva, O. D., and Noskova, R. I. (1990). The light curve of the symbiotic nova HM Sagittae, *Physics of Classical Novae*. A. Cassatella and R. Viotti (eds.), IAU Coll. No. 122. Springer-Verlag, Dordrecht, 437–440.

Audard, M., Gudel, M., and Skinner, S. L. (2003). Separating the X-ray emissions of UV Ceti A and B with Chandra. *Ap. J.*, **589**, 983–987.

Baratta, G. B. and Viotti, R. (1990). The spectrum of the symbiotic nova HBV 475 in 1969. II. *A. A.*, **229**, 104–116.

Bath, G. T. (1975). Dynamical instabilities and mass exchange in binary systems. *MNRAS*, **171**, 311–328.

Bauer, W. H. and Bennett, P. D. (2000). The ultraviolet spectrum of VV Cephei out of eclipse. *Publ. A. S. Pacific*, **112**, 31–49.

Bauer, W. H., Stencel, R. E., and Neff, D. H. (1991). Twelve years of IUE spectra of the interacting binary VV Cephei. *A. A. Suppl.*, **90**, 175–190.

Belczynski, K., Mikolajewska, J., Munari, U., Ivison, R. J., and Friedjung, M. (2000). A catalogue of symbiotic stars. *A. A. Suppl.*, **146**, 407–435.

Bloch, M. and Chalonge, D. (1965). Etude prelininaire du spectre de Nova Herculis 1963 dans le visible et l'ultraviolet. *Colloque Internatinal sur les Novae, Supernovae, Novoides*, CNRS, No. 121, Central National de la Recherche Scientifique, Paris, 78–82.

Bode, M. F. (ed.) (1987). *RS Ophiuchi (1985) and the recurrent nova phenomenon.* Proc. of Manchester Conference, M. F. Bode (ed.), VNU Science Press, Utrecht.

Bode, M. F. and Kahn, F. D. (1985). A model for the outburst of nova RS Ophiuchi in 1985. *MNRAS*, **217**, 205–215.

Bode, M. F., Roberts, J. A., Ivison, R. J., Meaburn, J., and Skopal, A. (1991). Echelle spectroscopy of the symbiotic star CH Cygni through quiescence. *MNRAS.*, **253**, 80–88.

Böhm-Vitense, E. (1992). Rotation and transition layer emission in cool giants. *A. J.*, **103**, 608–616.

Bopp, B. and Fekel Jr., F. (1977). Binary incidence among the BY Draconis variable. *A. J.*, **82**, 490–494.

Bowen, G. H. (1988). Dynamical modeling of long-period variable star atmospheres. *Ap. J.*, **329**, 299–317.

Bruevich, E. A., Katsova, M. M., and Livshits, M. A. (1990). Kinetics of hydrogen in the chromospheres of red dwarfs. *Sov. Astron.*, **34**, 60–65.

Budding, E. (1984). A catalogue of classical (evolved) Algol-type candidate stars. *Bull.de Centre des Donnees Stellaires*, **27**, 91–129.

Byrne, P. B., Abdul Aziz, H., Amado, P. J., Arevalo, M. J., Avgoloupis, S., Doyle, J. G., and 11 co-authors (1998). The photosphere and chromosphere of the RS CVn star. II Pegani. *A. A. Suppl.*, 127, 505–519.

Çakirli, Ö., Ibanoglu, C., Djurasevic, G., Erkapic, S., Evren, S., and Tas, G. (2003). Long-term photometric behaviur of the RS CVn binary RT Lacertae. *A. A.*, **405**, 733–745.

Chambers, H. L. II. (1995). The nature of the companion too Ceti. *PhD Thesis*, Purdue University, (Dissertation Abstracts International, Vol. 57–03, p. 1856).

Cheng, Q. Q., Engvold, O., and Elgaroy, O. (1997). A numerical simulation of the Wilson-Bappu relationship. *A. A.*, **327**, 1155–1163.

Chincarini, G. and Rosino, L. (1965). Spectral evolution of Nova Hercules (Dahlgren) from February to September 1963. *Novae, Novoides et Supernovae.* Coll.Int'l du CNRS, No. 121, Centre National de la Recherche Scientifique, Paris, 72–77.

Choi, H. J., Soon, W., Donahue, R. A., Baliunas, S., and Henry, G. W. (1995). A study of variability in a smaple of G and K giants. *PAS Pacific*, **107**, 744–750.

Contini, M. (2003). An analysis of the emission line spectra of AG Pegasi between phases 7.34 and 9.44. *MNRAS*, **339**, 125–132.

Contini, M., Orio, M., and Prialnik, D. (1995). RS Oph at day 201: a test for shocks in nova shells. *M. N. R. A. S.*, **275**, 195–208.

Cram, L. E. (1982). X-ray heating of the quiescent chromospheres of dMe stars. *Ap. J.*, **253**, 768–772.

Crowe, R. A. and Garrison, R. T. (1988). The visible spectra of southern hemisphere Mira variable stars. *Ap. J. Suppl*, **66**, 69–98.

D'Antona, F. and Mazzitelli, I. (1992). Stellar evolution, low mass stars. *The Astr and Astrophys. Encyclopedia*, S. P. Maran (ed.), Van Nostrand Reinhold, New York, 838–839.

de Jager, C., Heise, J., van Genderen, A. M., Foing, B. H., Ilyin, I. V., and 14 co-authors (1989). Coodinated Observations of a large impulsive flare on UV Ceti. *A. A.*, **211**, 157–172.

Dobrzycka, D., Kenyon, S. J., Proga, D., Mikolajewska, J., and Wade, R. A. (1996). The hot component of RS Ophiuchi. *A. J.*, **111**, 2090–2098.

Downes, R. A., Duerbeck, H. W., and Delahodde, C.E. (2001). Luminosities of [OIII]λ 5007 and Hydrogen Balmer lines in nova shells years and decades after outburst. *J. Astron. Data*, **7**, 6–53.

Doyle, J. G. (1989). Hα versus X-ray luminosity in dwarf M stars. *A. A.*, **218**, 195–198.

Drake, S. and Ulrich, R. (1980). The emission-line spectrum from a slab of hydrogen at moderate to high densities. *Ap. J. Suppl.*, **42**, 351–383.

Dufay, J., Bloch, M., Bertaud, Ch., and Dufay, M. (1965). Evolution du spectre de Nova RS Ophiuchi après l'explosion de 1958. *Novae, Supernovae, Novoides.* Coll. Internationaux du CNRS, No. 121, Centre National de la Recherche Scientifique, Paris, 18–49.

Dufay, J. and Bloch, M. (1965). Température de couleur, décrément de Balmer et rougissement interstellaire dans le spectre de Nova RS Ophiuchi. *Colloque International sur les Novae, Supernovae, Novoides.* CNRS, No. 121, Centre National de la Recherche Scientifique, Paris, 53–59.

Echevarria, J. (1988). A statistical analysis of the emission line ratios in cataclysmic variables. *M. N. R. A. S.*, **233**, 513–527.

Echevarria, J., Tovmassian, G., Shara, M., Tapia, M., Hohigas, J., and 17 co-authors (1996). Simultaneous multiwavelength observations of dwarf novae. I. SU UMa: Minihumps at a minioutburst? *Ap. J.*, **467**, 851–859.

Elitzur, M., Freland, G. J., Mathews, W. G., and Shields, G. A. (1983). Stimulated emission and the flat Balmer decrements of cataclysmic variables. *Ap. J.*, **272**, L55–L59.

Engvold, O. and Rygh, B. O. (1978). The CaII line width in late type stars. The Wilson–Bappu effect. *A. A.*, **70**, 399–407.

Formiggini, L., Contini, M., and Leibowitz, E. M. (1995). HM Sagittae: the shocked and photoionized interaction zone of two colliding stellar winds. *MNRAS*, **277**, 1071–1079.

Fox, M. W., Wood, P. R., and Dopita, M. A. (1984). Shock waves in Mira variables. I. Emission-line spectra. *Ap. J.*, **286**, 337–349.

Fox, M. W. and Wood. P. R. (1985). Shock waves in Mira variables. II. Theoretical models. *Ap. J.*, **297**, 455–475.

Garcia-Alvarez, D., Jevremovic, D., Doyle, J. G., and Butler, C. J. (2002). Observations and modelling of a large optical flare on AT Microscopii. *A. A.*, **383**, 548–557.

Gauzit, J. (1955). Le spectre de AX Persei et ses variation. *Ann. d'Ap.*, **18**, 354–374.

Gershberg, R. E. (1974). Balmer emission decrement and electron density in stellar chromopheres. *Sov. Astron.*, **18**, 326–330.

Gershberg, R. E. and Shnol, E. E. (1974). The Balmer decrement in spectra of moving medium. The case of collisional ionization and excitation. *Izv. Krim. Astrophys. Obs.*, **50**, 122–151.

Gershberg, R. E., Katsove, M. M., Lovkaya, M. N., Terebizh, A. V., and Shakhovskaya, N. I. (1999). Catalogue and bibliography of the UV Cet-type flare stars and related objects in the solar vicinity. *A. A. Suppl.*, **139**, 555–558.

Gill, C. D. and O'Brien, T. J. (1999). Emission-line profiles from model nova shell. *MNRAS*, **307**, 677–684.

Gillet, D. (1988). The Balmer emission profiles in Mira stars. *A. A.*, **192**, 206–220.

Gliese, W. (1969). *Catalogue of Nearby Stars.* Veroffentlichungen des Astronomishen Rechen-Instituts Heidelberg Nr. 22, Verlag G. Braun, Karlsruhe.

Gray, D. F. and Endal, A. S. (1982). The angular momemtum history of the Hyades K giants. *Ap. J.*, **252**, 162–167.

Gutiérrez-Morino, A, and Moreno, H. (1996). Spectroscopic observations of some D-type symbiotic stars. *PAS Pacific*, **108**, 972–979.

Gutierrez-Moreno, A., Moreno, H., and Costa, E. (1999). Spectroscopic observations of some S-type symbiotic stars. *PAS Pacific*, **111**, 571–586.

Hack, M., Engin, S., and Yilmaz, N. (1989). A study of the ultraviolet spectrum of VV Cephei. *A. A.*, **235**, 143–155.

Hack, M. (1992). Binary stars, atmospheric eclipses. *The Astronomy and Astrophysics Encyclopedia*, Maran, S. P. (ed.), Van Nostland Reinhold, New York, 58–61.

Hagen, W., Black, J. H., Dupree, A. K., and Holm, A. V. (1980). Ultraviolet spectroscopic observations of VV Cephei. *Ap. J.*, **238**, 203–209.

Haro, G. (1957). The possible connexion between T Taauri stars and UV Ceti stars. IAU Symp. No. 3, *Non-Stable Stars*, G. H. Herbig (ed.), Cambridge University Press, London, 26–30.

Haro, G. and Morgan, W. (1953). Rapid variables in the Orion Nebula. *Ap. J.*, **118**, 16–17.

Haro, G. and Parsamian, E. (1969). A new "slow" flare star in Orion. *Bol. Obs. Tonantzintla y Tacubaya*, **5**, 45–52.

Herbig, G. H. (1985). Chromospheric Hα emission in F8–G3 dwarfs, and its connection with the T Tauri stars. *Ap. J.*, **289**, 269–278.

Herbst, W. and Miller, J. R. (1989). Hα photometry of dwarf K and M stars : chromospheric activity. *A. J.*, **97**, 891–899.

Hoffmeister, C., Richter, G., and Wenzel, W. (1985). Pulsating stars (Chapter 2), Eruptive variables (Chapter 3). *Variable Stars*, Springer-Verlag, Berlin, Heiderberg, N.Y.

Horne, K. (1991). Variability and structure of accretion disks in cataclysmic variables. *Structure and Emission Properties of Accretion Disks*, C. Bertaux, S. Collin-Suffrin, and J. P. Lasota (eds.), IAU Coll. No. 129, Paris Editions Frontieres, 3–18.

Houdebine, E. R. Foing, B. H., and Rodono, M. (1990). Dynamics of flares on late-type dMe stars. I. Flare mass ejection and stellar evolution. *A. A.*, **238**, 249–255.

Houdebine, E. R., Doyle, J. G., and Koscielecki, M. (1995). Observation and modeling of main sequence star chromospheres III. Differential analysis of hydrogen lines versus activity level in M dwarfs. *A. A.*, **294**, 773–791.

Iben, I. and Tutukov, A. V. (1996). On the evolution of symbiotic stars and other binaries with accreting degenerate dwarfs. *Ap. J. Suppl.*, **105**, 145–180.

Ichimura, K., Shimizu, Y., Watanabe, E., and Okada, T. (1973). Continual photoelectric monitoring of flare stars, VIII : YZ CMi, AD Leo and UV Cet (1972). *Tokyo Ast. Bull.*, Second Ser. No. 224, 2607–2612.

Iijima, T, Strafella, F., Sabbadin, F., and Bianchini, A. (1994). High velocity and rapidly variable emission features in the spectra of RS Ophiuchi and CH Cygni. *A. A.*, **283**, 919–931.

Ikeda, Y. and Tamura, S. (2000a). A new mass function of a symbiotic star, HBV 475. (V1329 Cyg). Proc. Pacific Rim Conf., *Stellar Astrophysics*, L. S. Cheng, H. F. Chau, K. C. Leung (eds.), Kluwer, Dordrecht, pp. 423–427.

Ikeda, Y. and Tamura, S. (2000b). Spectroscopic observations of symbiotic stars. III. Radial velocity analysis of HBV 475. *PAS Japan*, 52, 589–599.

Illarionov, A. F., and Sunyaev, R. A. (1975). Why the number of galactic X-ray stars is so small? *A. A.*, **39**, 185–195.

Ivison, R. J., Seaquist, E. R., Schwarz, H. E., Hughes, D. H., and Bode, M. F. (1995). Millimetre continuum emission from symbiotic stars. I. The measurements. *M. N. R. A. S.*, **273**, 517–527.

Joy, A. H. (1954). Spectroscopic observations of Mira Ceti, 1934–1952. *Ap. J. Suppl.*, **1**, 39–61.

Jura, M. and Helfand, D. J. (1984). X-rays from accretion of red giant winds. *Ap. J.*, **287**, 785–792.

Kaitchuck, R. H. and Honeycutt, R. K. (1982). A survey for gaseous disks in short-period Algol systems during primary eclipse. *PASP*, **94**, 532–536.

Kaitchuck, R. H., Honeycutt, R. K., and Schlegel, E. M. (1985). A survey for transient accretion disks in short-period Algol systems. II. *PASP*, **97**, 1178–1185.

Karovska, M. (1992). Mira's companion(s). *Complementary Approaches to Double and Multiple Star Research*, H. A. McAlister and W. I. Hartkopf (eds.), IAU Coll. 135, ASP Conf. Ser., Vol. 32, San Francisco, CA, 558–560.

Karovska, M., Raymond, J., and Guinan, E. (1996). Coordinated ROSAT and speckle observations of the Mira AB system. *Technical report, Smithsonian Astrophys. Obs., Cambridge.*

Katsova, M. M. (1990). Balmer decrements of red dwarfs in the quiet state and during flares. *Sov. Astron.*, **34**, 614–620.

Kawabata, S., Saijo, K., Sato, H., and Saito, M. (1981). A spectroscopic study of VV Cephei during the 1976–78 eclipse. I. Observations of the Hα line. *PAS Japan*, **33**, 177–188.

Kawabata, S. and Saito, M. (1997). Expanding atmosphere of the M-type supergiant in VV Cephei. *PAS Japan*, **49**, 101–107.

Keenan, P. C. (1966). A catakogue of spectra of Mira variables of types Me and Se. *Ap. J. Suppl.*, **13**, 333–378.

Keenan, P. C., Garrison, R. F., and Deutch, A. J. (1974). Revised catalogue of spectra of Mia variables of types Me and Se. *Ap. J. Suppl.*, **28**, 271–307.

Kenyon, S. J. (1986). *The Symbiotic Stars*, Cambridge University Press, Cambridge, New York.

Kenyon, S. J. (1992). Symbiotic Stars. *The Astronomy and Astrophysics Encyclopedia*, S. P. Maran (ed.), Van Nostrand Reinhold, New York, 794–797.

Kenyon, S. J. and Truran, J. W. (1983). The outbursts of symbiotic novae. *Ap. J.*, **273**, 280–288.

Kharchenko, N., Kilpio, E., Malkov, O., and Schilback, E. (2003). Compiled catalog of stellar data of Miras. *VizeR On-line Data Catalogue*, Originally published in 2002, *A. A.*, **384**, 925.

Kippenhahn, R. and Weigert, A. (1967). Entwicklung in engen Doppelsternsystemen. I. Massenausyausch vor und nach Beendigung des zentralen Wasserstoff-Brennens. *Zs. f. Ap.*, **65**, 251–273.

Knapp, G. R. (1985). Mass loss from evolved stars. IV. The dust-to-gas ration in the envelopes of Mira variables and carbon stars. *Ap. J.*, **293**, 273–280.

Kneer, F. (1983). A possible explanation of the Wilson-Bappu relation and the chromospheric temperature rise in late-type stars. *A. A.*, **128**, 311–317.

Kopal, Z. (1959). *Close Binary Systems*, Chapman & Hall, London.

Kraft, R. P. (1959). The binary system nova DQ Herculis. II. An interpretation of the spectrum during the eclipse cycle. *Ap. J.*, **130**, 110–123.

Kukarkin, B. V. and Parenago, P. P. (1970). *General Catalogue of Variable Stars* (GCVS). Third edition, Akademia Nauk Moskva. (See Fouth edition 1988, Warren, Jr., W. H., National Space Science Data Center, NASA).

Kunkel, W. E. (1970). On the spectra of stellar flares. *Ap. J.*, **161**, 503–518.

la Dous, C. (1991). New insights from a statistical analysis of IUE spectra of dwarf novae and nova-like stars. *A. A.*, **252**, 100–122.

Lauterborn, D. (1969). White dwarf production in binary systems of large separation. *Mass Loss from Stars*, M. Hack (ed.), D. Reidel, Dordrecht, pp. 262–266.

Leedjärv, L. and Mikolajewski, M. (1995). The long-period symbiotic binary CH Cygni. IV. High velocity spectral features as the result of a propeller action. *A. A.*, **300**, 189–199.

Leung, K. C. (1992). Binary Stars, Contact. *The Astronomy and Astrophysics Encyclopedia*, Maran, S. P. (ed.), Van Nostrand Reinhold, New York, 63–65.

Livio, M. and Truran, J. W. (1992). Type I supernovae and accretion-induced collapses from cataclysmic variables. *Ap. J.*, **389**, 695–703.

López, J. A., Escalante, K., and Riesgo-Tirado, H. (2004). Links between symbiotic and planetary nebulae. *Rev. Mex. A. A.*, **20**, 226–227.

Luttermoser, D. G. and Bowen, G. H. (1992). NLTE synthetic spectra of the Mira-type variable S Car. *Seventh Cambridge Workshop on Cool Stars, Stellar Systems, and the Sun*, ASP Conf. Ser. Vol. **26**, San Francisco, CA, 558–560.

Luttermoser, D. G. and Castelaz, M. W. (2003). FUSE observations of a Mira variable star. *The Future of Cool-Star Astrophysics: 12th Cambridge Workshop*, A. Brown, G. M. Harper, and T. R. Ayres (eds.), University of Colorado, Boulder, 1042–1047.

Maeder, A. (1987). Changes of surface chemistry for standard massive star evolution: Cartography in the HR diagram. *A. A.*, **173**, 247–262.

Martinez-Pais, I. G., Giovannelli, F., Rossi, C., and Gaudenzi, S. (1994). An optical time-resolved spectroscopic study of SS Cygni. I. Quiescence. *A. A.*, **291**, 455–467.

Martinez-Pais, I. G., Giovannelli, F., Rossi, C., and Gaudenzi, S. (1996). An optical time-resolved spectroscopic study of SS Cygni. II. Outburst. *A. A.*, **308**, 833–846.

Mason, E., Skidmore, W., Howell, S. B., Ciardi, E. R., and Littlefair, S. (2000). Investigating the structure of the accretion disc in WZ Sge from multiwaveband timersolved spectroscopic observations. II. *M. N. R. A. S.*, **318**, 440–452.

McLaughlin, D. B. (1965). The behaviour of absorption systems in spectra of novae. *Colloque International sur Novae, Supernovae, Novoides.* CNRS, No. 121, Paris, 3–12.

Meier, S. R., Kafatos, M., Fahey, R. P., and Michalitsianos, A. G. (1994). A far-ultraviolet atlas of symbiotic stars observed with IUE. I. The SWP range. *Ap. J. Suppl.*, **94**, 183–220.

Merrill, P. W. (1958). Symbiosis in astronomy: Introductory report. *Etoiles a Raies D'emission. 8th International colloquium held in Liége*, Inst. of Astrophysics, Belgium, pp. 436–448.

Merrill, P. W. and Burwell, C. G. (1933). Catalogue and bibliography of stars of classes B and A whose spectra have bright hydrogen lines. *Ap. J.*, **78**, 87–140.

Mikolajewski, M., Mikolajewska, J., and Khudyakova, T. N. (1990). The long-period symbiotic binary CH Cyg. I. A hundred year's history of variability. *A. A.*, **235**, 219–233.

Mirzoyan, L. V. (1990). Flare stars in star clustiers, associations and the solar vicinity. *Flare Stars in Star Clusters, Associations and the Solar Vicinity*, L. V. Mirzoyan, B. R. Pettersen, and M. K. Tsvetkov (eds.), IAU Symp. No. 137, Kluwer, Dordrecht, 1–13.

Mirzoyan, L.V. (1993). Flare stars and the evolution of red dwarf stars. *Astrophysics*, **36**, 170–191.

Mirzoyan, L. V., Hambarian, V. V., Garibjanian, A. T., and Mirzoyan, A. L. (1995). Optical observations of flare stars in the Galaxy. *Astrophysics*, **38**, 276–279.

Montes, D., Fernandez-Igueroa, M. J., Cornide, M., and DeCastro, E. (1996). The behaviour of the excess CaII H and K and Hεemissions in chromospherically active binaries. *A. A.*, **312**, 221–233.

Morales-Rueda, L. and Marsh, T. R. (2002). Spectral atlas of dwarf novae in outburst. *MNRAS*, **332**, 814–826.

Munari, U., Yudin, B. F., Taranova, O. G., Massone, G., Marang, F., and 4 co-authors (1992). UBVRI-JHKL photometric catalogue of symbiotic stars. *A. A. Suppl.*, **93**, 383–390.

Munari, U. and Zwitter, T. (2002). A multi-epoch spectrophotometric atlas of symbiotic stars. *A. A.*, **383**, 188–196.

Nussbaumer, H., Schild, H. M., Vogel, M., and Schld, H. (1988). Relative C, N, O abundances in red giants, planetary nebulae, novae and symbiotic stars. *A. A.*, **198**, 179–186.

Oliverson, N. A. and Anderson, Ch. M. (1982). Spectra of individual symbiotic stars. *The Nature of Symbiotic Stars*, M. Friedjung and R. Viotti (eds.), D. Reidel, Dordrecht, 71–82.

Olson, E. C. and Etzel, P. B. (1995). Rapid Hα emission variationin accretion disks in long-period Algols. *A. J.*, **109**, 1308–1312.

Osaki, Y. (1974). An accretion model for the outburst of U Geminorum stars. *PAS Japan*, **26**, 429–436.

Osaki, Y. (1996). Dwarf—nova outbursts. *PASP*, **108**, 39–60.

Osten, R. A., Hawley, S. L., Allred, J., Johns-Krill, C. M., and Roark, Ch. (2005). From radio to X-ray: Flares on the dMe flare star EV lacertae. *Ap. J.*, **621**, 398–416.

Özeren, F. F., Doyle, J. G., and Jevremovic, D. (1999). The Wilson- Bappu relation for RS CVn stars. *A. A.*, **350**, 635–642.

Parsamian, É. S. (1971). Slow flares in Pleiades. *Astrophysics*, **7**, 326–330 (Astrofizika, **7**, 547–555, 1971).

Payne-Gaposchkin, C. (1957). *The Galactic Novae*, North-Holland Publishing Company, Amsterdum.

Peters, G. J. (1980). H-alpha observations of Algol-type interacting binary systems. *Close Binary Stars: Observations and Interpretation*, M. J. Plavec, D. M. Popper, and R. K. Ulrich (eds.). IAU Symp. No. 88, D. Reidel, Dordrecht, 287–290.

Petit, M. (1982). *Variable Stars* (English ed., 1987), John-Wiley & Son.

Pettersen, B. R. (1988). Atmospheric activity in the outer envelopes of cool dwarf stars. *Activity in Cool Star Envelopes,* O. Havnes, B. R. Pettersen, J. H. M. M. Schmitt, and J. E. Solheim (eds.), Kluwer, Dordrecht, 49–60.

Plavec, M. and Polidan, R. S. (1976). The Algols, red spectra, Be stars, and even neutrions. *Structure and Evolution of close Binary Systems,* IAU Symp. No. 73, P. Eggleton, S. Mitton, and J. Whelan (eds.), D. Reidel, Dordrecht, p. 289.

Ratering, C., Bruch, A., and Diaz, M. (1993). A spectroscopic study of the Z Camelopardalis type dwarf nova KT Persei. *A. A.*, **268**, 694–704.

Richards, M. T. and Albright, G. E. (1999). Morphologies of Hα accretion regions in Algol systems. *Ap. J. Suppl.*, **123**, 537–626.

Ritter, H. (1987). Catalogue of cataclysmic binaries, low-mass X ray binaries and related objects. *A. A. Suppl.*, **70**, 335–367.

Robinson, E. L. (1975). Pre-eruption light curves of novae. *A. J.*, **80**, 515–524.

Rodono, M. (1990). Prospects for studies of UV Ceti-type flare stars. *Flare Stars in Star Clusters, Associations and Solar Vicinity*, L. V. Mirzoyan, B. R. Pettersen, and M. K. Tsvetkov (eds.), IAU Symp. No. 137, Kluwer, Dordrecht, 371–391.

Rodono, M. (1994). RS CVn-type close binaries: a class of active stars suitable for MUSICOS campaigns. *Proc. 4th Workshop on Multi-site Continuous Spectroscopy*, Huang, L., Zhai, D. S., Catala, C., and Foing, B. H. (eds.), European Space Research and Technology Centre, Noordwijk, 69–80.

Rosino, L. and Iijima, T. (1987). The 1985 outburst of RS Ophiuchi. *RS Ophiuchi (1985) and the Recurrent Nova Phenomenon*, M. F. Bode (ed.), Utrecht VNU Science Press, Utrecht, 27–38.

Saijo, K. (1981). A spectroscopic study of VV Cephei during the 1976–78 eclipse. II. Structure of the Hα emission envelope around the early-type component. *PAS Japan*, **33**, 351–364.

Schmid, H. M. and Schild, H. (1990). Physical conditions and element abundances in the symbiotic novae V1016 Cyg, HM Sge and HBV 475. *MNRAS*, **246**, 84–97.

Schwank, M., Schmutz, W., and Nussbaumer, H. (1997). Irradiated red giant atmospheres in S-type symbiotic stars. *A. A.*, **319**, 166–175.

Seaquist, E. R., Taylor, A. R., and Button, S. (1984). A radio survey of symbiotic stars. *Ap. J.*, **284**, 202–210.

Seitter, W. C. (1990). Optical studies of classical novae in outburst. *Physics of Clasical Novae*, IAU Coll. No. 122, A. Cassatella and R. Viotti (eds.), Springer-Verlag, Dordrecht, 79–96.

Shafter, A. W. and Hessman, F. V. (1988). A time-resolved spectroscopic study of the SU Ursae Majoris dwarf nova YZ Cancri. *A. J.*, **95**, 178–189.

Shafter, A. W., Hessman, F. V., and Zhang, E. H. (1988). Photometric and spectroscopic observations of the eclipsing nova-like variable PG 1030-590 (DW Ursae Majoris). *Ap.J.*, **327**, 248–264.

Shara, M. M. (1989) Recent progress in understanding the eruptions of classical novae. *PASP*, **101**. 5–31.

Shawl, S. J. and Bord, D. B. (1991). A search for companions to Mira variables. *Bull. Am. Ast. Soc.*, **23**, 1380. Abstract 35.12.

Shore, S. N., Kenyon, S. J., Starrfield, S., and Sonneborn, G. (1996). On the interpretation of the ultraviolet spectraa of symbiotic stars and recurrent novae. II. The 1985 outburst of RS Ophiuchi. *Ap. J.*, **456**, 717–737.

Short, C. I. and Doyle, J. G. (1998). Chromospheric modelling of the Hα and Na I D lines in five M dwarfs of low to high activity level. *A. A.*, **336**, 613–625.

Short, C. I., Byrne, P. B., and Panagi, P. M. (1998). The chromosphere of II Pegasi: multi-line modelling of an RS CVn star. *A. A.*, **338**, 191–199.

Slavin, A. J., O'Brien, T. J., and Dunlop, J. S. (1995). A deep optical imaging study of the nebular remnants of classical novae. *MNRAS*, **276**, 353–371.

Starrfield, S., Truran, J. W., Sparks, W. M., and Kutter, G. S. (1972). CNO abundances and hydrodynamic models of the nova outburst. *Ap. J.*, **176**, 169–176.

Starrfield, S., Sparks, W. M., and Truran, J. W. (1974). CNO abundances and hydrodynamic models of the nova outburst. II. 1.00 M_{sun} models with enhanced carbon and oxygen. *Ap. J. Suppl.*, **28**, 247–270.

Stauffer, J. R. and Hartmann, L. W. (1986). Chromospheric activity, kinematics, and metalicities of nearby M dwarfs.(+) *Ap. J. Suppl.*, **61**, 531–568.

Stencel, E. R., Potter, D. E., and Bauer, W. H. (1993). Rapid mass-loss transients in VV Cephei. *A. J.*, **105**, 45–50.

Stickland, D. J., Kelly, B. D., Cooke, J. A., Coulson, I., Engelbrecht, C., Kilkenny, D., and Spencer-Johns, J. (1984). RZ Gru—A UX UMa 'disc star'. *MNRAS*, **206**, 819–831.

Strassmeier, K. G., Hall, D. S., Fekel, F. C., and Scheck, M. (1993). A catalog of chromospherically active binary stars. *A. A. Suppl.*, **100**, 173–225.

Strassmeier, K. G., Handler, G., Paunzen, E., and Rauth, M. (1994). Chromospheric activity in G and K giants and their rotation-activity relation. *A. A.*, **281**, 855–863.

Tappert, C., Mennickent, R. E., Arenas, J., Matsumoto, K., and Hanvschuk, R. W. (2003). An atlas of line profile studies for SU UMa type cataclysmic variables. *A. A.*, **408**, 651–661.

Vaughan, A. H., and Preston, G. W. (1980). A survey of Chromo- spheric CaII H and K emission in field stars of the Solar neighborhood. *PASP*, **92**, 385–391.

Viotti, R. (1990). The symbiotic novae. *Physics of Classical Novae*, IAU Coll. 122, A. Cassatella and R. Viotti (eds.), Springer-Verlag, Dordrecht, 416–421.

Vladilo, G., Moralo, P., Crivellari, L., Foing, B. H., Beckman, J. E., and Genova, R. (1987). Chromospheric Mg II h and k emissions free of interstellar contamination: Velocity structure in late-type dwarfs and giants. *A. A.*, **185**, 233–246.

Walker, M. F. (1954). Nova DQ Herculis (1934): An eclipsing binary with very short period. *PASP*, **66**, 230–232.

Walker, M. F. (1957). The extremely rapid light-variations of old novae and related objects. *Non-Stable Stars*. IAU Symposium, No. 3. G. H. Herbig (ed.), Cambridge University Press, London, 46–56.

Warner, B. (1995). Theories and Models of DN Outbursts. *Cataclismic Variable Stars*. Section 3.5.

Webbink, R. F., Livio, M., Truran, W., and Orio, M. (1987). The nature of the recurrent novae. *Ap. J.*, **314**, 653–672.

Williams, R. E. (1980). Emission lines from the accretion disks of calaclysmic variables. *Ap. J.*, **235**, 939–944.

Williams, R. E. (1990). The ionization of novae ejecta. IAU Coll. No. 122, *Physics of Classical Novae*, A. Cassatella and R. Viotti (eds.), Springer-Verlag, Dordrecht, 215–227.

Williams, R. E. (1992). The formation of novae spectra. *A. J.*, **104**, 725–733.

Wilson, O. C. (1959). Accuracy of absolute magnitudes derived from widths of H and K emission components. *Ap. J.*, **130**, 499–506.

Wilson, O. C. and Bappu, M. K. V. (1957). H and K emission in late type stars: Dependence of line width on luminosity and related topics. *Ap. J.*, **125**, 661–683.

Wood, B. E., Karovska, M., and Raymond, J. C. (2002). Analysis of H2 emission from Mira B in ultraviolet spectrum from the Hubble Space Telescope. *Ap. J.*, **575**, 1057–1077.

Wood, B. E. and Karovska, M. (2003). FUSE observations of molecular hydrogen emission from Mira B. *BAAS*, **203**, #84.09.

Yamashita, Y. and Maehara, H. (1978). Mass loss from Mira Ceti, *PAS Japan*, **30**, 409–417.

Yamashita, Y., and Maehara, H. (1979). A binary model for CH Cygni. *PAS Japan*, **31**, 307–316.

Young, A., Skumanich, A., Stauffer, J. R., Bopp, B. W., and Harlan, E. (1989). A study of excess H-alpha emission in chromospherically active M dwarf stars. *Ap. J.*, **344**, 427–436.

Zhai, D.S., Foing, B. H., Cutispoto, G., Zhang, R. X., Catala, C., Char, S., Zhang, X. B., and Jankov, S. (1994). Multisite continuous spectroscopy. III. Photometric analysis and spot modelling of the light curves of HR 1099 before and after the 1989 optical flares. *A. A.*, **282**, 168–178.

Zhai, D. S. and Zhang, X. B. (1996). Hα spectroscopy of the chromospherically active binary HR 1099 in 1993. *A. A.*, **309**, 530–543.

Chapter 7

Pre-main Sequence Stars

7.1 Herbig Ae/Be stars

7.1.1 Definition and catalogues

Herbig Ae/Be stars are the early-type pre-main sequence stars. Normally they are simply abbreviated as HES (Herbig emission-line stars) or HAEBE (Herbig Ae Be stars). Herbig (1960) originally defined HES by the following characteristics:

(i) The spectral type is A or earlier, with emission lines.
(ii) The star lies in an obscured region.
(iii) The star illuminates fairly bright nebulosity in its immediate vicinity.

These properties distinguish HESs from classical Be stars in the same spectral class. Later, Strom et al. (1972) added the following properties:

(a) Infrared excess—Conspicuous IR excess has been observed in many HESs, showing the existence of dust shells around the stars.
(b) Time variations—HESs generally exhibit irregular variations in brightness and spectral features. Light variations mainly occur in Ae stars, while spectral variations appear throughout Be and Ae in intensities and profiles of emission lines.
(c) Linear polarization—The degree of polarization in a continuum is around 1% in average as in the case of classical Be stars. In some HESs, however, the polarization is as high as 7–8%, implying that the origin may be different from that of classical Be stars.
(d) Association with star-forming regions—HESs are mostly associated with specific star-forming regions (molecular clouds, stellar associations, and/or young star clusters). Some stars are accompanied by molecular or optical bipolar flows.

These features will be discussed below in their respective sections. All of these properties show that HESs are the pre-main sequence stars of intermediate mass between OB stars (high-mass stars) and T Tauri stars (low-mass stars).

Herbig (1960) listed 26 HESs, selected under the first three characteristics, and Finkenzeller and Mundt (1984) extended the list to 57 stars by using the same properties. With the advance of infrared observations, and of emission-line star surveys in star forming regions, the number of HESs has been increased. In the Catalogue of Thé et al. (1994), 108 stars are listed as HES and HES candidates. Malfait et al. (1998) proposed the presence of broad infrared excess as the most important defining criterion for these stars. By this definition they counted 287 objects as HESs.

Hernández et al. (2004) analyzed the optical spectra of 75 early-type emission-line stars, using the 1.5 m telescope at the Whipple Observatory with the spectrograph of resolution 6 Å. Based on effective-temperature sensitive spectral features, including Fe I, He I, G band, and Balmer lines, they classified 39 HESs with an error of less than 2.5 spectral subtypes. Other types in their classification are "continuum stars" which are stars having essentially no absorption features with strong Hα emission (7 stars), and stars with spectral types later than F (12 stars) that are included as HES in previous catalogues. Some stars are classified as stars in uncertain evolutionary status. A selected list of HESs is given in Table 7.1.

7.1.2 Spectral features

Line spectra

Herbig Ae/Be stars generally exhibit conspicuous emission lines in their optical spectra. On one hand, their spectra are similar to the classical Be stars of the same spectral type, while on the other hand similar to the T Tauri stars (TTS) of the pre-main sequence type. The similarity with classical Be stars lies in the emission lines of Balmer series and some ionized metals such as FeII, TiII, and MgII. The similarity with TTS is the existence of low-excitation lines such as CaII, NaI, KI, and FeI.

The Hα line generally shows strong emission with various profiles. According to Finkenzeller and Mundt (1984), the types of profile are double peaks (50%), single peak (25%), and P Cygni profile (20%). The percentage in brackets denotes the rate of occurrence. The remaining 5% exhibit more complex profiles including inverse P Cygni profiles. Some examples of the profiles of Hα and NaI D lines are shown in Figure 7.1. The line profiles are often variable in the HESs. Spectral features of the Balmer lines will be discussed in Section 7.3 both for HESs and TTSs.

NaI D lines appear in emission, in absorption, or in more complex profiles in different stars, often accompanied by sharp interstellar absorption (see Figure 7.1). They are not necessarily parallel to the emission feature of the Hα line.

Forbidden lines are also observed in many Herbig Ae/Be stars. According to the high-resolution spectroscopic survey of Böhm and Catala (1994), [OI] λ6300.31 Å was detected in 17 out of 33 HESs. Among these, eight stars also showed detectable [OI] λ6363.79 Å which is weaker than λ6300.31 by a factor 3 as predicted by the radiative de-excitation rate.

Table 7.1: A list of selected Herbig Ae/Be stars

Name	HD/etc	$V^a_{\mathrm{mag.}}$	Sp. type[b]	IRAS[c]	Nebula[d]	IR-ex.[e]	Hα[f]	X-ray[g]
AB Aur	HD 31293	(6.9–8.4)	B9/A0e+sh	Y	N	I	P	Y
UX Ori	HD 293762	9.2–12.3	A2/3IIIe	Y	N	I	C	—
HK Ori	—	11.63	B8/A4ep	Y	N	I	D	—
T Ori	BD-5^{0}1329	(8.7–12.6)	A3/5eα	N	Y	I	InvP	—
V380 Ori	BD-6^{0}1253	(9.97–10.82)	B8/A1e	Y	Y	I	S	Y
BF Ori	BD-6^{0}1259	(9.60–13.5)	A1/F6e+sh	Y	N	I	D	—
ω Ori	HD 37490	(4.41–4.59)	B3III-IVe	Y	Y	III	D	N
R Mon	BD+8^{0}1427	10.37–11.89	B0e	Y	Y	II	—	—
GU CMa	HD 52721	(6.50–6.72)	B2Vne	Y	—	III	S	N
HR 5999	HD 144668	(6.73–8.47)	A7III-IVe	Y	N	—	D	Y
R CrA	CPD-37^0 8452	10.74–11.50	A5IIe+sh:/F0	Y	Y	II	—	N
MWC1080	—	(11.58–11.67)	B0eq	Y	Y	I	P	Y
V633 Cas	LkHα 198	14.29	B/Ae	Y	Y	II	—	N

[a] $V_{\mathrm{mag.}}$: In brackets the magnitudes in other bands such as pg = photographic, B, and R are given. In many case light variations are observed.
[b] Spectral type by different reference is distinguished by slash.
[c] IRAS source: Whether exists or not is shown by Y and N, respectively.
[d] Nebula: The association with nearby nebula is shown by Y or N.
[e] IR excess: The type of excess is shown by Hillenbrand et al. (1992) (see Table 7.2)
[f] Hα emission profile is distinguished as follows (Finkenzeller & Mundt, 1984): S = single peak, D = double peak, P = P Cyg, InvP = inverse P Cyg, C = complex
[g] X-ray source: Whether exists or not is shown by Y and N (Zinnecker and Preibisch 1994).
The third to sixth columns are taken from Thé et al. (1994).

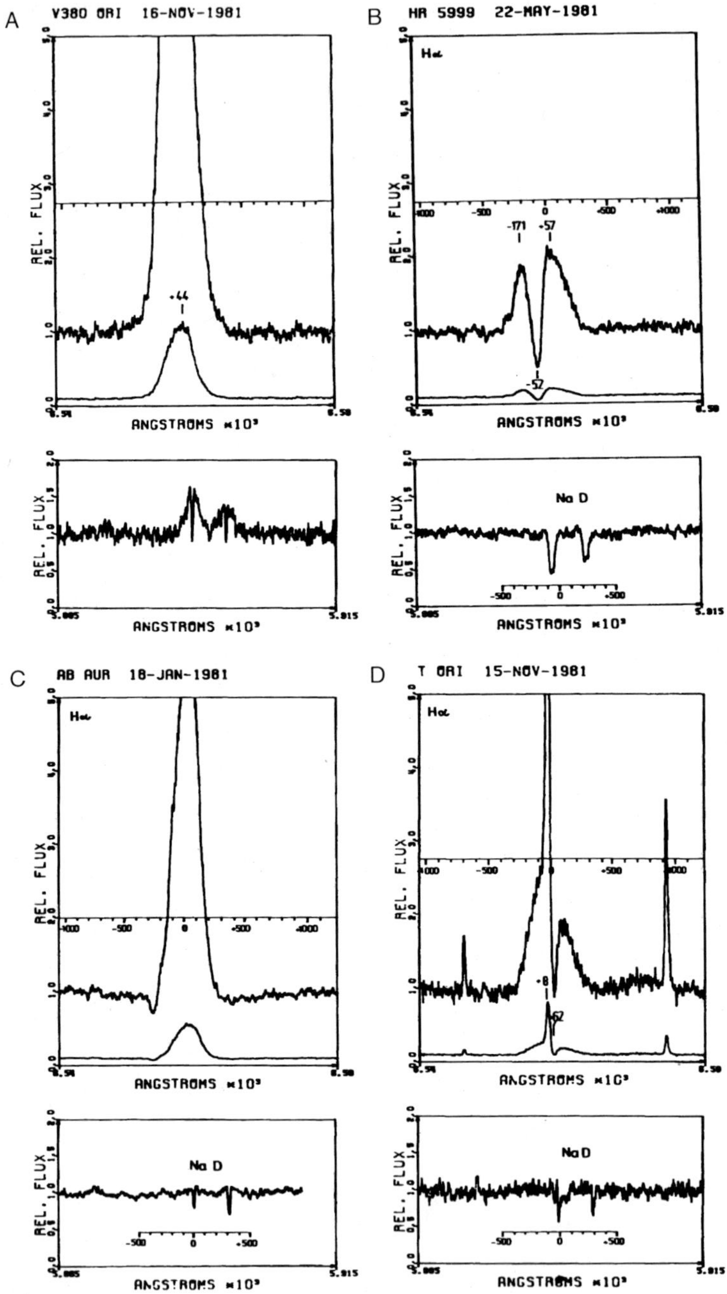

Figure 7.1: Samples of the line profile of Hα and NaI D in HESs. (A) V380 Ori (single peak); (B) HR5999 (double peaks); (C) AB Aur (P Cyg profile); (D) T Ori (complicated profile). In each panel, upper and lower parts exhibit Hα and Na ID line profiles, respectively. Hα lines are shown in two profiles with different intensity scales. The sharp absorption in Na ID lines denotes the interstellar lines. (From Finkenzeller & Mundt 1984)

The existence of these forbidden lines is somehow related to the profile of Hα emission. That is, stars with single-peaked Hα emission show no forbidden lines or less than detection limit (~40 mÅ), whereas stars with double-peaked or P Cygni profiles reveal the [OI] λ6300.31 Å emission. The strongest forbidden line (equivalent width: 0.6–0.8 Å, half-half width: 30–100 km s^{-1} for [OI] λ6300.31 Å), appears in stars with double-peaked Hα emission. The critical electron density defined by Equation (4.3.8) for this line is $N_e^c \sim 2 \times 10^7$ cm^{-3} for $T_e = 1000$ K, and $N_e^c \sim 3 \times 10^6$ cm^{-3} for $T_e = 10{,}000$ K. Forbidden lines of [OI] should be formed in the low-density HI region outside of the HII region in the stellar envelope, where the electron temperature may be nearer to 1000 K.

As an explanation for the absence of appreciable forbidden lines in the stars having single-peaked Hα emission, Böhm and Catala (1994) suggested that these stars have no stellar winds or insufficient mass loss rates. Alternatively, if we think of these stars as nearly pole-on stars as in case of classical Be stars, stellar winds from polar regions are of low gas density with a high ionization degree making the formation of [OI] lines difficult. For stars with double-peaked Hα emission, we can suppose that they have larger inclination angles, therefore, the stellar winds are strong with sufficient mass-loss rates and with suitable temperature gradient decreasing outward. This picture may explain the relationship between the existence of forbidden lines in [OI] and the profiles of Hα emission.

In the UV region, Herbig Ae/Be stars often exhibit the so-called superionization phenomena corresponding to the existence of regions hotter than 10^5 K where the emission lines of CIV and NV are produced. Examples are AB Aur (Bouret et al. 1997) and HR5999 (Pérez et al. 1993) among others. On the formation of such hot regions several mechanisms have so far been proposed. One is the inhomogeneous stellar-wind model, where collisions between high and low speed flows produce the shock-heated hot region (Bouret et al. 1997). The other is a chromospheric activity model analogous to classical TTSs (Catala 1989). A clumpy accretion model was also proposed as an alternative to the stellar wind model (Pérez and Thé 1994).

Infrared excess and infrared emission lines

A remarkable fraction of HESs exhibits infrared excess. Some examples of the spectral energy distributions (SED) observed by Hillenbrand et al. (1992) are shown in Figure 7.2 for the three groups I, II, and III, separately. This grouping represents different forms of the SED as summarized in Table 7.2. The infrared excess for individual stars is defined as the color excess $(V - 12\,\mu) - (V - 12\,\mu)_0$, where the latter term indicates the color excess for a standard star in the respective spectral type.

Hillenbrand et al. (1992) pointed out the following properties for each group in their 47 sample stars:

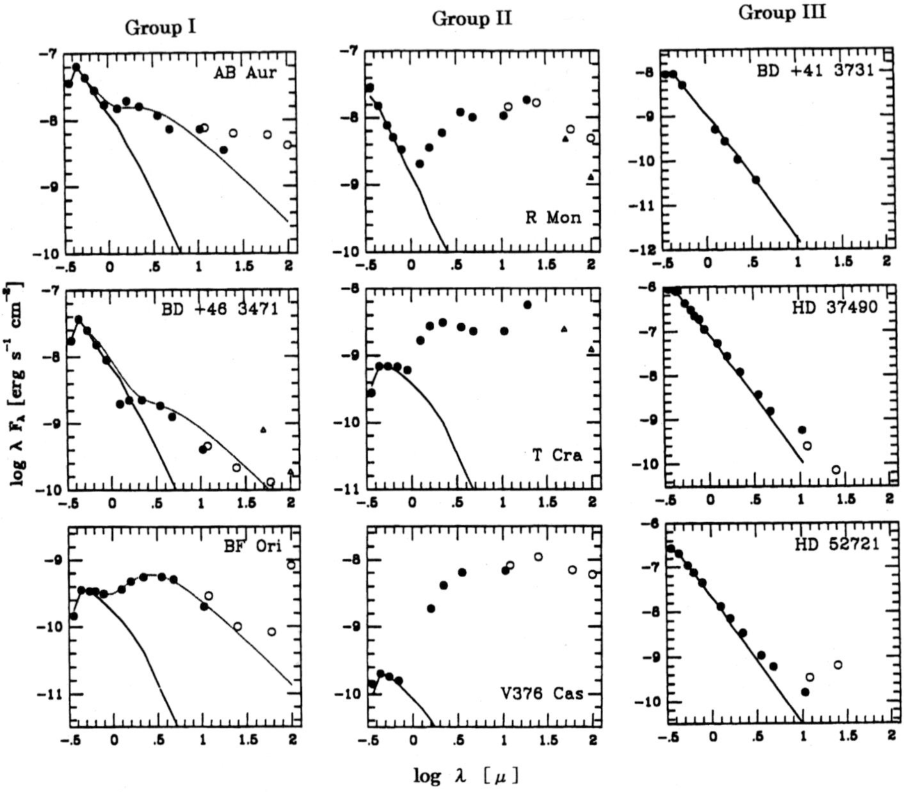

Figure 7.2: Samples of 0.30–100 μm spectral energy distributions (SEDs). (a) Group I: AB Aur; BD+46^{0}3471; BF Ori. (b) Group II: R Mon; T CrA; V376 Cas. (c) Group III: BD+41^{0}3731; HD 37490 (ω Ori); HD 52721 (GU CMa). Solid circles are ground-based observations; open circles represent IRAS observations; and triangles are KAO (Kuiper Airborn Observatory). The solid line is the SED appropriate to a standard star. The dotted line represents the best-fit disk model. (From Hillenbrand et al. 1992)

Group I contains 30 objects with large infrared excess. Infrared SED (>2.2 μm) can be well fitted by assuming that excess emission arises in an optically thick circumstellar accretion disk. Spectral types of stars in this group range from O9 to F2 (average type B8).

Table 7.2: Groups of IR spectra (adapted from Hillenbrand et al. 1992)

Group	IR excess (V–12 μ excess)	Spectral energy distribution (SED)	Number of stars
I	large (4–8 mag.)	$\lambda F_\lambda - \lambda^{-4/3}$ ($\lambda > 2.2$ μ)	30
II	large (7–11)	flat or rising spectrum out to 100 μm	11
III	small (0–2)	free-free emission from disk or wind	6

Group II comprises of 11 objects with large infrared excess showing flat or rising infrared spectra. The SED shows evidence of a circumstellar disk surrounded by gas and dust material not confined to the disk. From the form of SED these objects could be younger than the group I objects. Stars in this group range from B0 to F8 (average spectral type of A5) and masses lower than the stars of group I.

Group III contains six objects of early spectral types ranging from B0 to B3 (average B2). Their infrared excesses are far smaller than those of groups I and II objects, and the infrared SEDs appear similar to those of classical Be stars, whose small infrared exesses are supposed to originate from free-free emission in their gaseous envelopes. Although they lack optically thick disks, they are associated with reflection nebulosity and star-forming molecular clouds. Thus the group III objects may be young, intermediate-mass stars, analogous of diskless T Tauri stars located in the high-mass end of HESs.

Malfait et al. (1998) carried out photometric observations in multiband from U, B, V to K, L, M, for IRAS-selected 45 HES candidate stars, among which 33 stars were confirmed as HESs by the presence of broad infrared excess. In addition, they decomposed the infrared SED into two hot and cold components in the majority of HESs, and suggested the existence of a double structure of inner and outer shell/disks around the stars.

In near IR regions many emission lines are observable. Hamann and Persson (1992) carried out spectroscopic observations for 32 HESs in a spectral range of λ6500–9078 Å, and analyzed both permitted and forbidden emission lines. Besides hydrogen (Brα, Brγ, Pfβ, Pfγ, etc), the permitted emission lines are by neutral and ionized atoms such as KI, OI, MgI, MgII, FeII, and CaII. CaII triplet (λ8498.02, 8542.09, 8662.14 Å) is remarkable and observed in 84% of stars. The forbidden lines include [CaII], [NiII], [FeII], and [CrII]. Note that the permitted lines often show P Cygni line profile, indicating that mass outflow is the prevailing phenomena in HESs. In contrast, forbidden lines are usually of single peaked, broad, or blueshifted profiles, implying that these lines are formed in outer edges of the outflowing envelopes. Hamann and Persson also suggested that accretion disk models, successfully applied to the classical TTS stars, are not directly applicable to the hottest part of Herbig Be stars, where the stellar radiation plays an important role in the excitation of FeII and CaII triplet as in classical Be stars.

Linear polarization

The majority of Herbig Ae/Be stars exhibits linear polarization in the optical region (Breger, 1974; Vrba et al. 1979; Vink et al. 2002). The intrinsic polarization scatters in a range from 0.1% up to 7% or so. In contrast to classical Be stars, the polarization has been supposed to originate from the scattering of stellar light, mostly by aligned dust grains (Bastien and

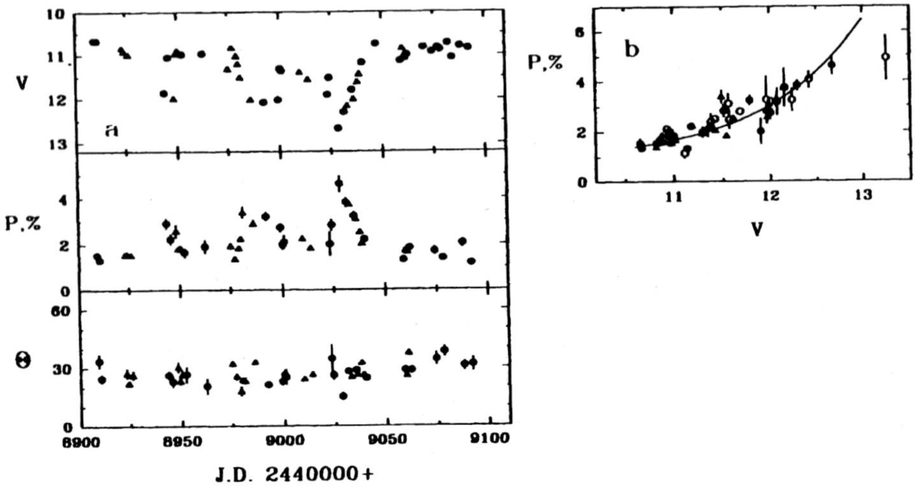

Figure 7.3: Linear polarization of the Herbig Ae star RR Tau. Panel (a): the light curve in V magnitude, linear polarization ($P\%$) and position angle, shown from top to bottom. Panel (b): observed linear polarization in V band as a function of V magnitude. The theoretical line corresponds to the variable circumstellar screening model. (From Grinin et al. 1994)

Landstreet 1979). This was based on the facts that (a) the wave-length dependence due to bound–free, free–free absorption of hydrogen atoms is usually not seen and (b) the depolarization effect for the Hα emission is not observed. Both of these are conspicuous in the classical Be stars (see Section 5.3.5 of Chapter 5). Recently, however, the polarization behavior in HESs has been found to be much more complex than previously supposed.

If the continuum polarization is produced by dust grains in circumstellar envelopes or in reflection nebulae in which the stars are embedded, the time variation may be related to the formation of dust envelopes in the vicinity of stars. Along this line, Grinin (1994) proposed a model of variable circumstellar screening which predicts that, if a star is screened by an opaque dust envelope, the star becomes faint and the linear polarization will increase. Figure 7.3 illustrates the observed variations in V magnitude, polarization $P(\%)$, and position angle (Θ) for the Herbig Ae star RR Tau in Panel (a). A comparison between the result of model calculation and observed correlation on the $V - P(\%)$ diagram is given in Panel (b). The model prediction shows good agreement with the observations.

Concerning the Hα polarization, it was found by Pontefract et al. (2000) that the Herbig Ae star AB Aur disclosed a linear-polarization in the Hα emission intrinsic to the source, instead of a depolarization effect as previously supposed. Vink et al. (2002) carried out Hα spectropolarimetry observations of a sample of 23 HESs and found a large variety of polarization effects inside the line profiles. Examples of Hα line profiles and polarization behaviors (linear polarization $P(\%)$, position angle θ°) are shown in Figure 7.4. As seen in this figure, Hα polarization behaviors can be classified into three types. According to Vink et al., the type, star numbers, and spectral subtype are as follows:

Type	Number of stars	Spectral subtype
Depolarization	4	B0–B2
Line polarization	14	B6–F2
No line effect	5	B0–A3

It seems that the depolarization mainly appears in early B stars. Vink et al. counted the depolarization frequency as 7/12 (=58%) of stars by combining other data. For these stars linear polarization can be attributed to electron-scattering circumstellar disks as in the case of classical Be stars. For the next group that show line polarization effect in late B to early F types, the polarization can be attributed to a compact Hα emission region that is itself polarized by magnetic field and located in an interior region of magnetic accretion flow. Because the depolarization effect tends to occur in early B type stars and the line polarization effect occurs in late B to A type stars, they supposed that there is a transition in the HR diagram from magnetic accretion at spectral type A to disk accretion at spectral type B. The stars with no line effect might expose some effect of inclination angle in disk accretion case, or some causes in the transition in the HR diagram in late B to A type stars. It is also remarkable that the polarization behaviors show almost no relationship to the profile of Hα emission.

Time variation of linear polarization is also prevalent among HESs. The time scale of variability ranges from a few hours to several years (Jain and Bhatt 1995, Beskrovnaya et al. 1995).

X-ray and radio spectra

The X-ray observations with ROSAT (energy range 0.1–2.4 keV) were carried out by Zinnecker and Preibisch (1994) for 21 HESs and X-ray sources were detected in 11 stars. These stars are characterized by the following properties:

(a) There is a positive correlation between X-ray luminosity and stellar bolometric luminosity as shown in Figure 7.5. As a comparison the line of $(L_X/L_{bol}) = 10^{-7}$, which nearly corresponds to the upper limit for the classical Be stars, is also indicated. It is apparent that the HESs are generally stronger X-ray emitters than classical Be stars.

(b) All known close binaries are detected as X-ray sources.

(c) Among the objects having circumstellar dust envelopes, no X-ray source is detected. This nondetection may be explained by strong extinction of X-rays in the dust envelope, and is not due to intrinsic low X-ray luminosity.

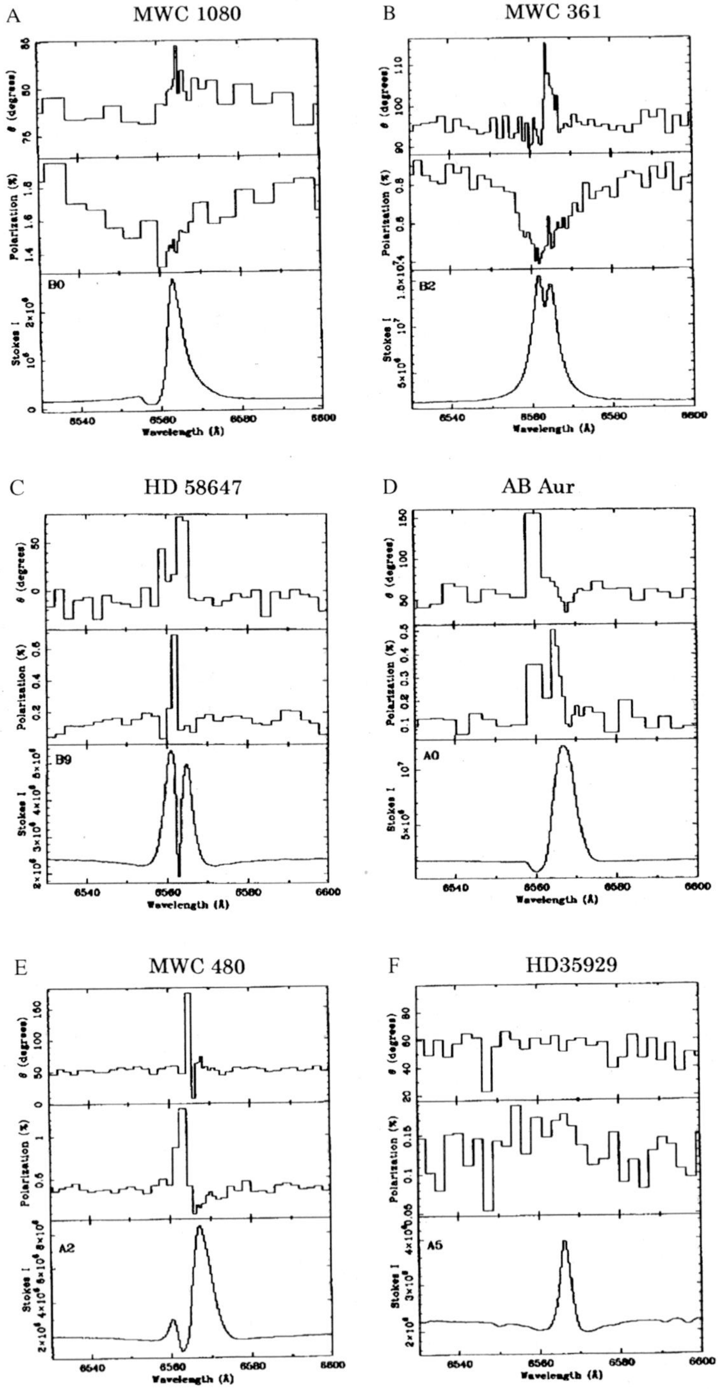
A
MWC 1080
B0
B
MWC 361
B2
C
HD 58647
B9
D
AB Aur
A0
E
MWC 480
A2
F
HD35929
A5
θ (degrees)
Polarization (%)
Stokes I
Wavelength (Å)

(d) The X-ray luminosities do not depend on the spectral types and rotational velocities ($V \sin i$) of stars, but weakly depend on mass-loss rates.

The above properties give us some constraints on the mechanism of X-ray emission. Since thermal winds with electron temperature around 10,000 K can not emit X-rays, some heating mechanism is needed. One possibility is the dynamic instability in a strong stellar wind producing shock waves as in the case of OB stars. Another possibility is collision of a fast wind ($v_\infty \approx 100$–500 km s^{-1}) with the remnant circumstellar material of the HESs. Figure 7.5 suggests the former possibility, since stellar winds must be stronger in stars with higher stellar luminosity.

Radio survey observations were carried out by Skinner et al. (1993) by using the VLA (Very Large Array, USA) and ATA (Australia Telescope Array) for 57 HESs in 3.6- and 6-cm wavelength bands. Radio sources were detected in 12 stars in 0.1 mJ level, and upper limits were found in 4 stars. For stars later than spectral type A2, no radio source was detected.

The detected radio emission is supposed to be predominantly thermal for several reasons: (1) the brightness temperature in partially resolved radio sources is less than around 10^4 K, (2) the radio spectral index $\alpha(S \propto \nu^\alpha)$ agrees with $\alpha = +0.6$ expected for free–free wind emission (see Section 5.1.2 of Chapter 5), (3) the circular polarization is not detected, and (4) the mass-loss signatures are present for most of radio detected stars. An exception is the eclipsing binary TY CrA, which reveals the nonthermal nature of radio sources based on the radio spectral index ($\alpha = -1.2$).

Figure 7.4: Examples of the Hα polarization behaviors in HESs. Position angle ($\theta°$), linear polarization (P(%)), and line profile (Stokes parameter I) are shown in this order for each panel. Stellar parameters and polarization behaviors for each star are given as follows. (Partly reproduced from Vink et al. 2002)

Star name	Sp. type	Hα Profile[a]	EW(Hα)[b]	Polarization[c]
MWC 1080	B0	P	−101 Å	D
MWC 361	B2	D	−63	D
HD 58647	B9	D (shell)	−8.9	L
AB Aur	A0	P	−40	L
MWC 480	A2	D (V ≪ R)	−21	L
HD 35929	A5	S	−3.2	N

[a] Hα profile: P = P Cyg, D = double peaked, S = single peak.
[b] EW (Hα): emission equivalent width in Å.
[c] Polarization: D = depolarization, L = line polarization, N = no effect.

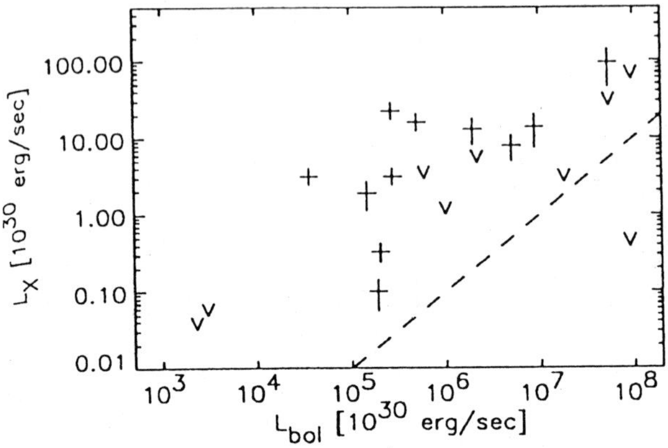

Figure 7.5: Correlation between X-ray (L_X) and bolometric ($L_{\rm bol}$) luminosities in HESs. The X-ray luminosity is shown with error bar (+) or upper limit mark (v). The broken line denotes the relation log ($L_X/L_{\rm bol}$) = −7. (From Zinnecker and Preibisch 1994)

Radio luminosity at 3.6 cm ($L_{\rm 3.6cm}$) suggests a positive correlation between bolometric luminosity ($L_{\rm bol}$) and effective temperature ($T_{\rm eff}$) of stars, and also the radio detection is limited to the stars of spectral types earlier than A2. These features infer that the radio emission is produced in the stellar winds. This is also supported by the positive correlation between 3.6 cm luminosity and Hα luminosity $L_{\rm H\alpha}/L_\odot$ as shown in Figure 7.6 (Skinner et al. 1993), where Hα luminosity is defined by $L_{\rm H\alpha}= 4\pi d\ ^2 W({\rm H}\alpha) F_R$ (d = distance to the star, W(Hα) = equivalent width, F_R = radiative flux in R band).

Skinner et al. (1993) assumed the free–free emission for radio waves and estimated mass-loss rate $\dot{M}$ ($M_\odot$ per year) by making use of terminal velocity v_∞ obtained from the P Cyg profile of optical emission lines. As a result they derived the mass-loss rate in a range from 1.1×10^{-8} (AB Aur), 1.4×10^{-8} (Elias 1) to 4.6×10^{-6} (Z CMa), 1.7×10^{-5} (MWC 300). It is nearly proportional to the bolometric luminosity $L_{\rm bol}$. These values are several orders of magnitude higher than those of classical Be stars in the same spectral type. Since it is known that HESs have accretion disks, Skinner et al. suggested the possibility of the coexistence of accretion flow and stellar wind as in the case of T Tauri stars (see Figure 7.30).

7.1.3 Rotational velocities and binarity

Rotational velocities

The projected rotational velocities (V sin$i$) of HESs can be estimated by the profiles of photospheric absorption lines of He λ4471 Å and/or MgII

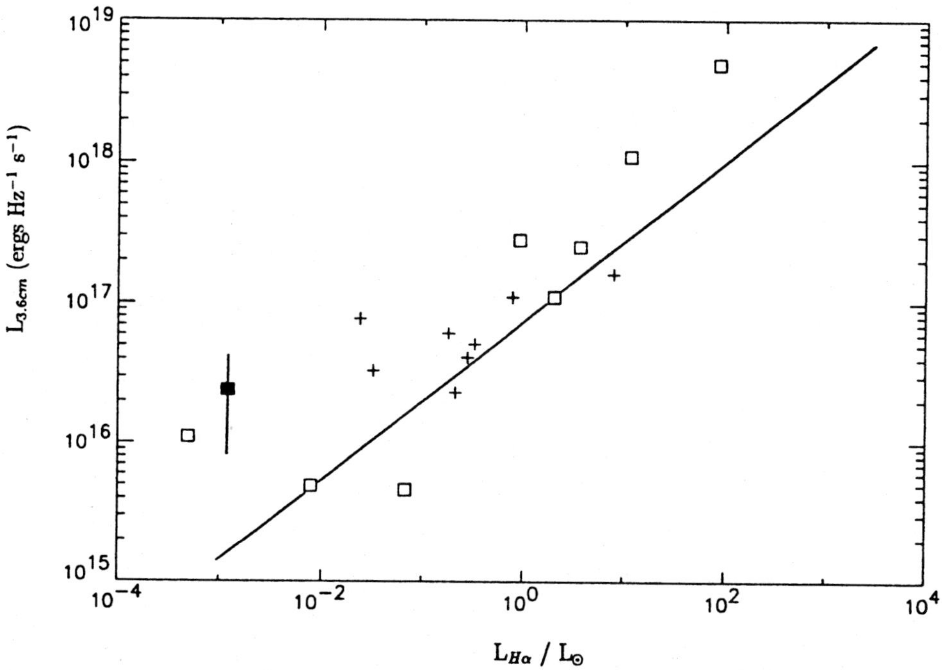

Figure 7.6: Radio luminosity (3.6 cm) vs. Hα luminosity ($L_{\mathrm{H}\alpha}$). Squares are radio detections, pluses are 3σ upper limits, and filled square is TY CrA. Solid line shows the maximum likelihood estimate $L_{3.6\mathrm{cm}} \propto (L_{\mathrm{H}\alpha})^{0.6\pm0.1}$. (From Skinner et al. 1993)

λ4481 Å. Finkenzeller (1985) for 19 stars, Böhm and Catala (1995) for 24 stars, respectively, measured V sin$i$ with an accuracy of around 20–30 km s^{-1}. Böhm and Catala's values are plotted in Figure 7.7, along with the average values of classical Be/Ae, and normal B/A stars for suitable ranges of spectral subtypes. The average values of Be/Ae and B/A stars are taken from Fukuda and Uesugi (1982). It is remarkable to see that HESs are generally much slow rotators as compared to classical Be/Ae stars or even to normal B/A stars. This implies that the effects of stellar rotation on the formation of emission lines may be quite restricted, if at all.

Binary systems

The binary frequency of Herbig Ae/Be stars has been estimated by many observers. Martin (1994) found 19 binary and multiple systems out of 108 stars in the catalogue of Thé et al. (1994). The binary frequency is thus 18% for the projected linear-separation range 200–2000 AU. Li et al. (1994) detected nine possible binary systems in 16 nearby HESs, by near infrared imaging

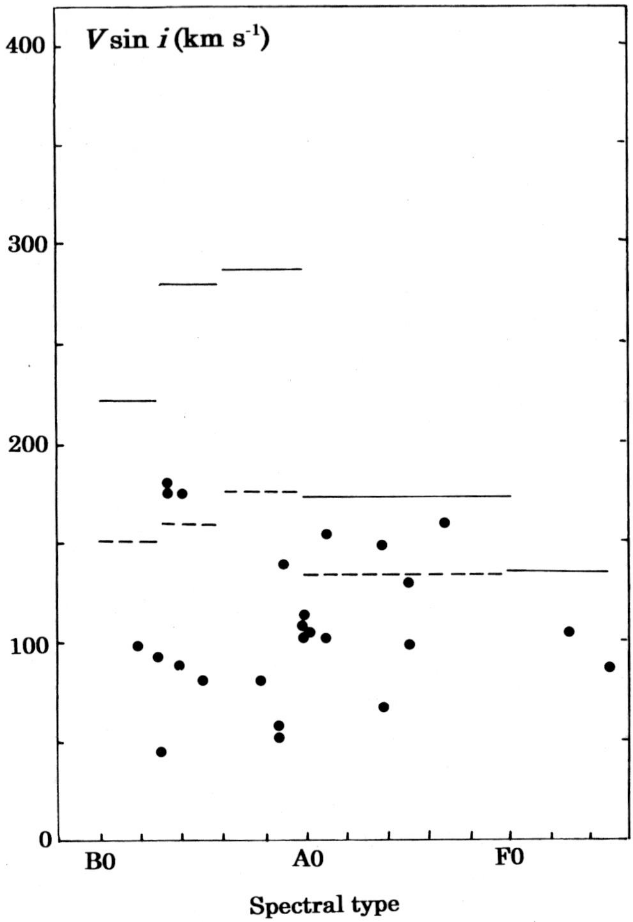

Figure 7.7: Rotational velocity $V \sin i$ vs. spectral type. The values of $V \sin i$ measured by Böhm and Catala (1995) are plotted by filled circles. The solid and dotted lines give the average values for the classical Be/Ae stars and normal B/A stars, respectively, adopted from Fukuda and Uesugi (1982).

observations with the spatial resolution range 1.3″–10″, which corresponds to the linear separation range 2500–9000 AU. The work of Li et al. was extended by Pirzkal et al. (1997), who carried out IR observations in K band with higher resolution of 0″.4 using 2.3-m telescope at the Wyoming IR Observatory. They detected seven binaries and two triple systems out of 39 bright HESs ($K \leqq 10.5$).

These observed binary frequencies are, however, constrained by observing conditions such as spatial resolution and photometric sensitivity. If we adopt some suitable distribution of mass, mass-ratio, and orbital parameters,

Table 7.3: List of binary and triple systems in HESs observed by Pirzkal et al. (1997)

Stellar system	Star	Angular distance (arc sec)	Projected linear distance (AU)	Distance (pc)	Magnitude difference ΔK (mag)
Binary	HD 200775	2.25	1350	600	4.9
	XY Per	1.2	192	160	0.0
	KK Oph	1.5	465	310	2.5
	HD 150193	1.1	176	160	2.2
Triple system	MWC 1080	4.69	11725	2500	6.2
		0.6	1500	—	3.1
	HD 141569	8	—	—	2.0
		6.8	—		2.7

Note: In triple system (A, B, C stars), upper and lower lines denote the angular distance and magnitude difference between A, B and A, C, respectively.

the observed frequency can be transformed to the intrinsic binary frequency. Pirzkal et al. (1997) made such correction for completeness and derived the binary rate 85% from their observed rate 9/39 = 0.23. This exceeds that of the solar-type main sequence stars (57%), but somewhat different from that of T Tauri stars (see Section 7.2.4).

The data for binary and triple systems in HESs, detected by Pirzkal et al. (1997) are summarized in Table 7.3.

The eclipsing binaries yield the information on stellar and orbital parameters. An example in HES is TY CrA (B9e), the double-lined eclipsing binary with the orbital period $P = 2.888777$ days. Though the Hα emission is weak, it shows the presence of circumstellar envelope and infrared excess as a pre-main sequence star. Corporon et al. (1994) carried out the spectroscopic analysis of TY CrA, and obtained the binary parameters as summarized in Table 7.4. It can be seen that the primary of this binary has the mass, radius, and luminosity typical of a Herbig Be star in late B type.

Table 7.4: Stellar parameters of the HES eclipsing binary TY CrA (adapted From Corporon et al. 1994)

	Mass ($M_\odot$)	Radius ($R_\odot$)	Luminosity ($L_\odot$)	Temperature (K)	Spectral Type
Primary	2.8	1.9	∼60	∼11200	B7/B9
Companion	1.5	1.5	∼4	∼6600	—

Note: The orbital parameters are derived by assuming the orbital inclination $i = 0$ and eccentricity $e = 0$. The semi-major axis is $a = 13.9\ R_\odot$, the orbital velocity amplitudes for the primary and the secondary are $K_1 = 86$ km s^{-1} and $K_2 = 157$ km s^{-1}, respectively.

7.1.4 Variability

Variations in brightness and colors

Many HESs exhibit variability in brightness and colors, usually in the form of a light decline from ordinary (maximum) levels. Bibo and Thé (1991) carried out monitoring observations for 2–8 years for 23 HESs in the Strömgren photometric system (y, b bands). Variability was detected for 16 stars with a large amplitude ($0.^{\rm m}65 \leqq \Delta y \leqq 3.^{\rm m}0$) for stars later than A0 and with a small amplitude ($\Delta y < 0.^{\rm m}20$) for stars earlier than A0.

Bibo and Thé (1991) classified the stars into three groups based on the behaviors in the color-magnitude diagram ($\Delta y, \Delta(b-y)$), as follows:

R = red behavior—When a star becomes fainter, its color becomes redder. The amplitude of variation is $\Delta y = 0.^{\rm m}4$–$1.^{\rm m}6$. This type contains eight stars including HR5999 (A7III-IVe), T Ori (A5e), and HD 250550 (A0e).

CR = color reversal—When a star becomes fainter, its color first becomes redder as an R type, but there is a turning point at which, with a further decline in brightness, the color index starts to decrease (bluer) until it is about the same as it was at the maximum brightness. The amplitude of brightness change is $\Delta y = 1.^{\rm m}6$–$3.^{\rm m}0$. Four stars belong to this type, including UX Ori (A2 IIIe), BF Ori (A5e), and CQ Tau (F2e). This color reversal is called the blueing effect.

RB = red-blue—The color index does not depend upon the variation in visual brightness. Amplitude of variation is generally small ($\Delta y = 0.^{\rm m}11$–$0.^{\rm m}18$), except NX Pup. Main stars are HD 53367 (B0 IVe), TY CrA (B9(e)), and NX Pup (A1 III, $\Delta y = 1.^{\rm m}7$).

Typical examples of these behaviors on the color-magnitude diagram are shown in Figure 7.8. The respective behavior is distinguished by its color when the star becomes faint. The irregular and sometimes rapid variability of HESs is generally supposed to be caused by the obscuration effect of dust clouds that have large optical depths and orbit in the outer region of the circumstellar disk. Thé (1994) suggested that the origin of the blueing effect was due to the existence of an edge-on disk. At the minimum light, the central star's blue light, scattered by small dust particles located backside of the star, becomes visible. Near the maximum light, such scattered light is not seen due to the strong light of the star.

The light variation is closely related with linear polarization as already stated in Section 7.1.2. In Figure 7.9 shows the light curve in V band, variations in linear polarization, and the position angle for UX Ori (CR group). A good correlation between brightness and polarization is seen, similar to the case of RR Tau (Figure 7.3). The polarization increases remarkably when stars deeply declined. This positive correlation again implies that HESs are surrounded by fully flattened dust disks (Thé 1994, Grinin 1994).

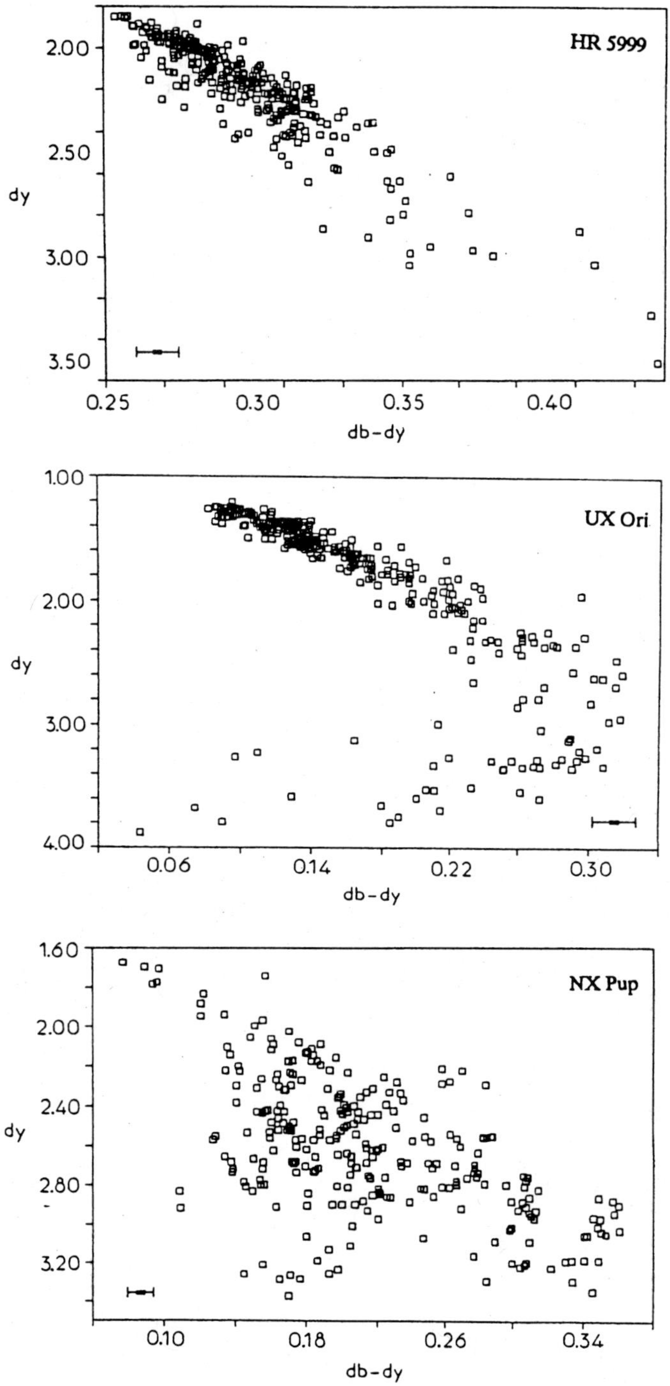

Figure 7.8: The variations on the color-magnitude diagram for three HESs. The abscissa is the color $(db - dy)$ and the ordinate is the magnitude difference from the standard star (dy). The stars are HR 5999 (R), UX Ori (CR) and NX Pup (RB). For the grouping R, CR, and RB, see text. (From Bibo and Thé 1991)

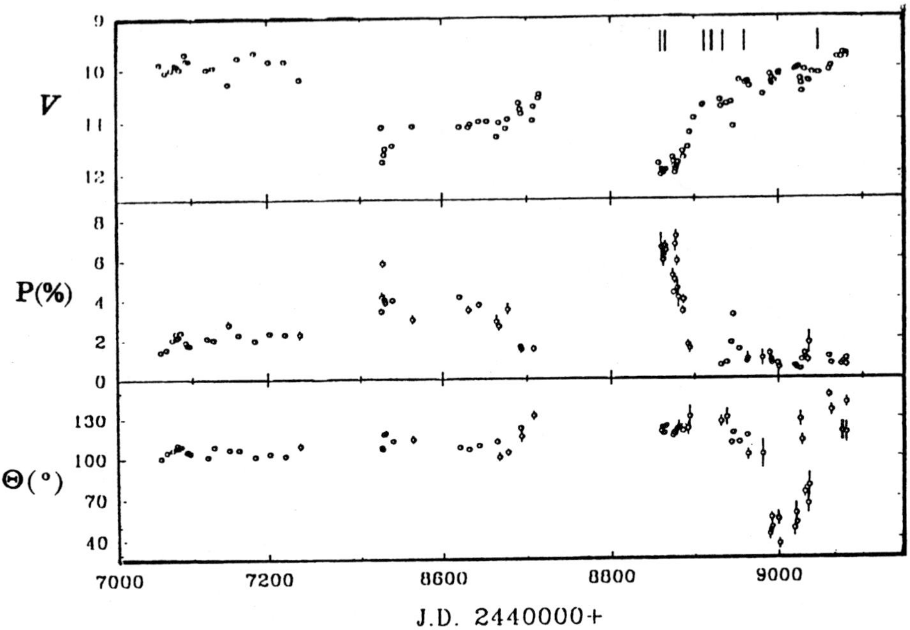

Figure 7.9: The light curve in V band and variation of linear polarization (percentage and phase angle) of the star UX Ori are given from top to bottom. Good correlation is seen between the decrease of brightness and increase of linear polarization. (From Thé 1994)

A group of HESs that exhibit light decrease with large amplitude, similar to the Algol binaries, is called the T Orionis variables. This name is originally used in GCVS (Kukarkin and Parenago 1970) as an A-type irregular variable. This was later changed by Glasby (1974) to the A-type HESs that show large-amplitude light decrease ($0.^{m}5$–$3.^{m}0$) in a short period just like Algol's eclipse. UX Ori (Figure 7.9) and WW Vul are the typical samples of this type, whose amplitudes (Δm) and length(Δt) of light decrease are as follows:

WW Vul: $\Delta m \sim 1.5^{\rm m}$, $\Delta t \sim$ 2–15 days.

UX Ori: $\Delta m \sim 2.0^{\rm m}$, $\Delta t \sim$ 4–30 days.

T Ori type stars generally exhibit Hα emission, and the Hα intensity appears to increase when the stellar light declines (Kolotilov 1977, Herbst et al. 1983).

Spectral variations

HESs generally exhibit remarkable spectral variations. While the light variations are mostly seen in Ae stars, spectral variations are widely observable for Be and Ae stars. In the optical region, variable emission is seen in the Balmer lines as well as in lines of FeII, HeI, NaI D, CaII H, K, etc., with various time scales from less than 1 day, to a month or years.

As an example of the Hα variation, a series of profile variation of WW Vul observed by Kolotilov (1977) is shown in Figure 7.10. The observations

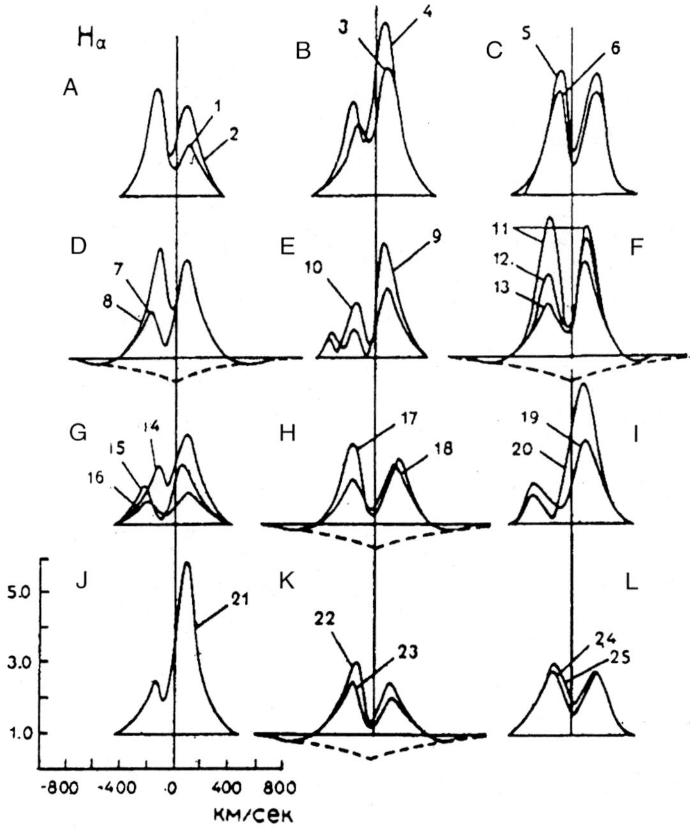

Figure 7.10: The variation of the Hα line profile in the Herbig Ae star WW Vul. The number for each profile indicates the epoch of observation, and each group from (*A*) to (*L*) gives the short-term variation within around 2 days. There is an time interval between neighboring groups from several days to several months showing some long-term variability. (From Kolotilov 1977)

were made using the spectrograph (dispersion 20 Å mm^{-1}) attached to the 125 cm reflector at the Crimean Observatory. The number for each profile gives the different epoch of observations, indicating 25 observations carried out in 1972–1974. Each group from (A) through (L) exhibits the short-term variability within 1 or 2 days, and the interval between neighboring groups denotes the long-term variability in a range from several days to several months. Remarkable changes in line intensity and profile are seen, mostly accompanied by V/R variation. The peak separation sometimes remains constant (Example: A, C, F, I, L), sometimes changes (B, D, G) and sometimes shows triple peaks (E). The V/R variation, which has been explained by the long-term one-arm oscillations in classical Be stars, may require a different explanation for the short-term change of 1 or 2 days.

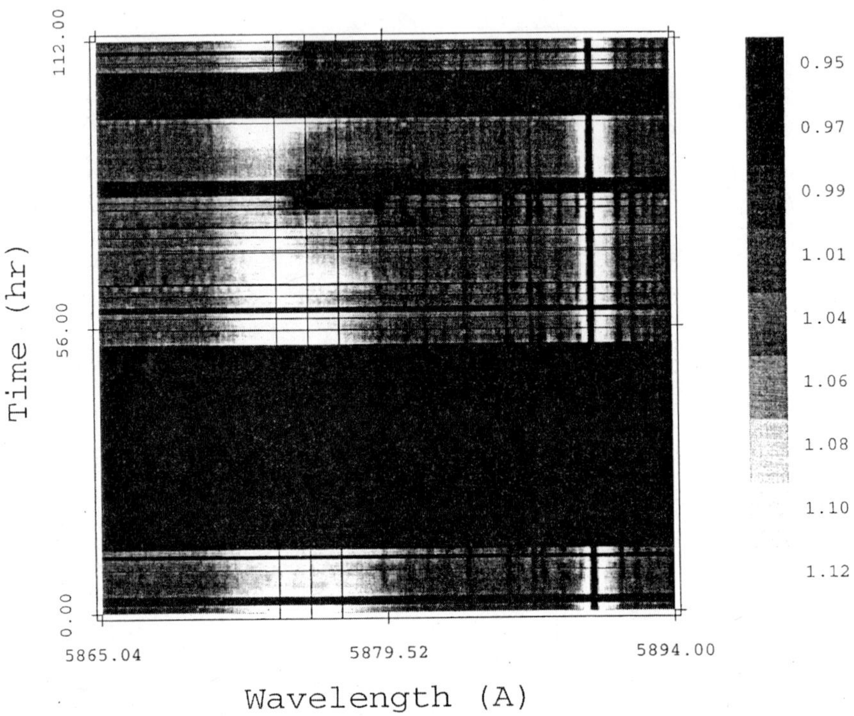

Figure 7.11: Dynamic spectrum of the HeI 5875.63 Å of the Herbig Ae star AB Aur. The ordinate gives the time series from the beginning of the campaign observations in 1992. The abscissa is wavelength. The three thin vertical lines in the He line correspond to the rest wavelength of the line and the corresponding maximum rotational velocities at $\pm V \sin i = 80.5$ km s^{-1}. (From Böhm et al. 1996)

We now proceed to the case of AB Aur (HD 31293, A0V/B9e) as an example of short-term spectral variation. This star is the brightest HES in the northern hemisphere and a high-dispersion spectral atlas in the spectral range λ3910–8740 Å was published by Böhm and Catala (1993). The Hα line basically exhibits weak P Cyg profile as seen in Figure 7.1, but shows profile variations in various time scales. Beskrovnaya et al. (1995) carried out simultaneous spectral and polarimetric observations at the Crimean Astrophysical Observatory in 1993 and 1993 and found a strong variation of the P Cyg profile at Hα in time scales from hours to months.

One of the remarkable behaviors of AB Aur is the existence of emission component in HeI D_3 λ5876 Å and its conspicuous rapid variation (see Section 4.4.4 of Chapter 4). Catala's group carried out international campaigns for the continuous high-dispersion spectroscopic observations in 1992 and 1996 (Böhm et al. 1996, Catala et al. 1999). The results are particularly interesting due to the short-term variation of HeI D_3 λ5876 Å. Figure 7.11 shows

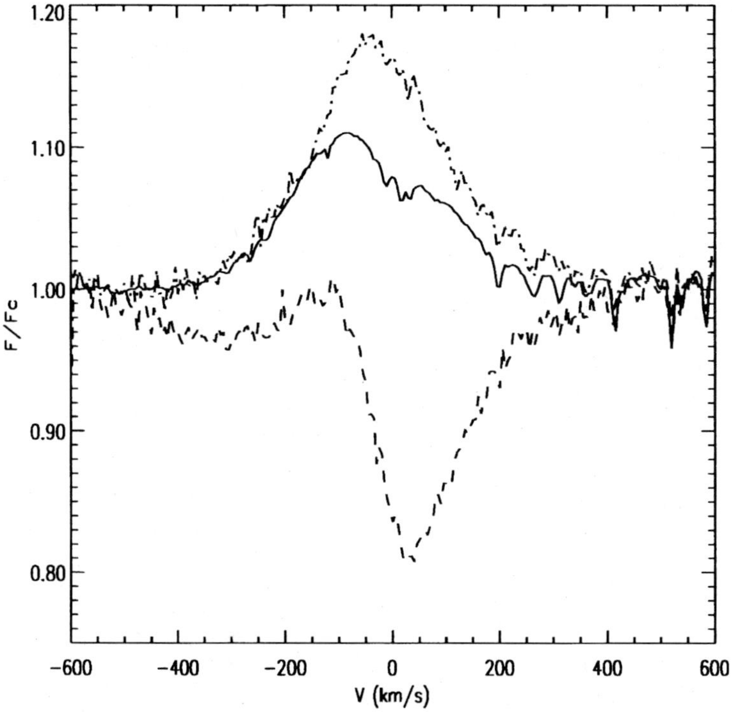

Figure 7.12: Line profiles of HeI 5876 Å of the Herbig Ae star AB Aur. Solid line: mean profile averaged over the whole series of campaign observations in 1996; Dashed line: profile near the minimum intensity, and dash-dotted line: profile near the maximum intensity. (From Catala et al. 1999)

the dynamic spectrum of HeI D_3 for a time over 112 hours in the 1992 campaign. White color denotes strong emission component. Three lines centered at $\lambda 5876$ Å correspond to the center of the line and the maximum rotational broadening at $\pm V \sin i$ ($=80 \pm 5$ km s^{-1}). Figure 7.12 is the line profiles of HeI D_3 observed in the 1996 campaign. The solid line is the mean profile, averaged over the whole period of observations. The two lines show the profiles at the time near the maximum and minimum intensities.

From these campaign observations the following behaviors of the HeI D_3 line were confirmed: (1) The line has two separate components, one blue and one red, as seen in Figure 7.12. (2) The blue component is always in emission. Its centroid varies in velocity modulated with a 45-hour period. (3) The red component is mostly in emission, but appears in absorption in several occasions. Its centroid does not change significantly, but its intensity is highly variable.

The period of velocity change differs by different lines. Catala et al. (1986) detected a variable component of 32 ± 4 hour in CaII K line and 45 ± 6 hour in MgII 2795 Å line. Praderie et al. (1986) suggested that the MgII emission is formed in the region more distant from the star than that of CaII, whereas Catala et al. (1986) attributed such period difference to a differential rotation in the outer atmosphere of the star, i.e., CaII and MgII lines are formed in different stellar latitudes, implying the possible existence of a surface magnetic field.

7.1.5 Toward the models of envelopes

Structure of the Hα envelopes

According to Finkenzeller and Mundt (1984), the profiles of Hα emission in 47 HESs are classified as follows:

Profile	Number of stars	Notes
Single peak	11	
Double peaks	14	
Be-shell type	5	Strong central absorption
P Cygni type	8	
Inv. P Cyg type	1	T Ori
Complex	3	Triple peaks
Absorption	1	TY CrA

For these profiles, if we apply the scheme of the classical Be stars, we can guess that the Hα envelopes are flattened and rapidly rotating. That is, the double peak is formed in the oblique rotating disk-like envelope, while the single peak is formed either in the nearly pole-on disk or in a sufficiently developed oblique envelope. The central shell absorption in Be-shell type indicates the existence of an optically thick envelope in front of the stellar disk. P Cyg type and inverse P Cyg type reveal the existence of global expansion or contraction of the Hα envelopes. In this case the structure of envelopes, whether they are spherically symmetric or disk-like, still remains unclear.

Stellar winds

Observational evidence for wind models, favorable at least to some fraction of HESs, may be summarized as follows:

P Cygni profiles—As seen above, around 20% of HESs reveal the P Cygni profiles in the Hα line. P Cygni profiles observed in Balmer lines, CaII, MgII, etc., and UV absorption lines with wide violet wings, are the indicators of the existence of stellar winds in these stars (Böhm and Catala 1995).

X-ray emission—Zinnecker and Preibisch (1994) detected 8 stars out of 14 HESs by ROSAT observations (energy range 0.1–2.4 keV) and

Damiani et al. (1994) detected 11 stars among 31 HESs by Einstein observations (2–3.5 keV). The X-ray emission is supposed to originate from some inhomogeneous winds such as the collision of fast wind (V_∞= 100–500 km s^{-1}) with the remnant circumstellar material (Zinnecker and Preibisch 1994).

Thermal radio emission—Thermal radio sources are detected by Skinner et al. (1993) in 21% of observed HESs. From the positive correlation of radio luminosity $L_{\rm rad}$ (3.6 cm) with $L_{\rm bol}$, $L_{\rm H\alpha}$, and $T_{\rm eff}$, they adopted the wind model for stars earlier than around A2 spectral type. The wind ionization rapidly decreases in stars of lower effective temperature.

Superionization—The possibility of superionization in clumpy structure of the wind is suggested by Bouret et al. (1997) in case of AB Aurigae. According to their model, fast and slow streams alternate and produce shocks where fast streams overtake slow ones. This model may be consistent with that of X-ray emission.

Profile of the forbidden lines [OI]—Forbidden lines of [OI] $\lambda\lambda$ 6300, 6364 Å show a broadening corresponding to velocities between 30 and 100 km s^{-1}. Böhm and Catala (1994) interpreted these broadening as the terminal velocities of a stellar wind, which form the low-excited forbidden lines in its outermost parts.

Accretion flows and accretion disks

Possible evidence for the disk accretion in HES's is enumerated by Ghandour et al. (1994) as follows:

Excess IR emission with spectral energy distribution (SED) consistent with origin in an optically thick accretion disk (see Section 7.1.2).

Excess optical and UV continuum emission arising in either a classical or a magnetospheric boundary layer and with a magnitude proportional to the accretion luminosity inferred from excess IR emission.

Forbidden line emission with strength proportional to both excess IR and optical/UV emission.

Kinematic signature of accretion, most notably inverse P Cygni profiles in the Balmer series and in other strong resonance lines (e.g., NaI, CaII, MgII).

Some additional diagnostics for the accretion process are given by Pérez and Thé (1994) including (1) strong correlation between the decrease in brightness and increase of linear polarization (see Figure 7.3) and (2) the blueing effect (Section 7.1.4).

Wind versus accretion

The rate of stars that disclose clear evidence for the wind or accretion flows is not high, around 20% or less. P Cygni profile appears more often than inverse P Cygni profile, showing that the wind prevails over accretion.

Stars sometimes show a change of state from wind to accretion or vice versa. For example, Pérez et al. (1993) reported that the Herbig Ae star HR 5999 (A7 III-IVe) showed a variation of the profile of MgII doublet from P Cyg type (1979–1990) to inverse P Cygni type (1992), suggesting the change of envelope structure from wind to accretion.

Some stars are also supposed to be in coexistence of wind and accretion such as the disk wind model proposed by Catala et al. (1999). In this case the wind originates from an accretion disk through the reprocessed energy flux sufficient to lift up disk matter to the wind (see also Section 7.2.5, Figure 7.30).

Chromospheric activity

The existence of stellar chromosphere is directly inferred from the emission line of HeI. Böhm and Catala (1995) detected HeI λ5876 Å in 13 HESs out of 29 observed stars. Among the detected stars, 10 stars are of spectral types later than B7, for which the stellar radiation is not sufficient to excite HeI lines. The equivalent widths of HeI λ5876, EW(HeI), are in a range of 25–230 mÅ, and there is no particular correlation with the equivalent with of the Hα line. It is also noticed that the strong HeI emission is sharply concentrated in stars with $T_{\rm eff} \sim 10{,}000$ K. This infers that He I emissions are formed in chromospheric region not related to the development of Hα emitting regions, but are favorably connected with stars of around A0 type.

Based on HeI line profile analysis of AB Aur, Catala (1988) derived a semi-empirical model of a stellar wind with $\dot{M} \sim 10^{-8}\ M_\odot$ per year and an extended chromosphere, located at the base of the wind, with a temperature of 17,000 K. Since many of these stars show violet shifts in HeI emission, the chromspheres are generally supposed to be expanding.

Magnetospheric models

Magnetospheric models of HESs have been proposed to explain observed infrared excess and some line profiles characteristic of accretion flows (Sorelli et al. 1996, Muzerolle et al. 2004). If a pre-main sequence star has a dipolar magnetosphere, the inner part of the accretion disk is destroyed and the material falls into the star along magnetic field lines producing hot regions of high density at the base of the accretion disk (see Figure 3.16 of Chapter 3). Magnetospheres will form optical lines such as Balmer lines and Na D, whereas cool accretion disks will produce infrared emission. Muzerolle et al. (2004) also pointed out that the accretion disk of HESs could be divided into inner gas disk and outer dust disk separated by a dust destruction wall at some radius due to stellar irradiation. Basic parameters of the magnetospheric model are the mass accretion rate, stellar luminosity, and stellar rotation velocity.

7.1.6 Optical jet flows

Molecular and/or optical outflowing phenomena, such as molecular flows, Herbig-Haro objects, and jets, are often observed in the vicinity of HESs and

Table 7.5: A list of optical jets associated with HESs (adapted from Mundt and Ray 1994)

Star	Sp. Type	Dist. (kpc)	Lumin. ($10^3 L_\odot$)	Proj. length (arcsec)	Proj. length (pc)	Max. outflow velocity (km s^{-1})	Full open angle (degree)
LkHα 198	B/Ae	0.95	0.26	23	0.11	$\leq$50	5
HK Ori ?	B8/A4ep	0.46	0.05	115	0.25	—	—
V380 Ori	B8/A1e	0.46	0.3	145	0.33	400	80
R Mon	B0e	0.8	0.7	400	1.6	75	10
LkHα 234	B5/B7e	1.0	1.3	40	0.2	200	10
MWC 1080	B0eq	2.2	17	20	0.25	300	70

other young stellar objects. Cantó et al. (1984) carried out radio observations of carbon monoxide, using the millimeter-wave telescope of the National Radio Observatory at Kitt Peak, for 11 HESs, and detected molecular clouds for all of them. Molecular outflow was found for two stars (LkHα 198 and MWC 1080). Mundt and Ray (1994) collected the data for optical outflows associated with HESs and other high-luminosity young stellar objects and listed 24 outflow objects including 15 HESs. Since optical outflows exhibit emission lines in Hα and in some forbidden lines, we can measure the physical parameters of the flows such as flow velocity and gas density, along with optical size and the opening angle of the jet-like flow. Some examples of outflows are shown in Table 7.5. The projected length of jets is in the range of 0.1–1.6 pc, much larger than the stellar envelope, and the jets are often connected to molecular outflows and Herbig-Haro objects.

We now consider the optical jet associated with LkHα 234, the luminous Herbig Be star embedded in the reflection nebula NGC 7129. The jet observed by Ray et al. (1990) is shown in Figure 7.13, where the right panel illustrates the narrow band image of the [SII] λ λ 6716, 6731 Å after subtracting the continuum emission. Highly collimated jet emanating from LkHα234 is seen to a projected distance of 40" corresponding to 0.19 pc. The jet is inhomogeneous containing several knots such as denoted by symbols A, B, C, . . . , but no hint of a counter-jet is found. The left panel is a long-slit spectrum of the LkHα 234 in the [SII] λ λ 6716, 6731 Å doublet region. The continuum is removed. The electron densities and heliocentric radial velocities measured for the individual knots are indicated. This jet is well collimated, but consists of many knots with different radial velocities, suggesting some intermittent ejection from the star.

If we assume a typical flow velocity of 50 km s^{-1} and the inclination angle of around 45°, i.e., the length of jet being $\sim 0.19\sqrt{2}$ pc, then the dynamical time scale is $\sim 5 \times 10^3$ years. It is also noted that LkHα 234 has a molecular (CO) jet in the direction opposite to the optical jet as observed by Mitchell and Matthews (1994). The inner $30''$ of the CO jet is seen to be aligned with the optical jet axis, and, in the outer part, the CO structure bends to the north, ending at a faint Herbig-Haro object HH105, separated around $80''$

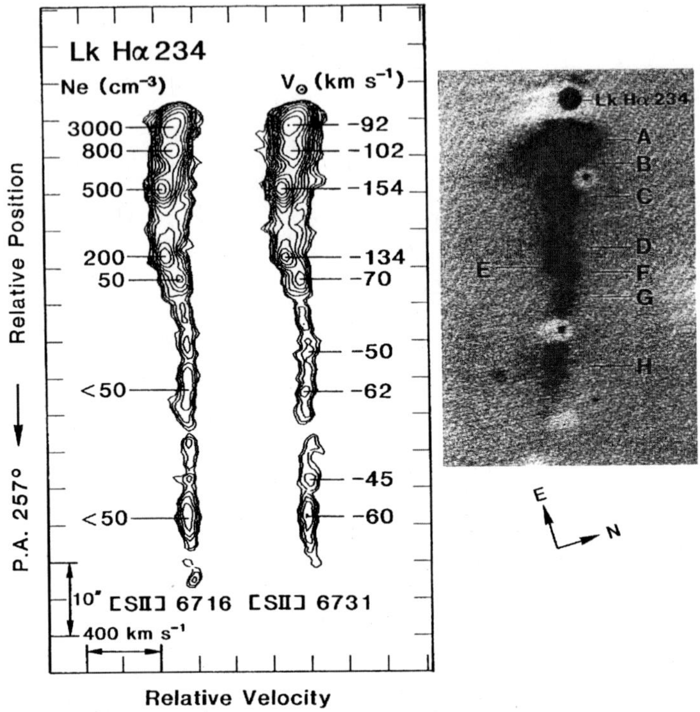

Figure 7.13: Optical jet and its physical structure of the Herbig Be star LkHα 234. The right panel shows the [SII] image after subtraction of the continuum emission. Main knots are designated by the symbols A, B, ..., H. The left panel indicates two position-velocity maps of [SII] doublet lines of λ6716 (left) and λ6731 (right). The equal electron density contours are given in the left, and the equal heliocentric velocity contours are given in the right image. The peak values in the main knots are also indicated. (From Ray et al. 1990)

from LkHα 234. The molecular flow velocity is typically 10 km s^{-1}, and then the dynamical time scale of the flow is $\sim$5 $\times$ 10^4 years, which is notably longer than the optical jet. The estimated mass loss rates are $\sim$3 $\times$ 10^{-8} $M_\odot$ per year for the blue-shifted optical jet and $\sim$3 $\times$ 10^{-3} $M_\odot$ per year for the red shifted molecular jet. The time scale is quite different for both jets.

7.2 T Tauri type stars

7.2.1 What are T Tauri type stars

T Tauri type stars (TTSs) were first noticed by Joy (1942) as irregular variables, existing near or embedded in dark clouds and showing strong emission in CaII H and K lines as well as in hydrogen and metallic lines. The pre-main-sequence nature of these stars has been pointed out by Ambartsumian

Table 7.6: A list of selected bright TTSs ($V < 12.0$)

Star	HRC[a]	Mag.$\langle V\rangle$[b]	Sp. type[b]	LC[c]	$\log L(L_\odot)$[d]	W(Ha) (Å)[d]	Notes[e]
V410 Tau	29	10.9	dK5(e)	—	0.38	2.1	WTTS
RY Tau	34	10.7	K1e	I, IV	1.24	21.0	CTTS
T Tau	35	10.4	K1e	I	1.45	38.0	CTTS
DF Tau	36	11.6	dM0e	II	0.73	53.9	CTTS
SU Aur	79	9.2	G2ne III	—	1.25	3.5	WTTS
RW Aur	80	10.8	dG5?e	II	>0.78	84.2	CTTS
CO Ori	84	10.6	Gpe	IV	1.68	10.2	CTTS
GW Ori	85	9.7	dK3e	I	1.82	27.6	CTTS
P1649	121	10.3	K0-2 III-IV	—	1.34	none	WTTS
FU Ori	186	8.9	F2:p I-II	—	2.43	none	FUO
R Mon	207	11.7	(B0)	IV	1.21	96.7	
Z CMa	243	8.8	eq	—	—	—	FUO
DI Cep	315	−11.2	dK3e	II	6.48	95.5	CTTS

[a]HRC = Herbig–Rao Catalogue (Herbig and Rao 1972).
[b]Data taken from HRC.
[c]LC = Form of light curve by Parenago (1954) (see Section 7.2.5).
[d]Values taken from Cohen & Kuhi (1969).
[e]CTTS = Classical T Tauri stars (G ~ M type); WTTS = weak-lined T Tauri stars; FUO = FU Ori type pre-main sequence star.

(1949) and Herbig (1952, 1962). The main evidence for their youth is as follows:

(a) Association with the star-forming regions—A star forming region is characterized as an aggregation of young nebulosity (molecular cloud, reflection/dark/emission nebula), OB association, and groups of infrared objects. TTSs coexist with these objects (Herbig 1977) and often form the so-called T association with or without accompanying O associations.
(b) Location on the HR diagram—TTSs are located in the upper part of the late-type main sequence and show good correspondence with the theoretical prediction of the pre-main sequence evolution of stars in a gravitational contraction (Kuhi 1965, Cohen and Kuhi 1979) (see Section 7.4).
(c) Spectroscopic characteristics. Strong stellar activities forming variable emission lines infer the existence of developed accretion disks or stellar winds. High Li abundance is also an indication of the youth of stars (see Section 7.2.2).

The catalogue of TTSs was first published by Herbig (1962) for 136 stars with data on brightness, variability type, and emission line strength. Herbig and Rao (1972) listed 323 HESs in the Orion population. Table 7.6 lists selected bright TTSs.

There are several subtypes so far proposed for the TTSs:

(a) **CTTS** (classical TTS)—TTSs having strong Hα emission (W(Hα) $\geqq$ 10 Å), and later than K0 type (Herbst et al. 1994).
(b) **WTTS** (weak-lined TTS)—TTSs with weak Hα emission (W(Hα) $<$ 10 Å), usually later than K0 type. CTTS and WTTS are rather clearly separated with each other (Herbst et al. 1994).
(c) **NTTS** (naked TTS)—Evolved TTS that look naked since they have losed the dust shell surrounding the star (Walter, 1986, Walter et al. 1988).
(d) **ETTS** (early-type TTS)—TTSs earlier than K0. The behaviors in colors and spectral variations are somewhat different from CTTSs (Herbst et al. 1994).
(e) **PTTS** (Post TTS)—Stars just before entering ZAMS, after ending the evolutionary stage as TTSs (Pallavicini et al. 1992). This type is not clearly separated from NTTs.

There are other types of TTS, called FU Ori type (Fuor or FUor), EXors and YY Ori type, in the early stage of TTS evolution. These stars will be considered in Section 7.2.6.

7.2.2 Spectroscopic features

Optical spectra

The optical spectra of TTSs are characterized by emission lines superposed on late-type spectra. Some examples of low-dispersion spectra given by Cohen and Kuhi (1979) are shown in Figure 7.14.

In the optical region, we particularly notice the following spectral properties:
(a) Existence of emission in CaII H, K, and Hα lines—Herbig et al. (1986) surveyed a large area of Tau-Aur dark cloud region and detected CaII H and K emission in all known TTSs. According to Walter et al. (1988), the surface flux of CaII K emission seems to peak in the early K stars and fall off into the M stars. On the other hand, the ratio of the Hα surface flux to that of CaII K increases with the advance of spectral type as shown in Figure 7.15. This suggests that the dominant source of radiative losses gradually changes from CaII and Mg II in early G type stars to the Hα emission in M type stars.

→

Figure 7.14: Examples of low-dispersion spectra of TTSs observed by Cohen and Kuhi (1979) at the Lick Observatory with the spectral resolution of 7 Å and wavelength range λ4270–λ6710 Å. The abscissa is the wavelength (Å) and the ordinate is the radiative flux (erg cm^{-2} s^{-1} Å^{-1}). The spectral type of each star is: (a) SU Aur (G2III), (b) RW Aur (G5), (c) T Tau (K1), (d) BP Tau (K7), (e) SU Ori (M1) and (f) Z CMa (P Cyg profile). Each spectrum is shown twice: The lower plot, which uses the right-hand ordinate scale, has been scaled to the peak of Hα emission as its maximum. (From Cohen and Kuhi 1979)

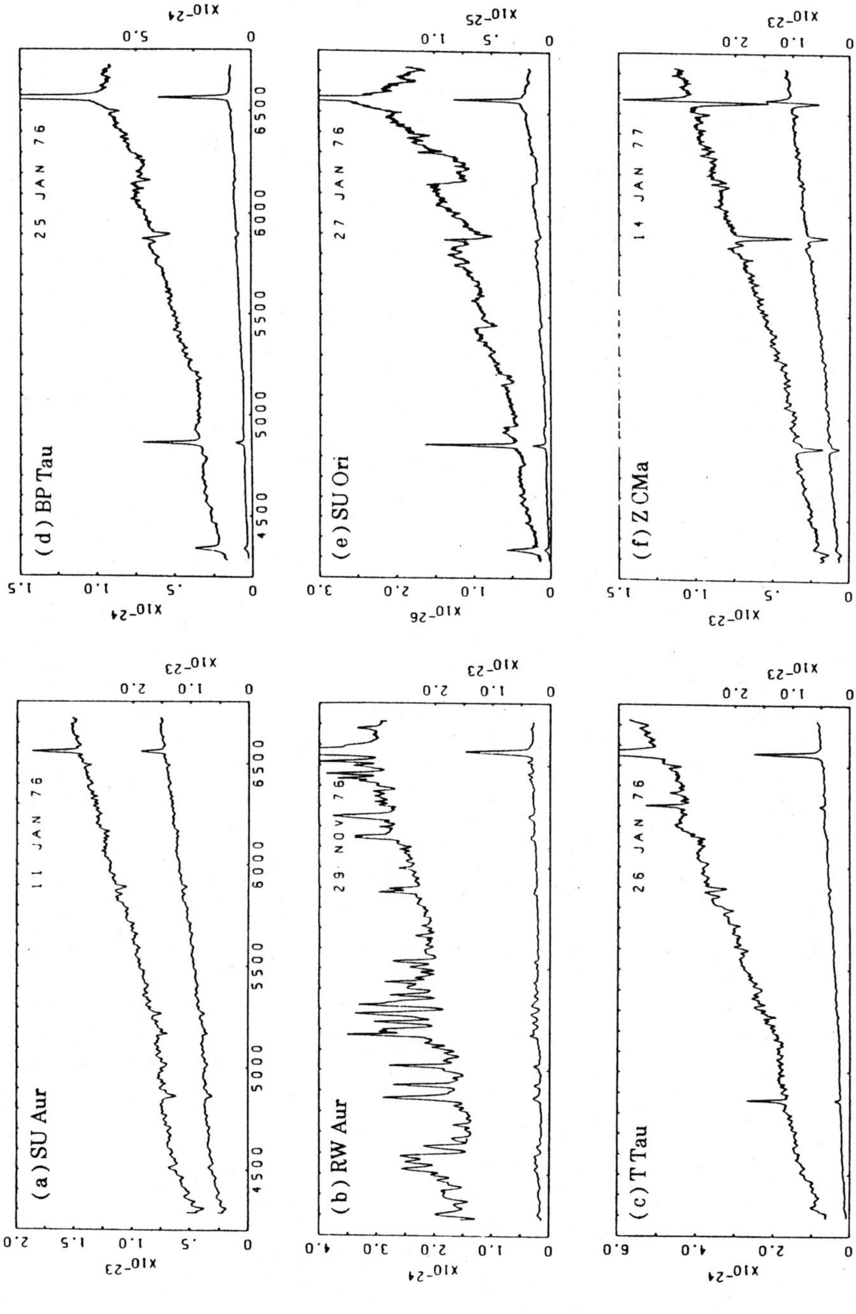

(a) SU Aur
11 JAN 76
(b) RW Aur
29 NOV 76
(c) T Tau
26 JAN 76
(d) BP Tau
25 JAN 76
(e) SU Ori
27 JAN 76
(f) Z CMa
14 JAN 77

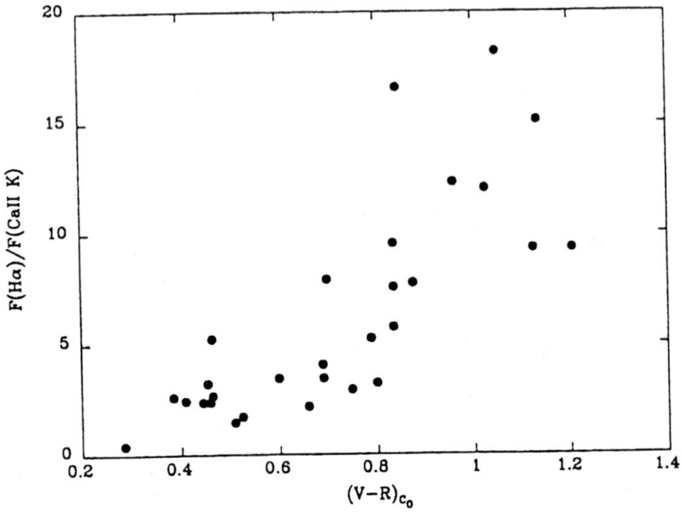

Figure 7.15: Ratio of the Hα to CaII K surface fluxes as a function of V – R color. The approximate correspondence between V – R color and spectral type is as follows: $(V-R)_0$: 0.4, 0.6, 0.8, 1.0, 1.2. Spectral type: G0, K0, K8, M0, M3. (From Walter et al. 1988)

(b) Existence of the fluorescence emission lines in FeI—Enhancement of the FeI emission lines λλ 4063, 4132 Å has been explained by the fluorescence mechanism (Herbig 1945). As seen in the partial Grotrian diagram of FeI given in Figure 7.16, FeI λ3969.261 Å (a^3F_4–$y^3F^0_3$) line is closely coincident to the

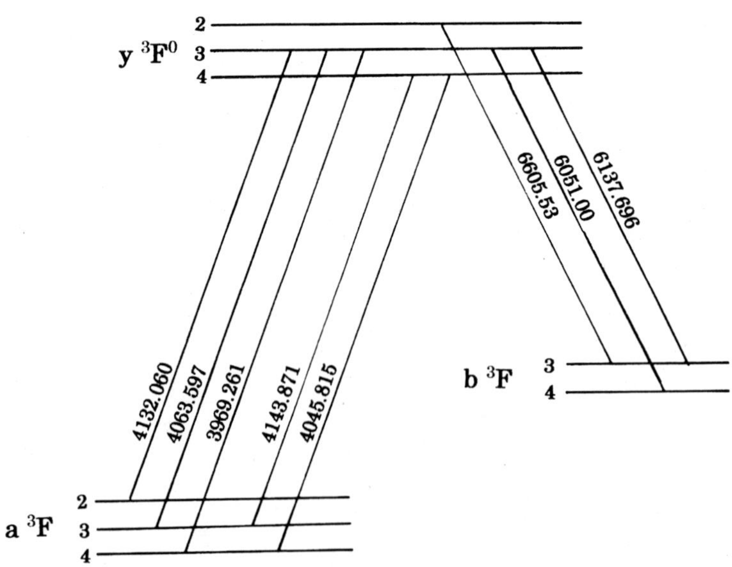

Figure 7.16: Schematic partial Grotrian diagram of three ^{3}F levels of FeI.

Table 7.7: Relative intensity of FeI fluorescence lines observed in RW Aur (adapted from Gahm 1970)

Wavelength (Å) transition line	4063.597 a^3F_3–$y^3F_3^0$ FL	4132.060 a^3F_2–$y^3F_3^0$ FL	4143.871 a^3F_3–$y^3F_4^0$ Non-FL
Date of observations			
19 Sep. 1962	56	44	4
4 Dec. 1968	48	27	3
30 Mar. 1969	38	27	2:

Notes: FL = fluorescence line; Non-FL = nonfluorescence, comparison line. Relative emission intensity is normalized at 4100 Å continuum level.

CaII H and Hε lines that are generally found in strong emission at around λ3969 Å. Due to this coincidence the $y^3F_3^0$ level is strongly overpopulated, leading to emission lines originating from this level. The relative intensities of fluorescence lines of RW Aur, observed at the Lick Observatory by Gahm (1970), are shown in Table 7.7, where one may see strong enhancement of fluorescence lines as compared to nonfluorescence lines. The intensity ratio of λ 4063/λ 4132 Å must be approximately equal to 3.7, which is the ratio of spontaneous transition probabilities from the common level $^3F_3^0$. But observed ratios in Table 7.7 are between 1.4 and 1.8, implying that the envelope of RW Aur is not optically thin for these lines. Wilson (1974) carried out a statistical equilibrium analysis for a 12-level iron atom, and derived the intensity of the fluorescence line λ4063 Å relative to the nonfluorescence line λ4143 Å in some suitable conditions of physical parameters for TTSs. As a result Wilson obtained an upper limit of electron density $N_e = 10^{16}\,W$ for the FeI line forming region, where the electron temperature is taken as 3000 K and the dilution factor W (see Equation (4.1.5)) is suggested to have a value between 10^{-3} and 10^{-6}. If we assume $W = 10^{-5}$, then N_e should be smaller than 10^{11} cm^{-3} at around 200 stellar radius.

(c) Forbidden line [OI] λ6300 Å prevailing in the spectra of TTSs—Cabrit et al. (1990) surveyed 36 TTSs and found [OI] λ6300 Å for 23 stars along with Hα emission with various intensities. Some stars also show forbidden lines of [SII] $\lambda\lambda$ 6717, 6731 Å, [OI] λ6363 Å, and [NII] $\lambda\lambda$ 6548, 6584 Å. The critical electron densities of these lines are at most 10^6 cm^{-3}, as shown in Table 4.5 in Chapter 4. This implies that the forbidden lines are formed in low-density outer parts of the stellar winds. Actually, Cabrit et al. (1990) found that the [OI] lines are blueshifted with average velocity of -42 km s^{-1} in a range from 0 to -160 km s^{-1}. The intensities of these lines well correlate with Hα luminosity over two orders of magnitude. They interpreted these features by the existence of energetic winds with a wide range of mass-loss rates among TTSs. Appenzeller et al. (1984) analyzed the profiles of forbidden lines and concluded that most observed profiles are not compatible with spherically symmetric radial winds but qualitatively agree with profiles predicted from axially symmetric velocity fields in filled or hollow cone winds.

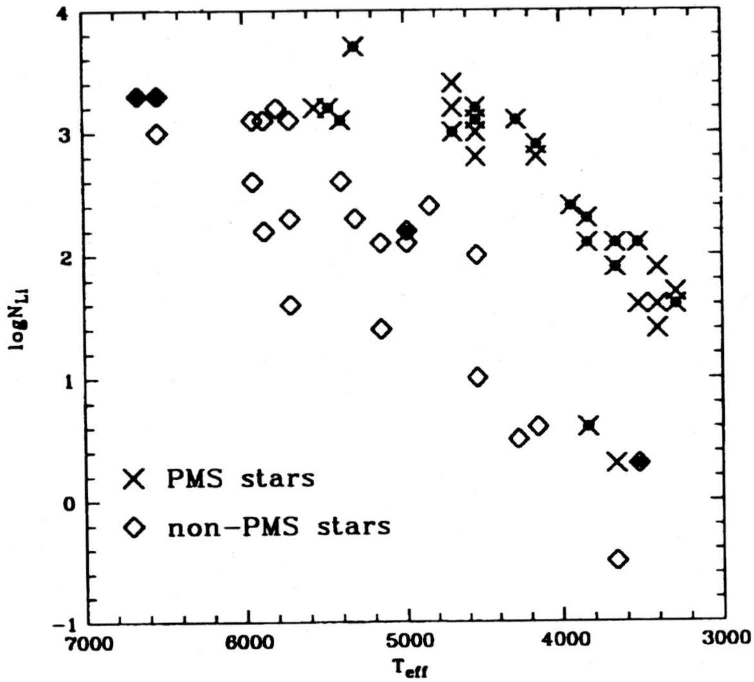

Figure 7.17: Lithium abundance log N(Li) vs. effective temperature $T_{\rm eff}$. Crosses represent pre-main sequence stars (PMS) and open diamonds indicate the non-PMS stars. Filled diamonds are the stars of the Chamaeleon star forming region. (From Magazzu et al. 1997)

(d) Strong absorption line in Li I λ6707 Å—According to Herbig (1965) Li abundance of TTSs is around two orders of magnitude higher than that of the main sequence stars and nearly the same order as the interstellar medium. He attributed this evidence as an indication of the youth of TTSs, since Li atoms can be destroyed by collision with protons in a deep convective layer. Along the same line of thought, Magazzu et al. (1992, 1997) measured the lithium abundance in order to see the effects of stellar effective temperature, mass, and age. In Figure 7.17, the lithium abundance log N (Li) (in the scale of log N (H) = 12) is plotted for $T_{\rm eff}$ for PMS (pre-main sequence stars = mostly TTSs) and non-PMS (mostly main-sequence field stars). It is apparent that the lithium abundance in PMS stars is around two orders of magnitude higher than that of non-PMS stars, and that both make a parallel decline toward red stars. This parallel declining may be explained by the development of convective layers in redder stars where the destruction of Li atoms takes place more efficiently. Martin et al. (1994) also showed that the TTSs with higher rotational velocities tend to have higher values of log N (Li). This can be understood by the effect of rotation, which prevents Li atoms from penetrating to the deeper convection layer.

Table 7.8: Distribution of linear polarization in TTSs

Number of stars	52
Linear polarization	
Less than 1%	51.92%
1–2%	17.31
2–4%	28.84
Higher than 4%	1.92

Note: The data are based on Bastien (1982, 1985).

Linear polarization

Most of the TTSs show intrinsic linear polarization in the optical continuum and in the Hα profile, in the similar manner as in Herbig Ae/Be stars.

In the continuum, Bastien (1982, 1985) carried out the polarization survey in the spectral region near λ5895 and λ7543 Å and found the degree of polarization as summarized in Table 7.8. Around half of TTSs show polarization lower than 1%, whereas a small part of stars (~2%) revealed polarization higher than 4%.

The intrinsic linear polarization is weakly correlated with the intensity of the Hα emission as shown in Figure 7.18, where the polarization data are taken from Bastien (1982) and the emission equivalent widths of the Hα are from Cohen and Kuhi (1979). Note that polarization higher than 1% appears only in stars with strong Hα emission (CTTS, EW(Hα) $>$ 10 Å), whereas weak emission stars (WTTS) show mostly weak polarization of less than 1%. The development of Hα emitting region seems to be an important factor in increasing linear polarization.

The wavelength dependence and its time variation have been widely investigated by Bastien and Landstreet (1979) and Bastien (1981). Since observations showed no depolarization effects in the Hα profile and no bound-free absorption in the Balmer continuum, it becomes clear that the linear polarization is caused by the scattering of dust particles rather than by the electron scattering. So the development of circumstellar dust envelope provides higher linear polarization.

Ménard et al. (2003) monitored the photometric and polarimetric variations of the classical T Tau star AA Tau and found cyclic variations coupled with the rotation of the system. The light curve is roughly constant in the bright state, with quasi-periodic fadings of amplitude up to 1.4 magnitudes in BVRI bands. A plot of polarimetric and photometric variatons of AA Tau is shown in Figure 7.19, which exhibits similar behavior with the Herbig Ae star RR Tau in Figure 7.3. That is, during the fading, the polarization increases. Because of the absence of color variations during the fadings, these polarization changes can be interpreted as an occultation phenomenon of AA Tau's photosphere by opaque circumstellar material.

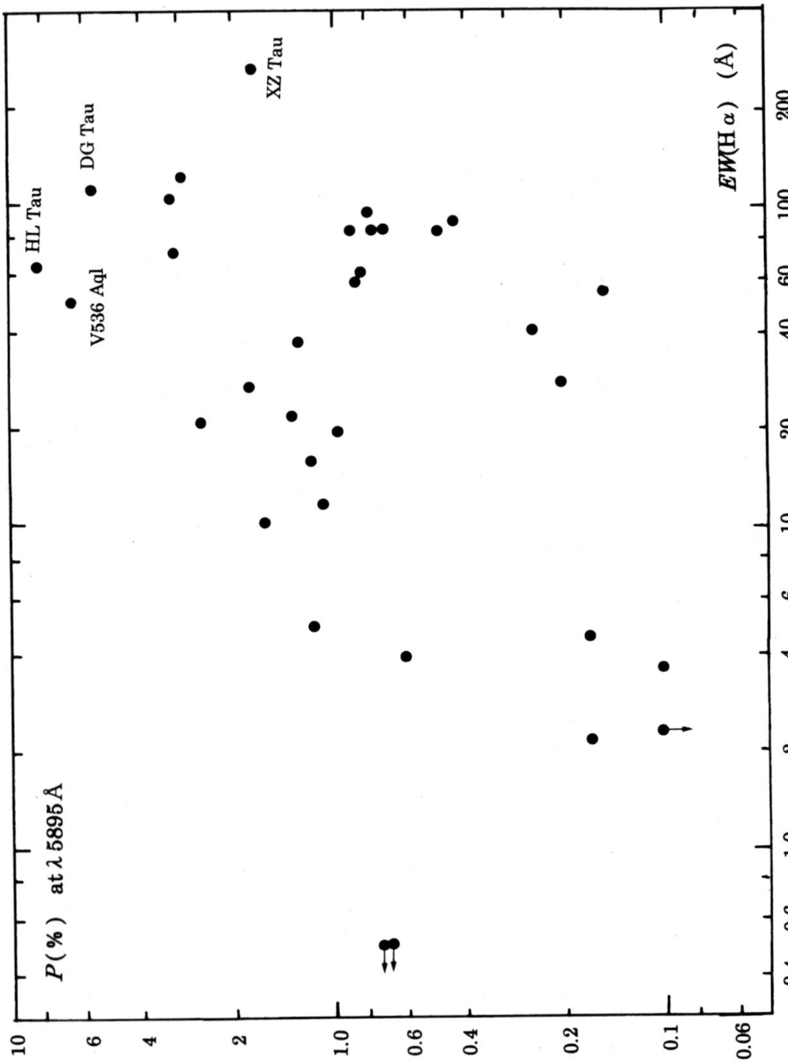

Figure 7.18: Linear polarization (at λ5894 Å) vs. Hα emission equivalent width in TTSs, plotted by the data of Bastien (1982) and Cohen and Kuhi (1979).

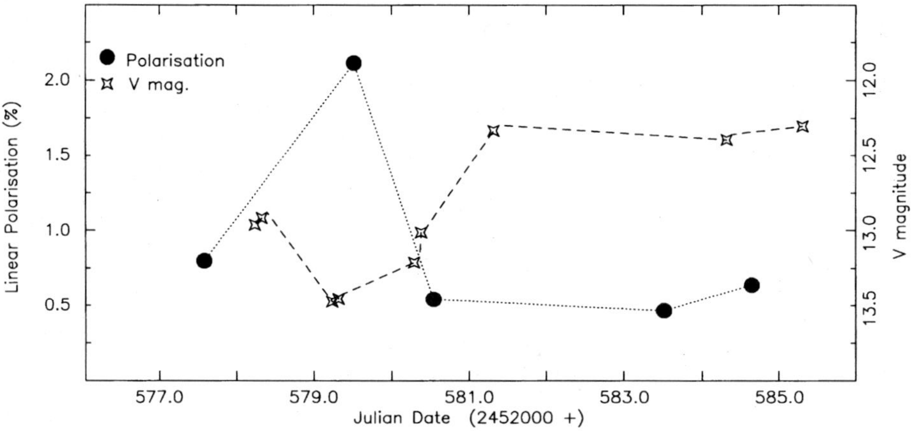

Figure 7.19: Light curve and linear polarization of AA Tau as a function of Julian days. Full circles and dotted lines denote I-band linear polarization, and open stars and dashed line gives the V-band photometry. (From Ménard et al. 2003)

Based on Hα spectropolarimetry observations, Vink et al. (2005) detected a change in the linear polarization across the Hα profile, that is, the line polarization, in most of TTSs (9 among 10 stars), just as seen in Figure 7.4 (L type polarization) in the case of HESs. They interpreted the line polarization in terms of a compact source of line photons that scattere off a rotating accretion disk. Furthermore, they found consistency between the position angle (PA) of the polarization and those of elongated disk image obtained from infrared and millimeter imaging and interferometric observations, though the latter imagings are probing much larger spatial scales.

Infrared spectra

Extensive infrared observations have been carried out by Mendoza (1968) for 33 TTSs in the spectral range 0.36–5 μm and then by Cohen and Kuhi (1979) for 207 pre-main sequence stars in three bands of H (1.26 μm), K (2.3 μm), and L (3.5 μm). The intrinsic two-color diagram ($H-K$, $K-L$) for the pre-main sequence stars in Taurus-Auriga clouds is shown in Figure 7.20. For comparison the loci of main-sequence stars and black bodies in various temperatures are illustrated, along with the locus of a K7 photosphere with increasing effects of superposed optically thin free-free, free–bound, and bound–bound continuum emission. The loci of a K7 photosphere, overlaid by thermal emissions from hot dust grains at 800, 1000, and 1500 K, are also indicated. On this diagram, one may see that the pre-main sequence stars (mainly TTSs) show a large infrared excess as compared to the main sequence stars, and that they are mostly scattered in a region bounded by the predicted curves of the dust temperature 1000 and 1500 K. This infers that the infrared excess of the TTSs is originated by a hot dust envelope surrounding the stellar photosphere.

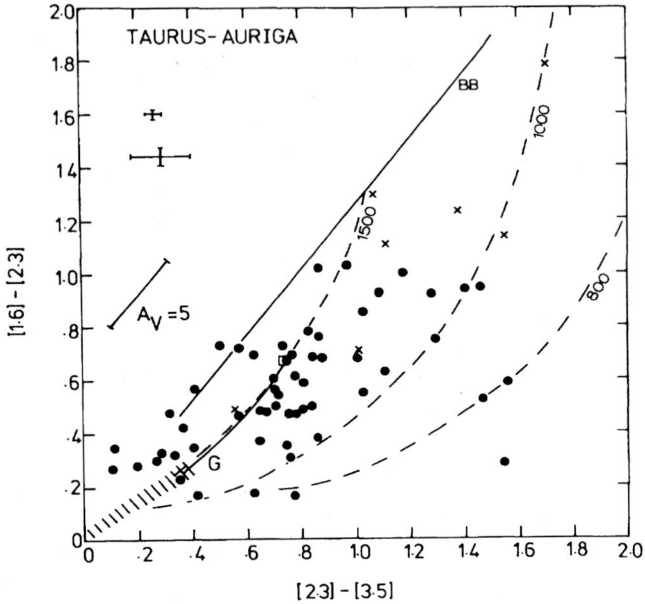

Figure 7.20: Two-color diagram ($H-K, K-L$) for pre-main sequence stars in the Taurus-Auriga clouds. Filled circles are stars for which continuum extinction can be determined (intrinsic color). Crosses are stars without extinction correction (observed color). The reddening correction vector for 5 magnitudes is shown in the lower left. The hatched area represents the locus of main-sequence stars. The solid line (BB) represents the locus of black bodies. The solid curve (G), terminated by an open square, indicates the locus of a K7 photosphere with increasing superposed optically thin free-free, free-bound, and bound-bound gas continuum emission. The dashed loci refer to a K7 photosphere overlaid by thermal emission from hot dust grains at 800, 1000, and 1500 K. Two error bars in the upper left are for bright stars (small bars) and faint stars (large bars). (From Cohen and Kuhi 1979)

The spectral form of the IR excess component was derived by Rydgren and Zak (1987). The spectral energy distributions (SED) and decomposition into stellar and IR components are exhibited in Figure 7.21 for three TTSs. The procedure of decomposition is as follows: The observed SED, over the wavelength range 0.36–20 μm, is first obtained from multiband photometry with a suitable correction for interstellar extinction. Then a fitting of a normal photosphere with the observed SED is made at I band (0.88 μm). By subtracting the photospheric component from the observed flux, we get the IR excess component for these stars. It is found that the IR excess components are remarkably similar from star to star, suggesting a common IR excess at wavelength shortward of 20 μm for T Tauri stars. The mean IR excess from 3.5 to 20 μm is well described by a power low of the form $\lambda F_\lambda \propto \lambda^{-0.75}$, with a peak at around 3 μm. From this spectral form one can estimate the dust

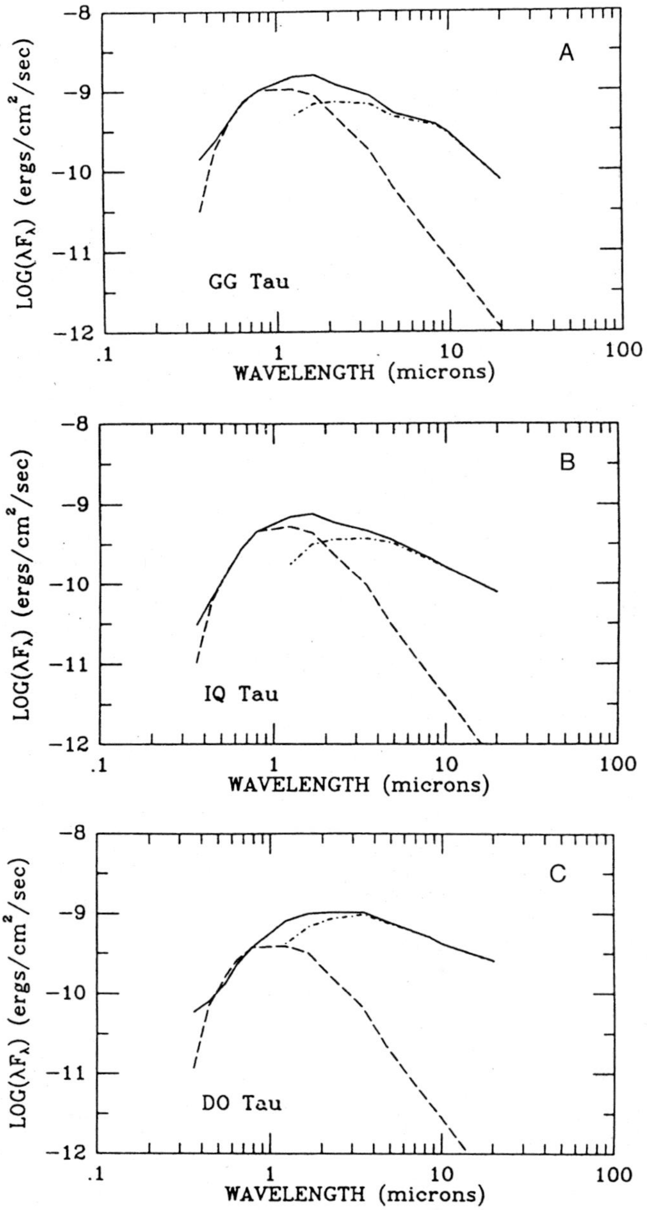

Figure 7.21: The SED for three TTSs (solid curves), corrected for interstellar extinction. The dashed lines are the fitted photospheric SED, while the dot-dashed line is the IR excess components. (From Rydgren and Zak 1987)

temperature of about 1300 K, which is in good agreement with the suspected temperature distribution between 1000 and 1500 K seen in Figure 7.20.

The structure of near IR excess sources of TTSs has become clear in 1990s, through (i) large infrared and millimeter continuum excess, (ii) blue-shifted emission-line profiles, and (iii) direct images of flattened, opaque circumstellar structure ~100 AU in size. Meyer et al. (1997) re-examined the near-infrared color–color diagrams, and derived accretion disk models that are characterized by a range of disk accretion rates ($10^{-8} < \dot{M}_{\rm disk} < 10^{-6}\, M_\odot$), inner disk holes sizes (1–6 R_*), and a random distribution of viewing inclination.

TTSs exhibit remarkable emission lines in the near-infrared region. Folha and Emerson (2001) carried out infrared observations using the United Kingdom Infrared Telescope (UKIRT) for 50 TTSs in 1994 and 1995. They found emission features in Paβ and Brγ lines in the most of observed stars. Examples of Paβ line profiles are shown in Figure 7.22, where the types of line profiles are defined as follows:

Type I—Symmetric profiles showing no or little evidence of absorption feature.
Type II—Profiles showing two peaks with the intensity of the second peak exceeding half the strength of the main peak.
Type III—Profiles showing two peaks with the intensity of the second peak being less than half the strength of the main peak.
Type IV—Profiles showing an absorption feature beyond which no emission is seen on the red or blue side.

To types II, III, and IV the letters B or R are appended, depending on the location of the secondary emission/absorption features relative to the main peak. If blue shifted a B is appended, and if red shifted an R is added. Note that Type IVB corresponds to P Cyg profiles, and Type IV R corresponds to inverse P Cyg profiles. The most conspicuous features in the line profiles among their observed stars is the almost complete absence of blueshifted absorption components and the relative high frequency of inverse P Cyg profiles. From these profile features, Folha and Emerson suggested the prevalence of infalling motion supporting accretion models.

Ultraviolet spectra

Like other late-type stars, TTSs also exhibit many emission lines in the far ultraviolet spectral region. Based on the IUE observations in 1978–1986, Imhoff and Appenzeller (1987) gave a list of 141 pre-main sequence stars, including 75 TTSs, 6 weak emission line stars, 28 HESs, and 2 FU Ori type stars.

Spectral features of the TTSs in the wavelength range of 2000–3000 Å are marked by strong emission of MgII h and k lines, along with other lines of MgI, FeII, CrII, etc, in emission or absorption. In the wavelength region 1200–2000 Å, remarkable emission generally appears in OI, SI, CII, CIII, CIV, SiII, SiIII, SiIV, and sometimes in higher excitation lines like HeII and NV. The relative intensity CIV ($I.P. = 64.5$ eV)/SiIV ($I.P. = 45.1$ eV) can be taken as

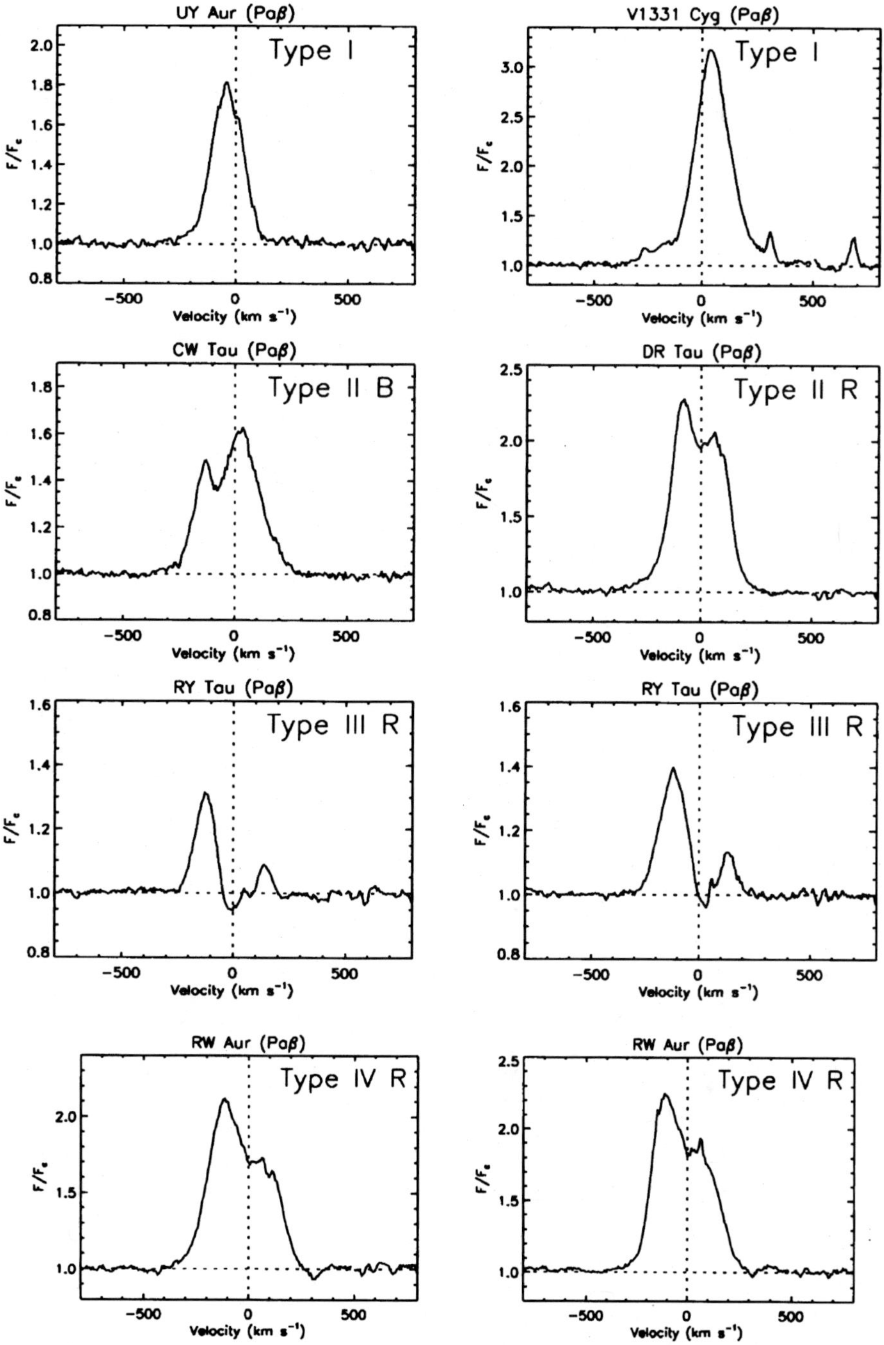

Figure 7.22: Examples of Paβ line profiles Types I–IV in TTSs. See text for the types of line profiles. (From Folha and Emerson 2001)

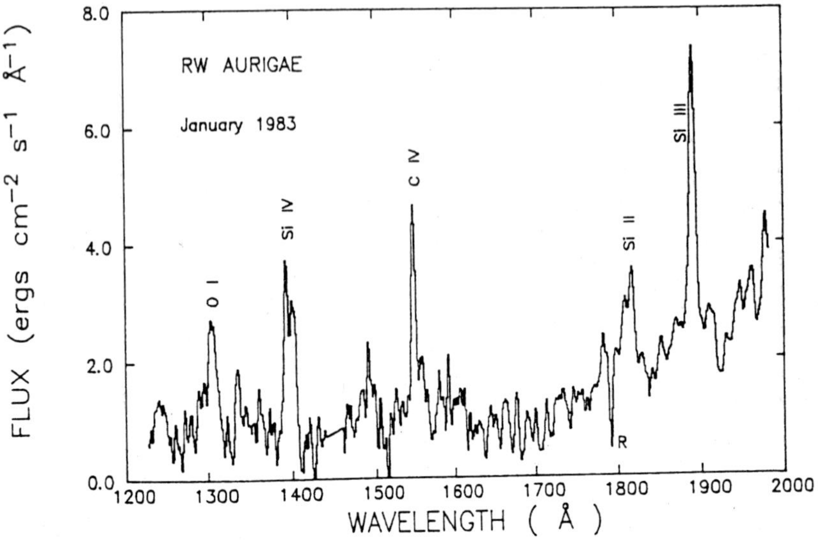

Figure 7.23: The far-UV spectrum of RW Aur observed by the IUE satellite. Observed flux is given in units of 10^{-14} erg cm^{-2} s^{-1} A^{-1}. The strong emission lines are indicated. Note the relative weakness of CIV line. (From Imhoff and Appenzeller 1987)

an index for the high-temperature lines. In active chromospheres in normal main sequence stars, this ratio is typically 2–3 and rarely less than 1.5. In contrast, TTSs show a wide range of this ratio from unity up to several (3–4). An example of low ratio of CIV/SiIV is shown in Figure 7.23 in the spectrum of RW Aur (K1 IV, V = 10.8) observed by Imhoff and Appenzeller (1997). The ratio is around unity. They suggested the existence of excess energy loss in this star.

Johns-Krull et al. (1998, 2000) measured the intensities of far-UV emission lines using the archive data of IUE low-dispersion spectra for 17 HESs. According to them the ratio CIV/SiIV takes a wide range as seen above and also the surface fluxes of CIV emission exhibit a good correlation with the mass-loss rate as shown in Figure 7.24. Based on this correlation Johns-Krull et al. suggested that the bulk of the UV emission seen in TTSs results from accretion related process and not from some dynamo-driven magnetic activity. The shock processes in the accretion flows may be the most likely origin of UV emission.

X-ray sources

The behaviors of X-ray sources are rather different in CTTSs (classical TTSs) and WTTSs (weak-lined TTSs). Based on the ROSAT unbiased X-ray surveys of TTSs in Taurus region, Neuhäuser et al. (1995) found that the detection

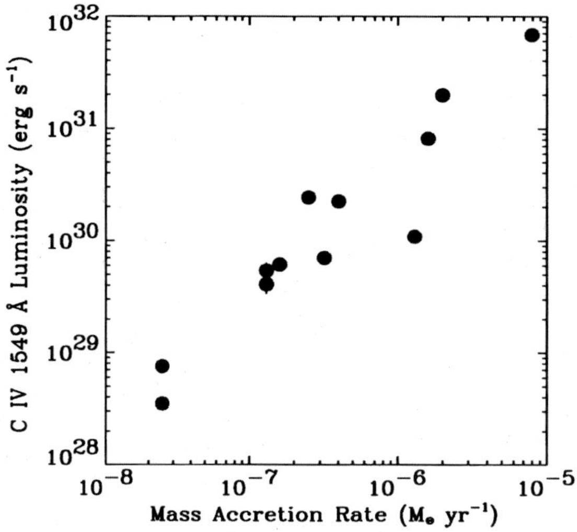

Figure 7.24: CIV emission flux versus mass accretion rate for CTTSs. (From Johns-Krull et al. 1998)

rates and mean X ray luminosities are several times higher in WTTSs than in CTTSs as shown in Table 7.9. The X-ray luminosity function shown in Figure 7.25 also exhibits the higher X-ray activity in WTTS than in CTTS. Neuhäuser et al. attributed such difference mainly to the difference in the rotational velocity and stellar age among other physical parameters such as stellar mass, bolometric luminosity, and effective temperature (see below). Stelzer and Neuhäuser (2001) carried out a systematic search for X-emission lines from pre-main sequence and young main sequence stars in the Taurus-Auriga-Perseus region, and derived the similar X-ray luminosity function for CTTS and WTTS as seen in Figure 7.25.

Correlation between X-ray surface flux and rotational period observed by Bouvier (1990) is shown in Figure 7.26. It is seen that X-ray flux exhibits a trend to be higher for rapid rotators and for high mass stars. Large scatter of observed points can be accounted for by intrinsic variability as indicated by vertical bars for some TTSs.

Table 7.9: X-ray sources in TTSs in the Taurus cloud region (adapted from Neuhäuser et al. 1995)

TTS type	Number of stars	Number of detection	Detection rate	Mean X-ray luminosity log L_X
CTTS	79	9	11%	29.09 erg s^{-1}
WTTS	65	43	66	29.70

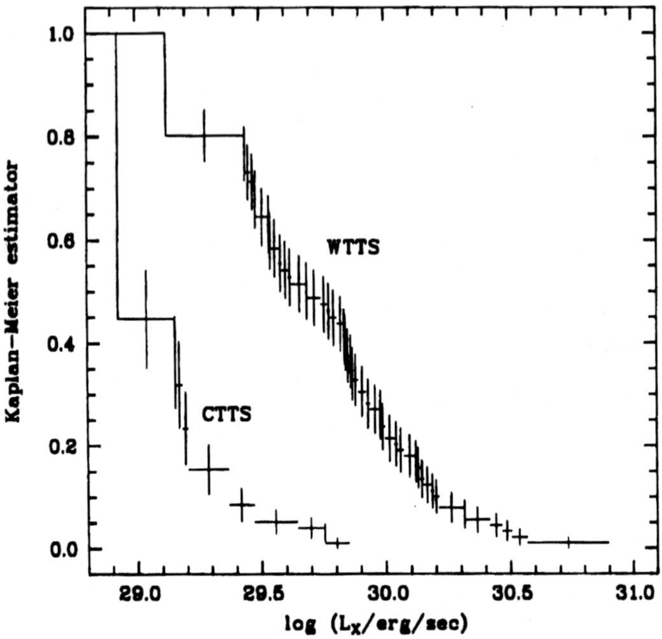

Figure 7.25: Normalized X-ray luminosity function. (From Neuhäuser et al. 1995)

The evolutionary scenario of TTSs in view of their X-ray emission and stellar rotation has been considered by Neuhäuser et al. (1995) as follows: If we presume that the TTSs take 10^8 years before entering the ZAMS, CTTSs mostly fall in younger ages of 10^4–$10^{6.5}$ years, whereas WTTSs concentrate in ages $10^{5.5}$~10^8 years. The average X-ray luminosity increases in the ages from ~$10^{4.5}$ years to ~$10^{6.5}$ years and then declines. On closer inspection, however, CTTSs show no increase of X-ray luminosity with age. This can be explained by the existence of a disk coupled with a stellar surface that prevents the star from spinning up due to impact of accreting flow. Given this, X-ray emission cannot increase with time, as long as disks are present. Contrarily, WTTSs which are supposed to have no disks, can spin up for stars younger than ~$10^{6.5}$ years, so that the rotation-supported magnetic activity increases, resulting in the enhancement of X-ray emission. At an age somewhere between ~10^6 and ~10^7 years, WTTSs reach their fastest rotation period, and after that their X-ray luminosity starts to decrease toward ZAMS. In this way we can understand that the X-ray luminosity depends upon the rotational velocity of stars.

7.2.3 Chromospheric structures

As late-type stars, TTSs possess developed convection layers and chromosphere-corona structure with inverted temperature distribution toward

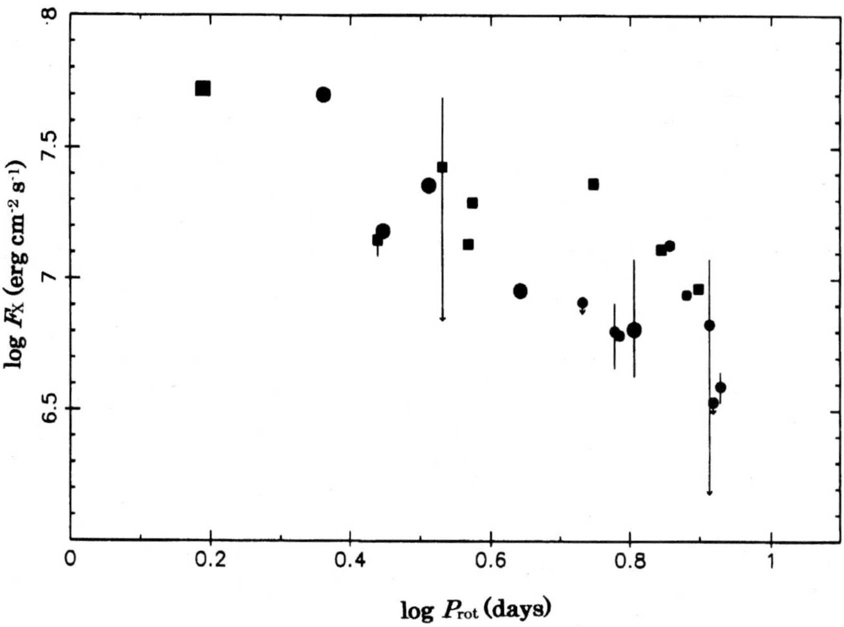

Figure 7.26: X-ray flux versus rotational period for TTSs. Filled squares denote WTTS and filled circles are CTTS. TTS with masses higher than 1.25 $M_\odot$ are represented by larger symbols. Vertical bars show the range of intrinsic variations. (From Bouvier 1990)

the outer layers. The layers of minimum temperature correspond to the surface of the photosphere, as in the case of solar atmospheres.

The models of chromosphere-corona structure have been widely developed since Cram (1979) and Brown et al. (1984). Batalha and Basri (1993) calculated a chromospheric model, which is connected to a standard photospheric model at its surface as the boundary condition. They first assumed the temperature distribution inside the chromosphere and solved the radiative transfer equations to calculate the emission intensities and profiles. In order that the results sufficiently match observations, successive approximations were repeated until a final result could be obtained. Figures 7.27 (a)–(d) exhibit the model calculations by Batalha and Basri for southern TTS stars designated by the Sz numbers. The spectral type in each panel is (a) G2, (b) K2, (c) K7 and (d) M0. The model calculations of some active main sequence stars are also shown for comparison. The abscissa is the column density of gas measured from outside, and the layer at which the column density becomes nearly unity corresponds to the photospheric surface with minimum temperature. Throughout the chromosphere, temperatures rise outward, but calculations ended at the temperature 10,000 K, so that the coronal regions were not shown. As seen in Figure 7.27, the temperature distributions in TTSs and

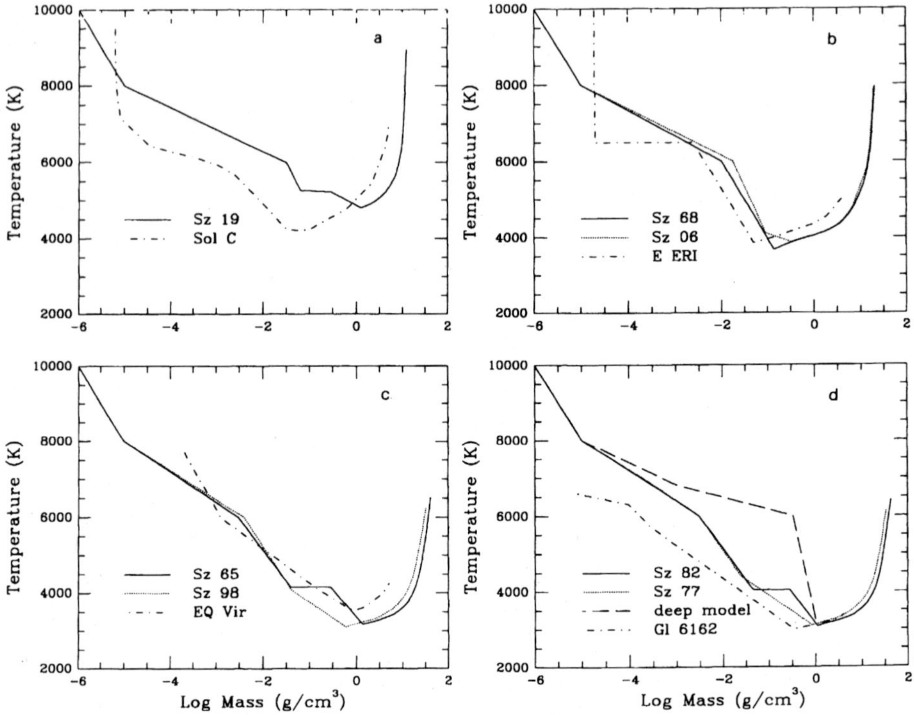

Figure 7.27: The models of chromospheric temperature structure. The abscissa is the column density of gas (g cm^{-2}) measured from outside. Panels: (a) G2 type, (b) K2, (c) K7 and (d) M0 type. The TTS stars are represented by Sz numbers, other stars are main-sequence stars. Sol C in Panle (a) denotes the solar model; deep model in panel (d) represents the TTS chromospheric model by Calvet et al. (1984). (From Batalha and Basri 1993)

in main-sequence stars are not markedly different. Using these models they explained the intensities and profiles of CaII λ3933 (K) and λ8542 Å lines as of chromospheric origin. For the enhanced emission lines like Hα, however, we need to take into account the contribution from outer envelopes.

7.2.4 Rotational velocities and binary systems

Rotational velocities and inclination angles

Periodic brightness variation due to surface structure like stellar spots gives us the rotational period $P_{\rm rot}$. The profiles of photospheric absorption lines allow us to get the projected rotational velocity $V \sin i$. If we can measure both $P_{\rm rot}$ and $V \sin i$ independently, we can separate the values of V and i, provided the

Table 7.10: Rotation period ($P_{\rm rot}$), projected rotation velocity ($V \sin i$), and inclination angle (i) for selected TTSs (adapted from Herbst 1986)

Star	$P_{\rm rot}$ (day)	$V \sin i$ (km s^{-1})	$R \sin i$ ($R_\odot$)	R ($R_\odot$)	i (degree)
T Tau	2.80	20.1	1.1	6.8	9.3
BP Tau	7.6	<10	<1.5	2.7	<33.7
DN Tau	6.6	8	1.0	2.8	20.9
AA Tau	8.2	11.0	1.8	2.5	46.1
HP Tau/G2	1.2	100	2.4	2.6	67.4
V410 Tau +	1.92	70.7	2.7	3.2	57.5
GK Tau +	4.6	16.9	1.5	3.9	22.6
GI Tau +	7.2	11.6	1.7	2.9	35.9
DI Tau +	7.7	10.6	1.6	2.2	46.6
SU Aur+	2.7 (1.6)	66.2	3.5 (2.1)	4.2	56.4 (30.0)

Note: Stars with + denote possible values

photospheric radius R is given. The formulae for R sin$i$ and i are given as

$$R \sin i = \frac{1}{2\pi} P_{\rm rot} V \sin i, \quad i = \sin^{-1}\left(\frac{P_{\rm rot} V \sin i}{2\pi R}\right). \tag{7.2.1}$$

The intrinsic rotational velocity V is due obtainable.

For TTSs, the values of $P_{\rm rot}$, $V \sin i$, and R are obtained by Herbst (1986) as shown in Table 7.10. We find that the rotation period is in the range of 1–8 days, while the projected rotation velocity is from 8 to 100 km s^{-1}. The inclination angles take various values as naturally expected. If we know the inclination angle of a star, we can estimate the position of equatorial plane and derive valuable information on the surface structure such as latitude distribution of star spots.

The frequency distribution of the rotation period was studied by Attridge and Herbst (1992) for the 35 TTSs in the Orion Nebula Cluster, and they found a bimodal distribution as given in Table 7.11. Bimodal nature is also seen among the stars in Table 7.10.

In a review on the rotation of pre-main sequence stars, Bouvier (1991) showed that the distribution of rotational period depends on the stellar mass. That is, while low-mass TTSs ($M_* < 1.2\ M_\odot$) show the bimodal shape with an apparent deficiency of stars with a period around 4 days, intermediate-mass

Table 7.11: Bimodal distribution of rotation period in TTSs (adapted from Attridge and Herbst 1992)

	Number of stars	Range of period (days)	Average period (days)
Rapid rotator	8	1.6–3.4	2.2±1
Slow rotator	28	5.7–17.2	8.5±2.5

TTSs ($M_* \geqq 1.2\,M_\odot$) exhibit single peaked distribution in a period range of 1.5–6 days. According to Bouvier, the rapid rotators of low-mass TTSs are mainly close binaries subjecting to strong tidal effects to synchronize the orbital and rotational periods, thus creating a rapid rotation. The slower rotators are then either single stars or members of wider binary systems. In the intermediate-mass TTSs the rotation velocity $V \sin i$ is likely increasing with increasing stellar mass.

Frequency of binary systems and emission lines

TTS stars as pre-main sequence stars mostly belong to specific star forming regions (SFR). The binary frequency seems to differ for different SFRs. When compared to that of nearby main-sequence stars, the variety is seen as follows:

Higher frequency in Tau-Aur SFR, and Oph-Sco SFR.
Lower frequency in Orion Trapezium cluster and Chamaeleon SFR.
Comparable frequency in IC 348 and Pleiades.

As an example of higher binary frequency, Leinert et al. (1993) carried out speckle observations (resolution $0.''13$) for 104 low-mass pre-main sequence stars in the Taurus-Aurigae region and found 44 multiple systems, a detection rate of 42%. After making a correction for the orbital distribution outside their observed range, they derived the value of $80 \pm 20\%$ for the total binary frequency, which is about factor two higher than that of the nearby main-sequence stars. Similar high binary frequency for the TTSs has also been derived by Reipurth and Zinnecker (1993) through CCD observations. They argued that around 80% are binaries. Ghez et al. (1993) derived them through speckle observations, and suggested that most, if not all, TTSs have companions.

Contrarily, low binary frequency was found by Köhler (2001) in his multiplicity survey among X-ray selected TTSs in Chamaeleon association. The binary frequency is lower by a factor of 0.61 ± 0.27 than that of nearby main sequence stars. In this case one should notice that the X-ray (ROSAT) selected TTSs are mostly weak-lined TTSs. There might be some difference in the statistical behaviors between CTTSs and WTTSs. Melo (2003) carried out high-resolution spectroscopic observations to estimate the short-period binary frequency ($P_{\rm orb} < 100$ days) for a sample of TTSs in some SFRs. As a result, Melo found that there is an excess of short-period binaries by a factor 2–2.5 in Oph-Sco SFR, whereas short-period binaries seem to be absent in Cha/Lup/CrA SFRs. This shows a similar frequency variety stated above.

Mathieu (1994) compiled the binary and stellar data for 25 stars, among which 12 stars are double-lined SBs (orbital period $P = 1.68-35.9$ days) and 13 stars are single-lined SBs ($P = 2.4-2530$ days). Stars with P > 50 days are all single-lined SBs. All SBs listed are weak-lined TTSs (EW(Hα) < 10 Å) except two stars (HBC 662 = V4066 Sgr: EW(Hα) > 100 Å, HBC 85 = GW Ori: EW(Hα) = 27.6 Å). Five stars exhibit weak absorption in Hα. The fact that there is no correlation between the intensity of the Hα emission and orbital period gives a picture, different from the case of Algol systems (see Section 6.4.2), implying that binary systems do not contribute markedly to the formation of emission lines in the TTS binary systems.

7.2.5 Variabilities and activities

Variations in brightness and colors

TTSs generally exhibit irregular variations. Herbig (1962) classified the forms of light curves into four classes following Parenago's (1954) scheme (see column LC in Table 7.6)

Class I—The variable is more frequently bright than faint.
Class II—The variable most frequently has the mean brightness
Class III—The variable is more frequently faint than bright.
Class IV—The variable shows no preference for any level in its range.

Time scales of these light variations are generally from several 10 to several 100 days. As an example of light curves, Figure 7.28 illustrates the light curve of RY Tau for around 20 years. It looks very irregular, but, according to Parenago's system, this star is a combination of Class I and IV.

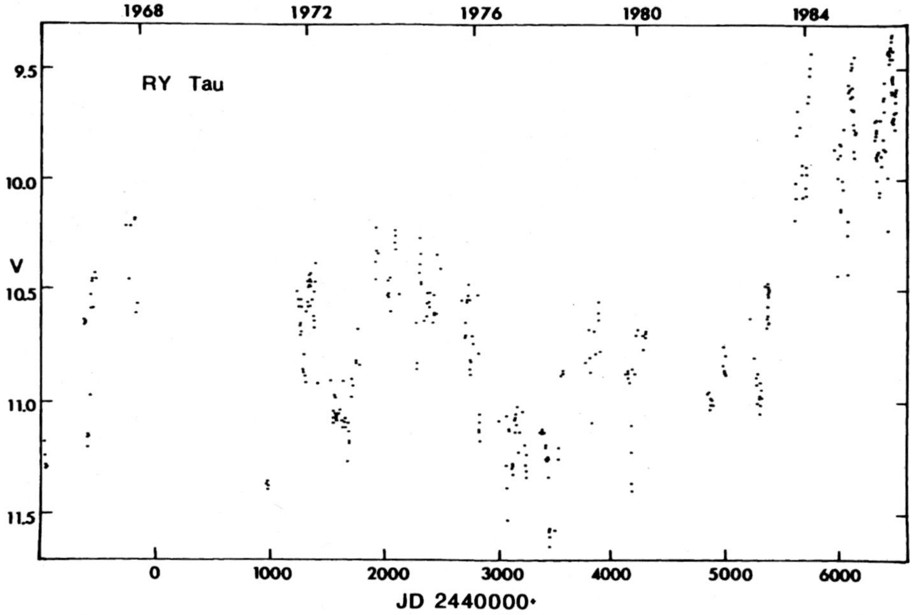

Figure 7.28: The light curve of RY Tau (CTTS) in V magnitude spanned for 20 years. Parenago's type of the light curve is a combination of I and IV. (From Herbst 1986)

On the other hand, Herbst et al. (1994) catalogued *UBVRI* photometry for 79 TTSs (43 CTTSs, 18 WTTSs, and 18 ETTSs), and classified them into four types based on the color and spectral variations. Each of the types has the following characteristic features.

Type 1—Periodic variations in *VRI*, undoubtedly caused by rotational modulation of a star with an asymmetric distribution of cool spots on its surface. Irregular flare activity is sometimes seen in U and B. This type mainly appears in WTTSs.

Type 2—Variations caused by hot spots or zones, and the "veiling" effect (shallowing of absorption lines by the rising of continuum) is commonly attributed to an accretion boundary layer or impact zone of a magnetically channeled accretion flow. This type is predominantly seen in CTTS.

Type 2p (subtype of Type 2)—Periodic variations caused by hot spots. It is supposed that both unsteady accretion and rotation of the star contribute to the variation of this type. This type exists among ETTSs, including stars as early as A type.

Type 3 (UXors)—UX Ori is a typical example of UXors. Variations are of very large amplitudes (exceeding 2.8 mag in V) without showing a veiling continuum or any substantial change in their photospheric spectra. The leading hypothesis for explaining Type 3 variations is variable obscuration by circumstellar dust.

The relationship between Herbig's "class" and Herbst et al.'s "type" is not clear, but may reflect the different time scales of variability, i.e., longer for the "class" (10–100 days) and shorter for the "type" (0.5–30 days).

In some TTSs, variations with very long time scales are known. Bellingham and Rossano (1980) observed R CrA, S CrA, T CrA, R Mon, and other T Tauri-like irregular variables. They found long-term variations with time spans from 1000 to 10,000 days. The magnitude change is a linear increase or decrease in the range 0.2–1.0 mag in V band. These time scales are much longer than those of Mira variables (less than around 500 days) so that they cannot be attributed to the stellar origin such as radial pulsation. Instead they suggested the changes in circumstellar shells or the existence of variable reflection nebulae.

Surface activities

Surface activities of TTSs can be observed as short-term variability and are classified into two types: rapid variations, less than around 1 day, and slow variations, taking place for 2–8 days.

For rapid variations, Gahm et al. (1995) found two different types of events, in six stars (two CTTSs, three WTTSs, and one PTTS), as follows:

(a) Balmer flare events—Rapid bursts less than 1 hour and solely visible in the increase of Balmer continuum and the Balmer line emission. This type of events only occurs in WTTSs. For the origin of the events, Gahm et al. proposed surface flares connected with the dissipation of magnetic energy. Though the detailed mechanism is not known, the flares are extremely powerful (10^{33}–10^{34} ergs) compared to ordinary flare stars.

(b) Continuum elevation—Slow and smooth increase of the continuum level over several hours. Inhomogeneous mass accretion from a circumstellar envelope (or disk) is proposed as its origin.

Among the stars they observed, the star which showed most frequent activities was VW Cha (CTTS), while HD 70309B (PTTS) did not show any flare events.

The slow variations in 2–8 days are supposed to be surface activities modulated by the rotation of stars. Bouvier and Bertout (1989) detected the periodic variation for 15 TTSs. The periods are in a range of 1.9–8.5 days. Most of those detected stars are supposed to have dark spots that are cooler than the photospheres around 500–1000 K and occupy the areas of 3–17% of the photospheric surfaces. Hot spots are also seen though the number is small. They are hotter than the photosphere up to 7000 K, but their areas of coverage are generally less than 1%. Bouvier and Bertout suggested that the hot spots are the points of accretion flows hitting the stellar surface.

Stellar winds, accretion flows, and jets

The relationship between accretion flow and outflowing wind was attributed to an accretion-powered disk wind by Cabrit et al. (1990). Based on the spectroscopic survey of the Hα and forbidden lines for 36 TTSs at Kitt Peak Observatory (resolution 1.8 Å), and combining with published data on optical and infrared photometry, they argued as follows:

(i) All of surveyed stars showed the Hα emission covering a range of the equivalent widths from 1 to 140 Å. In 23 stars, blueshifted forbidden line [OI] λ6300 Å was detected. Thus most of TTSs possess energetic winds and opaque circumstellar disks.

(ii) [OI] λ6300 Å line is supposed to form in the outer part of stellar wind, since the [OI] λ6300 Å and Hα line luminosities exhibit a positive correlation over two orders of magnitudes. This correlation indicates that the winds have a wide range of mass loss rates among the stars.

(iii) Infrared excess is usually attributed to the thermal emission from the dust in an accretion disk. There exists a good correlation between infrared excess and the intensities of Hα and [OI] emission. This correlation represents a notable relationship between accretion rate of gas and the mass-loss rate.

(iv) There exists no correlation between stellar luminosity L_* and other parameters, such as emission intensities of Hα and [OI], and infrared

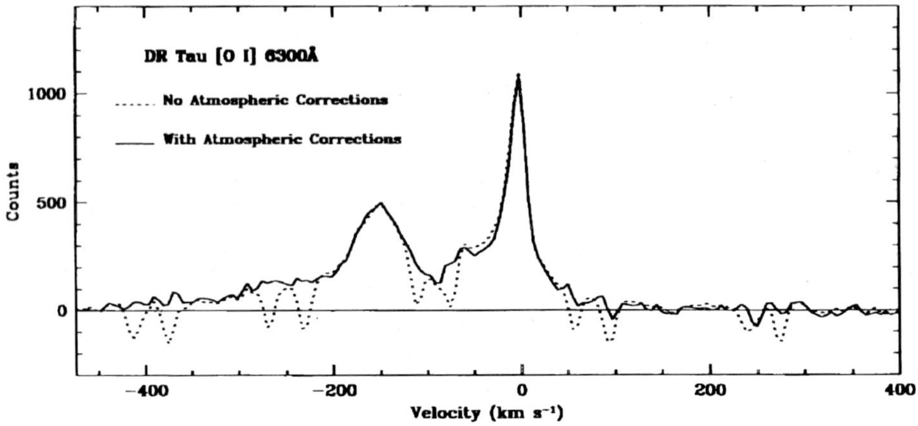

Figure 7.29: The line profiles of [OI] λ6300 Å for DR Tau, with no correction for terrestrial water vapor absorption (dotted line) and after atmospheric correction (solid line). The photospheric line and continuum are subtracted. The velocity scale is presented with respect to the rest velocity of the star. (From Hartigan et al. 1995)

excess, implying that it is the disk, not the star, that primarily determines the strength of wind in TTSs.

Summarizing the above correlation, we get a picture that the accretion disk is formed by the infall of dust gas, and the disk wind is powered by the release of gravitational energy of infalling gas. This is the principle of the accretion-powered disk wind model of Cabrit et al. (1990).

The existence of disk wind is supported by Hartigan et al. (1995), based on their high-dispersion spectroscopic observations at Kitt Peak Observatory. They have resolved two components in the profiles of the forbidden lines [OI], [SII], and [NII] with different radial velocities. Figure 7.29 exhibits an example of line profiles in [OI] λ6300 Å for DR Tau (K7), clearly resolved into double peaks. The first component with high radial velocity (~ -150 km s^{-1}) represents a gas flow with low gas density (electron density $\sim 10^4 - 10^5$ cm^{-3}), and is supposed to be the stellar wind from the star itself. The second component with low radial velocity ($\sim$5 km s^{-1}) represents the gas flow with high density (electron density $>10^5$ cm^{-3}) and low excitation level ([NII] line is absent), and this component is supposed to be the disk wind from the accretion disk. Figure 7.29 is direct evidence of the existence of two component gas flows, supposedly from the stellar surface and from the accretion disk.

If we accept this two-flow model, the envelope around a TTS may be depicted as in Figure 7.30. In addition to the high velocity wind from the star, there appear to be the accretion-powered low velocity winds from both sides of the disk. Note that, when a magnetospheric dipole exists, inner parts of accretion flow and stellar wind will be disturbed such as shown in Figure 3.16.

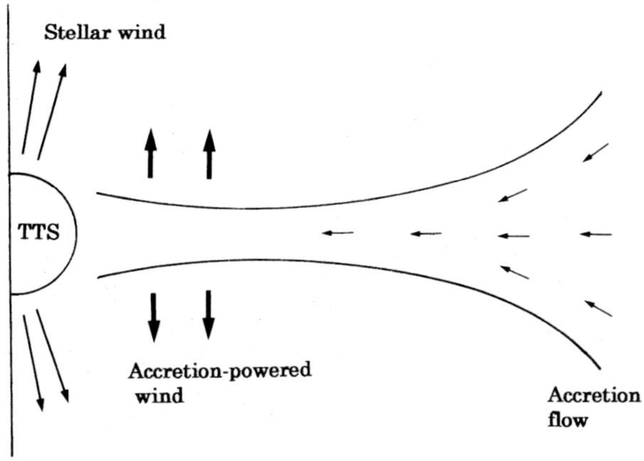

Figure 7.30: Schematic picture of TTS winds. The accreting flow produces a rotating disk around the star, and a part of gas bounces up as low-velocity wind due to the gravitational energy of infalling gas. A high-velocity wind from the star is also formed in the polar regions.

The jet streams around the TTSs are actually observed in the optical region. Hirth (1994) has carried out long-slit spectroscopic observations for CW Tau and DO Tau (both TTS) at Calar Alto Observatory. He obtained the position-velocity maps with spatial resolution of 1.″4 to 2.″0, and velocity resolution of 50–20 km s^{-1}. He observed forbidden emission lines [OI] λ6300, 6363 Å, [NII] λ6583 Å, [SII] λ6716, 6731 Å and analyzed the spatial and kinematic properties of rarefied gas within a few arc seconds from the star. Figure 7.31 is the case of CW Tau (K5e), where we can see that the forbidden lines exhibit the double-peaked profiles with high velocity component (around -100 km s^{-1}) and low velocity component (-5 km s^{-1}), except [NII] which only shows high-velocity component. Hirth supposed that the high-velocity component represents the stellar wind from the star, while the low-velocity component represents the disk wind. The gas density of the flow is expressed by a set of contours in logarithmic scale.

If we assume that the distance to CW Tau is 160 pc and the spatial extension of the jet is around 5" which corresponds to 800 AU, then the dynamic time scale for the high-velocity component is only 800 AU/100 km $s^{-1} \sim$ 40 years. This is very short, and implies that the accretion process may be a short-lived, intermittent phenomenon in this star.

7.2.6 FU Orionis and YY Orionis type stars

FU Orionis object (FUor)

The group of TTSs showing a sudden and persistent increase of several magnitudes and decreasing very slowly is called FU Orionis objects (or FUors).

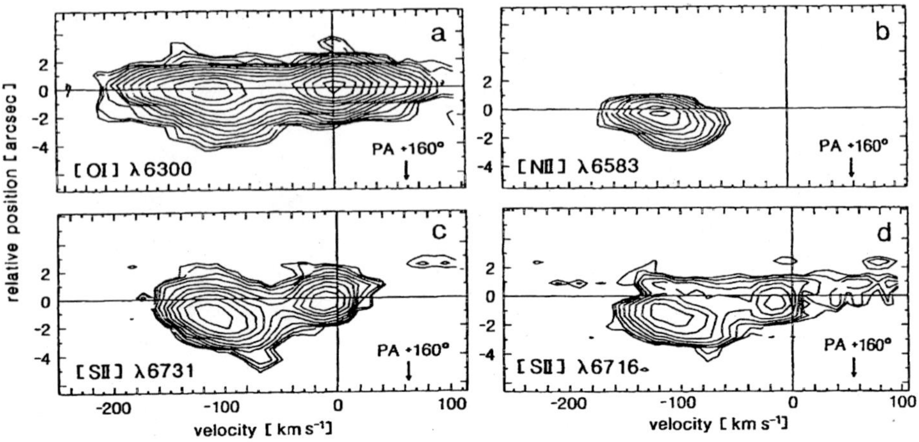

Figure 7.31: Position-velocity maps of several forbidden emission lines of CW Tau (K5e) at position angle 160°. (a) [OI] λ6300 Å, (b) [NII] λ6583 Å, (c) [SII] λ6716 Å, and (d) [SII] λ6731 Å. The continuum is removed. The spacing of the contours is logarithmic, corresponding to a factor $2^{0.5}$. Relative positions and velocities are quoted with respect to the stellar position and stellar rest velocity, respectively. (From Hirth 1994)

They are all spatially and kinematically associated with the star-forming regions. FU Ori, the prototype star, exhibited a sudden increase of brightness up to 6 magnitudes during a 4-month period in 1936–1937, and then its light gradually declined by only one magnitude in 60 years. Figures 7.32 illustrates the light curves of three FU Ori objects, and a selected list is given in Table 7.12.

Rapid brightening of FUors is generated by a large scale outburst due to the sudden increase of mass accretion rate (from $\sim 10^{-7}$ to $\sim 10^{-4}\,M_\odot$ per year onto the star. A high brightness level is retained as long as the high accretion rate continues. Since high-velocity stellar winds (>300 km s^{-1}) are also observed, some outbursts inside the accretion disks might have triggered the high-velocity winds. FUors are supposed to be in a very active stage in the early phase of the TTS evolution.

Herbig et al. (2003) carried out high-dispersion spectroscopic observations of FU Ori and V1057 Cyg between 1995 and 2002, after the period given in Figure 7.32. During these days FU Ori remained about 1 magnitude (in B) below its 1938–1939 maximum brightness, while V1057 recovered slightly after reaching minimum light in about 2000. In this period, both stars showed P Cyg structure in the Hα indicating the existence of powerful winds.

There is another class of eruptive variables, called EXors after Herbig (1989) with the prototype EX Lupi. They also experience sudden outbursts like FUors, but on a somewhat smaller scale. EXors are thought to be objects

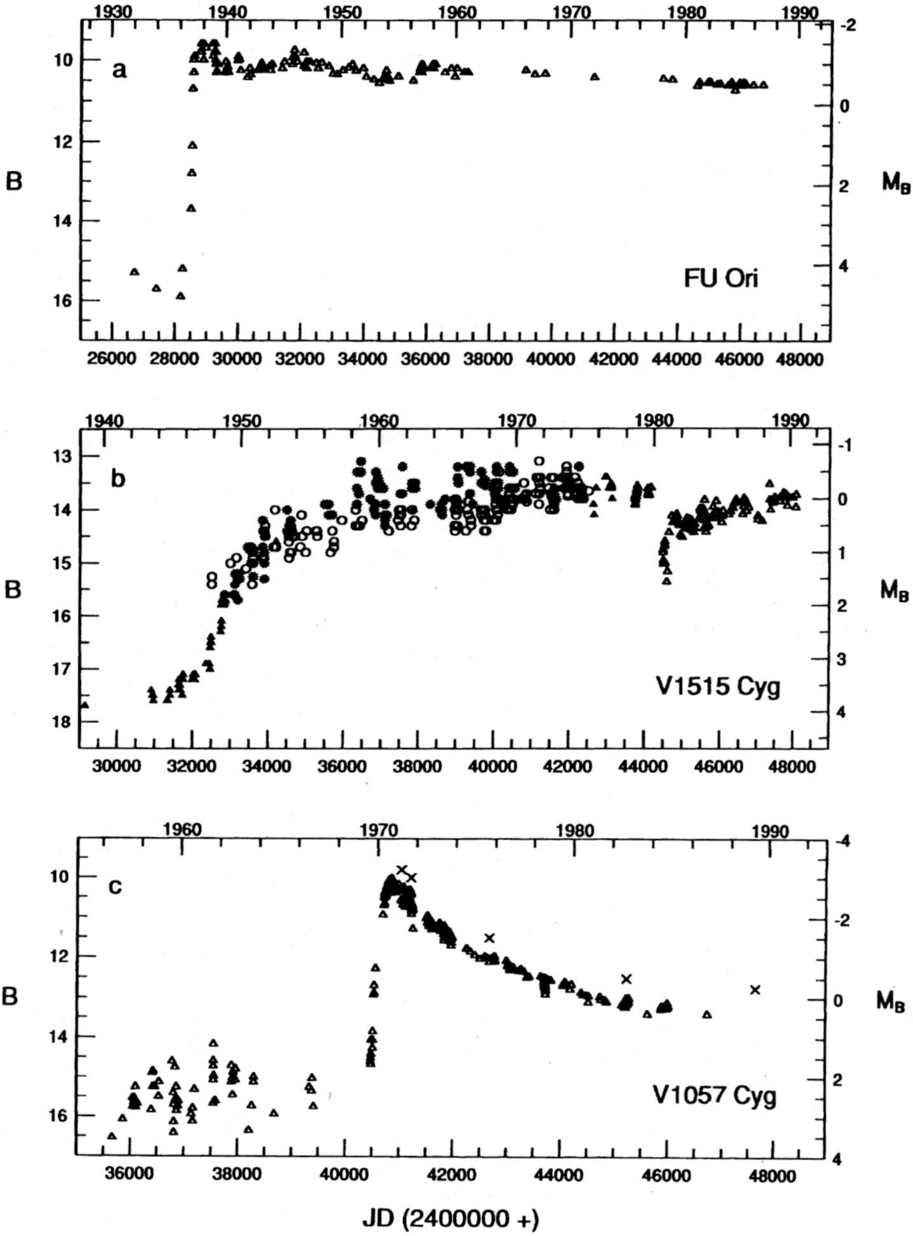

Figure 7.32: Optical light curves (B band photographic magnitude) of three FU Ori objects, with estimated $M_{\rm B}$ on the right-hand axis. Different symbols are by different observers. (From Bell et al. 1995)

Table 7.12: List of selected FU Ori objects (FUors) (adapted from Hartmann and Kenyon 1996)

Star	outburst	Brightening duration	Declining duration	Distance (kpc)	Luminosity ($L/L_\odot$)
FU Ori	1937	~1 yr	~100 yr	0.5	500
V1057 Cyg	1970	~1	~10	0.6	800–250
V1515 Cyg	1950s	~20	~30	1.0	200
V1735 Cyg	1957–65	<8	<20	0.9	>75
Z CMa	?	?	>100	1.1	600

lying on the evolutionary track from FUors to the classical TTSs. Herbig listed nine stars as the members of EXors.

YY Orionis type stars

Among TTSs, stars that show the inverse P Cygni profiles in the Balmer lines and CaII H, K lines are called the YY Orionis type. Walker (1972) carried out spectroscopic observations for 25 young stars with remarkable ultraviolet excess in the young clusters of Orion and Monoceros and found YY Orionis characteristics in 13 stars. Since then the name of YY Orionis type stars has become widely used.

The examples of optical spectra of YY Ori taken by Aiad et al. (1984) are shown in Figure 7.33. One can see the inverse P Cygni profiles in the Balmer lines and CaII H, K lines. The spectra are arranged from top to bottom in the increasing order of the strength of the Balmer continuum emission. YY Ori stars generally exhibit the Balmer discontinuity as a continuum emission with different magnitudes from star to star, or from time to time. In Figure 7.33, the spectrum of DS Tau is shown as an example of weak Balmer continuum. One may also see in this figure the remarkable profile variations of YY Ori during 4 days from late December 1980, to early January 1981. These short-term variations suggest the existence of inhomogenaity and a variable nature of the infalling materials. For the long-term variations, Walker (1972) reported a change of profile from inverse P Cygni in October 1962 to a single-peaked profile in December 1963.

Bertout et al. (1996) found the existence of a periodic component in the light curve of YY Ori with period 7.58 ± 0.15 days. They attributed this period to rotational modulation of the light curve by a large hot spot (or group of spots) located on the stellar surface. Combining with the inverse P Cygni profiles they estimated the mass accretion rate ($M_{\rm acc} \sim 3 \times 10^{-5}\ M_\odot$ per year) much higher than that of ordinary CTTSs ($M_{\rm acc} = 10^{-6} \sim 10^{-7}\ M_\odot$ per year).

From the spectroscopic behaviors it becomes evident that YY Ori stars are the members of young TTSs, still surrounded by infalling envelopes and accretion disks, and in the early stage of TTS evolution.

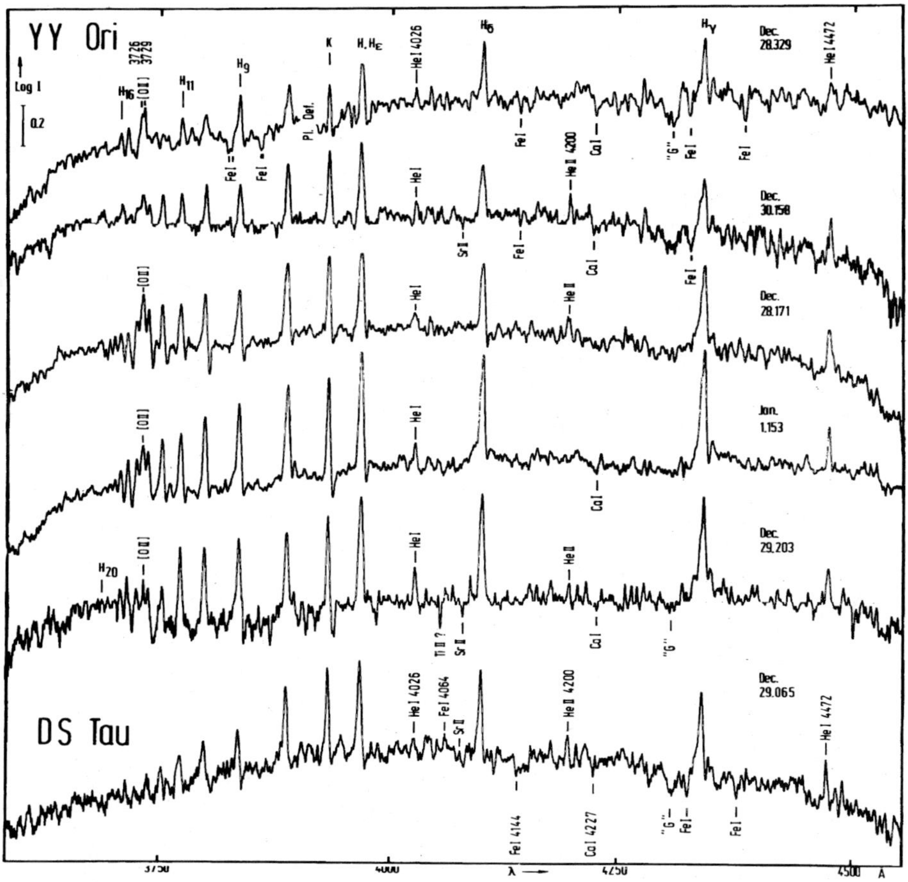

Figure 7.33: Examples of YY Ori spectra illustrating the variability of the Balmer continuum emission and spectral lines in Hα, CaII H, K, and others. The spectrograms are arranged in the order of increasing Balmer continuum emission from top to bottom. For comparison a spectrum of DS Tau with weak Balmer continuum is also presented. (From Aiad et al. 1984)

7.3 Pre-main sequence stars and hydrogen spectra

7.3.1 Emission-line intensities and Balmer decrements

Hα emission-line intensities

Pre-main sequence stars having emission lines are mostly concentrated in early type (B–A) and late type (K–M). In Figure 7.34 we show the spectral distribution of Hα intensity, based on the data of Fernandez et al. (1995) for

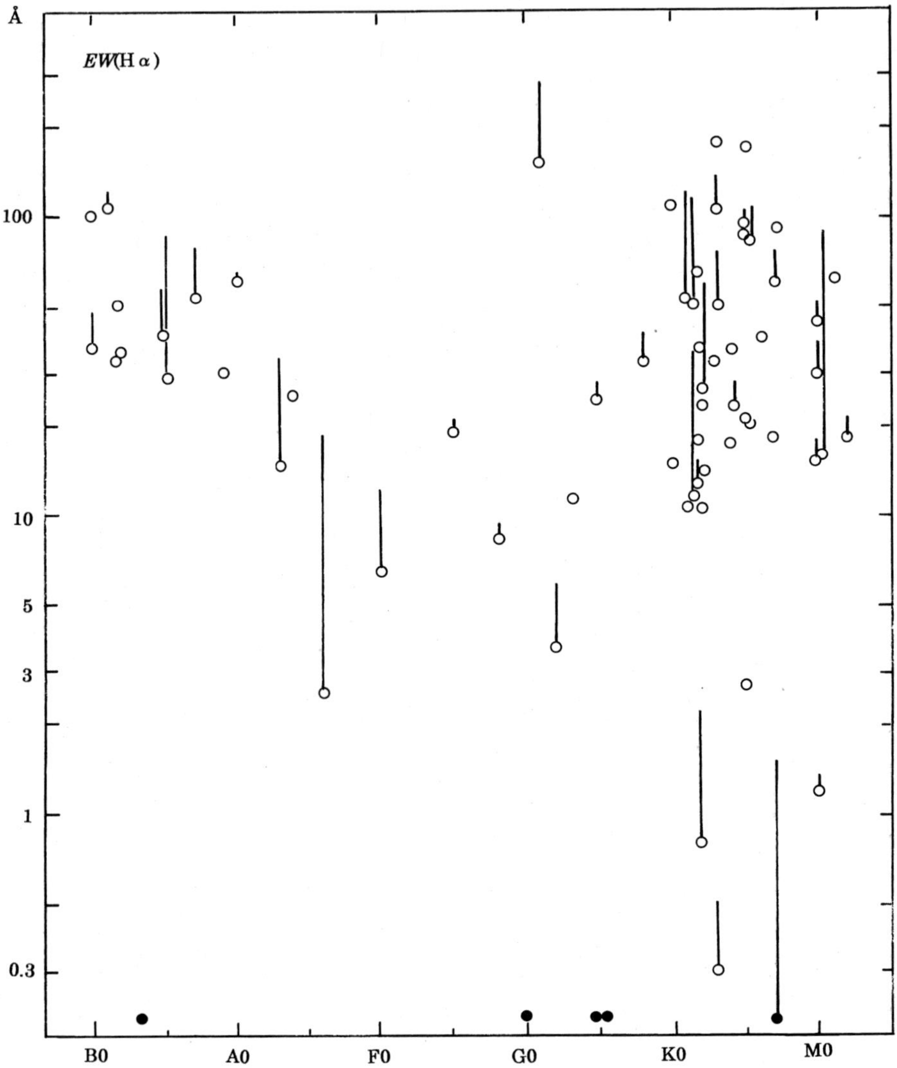

Figure 7.34: Hα emission equivalent width versus spectral type for HES and TTS. Open circle denotes the observed value or its minimum value when variable. Vertical line shows the range of variability. Filled circle denotes the disappearance of emission component. (Based on the data of Fernandez et al. 1995)

75 pre-main sequence stars (17 HESs, 47 CTTSs, and 11 WTTSs). They observed these stars with an intermediate dispersion spectrograph (dispersion 50, 22 Å mm^{-1}) at Calar Alto Observatory. The range of variability for each star is also designated.

Table 7.13: The number and rate of CTTS and WTTS in some star-forming regions (adapted from the table of Cohen and Kuhi (1979))

Star-Forming region	Distance (pc)	Number of TTSs	Number of		
			CTTS	WTTS	No emission
Tau-Aur region	160	83	61 (73.5)	20 (24.1)	2 (2.4)
Orion region	450–500	110	76 (69.1)	23 (20.9)	11 (10.0)
NGC 2264 region	800	97	81 (83.5)	10 (10.3)	6 (6.2)
Other regions	—	156	120 (76.9)	23 (14.7)	13 (8.3)
Total		446	338 (75.8)	76 (17.0)	32 (7.1)

Note: The numbers in brackets denote percentage of frequency.

One may see in Figure 7.34 that the Hα intensity gradually declines toward the later spectral types from B to G (HES), but clearly separates into strong emission (CTTS) and weak emission (WTTS) stars in K and M types.

The emission equivalent widths of Balmer lines (Hα, Hβ, Hγ) for TTSs have been measured by Cohen and Kuhi (1979). Table 7.13 gives the number and rate of CTTS and WTTS in some star forming regions. The total number of TTSs is 446. The number rate of CTTSs is around 76%, and that of WTTSs is 17%, almost regardless the star-forming regions. Around 7% of TTSs did not show the emission.

The Balmer decrements: observational data

(a) HES—In the calatogue of Cohen and Kuhi (1979) for the pre-main sequence stars there are around 14 HESs, with steep decrements D_{34} measured in a range of 10 to higher than 100. Kolotilov (1977) also found very steep decrement D_{34} in three stars as shown in Table 7.14.

Table 7.14: Very steep Balmer decrement D_{34} in HESs (adapted from Kolotilov 1977)

Star	Spectral Type	EW(Hα)	D_{34}
WW Vul	A0–3 Ve	35.8 Å	~50
VX Cas	A0–3e	15–25	>50
UX Ori	A2–3 IIIe	10–20	>50

Table 7.15: The Balmer decrement of HD 200775 (adapted from Köppen et al. 1982)

Line	Equivalent Width	Decrement
Hα	40 Å	2.65
	36	2.36
Hβ	6.0	1.0
Hγ	2.5	0.6
Hδ	1.3	0.36

Among them WW Vul displayed large profile variations as seen in Figure 7.10. In spite of strong Hα emission, Hβ emission is nearly absent or very weak, so that the decrements become very steep.
On the other hand, Köppen et al. (1982) found a rather flat decrement in the early spectral type star HD 200775 (B3Ve) nearer to the nebular decrement as shown in Table 7.15. However, this decrement seems rather exceptional, even among early type stars, since Cohen and Kuhi's catalogue contains several early type B stars, which exhibit a steep decrement. In classical Be stars (Section 5.3.4), the decrements gradually steepen with the advance of spectral subtype (see Figures 5.21 and 5.22), but in HESs, no such dependence is seen along the spectral type.

(b) TTS—The decrements D_{34} measured by Cohen and Kuhi (1979) are plotted in Figure 7.35 for the stars of Ophiuchus and Tau-Aur star-forming regions. It is seen that the decrements are generally steeper than that of standard nebular decrement, particularly in K type stars. According to Katysheva (1981), the average decrements for Cohen and Kuhi's stars after the correction for the interstellar extinction are as follows:

$$\langle D_{34} \rangle = \begin{cases} 6.9 \pm 3.2 & \text{for G5} - \text{K5} \\ 5.2 \pm 2.4 & \text{for } K5 - M5 \end{cases} \tag{7.3.1}$$

$$\langle D_{54} \rangle = \begin{cases} 0.4 \pm 0.2 & \text{for G5} - \text{K5} \\ 0.6 \pm 0.2 & \text{for K5} - \text{M5} \end{cases} \tag{7.3.2}$$

(c) NTTS—In the observation of NTTSs associated with X-ray sources, Walter (1986) cited Elias 12 as a possible TTS. The pre-main sequence nature of this star is clear by the presence of Li absorption and strong emission of CaII H and K. The optical spectrum is shown in Figure 7.36. The Balmer lines are seen from Hα (equivalent width 6 Å) up to at least H11. Though Walter did not measure it, the Balmer decrement should be very flat in contrast to ordinary TTSs.

(d) YY Ori type stars—As seen in Figure 7.33, the Balmer lines of YY Ori and DS Tau exhibit emission up to their high series members

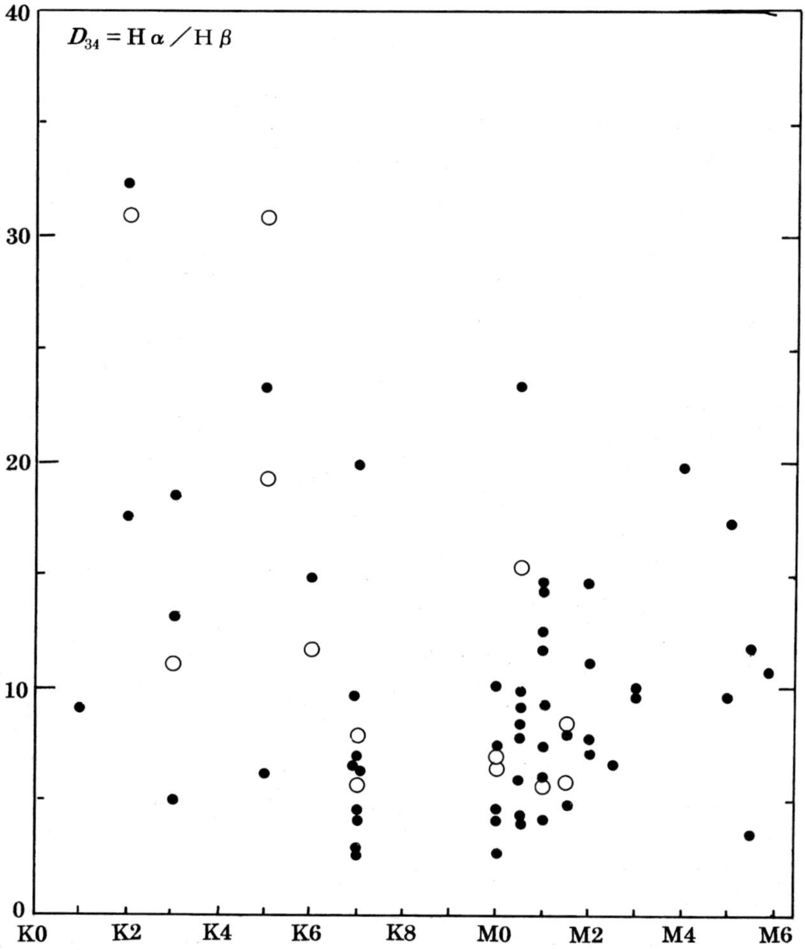

Figure 7.35: Balmer decrements D_{34} of nearby TTSs. Open circle denotes the stars in Ophiuchus region and filled circles those of Tau-Aur region. (Based on the data of Cohen and Kuhi 1979)

along with inverse P Cygni profiles. DR Tau, the other YY Ori type star, also show similar spectrum. Though not measured, the Balmer decrements of these stars should be very flat as in case of NTTS cited above.

We have surveyed the steepness of Balmer decrements in the literatures, but the statistical relationship of the steepness among the different types of HES and TTS remained somewhat unclear, mainly because observational data are still not sufficiently accumulated.

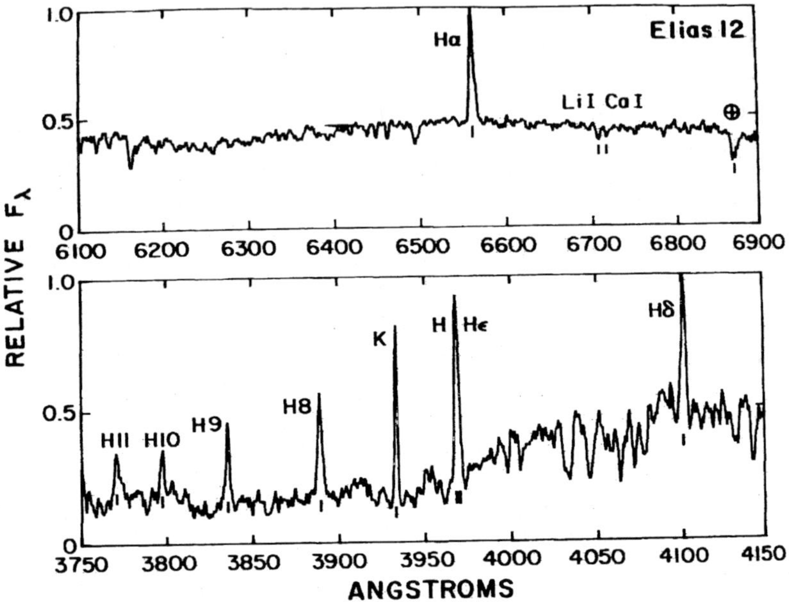

Figure 7.36: Spectra of Elias 12 (NTTS) in red and blue spectral regions. (From Walter 1986)

Theoretical decrements

(a) Extension of nebular approximation—In order to obtain steeper decrement D_{34}, Grinin (1969) adopted the nebular approximation by taking into account the finite optical thickness for the Hα line. He considered a plane–parallel layer, assuming that the layer is completely opaque for the Lyman lines and fully transparent for the high-level lines such as Paschen and Bracket series. By taking the optical thickness for the Hα, τ (Hα), as a free parameter, he calculated the escape probabilities by scattering for the Balmer lines and solved the equations of statistical equilibrium in this gas layer. If τ (Hα) $= 0$, the solution reduces to the standard nebular approximation (Case B). When τ (Hα) becomes large, the decrement D_{34} gets steeper, deviating from the nebular approximation. A part of the results of his calculation is shown in Table 7.16. A remarkable feature of his calculation is that while D_{34} is getting steeper as τ (Hα) increases, the decrements D_{54}, D_{64} are getting flatter. This feature seems not to conflict with observed decrements of TTSs given by Equations (7.3.1) and (7.3.2). Kolotilov (1977) also suggested that the very steep decrements of HESs ($D_{34} > 50$) seen in Table 7.14 could be explained by taking a large value of τ (Hα)($>$100) in Grinin's calculations.

Table 7.16: Tbe Balmer decrements calculated by the nebular approximation with finite optical thickness for Hα (adapted from Grinin 1969)

Te	10,000 K				20,000 K			
τ (Hα)	0	10	50	100	0	10	50	100
D_{34}	2.72	4.07	9.08	15.20	2.81	4.24	9.32	15.60
D_{54}	0.50	0.57	0.74	0.84	0.49	0.56	0.73	0.82
D_{64}	0.29	0.36	0.53	0.67	0.28	0.34	0.51	0.64

(b) Stellar envelope approximation—Balmer decrement of the Herbig Be star HD 200775 (B3V) is slightly flatter than the nebular decrement (Case B) as shown in Table 7.15. To explain this decrement, Köppen et al. (1982) considered the radiation field of a volume of hydrogen gas with the dilution factor W. Hydrogen atoms are assumed to have eight levels plus a continuum. All radiative and collisional processes for excitation and ionization are taken into account. For the lines, self-absorption is treated by the escape probability by motion. The parameters in this model are electron temperature (T_e), stellar temperature (T_*), dilution factor (W), and optical thickness for the Hα. The results of their model calculations are partly shown in Table 7.17. As Köppen et al. argued, a parameter set of $T_* = 20{,}000$ K, $T_e = 10.000$ K, $W = 10^{-2}$, τ (Hα) $= 2$, agrees well with observed values, though the decrement is rather insensitive to the parameters. The star HD 200775 has a strong Hα emission (equivalent width = 33.6 Å) with a deep and sharp central depth, suggesting the existence of a gas shell around the star. If so, the optical thickness τ (Hα) should be larger than around 100. Another set of parameters with large optical thickness is to be searched for among the same model.

Table 7.17: Balmer decrements in stellar-envelope approximation (adapted from Köppen et al. 1982)

Model parameters				Balmer decrement		
T^* (K)	T_e (K)	W	τ (Hα)	D_{34}	D_{54}	D_{64}
15,000	10,000	10^{-2}	6.7	2.50	0.42	0.23
			330	2.21	0.53	0.37
20,000	10,000	10^{-2}	2	2.50	0.55	0.29
		10^{-4}	0.04	2.79	0.49	0.28
20,000	20,000	10^{-2}	2.5	2.47	0.49	0.28
		10^{-2}	10^4	2.55	0.66	0.57

(c) Balmer decrement in moving envelopes—The radiative processes in a nonthermally heated and moving gas layer were examined by Gershberg and Shnol (1974) as seen in Section 7.6.6 of Chapter 6. They derived the Balmer decrement D_{34} as a function N_e, T_e, and β_{12°, where β_{12° is the escape probability by motion for the Lyman α line. The results of their calculation are partly given in Table 6.20 of Chapter 6, where a large variety of the Balmer decrement from very flat to very steep can be seen. Their models may also be applicable to the envelopes of TTSs by adopting suitable values of parameters for both cases of steep and flat decrements. This moving envelope model seems most promising to explain the large variety of Balmer decrements in TTSs.

7.3.2 Hydrogen infrared emission lines and mass-loss rates

In the infrared region (1–10 μm), hydrogen spectrum shows Paschen, Bracket, and Pfund series as seen in Table 2.4. In ground based observations, atmospheric extinction, mainly due to water vapor, obscure most of these lines, so that the number of observable lines is limited to include Pα, Brα, Brγ, Pfα. In 1995, the ISO (Infrared Space Observatory) was launched, and thereafter, a full range of IR line spectra became observable. Based on IR line intensities, mass-loss rates of TTSs (Givanardi et al. 1991) and of HESs (Nisini et al. 1995, Benedettini et al. 1998) have been derived as follows.

(i) TTS—Giovanardi et al. (1991), assuming the wind geometry and velocity law in the winds, found that the IR line luminosities of Pα, Brα, Brγ, and their ratios strongly depend upon the kinetic temperature T_k of wind gas and on the mass-loss rate $\dot{M}$ from the star. In order to separate these two parameters, they used the method of hydrogen-sodium diagnostics, which is based on the fact that the optical depth τ (NaI) in the neutral sodium resonance-doublet is more strongly determined by the $\dot{M}$ value. The optical depth τ (NaI) can be obtained from the depth of the NaI absorption line. In principle, this fact allows one to estimate both $\dot{M}$ and T_k, once an H line luminosity and a Na D2 (λ5890 Å) spectrum are available for a star. As an example of the hydrogen-sodium diagnostic diagrams is shown in Figure 7.37, where the curves of constant luminosity of Brγ line, log L(Brγ), and the curves of equal optical depth τ (NaI) are delineated on the plane log$\dot{M} - T_k$. By plotting the stars with observed Brγ line luminosity and NaI optical depth, one may estimate the values of $\dot{M}$ and T_k. In this way Giovanardi et al. derived the wind temperature (T_k = 5000–7000 K) and mass-loss rate ($\dot{M} = 10^{-6.5}$–10^{-8} $M_\odot$ per year) for 13 TTSs. The mass-loss rates thus estimated are shown in the low-luminosity region of Figure 7.39.

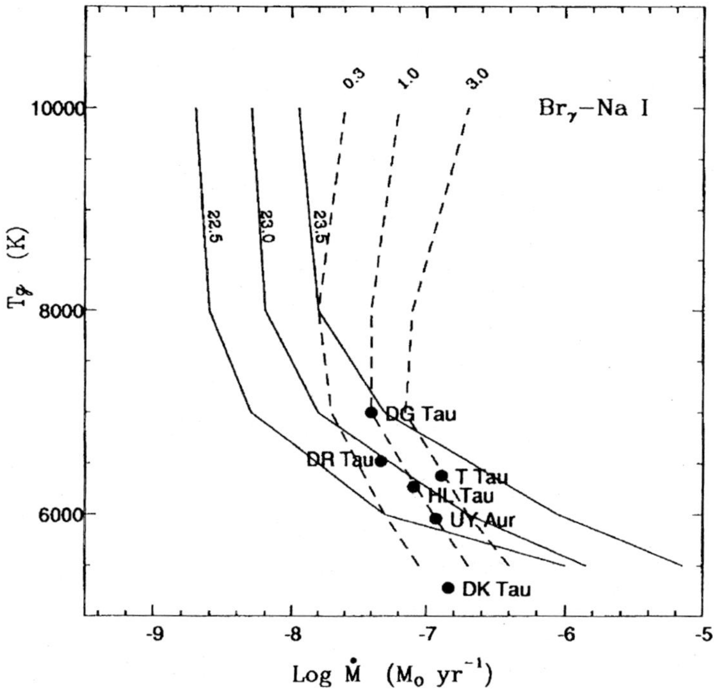

Figure 7.37: Hydrogen-sodium diagnostics for TTSs. The curves of constant luminosity of Brγ line and the curves of equal optical depth in NaI line are delineated on the plane $\log\dot{M} - T_k$. Stars with observed line luminosity $\log L(\text{Br}\gamma)$ (in unit of W) and τ (NaI) are plotted on this plane and the values of $\dot{M}$ and T_k. are estimated. (From Giovanardi et al. 1991)

(ii) HESs—Nisini et al. (1995) adopted a spherically symmetric, isothermal (T = 10,000 K) wind with the velocity law

$$V(r) = V_i + (V_{\max} - V_i)\left(1 - \left(\frac{r_i}{r}\right)^{\alpha}\right), \tag{7.3.3}$$

where r is the distance from the star's center and V_i and r_i denote the velocity and radius at the base of stellar winds, respectively. $V_{\max}$ is the terminal velocity of the wind and is estimated from the profile of the Hα emission in the range of 200–400 km s^{-1}. For these parameters Nisini et al. adopted the values of V_i = 20 km s^{-1}, r_i = stellar radius, $V_{\max}$ = 250 km s^{-1}, and $\alpha = 4$. Instead of solving equations of radiative transfer, they used the escape probability by motion in the above velocity law. By taking the mass-loss rate as a free parameter, they calculated the emission line intensities of Paβ, Brγ, Pfβ, and Pfγ relative to Brα, and compared them with observed line ratios. Thus

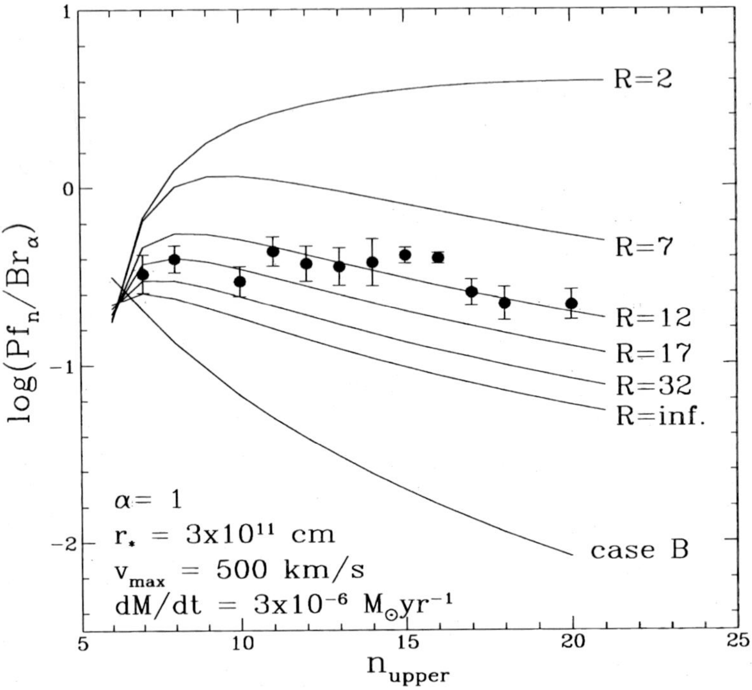

Figure 7.38: CoD-42°11721. Comparison between the Pfund line ratios relative to the Brα line (filled dots) and the predicted line ratios (solid lines) for different dimensions of the ionized region ($R = r_{\rm rec}/r_*$), where $r_{\rm rec}$ represents the radius at which the hydrogen atoms recombine. The model parameters and mass-loss rate are indicated in the figure, $\alpha = 1$ indicated the index parameter in equation (7.3.3). The line of pure recombination spectrum (case B) is also indicated. (From Benedettini et al. 1998)

they estimated the mass-loss rates in the range of 10^{-8}–10^{-6} $M_\odot$ per year for HESs.

Based on the ISO spectrometer Benedettini et al. (1998) obtained the infrared spectra showing hydrogen recombination lines of Bracket, Pfund and Humphreys series for two HESs, MWC 1080 and CoD-42°11721. The observed HI line intensities have been compared with a wind model, similar to Nisini et al.'s (1995) model. Figure 7.38 exhibits the comparison between the Pfund line ratios relative to Brα line and the predicted line ratios in the case of CoD-42°11721. The best fit parameters are indicated in the figure. Thus the mass-loss rate of $(3 \pm 1) \cdot 10^{-6}$ $M_\odot$ per year has been estimated.

(iii) Mass loss rate—The results are shown in Figure 7.39 as a function of stellar luminosity for TTSs (Giovanardi et al. 1991) and HESs (Nisini et al. (1995). For comparison, mass loss rates of classical Be stars

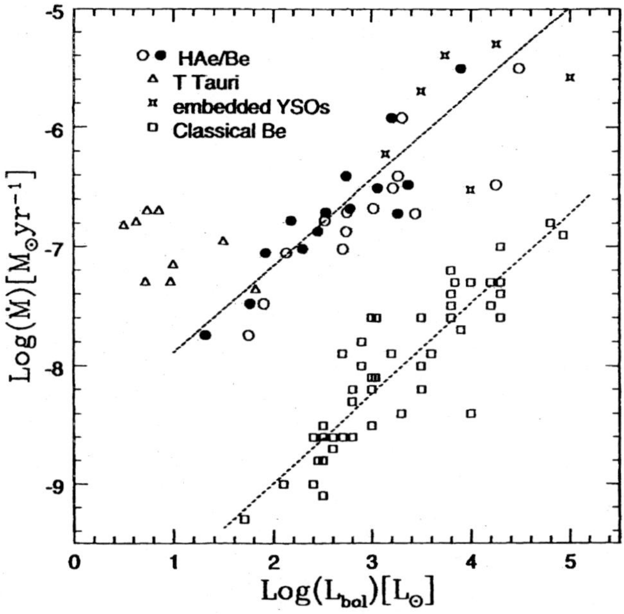

Figure 7.39: The mass-loss rates of HES (Nisini et al. 1995), TTS (Giovanardi et al. 1991), YSO (young stellar object, Höfflich and Wehrse 1987) and classical Be stars (Waters et al. 1987), as a function of stellar luminosity. The filled and open circles in HESs denote the minimum and maximum estimation for the stellar luminosity, respectively. The dashed lines are the best-fit linear relationship through the HESs and classical Be stars. (From Nisini et al. 1995)

(Waters et al. 1987) are also plotted. It is seen in this Figure that the mass-loss rates for HES reveal a good correlation with stellar luminosity in parallel with classical Be stars, but around 1.5 orders of magnitude higher in HESs. In contrast, mass-lass rates of TTS are rather insensitive to the stellar luminosity and mostly higher than the best-fitted line of HES extrapolated toward low luminosity stars. From this it may be inferred that the mass-loss mechanism of TTSs should be different from that of HESs.

7.3.3 Shell absorption lines

Shell absorption lines (see Chapter 5, Section 5.3.4) in HESs can be seen in the spectral atlas of Finkenzeller and Jankovics (1984), which gives the spectral line profiles from Hβ up to higher members of the Balmer series with the intermediate dispersion (29–8.8 Å mm^{-1}). Among 27 HESs, the characteristics of shell absorption are found in 7 stars, for which the highest series number

n_m is traceable as Balmer shell lines. The value of n_m and the property of Hβ profile (emission/absorption, shell/non-shell, V/R) are given as follows:

HD 37490	$n_m = 19$	Hβ (emission, shell, V = R)
CoD $-44°3318$	$n_m = 19$	Hβ (weak emission, shell, V > R)
HR 5999	$n_m = 17$	Hβ (weak emission with inverse P Cygni profile, shell, V ≫ R)
HD 150193	$n_m = 14$	Hβ (weak emission, shell, V ∼ R)
HD 163296	$n_m = 17$	Hβ (weak emission with P Cygni profile, shell, V ≪ R)
BD $+40°4124$	$n_m =$ not clear	Hβ (strong emission, shell, V ≪ R)
HD 200775	$n_m \sim 14$	Hβ (strong emission, shell, V ∼ R)

Among these, HR 5999 (HD 144668 = V856 Sco) shows the shell lines from the Hα to H20 (Tjin et al. 1989). As seen above, the values of n_m is less than around 20 in HESs, in contrast to the classical Be stars for which we meet with developed gas shell of $n_m \sim 30$–40. It is then supposed that the optical depth for the Hα, τ (Hα), is no higher than the shells of classical Be stars.

When the spectrum of the blue region is not available, we can judge the shell characteristic from the profiles of Hα and/or Hβ. The stars showing a sharp absorption component deeper than the photospheric line profile can be picked up as the candidates for shell line stars. In the atlas of Reipurth et al. (1996), we select those stars as follows:

TTS—4 stars among 43 TTSs
UX Tau (K0), GQ Lup (K7), RY Lup (K1), SZ Cha (K0)
HES—6 stars among 18 HESs
NX Pup (F1), ESO Hα28 (B8), HR 5999(A7), AK Sco(F5), VV Ser (A2), WW Vul (A0)
FUor—No candidate in two stars.

Alencar and Basri (2000) carried out spectroscopic observations for 30 TTSs and measured the profiles of strong permitted lines including Hα, Hβ, Hγ, Hδ, and other metallic lines. Though the profiles of Balmer lines showed large variety such as single-, double-peaked emission or P Cyg profile, nine stars revealed clear evidence of strong shell absorption in the near centers of the Balmer lines.

Thus, the candidates for shell line stars can be found in early to late type stars. The existence of shell line stars implies that the envelopes are optically thick in the Hα and if we can see the profiles of higher Balmer members, we can get the information on the structure of the envelope as in case of classical Be stars.

Three edge-on TTSs, HH30*, HK Tau B, and HV Tau C, observed by Appenzeller et al. (2005), are expected to be shell stars by their suspected large inclination angles. However, these stars did not show any shell-absorption features in Balmer lines, and showed the existence of photospheric absorption lines. These spectral features imply that the disks of these TTSs might be exceptionally optically thin for the Hα line.

7.3.4 Magnetospheric accretion models and line profiles

The concept of magnetically controlled accretion in CTTSs was developed in 1980's. Since then many models have been proposed and applied to the TTS phenomena. In the study of formation of emission-line profiles, Muzerolle et al. (2001) presented models of an axisymmetric, dipolar magnetic field geometry for the accretion flows. Schematic geometry of the dipole magnetic field is shown in Figure 7.40, where the inner and outer radii of the disk are taken as free parameters. For modeling they solved the radiative transfer problem by using the method of escape probability by motion, and calculated the profiles of Hα, Hβ and Na I D lines. Parameters are accretion rate, gas temperature, inclination angle, and magnetospheric size. A grid of Hβ model profiles, showing the effects of accretion rate and gas temperature, are given in Figure 7.41 in the case of fixed magnetospheric dimension, and inclination angle. Large variations of the profile depending on the parameters are seen and this enables model calculations to use as the emission-line diagnostics of T Tauri mangetospheric accretion disks. Muzerolle et al. (2004) applied this

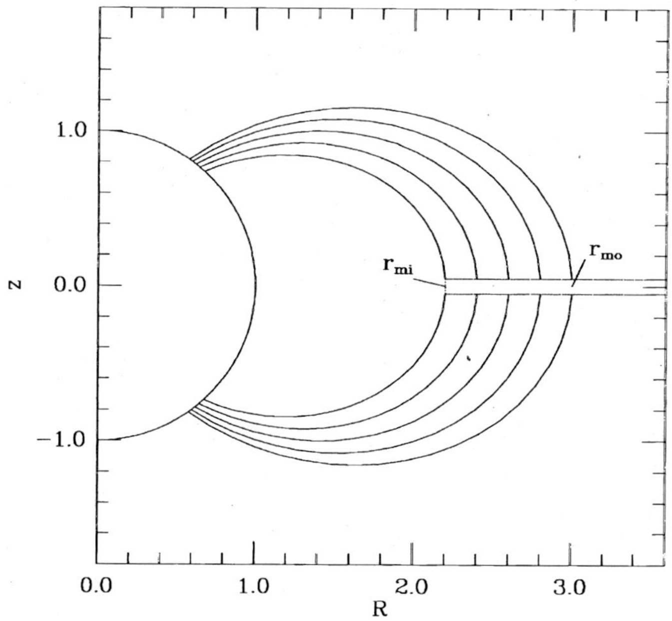

Figure 7.40: Schematic geometry of the dipole magnetic field and accretion disk, used in the models of Muzerolle et al. (2001).

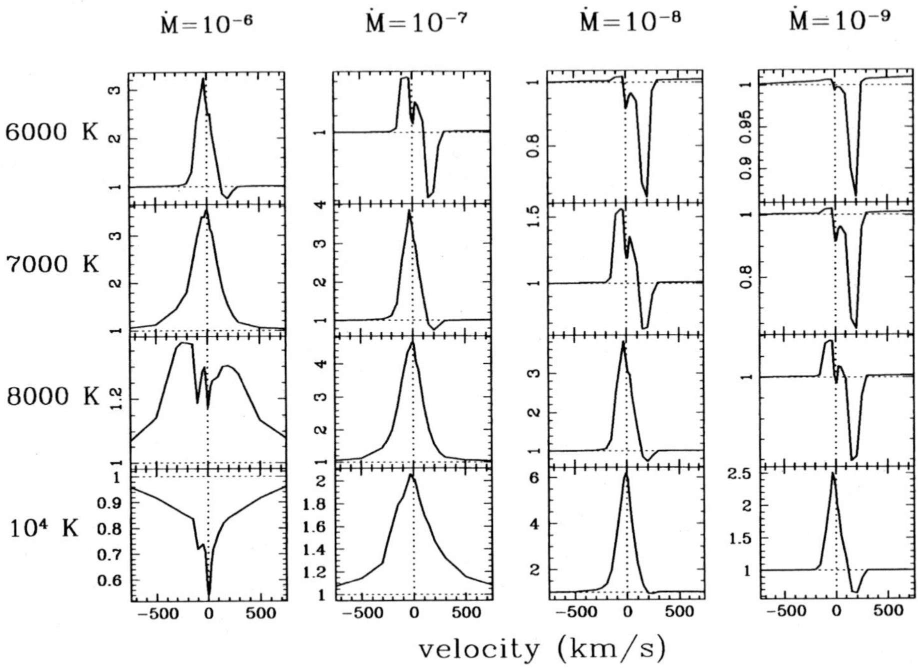

Figure 7.41: A grid of Hβ model profiles, showing the effects of mass accretion rate (in $M_\odot$ yr-1) vs. temperature (characterized by the maximum temperature inside the disk). All models are calculated with the fiducial magnetospheric dimensions, $r_m = 2.2$–$3R_*$, and inclination $i = 60°$. (From Muzerolle et al. 2001)

diagnostics to the Herbig Ae star UX Ori, and derived the mass accretion rate and other plausible disk parameters.

Beristan et al. (2001) analyzed emission-line profiles of HeI and HeII in 31 TTSs, from the viewpoint of accretion and wind flow in magnetospheric structure. These lines should be formed in a region either of high temperature or close proximity to a source of ionizing radiation, both of which are related to the stellar magnetosphere. They particularly analyzed the HeI λ5876 Å line which is decomposed into narrow and broad components. Some stars show only broad components (3 stars), some others show combined narrow and broad components (19 stars), or only narrow components (9 stars). The average line profiles for these three types among their observed stars are exhibited in Figure 7.42. A narrow component is characterized by relatively uniform line widths and centroid velocities, suggesting their formation in the decelerating postshock gas at the magnetosphere foot point. A broad component displays a diversity of kinematic properties, showing blueshifted profiles as seen in Figure 7.42. This component showed maximum blue wing velocities exceeding –200 km s^{-1} in 14 stars, suggesting that these lines are formed in the inner hot regions of winds in T Tauri accretion disk systems.

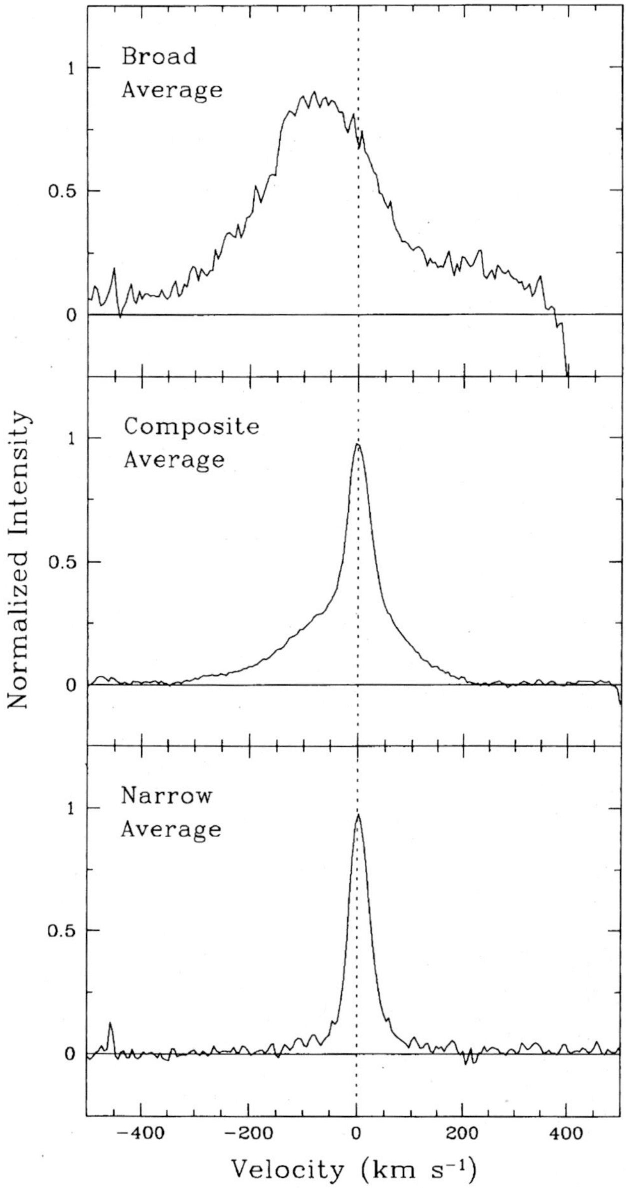

Figure 7.42: Average profiles of He Iλ5876 in TTSs in three types: broad component only (upper panel), narrow and broad composite (middle panel), and narrow component only (bottom panel). (From Beristain et al. 2001)

7.4 Evolution of pre-main sequence stars

The protostars formed in molecular clouds are accessible by infrared observations because of their low surface temperature. On the way toward the main sequence, the protostellar surface is getting hotter through continuing contraction, and at some point, the star becomes optically observable. The protostars with different masses become visible at different points in the HR diagram. The locus of such points is called the "birthline" (Stahler 1983).

The evolutionary tracks in the HR diagram, starting at the birthline and ending at the ZAMS, are calculated by Palla and Stahler (1993) and illustrated in Figure 7.43 for stars with masses from 0.6 to 6.0 $M_{\odot}$. The birthline has been

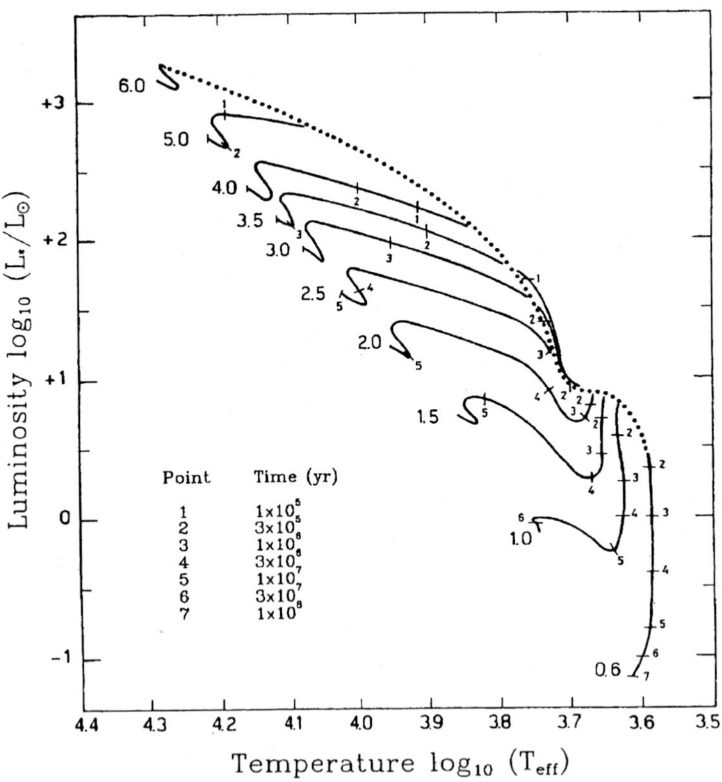

Figure 7.43: Evolutionary tracks in the HR diagram for stars with masses from 0.6 to 6 $M_{\odot}$. Each track is labeled by the corresponding mass and the tick marks indicate evolutionary times measured from the birthline (dotted curve). (From Palla and Stahler 1993)

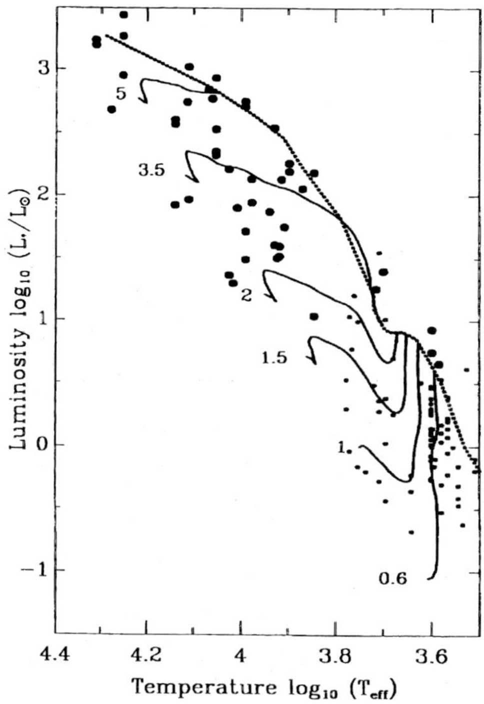

Figure 7.44: Observed distribution of low- and intermediate-mass pre-main sequence stars in the HR-diagram. Large filled circles denote the HESs, and small dots the TTSs. Theoretical tracks and birthline are the same as in Figure 7.39. (From Palla and Stahl 1993)

calculated under the assumption of a constant mass-accretion rate of 10^{-5} $M_\odot$ per year in the protostar stage. Observed distribution of HES and TTS in the same HR diagram is shown in Figure 7.44 plotted by Palla and Stahler (1993). As theoretically predicted, these pre-main sequence stars are well distributed between the birthline and the ZAMS line. If we extrapolate the birthline to the more massive stars, it intersects with ZAMS at $M = 8$ $M_\odot$, implying that stars more massive than this value might have no pre-main sequence phase. That is, ZAMS is effectively the birthline for early B and O stars. In addition, evolutionary time, $t_{\rm evo}$, also depends on stellar mass. The time $t_{\rm evo}$ is less than 1 million years for massive stars higher than 4 $M_\odot$, whereas $t_{\rm evo}$ is around 10^8 years or longer for stars with mass smaller than 1 $M_\odot$.

An evolutionary scenario for a low-mass stellar accretion has been proposed by Hartmann (1998) as schematically shown in Figure 7.45. Suppose a molecular cloud core of a few solar masses evolving into self-gravitating collapse at nearly free-fall. This collapse lasts approximately 0.1–0.2 Myr, forming a stellar core and surrounding circumstellar disk. Most of the infalling

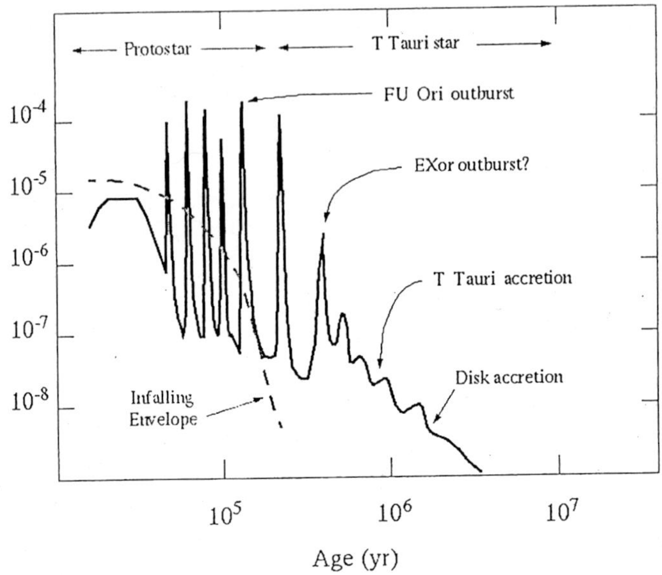

Figure 7.45: Outline of mass-accreting rates during the formation of a typical low-mass star. The dotted curve denotes the infalling rate of gas onto the circumstellar disk, and the solid curve is the accretion rate from disk to the central star. (From Hartmann 1998)

mass lands on the disk, and its infalling rate may be something like as shown by the dotted curve in Figure 7.45. Eventually, the disk mass is gradually or intermittently accreted onto the central star.

The accretion rate onto the star is shown by a solid curve in Figure 7.45. During or immediately after the protostellar phase, disk masses are likely to be relatively large, so that the disks could be subject to gravitational instabilities, which cause rapid accretion and give rise to the FU Ori type outbursts. After repeating FU Ori outbursts several times, the star enters the EXor and/or T Tauri star phase, and gradually evolves into the main sequence star.

The lifetime of accretion disks and accretion rates show a large dispersion among TTSs. According to Armitage et al. (2003), around 30% of stars lose their disks within 1 Myr, while the remainder has disk lifetimes typically in the range of 1–10 Myr, mostly depending on the initial conditions of disk formation.

Further reading

Bertout, C. (1989), T Tauri stars: Wild as dust. *Ann. Rev. A. & A*, **27**, 351–395.

Hartmann, L. (1998), *Accretion Processes in Star Formation.* Cambridge University Press, Cambridge.

Reipurth, B. (ed.) (1989), *Low Mass Star Formation and Pre-main Sequence Objects*, ESO Workshop Proceedings No. 33, ESO.

Thé, P. S., Pérez, M. R., and van den Heuvel, E. P. J. (eds.) (1994), *The Nature and Evolutionary Status of Herbig Ae/Be Stars.* First International Meeting in Amsterdam, ASP Conf. Series, Vol. 62, San Francisco, CA.

References

Aiad, A., Appenzeller, I., Bertout, C., Stahl, O., and 4 co-authors (1984), Coordinated spectroscopic observations of YY Orionis stars. *A.A.*, **130**, 67–78.

Alencar, S. H. P. and Basri, G. (2000), Profiless of strong permitted lines in classical T Tauri stars. *A.J.*, **119**, 1881–1900.

Ambartsumian, V.A. (1949), Stellar associations. *Astr. Zhur.*, **26**, 3–12.

Appenzeller, I., Jankovics, I., and Östreicher, R. (1984), Forbidden-line profiles of T Tauri stars. *A. A.*, **141**, 108–115.

Appenzeller, I., Bertout, C., and Stahl, O. (2005), Edge-on T Tauri stars. *A.A.* **434**, 1005–1019.

Armitage, P. J., Clarke, C. J., and Palla, F. (2003), Dispersion in the lifetime and accretion rate of T Tauri stars. *M. N. R. A. S.*, **342**, 1139–1146.

Attridge, J. M. and Herbst, W. (1992), Rotation priods of T Tauri stars in the Orion Nebula cluster: A bimodal frequency distribution. *Ap. J.*, **398**, L61–L64.

Bastien, P. (1981), The wavelength dependence of linear polarization in T Tauri stars. *A.A.*, **94**, 294–298.

Bastien, P. (1982), A linear polarization survey of T Tauri stars. *A. A. Suppl.*, **48**, 153–164.

Bastien, P. (1985), A linear polarization survey of southern T Tauri stars. *Ap. J. Suppl.*, **59**, 277–291.

Bastien, P. and Landstreet, J. D. (1979), Polarization observations of the T Tauri stars RY Tauri, T Tauri, and V866 Scorpii. *Ap. J.*, **229**, L134–140.

Batalha, C. C. and Basri, G. (1993), The atmospheres of T Tauri stars. II. Chromospheric line fluxes and veiling. *Ap. J.*, **412**, 363–374.

Bell, K. R., Lin, D. N. C., Hartmann, L. W., and Kenyon, S. J. (1995), The FU Orionis outburst as a thermal accretion event: Observational constraints for protostellar models. *Ap. J.*, **444**, 376–395.

Bellingham, J. G. and Rossano, G. S. (1980), Long-period variations in R CrA, S CrA, T CrA, and R Mon. *A.J.*, **85**, 555–559.

Benedettini, M., Nisini, B., Giannini, T., Lorenzetti, D., Tommasi, E., Saraceno, P., and Smith, H. A. (1998), ISO-SWS observations of Herbig Ae/Be stars: HI recombination lines in MWC 1080 and CoD-42°11721. *A.A.* **339**, 159–164.

Beristain, G., Edwards, S., amd Kwan, J. (2001), Helium emission from classical T Tauri stars. Dual origin in magnetospheric infall and hot wind. *Ap. J.*, **551**, 1037–1064.

Bertout, C., Harder, S., Malbet, F., Mennesier, C., and Regev, C. (1996), Photometric observations of YY Orionis: New insight into the accretion process. *A.J.* **112**, 2159–2167.

Beskrovnaya, N. G., Pogodin, M. A., Najdenov, I., and Pomanyuk, I. (1995), Short-term spectral and polarimetric variability in the Herbig Ae star AB Aur as an indicator of the circumstellar inhomogeneity. *A.A.*, **298**, 585-593.

Bibo, E. A. and Thé , P. S. (1991), Type of variability of Herbig Ae/Be stars. *A. A. Suppl.*, **89**, 319–334.

Böhm, T. and Catala, C. (1993), A spectral atlas of the Herbig Ae star AB Aurigae. The visible domain from 391 to 874 nm. *A. A. Suppl.*, **101**, 629–672.

Böhm, T. and Catala, C. (1994), Forbidden lines in Herbig Ae/Be stars: the [OI] (IF) 6300.31and 6363.79 Å. *A.A.*, **290**, 167–175.

Böhm, T. and Catala, C. (1995), Rotation, winds and active phenomena in Herbig Ae/Be stars. *A.A.* **301**, 155–169.

Böhm, T., Catala, C., Donati, J. F., Welty, A., Baudrand, J., and 17 co-authors (1996), Azimuthal structure in the wind and chromosphere of the Herbig Ae star AB Aurigae. Results from the MUSICOS 1992 compaign. *A. A. Suppl.*, **120**, 431–450.

Bouret, J. C., Catala, C., and Simon, T. (1997), Nitrogen V in the wind of the pre-main sequence Herbig Ae star AB Aurigae. *A.A.*, **328**, 606–616.

Bouvier, J. (1990), Rotation in T Tauri stars. II. Clues for magnetic activity. *A.J.*, **99**, 946–964.

Bouvier, J. (1991), Rotation in pre-main sequence stars—Properties and evolution. *Proc. NATO Advanced Res. Workshop*, Catalano, S. and Stauffer, J. R. (eds.), Kluwer, Dordrecht, p. 41.

Bouvier, J. and Bertout, C. (1989), Spots on T Tauri stars. *A.A.*, **211**, 99–114.

Breger, M. 1974, Pre-main sequence stars. III. Herbig Be/Ae Stars and other selected objects. *Ap. J.*, **188**, 53–58.

Brown, A., Ferraz, M. C. de M., and Jordan, C. (1984), The chromosphere and corona of T Tauri. *M. N. R. A. S.*, **207**, 831–859.

Cabrit, S., Edwards, S., Strom, S. E., and Strom, K. M. (1990), Forbidden-line emission and infrared excess in T Tauri stars: Evidence for accretion-driven mass loss. *Ap. J.*, **354**, 687–700.

Calvet, N., Basri, G., and Kuhi, L. V. (1984), The chromospheric hypothesis for the T Tauri phenomenon, A. *p. j.*, **277**, 725–737.

Cantó, J., Rodriguez, L. F., Calvet, N., and Levreault, R. M. (1984), Stellar winds and molecular clouds: Herbig Be and Ae type stars. *Ap. J.*, **282**, 631–640.

Catala, C. (1988), Line formation in the winds of Herbig Ae/Be stars. The CIV resonance lines. *A.A.*, **193**, 222–228.

Catala, C. (1989), Herbig Ae and Be stars. *Low Mass Star Formation and Pre-main Sequence Objects*, E SA Workshop, Reipurth, B. (ed.), European Southern Observatory, Garching bei Munchen, pp. 471–489.

Catala, C., Felenbok, P., Czarny, J., Talavera, A., and Boesgaard, A. M. (1986), Sjhort-term variability in AB Aur: Clues for acitivity in Herbig Ae stars. II. The CaII K line. *Ap. J.*, **308**, 791–804.

Catala, C., Donati, J. F., Böhm, T., Landstreet, J., Henrichs, H. F., Unruh, Y., and 21 co-authors (1999), Short-term spectroscopic variability in the pre-main sequence Herbig Ae star AB Aurigae during the MUSICOS 96 campaign. *A.A.*, **345**, 884–904.

Cohen, M. and Kuhi, L. V. (1979), Observational studies of pre-main- sequence evolution. *Ap. J. Suppl.*, **41**, 743–843.

Corporon, P., Lagrange, A. M., and Bouvier, J. (1994), Direct determination of stellar and orbital parameters of the spectroscopic binary TY CrA. *A.A.*, **284**, L21–L24.

Cram, L. E. (1979), Atmospheres of T Tauri stars: The photosphere and low chromosphere. *Ap. J.*, **234**, 949–957.

Damiani, F., Micela, G., Sciortino, S., and Farata, F. (1994), Einstein X-ray observations of Herbig Ae/Be stars. *Ap. J.*, **436**, 807–811.

Fernandez, M., Ortiz, E., Eiroa, C., and Miranda, L. F. (1995), Hα emission from pre-main sequence stars. *A. A. Suppl.*, **114**, 439–464.

Finkenzeller, U. 1985, Rotational velocities, spectral types, and forbidden lines of Herbig Ae/Be stars. *A.A.*, **151**, 340–348.

Finkenzeller, U. and Mundt, R. (1984), The Herbig Ae/Be Stars associated with nebulosity. *A. A. Suppl.*, **55**, 109–141.

Finkenzeller, U. and Jankovics, I. (1984), Line profiles and radial velocities of Herbig Ae/Be stars. *A. A. Suppl.*, **57**, 285–326.

Folha, D. F. M. and Emerson, J. P. (2001), Near infrared hydrogen lines as diagnostics of accretion and winds in T Tauri stars. *A. A.*, 365, 90–109.

Fukuda, I. and Uesugi, A. (1982), *Revised Catalogue of stellar rotational velocities.* Department of Astronomy, Kyoto University.

Gahm, G. F. (1970), The spectrum of RW Aurigae, 3250–4900 Å, *Ap. J.* 160, 1117–1125.

Gahm, G. F., Lodén, K., Gullbring, E., and Hartstein, D. (1995), Activity on young stars. *A.A.*, **301**, 89–104.

Gershberg, R. E. and Shnol, E. E. (1974) The Balmer decrement in spectra of moving medium. The case of collisional ionization and excitation. *Izv. Krim Astrophys. Obs.*, **50**, 122–151.

Ghandour, L., Strom, S. E., Edwards, S., and Hillenbrand, L. A. (1994), Spectroscopic diagnostics of disk accretion in Herbig Ae/Be stars. *The Nature and Evolutionary Status of Herbig Ae/Be Stars.* P. S. Thé, M. R. Perez, and E. J. van den Heuvel (eds.). ASP Conf. Ser. Vol. 62, San Francisco, CA, pp. 223–226.

Ghez, A. M., Neugebauer, G., and Matthews, K. (1993), The multiplicity of T Tauri stars in the star forming regions Taurus—Auriga and Ophiuchus—Scorpius: A 2.2 micron speckle imaging survey. *A.J.*, **106**, 2005–2023.

Giovanardi, C., Gennari, S., Natta, A., and Stanga, R. (1991), Infrared spectroscopy of T Tauri stars. *Ap. J.*, **367**, 173–181.

Glasby, J. S. (1974), *The Nebular Variables.* Int'l Ser. Monographs in Natural Philosophy, Vol. 69. Pergamon Press, Oxford, pp. 1–201.

Grinin, V. P. (1969), Hydrogen-line intensities in the spectrum of an optically thick plasma. *Astrophysics*, **5**, 172–177.

Grinin, V. P. (1994), Polarimetirc activities of Herbig Ae/Be stars. *The Nature and Evolutionary Status of Herbig Ae/Be Stars.* P. S. Thé, M. R. Perez, and E. J. van den Heuvel (eds.). ASP Conf. Ser. Vol. 62, San Francisco, CA, pp. 63–71.

Grinin, V. P., Rostopchina, A. N., Okazaki, A., Kikuchi, S., and Minikhulov, N. H. (1994), Intrinsic polarization of classical Ae Herbig star RR Tau. *The Nature and Evolutionary Status of Herbig Ae/Be Stars*, P. S. Thé, M. R. Perez, and E. J. van den Heuvel (eds.). ASP Conf. Ser. Vol. 62, San Francisco, CA, pp. 86–87.

Hamann, F. and Persson, S. E. (1992), Emission-line studies of young stars. II. The Herbig Ae/Be stars. *Ap. J. Suppl.*, 82, 285–309.

Hartigan, P., Edwards, S., and Ghandour, L. (1995), Disk accretion and mass loss from young stars. *Ap. J.*, **452**, 736–768.

Hartmann, L. (1998), Overview. *Accretion Processes in Star Formation*, Cambridge Astrophysics, Series 32. Cambridge University Press, Cambridge, pp. 1–15.

Hartmann, L. and Kenyon, S. J. (1996), The FU Orionis phenomenon. *Ann. Rev. A. A.*, **34**, 207–240

Herbig, G. H. (1945), Emission lines of FeI in RW Aurigae. *P.A.S. Pacific*, **57**, 166–168.

Herbig, G. H. (1952), Emission-line stars in galactic nebulosities. *J. R. A. S. Canada*, **46**, 222–233.

Herbig, G. H. (1960), The spectra of Be- and Ae-type stars associated with nebulosity. *Ap. J. Suppl.*, **4**, 337–351.

Herbig, G. H. (1962), The properties and problems of T Tauri stars and related objects. *Adv. Ast. & Astrphys.*, **1**, 47–103.

Herbig, G. H. (1965), Lithium abundances in F5–G8 stars. *Ap. J.*, **141**, 588–609.

Herbig, G. H. (1977), Radial velocities and spectral types of T Tauri stars. *Ap. J.*, **214**, 747–758.

Herbig, G. H. (1989), FU Ori eruption, *Proc. ESO Workshop on Low Mass Star Formation and Pre-Main Sequence Objects* B Reipurth (ed.), ESO, Garching bei Munchen, pp. 233–246.

Herbig, G. H. and Rao, N. K. (1972), Second catalog of emission-line stars of the Orion population. *Ap. J.*, **174**, 401–423.

Herbig, G. H., Vrba, F. J., and Rydgren, A. E. (1986), A spectroscopic survey of the Taurus-Auriga dark clouds for pre-main sequence atars having Ca II H, K emission. *A.J.*, **91**, 575–582.

Herbig, G. H., Petrov, P. P., and Duemmler, R. (2003), High-resolution spectroscopy of FU Orionis stars. *Ap. J.*, **595**, 384–411.

Herbst, W., (1986), T Tauri variables. *P. A. S. P.*, **98**, 1088–1094.

Herbst, W., Holtzman, J. A., and Klasky, R. S. (1983), Photometric variations of Orion population stars. II. Ae-irregular variables and T Tauri stars. *A.J.*, **88**, 1648–1663.

Herbst, W., Herbst, D., Grossman, E. J., and Weinstein, D. (1994), Catalogue of UBVRI photometry of T Tauri stars and analysis of the causes of their variability. *A.J.*, **108**, 1906–1923.

Hernández. J. Calvet, N., Briceño, C., Hartmann, L., and Berlind, P. (2004), Spectral analysis and classification of Herbit Ae/Be stars. *A.J.*, **127**, 1682–1701.

Hillenbrand, L. A., Strom, S., Vrba, F. J., and Keene, J. (1992), Herbig Ae/Be stars: Intemediate-mass stars surrounded by massive circumstellar disks. *Ap. J.*, **397**, 613–643.

Hirth, G. A. (1994), Forbidden emission lines of T Tauri stars as tracers of their jets and disk winds. *Stellar and Circumstellar Astrophysics*, G. Wallerstein and A. Noriega-Crespo (eds.), ASP Conf. Ser. Vol. 57, San Francisco, CA, pp. 32–39.

Höfflich, P. and Wehrse, R. (1987), NLTE models for cocoon stars. *A.A.* **185**, 107–116.

Imhoff, C. L. and Appenzeller, I. (1987), Pre-main sequence stars. *Exploring the Universe with the IUE Satellite,* Y. Kondo (ed.), D. Reidel Pub. Comp. Dordrecht, pp. 295–319.

Jain, S. K. and Bhatt, H. C. (1995), *A. A. Suppl.*, **111**, 399–405, Study of variability of The polarization in Herbig Ae/Be stars.

Johns-Krull, C. M., Vatenti, J. A., and Linsky, J. L., (1998), A short wavelength IUE atlas of pre-main sequence stars. *Cool Stars, Steller Systems and the Sun.* R. A. Donahue and J. A. Bookbinder (eds.), ASP Conf. Ser. Vol. 154, San Francisco, CA, pp. 1724–1734.

Johns-Krull, C. M., Vatenti, J. A., and Linsky, J. L., (2000), An IAU atlas of pre-main-sequence stars. II. Far-ultraviolet accretion diagnostics in T Tauri stars. *Ap. J.*, **539**, 815–833.

Joy, A. H. (1942), Spectral criteria in the classification of Variable stars. *P. A. S. P.*, **54**, 15–18.

Katysheva, N. A. (1981), Balmer decrements of T Tau stars. *Astrophysics*, **17**, 165–169.

Köhler, R. (2001), Multiplicity of X-ray selected T Tauri stars in Chamaeleon. *A.J.*, **122**, 3325–3334.

Kolotilov, E. A. (1977), Spectral and photometric observations of the fast irregular variables. II. Hα and Hβ lines in the spectrum of WW Vul, VX Cas, and UX Ori. *Astrophysics*, **13**, 17–25.

Köppen, J., Finkenzeller, U., Mundt, R., and Beltrametti, M. (1982), On the Balmer emission lines of the Herbig Be star HD 200775. *A.A.*, **112**, 174–177.

Kuhi, L. V. (1965), T Tauri stars: A short review. *J. R. A. S., Canada*, **60**, (1), 1–14.

Kukarkin, B. V. and Parenago, P. P. (1970), *General Catalouge of Variable Stars (GCVS).* Third edition, Akademia Naulc Moskva.

Leinert, Ch., Zinnecker, H., Weitzel, N., Christou, J., and 4 co-authors (1993), A systematic search for young binaries in Taurus. *A.A.*, **278**, 129–149.

Li, W., Evans, H. J. II, Harvey, P. M., and Colome, C. (1994), Near-infrared (J. H. K) imaging of Herbig Ae/Be stars. *Ap. J.*, **433**, 199–215.

Magazzu, A. Rebolo, R., and Pavlenko, Ya. V. (1992), Lithium abundance in classical and weak T Tauri stars. *Ap. J.*, **392,** 159–171.

Magazzu, A., Alcala, J. M., and Coving, E. (1997), Lithium, rotation, and activity in Chamaeleon T Tauri stars. *Mem. Soc. A. Italia*, **65**, 925–928.

Malfait, K., Bogaert, E., and Waelkens, C. (1998), An ultraviolet, optical and infrared study of Herbig Ae/Be stars. *A.A.*, **331**, 211–223.

Martin, E. L. (1994), Binarity among Ae/Be stars. *The Nature and Evolutionary Status of Herbig Ae and Be stars*, P. S. Thé, M. R. Perez, and E. J. van den Heuvel (eds.). ASP Conf. Ser. Vol. 62, San Francisco, CA, pp. 315–318.

Martin, E. L., Rebolo, R., Magazzu, A., and Pavlenko, Ya. V. (1994), Pre-main sequence lithium burning. I. Weak T Tauri stars. *A.A.*, **282**, 503–517.

Mathieu, R.D. (1994), Pre-main-sequence binary stars. *Ann. Rev. A. & A*, **32**, 465–530.

Melo, C. H. F. (2003), The short-period multiplicity among the T Tauri stars. *The Future of Cool-Star Astrophysics: 12th Cambridge Workshop on Cool Stars. Stellar Systems and the Sun*, A. Brown, G. M. Harper, and T. R. Ayres (eds.)., U. Colorado, Colorado, pp. 741–746.

Ménard, F., Bouvier, J., Dougados, C., Mel'nikov, S. Y., and Grankin, K. N. (2003), Constrains on the disk geometry of the T Tauri star AA Tau from linear polarimetry. *A.A.*, **409**, 163–167.

Mendoza, E. E. (1968), Infrared excess in T Tauri stars and related objects. *Ap. J.*, **151**, 977–987.

Meyer, M. R., Calvet, N., and Hillenbrand, L. A. (1997), Intrinsic near-infrared excesses of T Tauri stars: Understanding the classical T Tauri star locus. *A.J.*, **114**, 288–300.

Mitchell, G. F. and Mathews, H. E. (1994), A molecular jet from LkHα 234. *Ap. J.*, **423**, L55–L58.

Mundt, R. and Ray, T. P. (1994), Optical outflows from Herbig Ae/Be stars and other high luminosity young stellar objects. *The Nature and Evol. Status of Herbig Ae and Be stars*, P. S. Thé, M. R. Perez, and E. J. van den Heuvel (eds.), ASP Conf. Ser. Vol. 62, San Francisco, CA, pp. 237–252.

Muzerolle, J., Calvet, N., and Hartmann, L. (2001), Emission-line diagnostics of T Tauri magnetospheric accretion. II. Improved model tests and insights into accretion physics. *Ap. J.*, **550**, 944–961.

Muzerolle, J., D'Alesso, P., Calvet, N., and Hartmann, L. (2004), Magnetospheres and disk accretion in Herbig Ae/Be stars. *Ap. J.*, **617**, 406–417.

Neuhäuser, R., Sterzik, M. F., Schmidt, J. H., Wichmann, R., and Krautter, J. (1995), ROSAT survey observation of T Tauri stars in Taurus. *A.A.*, **297**, 391–417.

Nisini, B., Milillo, A., Saraceno, P., and Vitali, F. (1995), Mass loss rates from HI infrared lines in Herbig Ae/Be stars. *A.A.*, **302**, 169–183.

Palla, F. and Stahler, S. W. (1993), The pre-main-seequence evolution of intermediate-mass stars. *Ap. J.*, **418**, 414–425.

Pallavicini, R., Pasquini, L., and Randich, S. (1992), Optical spectroscopy of post-TTS star candidates. *A.A.*, **261**, 245–254.

Parenago, P. (1954), Study of stars in Orion Nebula Region. *Publ. Sternberg Astr. Inst.* (Trudhi Astronomicheskii Instituta im P.K. Schternberga), **25**, 222–233.

Pérez, M. R., Grady, C. A., and Thé, P.S. (1993), UV spectral variability in the Herbig Ae star HR 5999 XI. The accretion interpretation. *A.A.*, **274**, 381–390.

Pérez, M. and Thé, P. S. (1994), The clumpy accretion in Herbig Ae/Be stars. *Rev. Mex. A. A.*, **29**, 54–58.

Pirzkal, N., Spillar, E. J., and Dyck, H. M. (1997), A search for close bright companions to AeBe stars. *Ap. J.*, **481**, 392–395.

Pontefract, M., Drew, J. E., Harris, T. J., and Oudmaijer, R. D. (2000), Hα spectropolarimetry of the Herbig Ae star AB Aurigae. *M. N. R. A. S.,* **319**, L19–L23.

Praderie, F., Simon, T., Catala, C., and Boesgaad, A. M. (1986), Short-term spectral variability in AB Aurigae : Clue for activity in Herbig Ae stars. I. The ultraviolet lines of MgII and FeII. *Ap. J.*, **303**, 311–326.

Ray, T.P., Poetzel, R., Solf, J., and Mundt, R. (1990), Optical jets from the high-luminosity young stars LkHα 234 and AFGL 4029. *Ap. J.*, **357**, L45–L48.

Reipurth, B. and Zinnecker, H. (1993), Visual binaries among pre-main sequence stars. *A.A.*, **278**, 81–108.

Reipurth, B., Pedrosa, A., and Lago, M. T. V. T. (1996), Hα emission in pre-main sequence stars. I. An atlas of line profiles. *A. A. Suppl.*, **120**, 229–256.

Rydgren, A.E. and Zak, D. S. (1987), On the spectral form of the infrared excess component in T Tauri systems. *P. A. S. P.*, **99**, 141–145.

Skinner, S. L., Brown, A., and Stewart, R. T. (1993), A high-sensitivity survey of radio continuum emission from Herbig Ae/Be stars. *Ap. J. Suppl.*, **87**, 217–265.

Sorelli,C., Grinin, V. P., and Natta, A. (1996), Infall in Herbig Ae/Be stars: what Na D lines tells us. *A.A.*, 309, 155–162.

Stahler, S. W. (1983), The birthline for low-mass stars. *Ap. J.*, **274**, 822–829.

Stelzer. R. and Neuhäuser, R. (2001), X-ray emission from young stars in Taurus-Auriga-Perseud: Luminosity function and the rotation- activity relation. *A. A.* **377**, 538–556.

Strom, S. E., Strom, K. M., Yost, J., Carrasco, L., and Grasdalen, G. (1972), The nature of the Herbig Ae and Be-type stars associated with nebulosity. *Ap. J.*, **173**, 353–366.

Thé, P. S. (1994), Evidence for circumstellar disks around variable Herbig Ae/Be stars from long-term photometry. *The Impact of Long-Term Monitoring on Variable Star Research*, C. Sterken and M. de Groot (eds.), Kluwer, Dordrecht, pp. 31–40.

Thé, P. S., de Winter, D., and Perez, M. R. (1994), A new catalogue of members and candidate members of the Herbig Ae/Be (HAEBE) stellar group. *A. A. Suppl.*, **104**, 315–339.

Tjin, A., Djie, H. R. E., Thé, P. S., Anderson, J., Nordstrom, B., Finkenzeller, U., and Jankovics, I. (1989). The variable Herbig Ae star HR 5999 VIII – Spectrosopic observations 1975–1985 and correlations with simultaneous photometry, *A. A. Suppl.* **78**, 1–24.

Vink, J. S., Drew, J. E., Harris, T. J., and Oudmaijer, R. D. (2002), Probing the circumstellar structure of Herbig Ae/Be stars. *M. N. R. A. S.* **337,** 356–368.

Vink, J. S., Drew, J. E., Harris, T. J., Oudmaijer, R. D., and Unruh, Y. (2005), Probing the circumstellar structures of T Tauri stars and their relationship to those of Herbig stars. *M. N. R. A. S.*, **359**, 1049–1064.

Vrba, F., Schmidt, G. D., and Hinzen, P. M. (1979), observations and evaluation of the polarization in Herbig Ae/Be stars. *Ap. J.*, **227**, 185–196.

Walker, M. F. (1972), Studies of extremely young cluster. VI. Spectroscopic observations of the ultraviolet-excess stars in the Orion Nebula Cluster and NGC 2264. *Ap. J.*, **175**, 89–116.

Walter, F. M. (1986), X-ray sources in regions of star formation. I. The naked T Tauri stars. *Ap. J.*, **306**, 573–586.

Walter, F. M., Brown, A., Mathieu, R. D., Myers, P. C., and Vrba, F. J. (1988), X-ray sources in regions of star formation. III. Naked T Tauri stars associated with the Taurus-Auriga Complex. *A.J.*, **96**, 297–325.

Waters, L. B. F. M., Coté. J., and Lamers, M. J. G. L. M. (1987), IRAS observations of Be stars. II. Far-IR characteristics and mass loss rates. *A.A.*, **185**, 206–224.

Wilson, L. A. (1974), Fe Fluorescence in T Tauri stars. *Ap. J.* **191**, 143–155.

Zinnecker, H. and Preibisch, Th. (1994), X-ray emission from Herbig Ae/Be stars: a ROSAT survey. *A.A.*, **292**, 152–164.

Supplement

Since the draft of this book was submitted, remarkable progress has been achieved in the field of the physics of emission-line stars. In this supplement, selected papers (published mostly in 2005 and 2006) are presented with some notes focusing into two topics: fine structure of emission-line forming regions (envelope, wind, and disk) and magnetic fields of early-type stars (February, 2007).

Structure of emission-line forming regions

With the advancement of optical and infrared interferometry and other sophisticated observational techniques, dimensions and internal structure of the emission-line forming regions have been markedly unveiled recently and compared with theoretical models. Many types of interferometer systems have been developed and used for observations. They include Very Large Telescope Interferometer (VLTI, ESO), Infrared and Optical Telescope Array (IOTA, Mt. Hopkins), Navy Prototype Optical Interferometer (NPOI, US Naval Observatory), Stellar Interferometer (Sydney University), and Center for High Angular Resolution Astronomy (CHARA Array, Mt. Wilson). Coronagraphic Imaging system with Adaptive Optics (CIAO, Subaru telescope) also yields high spatially resolved infrared images of stellar envelopes.

LBV and central stars of planetary nebulae

Near-infrared observations with the VLTI have been carried out for Eta Carinae and the central star of planetary nebula CPD-56°8032 (Chesneau et al. 2006). Weigelt et al. (2006) measured different disk diameters of Eta Carinae in the continuum (4.3 mas), in HeI emission (6.5 mas), and in Brγ emission (9.6 mas) in K band. Line emissions showed a larger diameter as compared to that in the continuum. Chesneau et al. (2005) also derived the sub-arcsecond structure of the Eta Carinae envelope in the narrow-band images at 3.74 and 4.05 μm. A butterfly-shaped dusty environment and a void around the central star were found. Through spectropolarimetric observations, Davies et al. (2005) found an aspheric and clumpy structure in the winds of LBVs, which is more apparent in stars of strong Hα emission.

Be stars

Stee et al. (2005) reviewed the methods and techniques of interferometric observations of hot star disks with application to Be and B[e] stars.

In the optical region, Tycner et al. (2005, 2006) carried out narrow-band Hα interferometry using NPOI and found the intensity distribution in the envelopes for γ Cas and ϕ Per, and a relationship between Hα emission and linear size of emission-line forming region for η Tau and β CMi. They attributed this relationship to the large optical thickness of Hα radiation. Grundstrom and Gies (2006) calculated numerical model of disks for the Hα emission line, and found that the Hα disk radii as theoretically predicted are consistent with Hα interferometric observations.

In the infrared spectral regions, various types of long baseline interferometers and interferometric array systems were used. Gies et al. (2007) in K band, Kervella and Domiciano de Souza (2006) in H, K bands, and Meilland and Stee (2006) determined the size and geometrical structure of the envelopes of some Be stars. Chesneau et al. (2005) carried out interferometric observations of the Be stars Alpha Arae in the N band at VLT, and derived the upper limits of the envelope size to be approximately 4 mas, corresponding to 14 stellar radius.

The formation and dissipation of the envelopes of Be stars are considered by Meilland et al. (2006) and Rivinius (2005). Meilland et al. suggested two scenarios: one is the successive outbursts of central stars to form disks and rings, and the other is the slowly decreasing mass loss until the disks vanish. Rivinius considered the lifecycles of classical Be stars, similar to the successive outburst scenario of Meilland. If not replenished by subsequent outbursts, the ring will finally dissipate and Be stars will become B star.

Mira variables

Stratified structure of the circumstellar envelopes of Miras, such as the difference in the radii of optical and radio photospheres and of inner dust shell, has been depicted by combined optical, infrared, and radio interferometers (Cotton et al. 2005, Whittkowiski and Boboltz 2005). Stellar diameters in the optical (Ireland and Scholz 2006, Ireland et al. 2005) and infrared (Millan-Gabet et al. 2005, Ohnaka et al. 2005) spectral regions revealed the marked dependence on the wavelength and pulsational phase of Mira stars, where we can see the effects of dust formation and pulsational shock propagation. Dynamic models have been calculated and compared with observations (Ohnaka et al. 2006).

Herbig Ae/Be stars

Highly spatially resolved observations of HESs in the optical and infrared spectral regions have been carried out mostly by three groups: VLTI

(Benisty et al. 2005, Preibisch et al. 2006), IOTA (Millan-Gabet et al. 2006, Monnier et al. 2005, 2006), and Subaru CIAO (Tamura and Fukagawa 2005, Fujiwara et al. 2006, Fukagawa et al. 2006, Honda et al. 2005, Lin et al. 2006, Okamoto et al. 2005). Far-UV long-slit spectrograph with the HST is also used to resolve the inner cavity of a disk (Grady et al. 2005). Complicated structure of circumstellar disks of dust or molecular gases, such as central cavity, asymmetric disk, spiral arms, etc., are elucidated, along with some relationship with the Hα emission intensity.

T Tau stars

As in the case of HES, recent observations of the structure of accretion disks have been made in near- and mid-infrared spectral regions mainly at Mauna Kea (Subaru, Keck telescopes) and ESO (VLT). Particular attention has been paid to the imaging of the inner part of the disks. Akeson et al. (2005) confirmed the existence of inner edge of dust disk using the Keck interferometer. Mayama et al. (2006) using the CIAO of Subaru telescope, and Duchêne et al. (2005) combining Keck telescope II, resolved a complex circumstellar structure around the multiple system of T Tau. Quanz et al. (2006) and Millan-Gabet et al. (2006) observed the structure of optically thick accretion disk of FU Ori in mid-infrared band using the VLT interferometer.

Magnetic fields of early-type stars

It has long been supposed that early-type stars are lacking magnetic fields because of the absence of convection layers theoretically predicted. Recently, however, magnetic fields have been detected in early-type stars, particularly in Be and Herbig Ae/Be stars. Its significant effects on the structure and evolution of envelopes have become widely recognized.

Be stars

Neiner and Hubert (2005) reviewed the indirect and direct methods of detection based on oblique rotator models. Rotational modulation of spectral lines and X-ray fluxes provide a promising method. Smith and Balona (2006) and Smith et al. (2006) suggested the existence of strong magnetic fields on the surface of Be stars by analyzing short-term variabilities in B, V bands, line emissions, and X-ray fluxes.

Several theoretical models for magnetic winds and disks are proposed, generally based on the oblique rotator scheme with dipole-like magnetic fields (Brown and Cassinelli 2005, Maheswaran 2005, Ud-Doula et al. 2005). Cassinelli and Neiner (2005) presented a broad discussion on the origin and dissipation of magnetic fields in Be stars. On the origin, two possible mechanisms were proposed: one is the dynamo action in the convection core and its transportation to the surface and envelope, and the other is that the fossil fields remained from the initial stage of star formation.

HES

Detection and measurements of magnetic fields in HESs have been performed mainly at VLT, ESO, and at CFHT, Mauna Kea, using the spectropolarimeters. Hubrig et al. (2005, 2006) measured the magnetic fields for several HESs and found a large magnetic field of around 450 G for HD 139614 as the largest case among Herbig Ae stars. Yudin et al. (2006) detected Zeeman features in CaII doublet and in metallic lines, whereas Yudin (2005) suggested the existence of localized magnetic fields generated during the evolution of circumstellar envelopes.

Catala et al. (2007) observed the Herbig Ae star HD 190073 with the echelle spectropolarimetric device attached to the CFH Telescope, and detected the magnetic field in the photosphere of this star. Drouin et al. (2005) used both VLT and CFHT to detect the magnetic fields and chemical peculiarities in two HESs that are supposed to be the progenitors of the magnetic Ap/Bp stars. Hamaguchi et al. (2005) showed that the properties of thermal X-rays observed by the ASCA satellite are well explained by magnetic activity in the circumstellar disk of HESs.

References

LBV and central stars of planetary nebulae

Chesneau, O., Colliud, A., de Marco, O., and 7 co-authors (2006). A close look into the carbon disk at the core of the planetary nebula CPD-56°8032. *A. A.*, **455**, 1009–1018.

Chesneau, O., Min, M., Herbst, T., and 15 co-authors (2005). The sub-arcsecond dusty environment of Eta Carinae. *A. A.*, **435**, 1043–1061.

Davies, B., Oudmaijer, T. D., and Vink, J. S. (2005). Asphericity and clumpiness in the winds of Luminous Bleu Variables. *A. A.*, **439**, 1107–1125.

Weigelt, G., Petrov, R. G., Chesneau, O., Davidson, K., and 21 co-authors (2006). VLTI-AMBER observations of Eta Carinae with high spatial resolution and spectral resolutions of 1500 and 10,000. *Advances in Stellar Interferometry*. Monnier, J. D., Schöller, M., and Danchi W. C. (eds.), Proc. The SPIE, Soc, Photo-Optical Instrumental Engineering. Bellington, WA, Vol. 6268.

Be stars

Chesneau, O., Meilland, A., Rivinius, T., and 12 co-authors (2005). First VLTI/MIDI observations of a Be star: Alpha Arae. *A. A.*, **435**, 275–287.

Gies, D. R., Bagnuolo, W. G., Baines, E. K., ten Brummelaar, T. A., and 21 co-authors (2007). CHARA Array K'-band measurements of the angular dimensions of Be star disks.*Ap. J.*, **654**, 527–543.

Grundstrom, E. D. and Gies, D. R. (2006). Estimating Be star disk radii using Hα emission equivalent widths. *Ap. J.*, **651**, L53–L56.

Kervella, P. and Domiciano de Souza, A. (2006). The polar wind of the fast rotating Be star Scherner. VINCI/VLTI interferometric observations of an elongated polar envelope. *A. A.*, **453**, 1059–1066.

Meilland, A. and Stee, Ph. (2006). Recent results from the SIMECA code and VLTI observations. *EAS Pub. Ser.*, **18**, 273–290.

Meilland, A., Stee, Ph., Zorec, J., and Kanaan, S. (2006). Be stars: one ring to rule them all? *A. A.*, **455**, 953–961.

Rivinius, T. (2005). Links between hot stars and their disks. *The Nature and Evolution of Disks around Hot Stars.* Ignace R. and Gayley K. G. (eds.), ASP Conf. Ser. Vol. **337**, Ast. Soc. Pacific, San Francisco, CA, 178–189.

Stee, Ph., Meilland, A., Berger, D., and Gies, D. (2005). Interferometic study of hot star disks. *The Nature and Evolution of Disks around Hot Stars.* ASP Conf. Ser. Vol. **337**, Ast. Soc. Pacific, San Francisco, CA, 211–222.

Tycner, Ch., Lester, J. B., Hajian, A. R., Armstrong, J. T., and 5 co-authors (2005). Properties of the Halpha-emittiong circumstellar regions of Be stars. *Ap. J.*, **624**, 359–371.

Tycner, Ch., Gilbreath, G. C., Zavala, R. T., Armstrong, J. T., and 5 co-authors (2006). Constraining disk parameters of Be stars using Norrowband Halpha Interferometry with the Navy Prototype Optical Interferometer. *A. J.*, **131**, 2710–2721.

Mira variables

Boboltz, D. A. and Wittkowski, M. (2005). Joint VLBA/VLTI observations of the Mira variable S Orionis. *Ap. J.*, **618**, 953–961.

Cotton, W. D., Mennesson, B., Diamond, P. J., Perrin, G., and 10 co-authors (2005). IR and SiO maser observation of Miras. *Future Direction in High Resolution Astronomy.* Romney J. D. and Reid M. J. (eds.), ASP Conf. Ser. Vol. **340**, Ast. Soc. Pacific, San Francisco, CA, 380.

Ireland, M. J. and Scholz, M. (2006). Observable effects of dust formation in dynamic atmospheres of M-type Mira variables. *M. N. R. A. S.*, **367**, 1585–1593.

Ireland, M. J., Tuthill, P. G., Davis, J., and Tango, W. (2005). Dust scattering in the Miras resolved by Optical interferometric polarimetry. *M. N. R. A. S.*, **361**, 337–344.

Millan-Gabet, R., Pedretti, E., Monnier, J. D., Scholerb, F. P., and 4 co-authors (2005). Diameters of Mira stars measured simultaneously in the J, H, K' near-infrared bands. *Ap. J.*, **620**, 961–969.

Ohnaka, K., Scholz, M., and Wood, P. R. (2006). Comparison of dynamical model atmospheres of Mira variables with mid-infrared interferometric and spectroscopic observations. *A. A.*, **446,** 1119–1127.

Ohnaka, K., Bergeat, J., Driebe, T., Graser, U., Hofmann, H.-H., and 14 co-authors (2005). Mid-infrared interferometry of the Mira variable RR Sco with the VLTI MIDI instumemt. *A. A.*, **429**, 1057–1067.

Wittkowski, M. and Boboltz, D. A. (2005). Joint VLTI/VLBA observations of Mira stars. *Future Directions in High Resolution Astronomy.* Romney J. D. and Reid M. J. (eds.), ASP Conf. Ser. Vol. **340**, Ast. Soc. Pacific, San Francisco, CA, 389–393.

Herbig Ae/Be stars

Benisty, M., Malbet, F., de Wit, W. J., Kraus, S., Meilland, A., and 12 co-authors (2005). MWC 297: disk and wind spatially resolved with VLTI/AMBER. *Protostars and Planets V.* Proceedings of the Conf. held Oct. 24–28, in Hilton Wailoloa Village, Hawaii, LPI Contr. No. 1286.

Cabrit, S., Pety, J., Pesenti, N., and Dougados, C. (2006). Tidal stripping and disk kinametics in the RW Aurigae system. *A. A.*, **452**, 897–906.

Fujiwara, H., Honda, M., Kataza, H., Yamashita, T., Onaka, T., and 6 co-authors (2006). The asymmetric thermal emission of the protoplanetary disk surrounding HD 142527 seen by SUBARU/COMICS. *Ap. J.*, **644**, L433–L136.

Fukagawa, M., Tamura, M., Itoh, Y., Kudo, T., Imaeda, Y., and 3 co-authors (2006). Near-infrared images of protoplanetary disk arounding HD 142527. *Ap. J.*, **636**, L153–L156.

Grady, C. A., Woodgate, B., Heap, S. R., Bowers, C., and 3 co-authors (2005). Resolving the inner cavity of the HD 100546 disk: A candidate young planetary system? *Ap. J.*, **620**, 470–480.

Honda, M., Kataza, H., Okamoto, Y. K., Yamashita, T., and 7 co-authors. SUBARU/COMICS 24.5 micron imaging of neaby Herbig Ae/Be disks. *Protostars and Planets V.* Proceedings of the Conf. held Oct. 24–28, in Hilton Wailoloa Village, Hawaii, LPI Contr. No. 1286.

Lin, S. Y., Ohashi, N., Lim, J., Ho, P., Fukagawa, M., and Tamura, M. (2006). Possible molecular spiral arms in the protoplanetary disk of AB Aurigae. *Ap. J.*, **645**, 1297–1304.

Millan-Gabet, R., Monnier, J. D., Berger, J. P., Traub, W. A., and 12 co-authors (2006). Bright localized near-infrared emission at 1–4 AU in the AB Aurigae disk revealed by IOTA Closure Phases. *Ap. J.*, **645**, L77–L80.

Monnier, J. D., Berger, J. P., Millan-Gabet, R., Traub, W. A., and 12 co-authors (2006). Few skewed disks found in first closure-phase survey of Herbig Ae/Be stars. *Ap. J.*, **647**, 444–463.

Monnier, J. D., Millan-Gabet, R., Billmeier, R., Akeson, R. L., and 30 co-authors (2005). The near-infrared size-luminosity relations for Herbig Ae/Be disks. *Ap. J.*, **624**, 832–840.

Okamoto, Y. K., Kataza, H., Honda, M., Yamashita, T., and 6 co-authors (2006). Extended mid-infrared dust emission survey toward Herbig Ae/Be stars. *Protostars and Planets V.* Proceedings of the Conf. held Oct. 24–28, in Hilton Wailoloa Village, Hawaii, LPI Contr. No. 1286.

Preibisch, Th., Kraus, S., Driebe, Th., van Boekel, R., and Weigelt, G. (2006). A compact dusty disk around the Herbig As star HR 5999 resolved with VLTI/MIDI. *A. A.*, **458**, 235–243.

Raman, A., Lisanti, M., Wilner, D. J., Oi, C., and Hogerheijde, M. (2006). A Keplelrian disk around the Herbig Ae star HD 169142. *A. J.*, **131**, 2290–2293.

Tamura, M. and Fukagawa, M. (2005). Circumstellar disks in PMS and T Tauri stars – Herbig Ae/Be stars, Vega-like stars, and submillimeter polarizations. *Astronomical Polarimetry: Current Status and Future Directions.* Adamson, A., Aspin, C., Davis, C. J., and Fujiyoshi, T. (eds.), ASP Conf. Ser. Vol. **343**, Ast. Soc. Pacific, San Francisco, CA, 215.

T Tau stars

Akeson, R. L., Walker, C. H., Wood, K., Eisner, J. A., and 6 co-authors (2005). Keck interferometer observations of classical and weak-line T Tauri stars. *Ap. J.*, **635**, 1173–1181.

Duchêne, G., Ghez, A. M., McCabe, C., and Ceccarelli, C. (2005). The circumstellar environment of T Tauri S at high spatial and spectral resolution. *Ap. J.*, **628**, 832–846.

Mayama, S., Tamura, M., Hayashi, M., Itoh, Y., Fukagawa, M., and 16 co-authors (2006). Subaru near infrared coronagraphic images of T Tauri. *P. A. S. Japan*, **58**, 375–382.

Millan-Gabet, R., Monnier, J. D., Akeson, R. L., Hartmann, L. and 30 co-authors (2006). Keck interferometer observations of FU Orionis objects. *Ap. J.*, **641**, 547–555.

Quanz, S. P., Henning, Th., Bouwman, J., Ratzka, Th., and Leinert, Ch. (2006). FU Orionis: The MIDI VLTI perspective. *Ap. J.*, **648**, 472–483.

Magnetic fields of early-type stars

Be stars

Brown, J. C. and Cassinelli, J. P. (2005). Effects of magnetic fields on winds and disks. *The Nature and Evolution of Disks around Hot Stars.* Ignace R. and Gayley K. G. (eds.), ASP Conf. Ser. Vol. **337**, Ast. Soc. Pacific, San Francisco, CA, 88–99.

Cassinelli, J. P. and Neiner, C. (2005). Magnetic fields and Be stars. *The Nature and Evolution of Disks around Hot Stars.* ASP Conf. Ser. Vol. **337**, Ast. Soc. Pacific, San Francisco, CA, 43–55.

Maheswaran, M. (2005). A magnetic rotator wind disk model for Be stars. *The Nature and Evolution of Disks around Hot Stars.* ASP Conf. Ser. Vol. **337**, Ast. Soc. Pacific, San Francisco, CA, 259–263.

Neiner, C. and Hubert, A. M. (2005). Magnetic fields in Be stars. *The Nature and Evolution of Disks around Hot Stars.* ASP Conf. Ser. Vol. **337**, Ast. Soc. Pacific, San Francisco, CA, 275–278.

Smith, M. A. and Balona, L. (2006). The remarkable Be star HD 110432 (BZ Crucis). *Ap. J.*, **640**, 491–504.

Smith, M. A., Henry, G. W., and Vishniac, E. (2006). Rotational and cyclical variability in gamma Cassiopeia. *Ap. J.*, **647**, 1375–1386.

Ud-Doula, A., Townsend, R., and Owocki, S. (2005). Centrifugal breakout of magnetically confined line-driven stellar winds. *The Nature and Evolution of Disks around Hot Stars.* ASP Conf. Ser. Vol. **337**, Ast. Soc. Pacific, San Francisco, CA, 319–323.

HES

Catala, C., Alecian, E., Donati, J. F., Wade, G. A., and 6 co-authors (2007). The magnetic field of the pre-main sequence Herbig Ae star HD 190073. *A. A.*, **462**, 293–301.

Drouin, D., Wadem G. A., Landstreet, J. D., Mason, E., and 6 co-authors (2005). Seeking the progenitors of magnetic Ap stars: A search for magnetic fields in HAeBe stars using FORS 1 and ESPaDOnS. *EAS Publ.*, Ser. **17**, 309–312.

Hamaguchi, K., Yamauchi, S., and Koyama, K. (2005). X-ray study of Herbig Ae/Be stars. *Ap. J.*, **618**, 360–384.

Hubrig, S., Szeifert, T., North, P., Schöller, M., and Yudin, R. V. (2005). Magnetic fields of B and Herbig Ae stars measured with FORS 1 at VLT. *The Nature and Evolution of Disks around Hot Stars.* Ignace R. and Galey K. G. (eds.), ASP Conf. Ser. V. **337**, Ast. Soc. Pacific, San Francisco, CA, 236–239.

Hubrig, S., Yudin, R. V., Schöller, M., and Pogodin, M. A. (2006). Accurate magnetic field measurement of Vega-like stars and Herbig Ae/Be stars. *A. A.*, **446**, 1089–1094.

Yudin, R. V., Pogodin, M. A., Hubrig, S., and Schöller, M. (2006). Magnetic fields in Herbig Ae/Be stars. Convection in Astrophysics. *IAU Symp.* **239**, held on 21–25. August, 2006 in Prague, Czech Republic, S239, 51.

Yudin, R. V. (2005). Circumstellar discs around Ae/Be and Vega-type stars and local magnetic fields. *The Nature and Evolution of Disks around Hot Stars.* Ignace R. and Galey K. G. (eds.), ASP Conf. Ser. V. **337**, Ast. Soc. Pacific, San Francisco, CA, 342–345.

Author Index

Letter T or F after the page number indicates the author found in the table or figure of the page.

Subject Index

The letter T or F after the page number denotes the subject found in Table or Figure of the page.

Y

Z

Index of Star Name

Stellar objects that appear only in the selected lists of respective types of stars are not included. The types of stars are indicated in the brackets in an abbreviated form after the star name. Some stars are classified into two types. Letter T or F after the page number indicates the star found in Table or Figure. The Gothic letter shows the main page of the star.

Proper name

Constellation name

A

L

M

O

P

Catalogue number

Printed in the USA